International Series in Operations Research & Management Science

Volume 137

For further volumes:
http://www.springer.com/series/6161

Katta G. Murty

Optimization for Decision Making

Linear and Quadratic Models

Springer

Katta G. Murty
University of Michigan
Dept. Industrial and Operations Engineering
1205 Beal Avenue
Ann Arbor MI 48109-2117
USA
murty@umich.edu

and

Professor, Systems Engineering Department
King Fahd University of Petroleum and Minerals
Dhahran
Saudi Arabia

ISSN 0884-8289
ISBN 978-1-4614-2517-5 e-ISBN 978-1-4419-1291-6
DOI 10.1007/978-1-4419-1291-6
Springer New York Dordrecht Heidelberg London

Printed on acid-free paper

Springer is part of Springer Science+Business Media (www.springer.com)

To the memory of my thesis advisor David Gale (my last teacher in the formal education system) who inspired me into doing research in Optimization, and to my mother Katta Adilakshmi (my first teacher in childhood at home)

Preface

I was fortunate to get my first exposure to linear programming in a course taught by the father of the subject, George Dantzig, at the University of California, Berkeley, in the fall of 1965. It was love at first sight! I fell in love with linear programming (LP), optimization models, and algorithms in general right then, and became inspired to work on them. Another of my fortunes was to have as my thesis advisor David Gale, who along with Harold Kuhn and Albert Tucker contributed to the development of optimality conditions. Eventually, I started teaching optimization in the IOE Department at the University of Michigan, Ann Arbor, and using it in applications myself, and I would now like to share this background with future generations of students.

Level of the Book and Background Needed

This is a first-year graduate (Master's) level textbook on optimization models, linear and quadratic, for decision making, how to formulate real-world problems using these models, use efficient algorithms (both old and new) for solving these models, and how to draw useful conclusions, and derive useful planning information, from the output of these algorithms.

It builds on the undergraduate (Junior) level book *Optimization Models for Decision Making Volume 1* on the same subject (Murty (2005) of Chap. 1), which I posted at the public access website:

http://ioe.engin.umich.edu/people/fac/books/murty/opti_model/,

from which you can download the whole book for a small contribution. Readers who are new to the subject should read this Junior-level book to acquire the background for reading this graduate-level book.

Why Another Book on Linear Programming

When friends learned that I was working on this book, they asked me, "Why another book on linear programming (LP)?" There are two reasons:

1. Almost all the best-known books on LP are mathematics books, with little discussion on how to formulate real-world problems as LPs and with very simple modeling examples. Within a short time of beginning work on applications, I realized that modeling could actually be as complex as proving mathematical results and requires very special skills. To get good results, it is important to model real-world problems intelligently. To help the reader develop this skill, I discuss several illustrative examples from my experience, and include many exercises from a variety of application areas.

2. All the available books on LP discuss only the simplex method (developed based on the study of LP using the simplex, one of the solids in classical geometry) and perhaps existing interior point methods (developed based on the study of LP using the ellipsoid). All these methods are based on matrix inversion operations involving every constraint in the model in every step, and work well for LPs in which the coefficient matrix is very sparse. We discuss also a new method being developed based on the study of LP using the sphere, which uses matrix inversion operations sparingly and seems well suited to solve large-scale LPs, and those that may not have the property of being very sparse.

Contents of the Book

Chapter 1 contains a brief account of the history of mathematical modeling, the Gasuss–Jordan elimination method for solving linear equations; the simplex method for solving LPs and systems of linear constraints including inequalities; and the importance of LP models in decision making.

Chapter 2 discusses methods for formulating real-world problems, including those in which the objective function to be optimized is a piecewise linear convex function and multiobjective problems, as linear programs. The chapter is illustrated with many examples and exercises from a variety of applications.

Chapter 3 explains the need for intelligent modeling in order to get good results, illustrated with three case studies: one from a container terminal, the second at a bus-rental company, and the third at an international airport.

Chapter 4 discusses the portion of the classical theory of polyhedral geometry that plays an important role in the study of linear programming and in developing algorithms for solving linear programs, illustrated with many numerical examples.

Chapter 5 treats duality theory, optimality conditions for LP, and marginal analysis; and Chap. 6 discusses the variants of the revised simplex method. Both chapters deal with traditional topics in linear programming. In Chap. 5 we discuss also optimality conditions for continuous variable nonlinear programs and their relationship to optimality conditions for LP.

Chapter 7 discusses interior point methods (IPMs) for LP, including brief descriptions of the affine scaling method, which is the first IPM to be developed, and the primal-dual IPM, which is most commonly used in software implementations.

Chapter 8 discusses the sphere methods, new IPMs that have the advantage of using matrix inversion operations sparingly, and thus are the next generation of methods for solving large-scale LPs.

Chapter 9 discusses extensions of the sphere methods – to convex and nonconvex quadratic programs, and to 0–1 integer programs through quadratic formulations.

Additional Exercises

Exercises offer students a great opportunity to gain a deeper understanding of the subject. Modeling exercises open the student's mind to a variety of applications of the theory developed in the book and to a variety of settings where such useful applications have been carried out. This helps them to develop modeling skills that are essential for a successful career as a practitioner. Mathematical exercises help train the student in skills that are essential for a career in research or a career as a higher-level practitioner who can tackle very challenging applied problems.

Because of the limitations on the length of the book, not all exercises could be included in it. These additional exercises will be included in the website for the book at springer.com in the near future, and even more added over time. Some of the formulation exercises at the website deal with medium-size applications; these problems can be used as computational project problems for groups of two or three students. Formulating and actually solving such problems using an LP software package gives the student a taste of real-world decision making.

Citing References in the Text

At the end of each chapter, we list only references that are cited in the text. Thus the list of references is actually small; it does not provide extensive bibliographies of the subjects. For readers who are interested, we refer them to other books available that have extensive bibliographies.

We use the following style for citing references: A citation such as "Wolfram (2002)" refers to the paper or book of Wolfram of year 2002 listed among references at the end of the current chapter where this citation appears. Alternately, a reference such as "(Dikin (1967) of Chap. 1) refers to the document of Dikin of year 1967 in the list of references at the end of Chap. 1.

Solutions Manual

Springer will host the solutions manual at springer.com, allowing token access to registered adopting faculty.

Acknowledgments

I received comments, encouragement, and other help from many people in preparing this book, including Richard Chen, Jose Dula, Stein-Erik Fleten, Santosh Kabadi, Shantisri Katta, Justin Lin, Mohammad Oskoorouchi, A. Ravi Ravindran, Romesh Saigal, Arvind Sharma, Eric Svaan, Volodymyr Babich. I am grateful to all of them. I am also grateful to my editors and supporters, Fred Hillier, Camille Price, and the Springer team for constant encouragement. Finally I thank my wife Vijaya Katta for being my companion in all these years.

Conclusion

Optimum decision making is all about improving lives. As the Sanskrit proverb (jiivaa ssamastaa ssukhinoo bhavamtu) shown in Telugu script says:

జీవా సమస్తా సుఖినో భవంతు

I hope readers will use these methods to improve the lives of all living beings!

9 April 2009 *Katta Gopalakrishna Murty*

Other Textbooks by Katta G. Murty

Linear and Combinatorial Programming, first published in 1976, available from R.E. Krieger, Inc., P. O. Box 9542, Melbourne, FL 32901.

Linear Programming, published in 1983, available from John Wiley & Sons, 111 River Street, Hoboken, NJ 07030-5774.

Linear Complementarity, Linear and Nonlinear Programming, published in 1988 by Heldermann Verlag, Germany; now available as a download for a voluntary contribution at: http://ioe.engin.umich.edu/people/fac/books/murty/linear_complementarity_webbook/

Operations Research: Deterministic Optimization Models, published in 1995, available from Prentice-Hall Inc. Englewood Cliffs, NJ 07632.

Network Programming, published in 1992, available from Prentice-Hall Inc. Englewood Cliffs, NJ 07632; also as a download for a voluntary contribution at: http://ioe.engin.umich.edu/people/fac/ books/murty/ network_programming/

Optimization Models for Decision Making: Volume 1, Junior Level, available as a download for a voluntary contribution at: http://ioe.engin.umich.edu/people/fac/ books/murty/opti_model/

Computational and Algorithmic Linear Algebra and n-Dimensional Geometry, Sophomore level, available as a download for a voluntary contribution at: http://ioe. engin.umich.edu/people/fac/books/murty/ algorithmic_linear_algebra/

Contents

Glossary of Symbols and Abbreviations

Equations, Results, Theorems, Examples, Tables, and within-section Exercises are numbered serially in each section; so an $i.j.k$ refers to the kth item in Sect. $i.j$. End-of-chapter exercises are numbered serially in each chapter. So Exercise $i.j$ refers to the jth exercise at the end of Chap. i. Similarly, figures are numbered serially in each chapter, so Fig. $i.j$ refers to the jth figure in Chap. i. In tableaus in which there is no constraint symbol ($=, \geq, \leq$) mentioned corresponding to the rows, it implies that each of those rows corresponds to an equation.

Abbreviations in Alphabetical Order

BFS	Basic feasible solution for a linear program
(B, L, U), B-N	Basic–nonbasic partitions of variables in the bounded variable simplex method
BV	Basic vector or Basic variables
CQ	Constraint qualifications for an NLP
CQP	Convex QP
CS	Complementary slackness optimality conditions
D1, D2, D3, D4, D5.1, D5.2	Descent steps used in the sphere methods for LP
DQ1, DQ2	Descent steps used in the sphere method for QP
GJ	Gauss–Jordan (pivot step, algorithm)
GPTC	Gradient projections on touching constraints used in the sphere methods
I/O	Input–output coefficients in a linear program
IPM	Interior point method
IT	Inverse tableau with respect to a basis for an LP in standard form
KKT conds.	Necessary optimality conditions for an NLP
KKT point	A point satisfying the KKT conditions
LCP	Linear complemetarity problem, a type of QP

LP	Linear program
LSCPD	A centering subroutine used in the sphere methods
LSFN	A centering subroutine used in the sphere methods
MDR	Minimum daily requirement for a nutrient in a diet model
MIP	Mixed integer program
MV	Marginal values in an LP
ND, NSD	Negative definite, semidefinite, respectively
NLP	Nonlinear programming problem
NTP	Near touching points used in the sphere methods
OR	Operations Research
PC	Pivot column (for a GJ pivot step or in a pivot step of the simplex method)
PD, PSD	Positive definite, semidefinite, respectively
PE	Pivot element in a GJ pivot step
PFI	Product form of the inverse of a basis
PL	Piecewise linear
PR	Pivot row (for a GJ pivot step or in a pivot step of the Simplex method)
QP	Quadratic program
RC	Redundant constraint identified in a system of linear constraints
RHS	Right hand side (constants or vector of constants in an LP)
RM_i	Raw material i in a product mix problem
SY	Storage yard in a container terminal at a port
TEU	A unit for measuring container volume in container shipping
TG	The terminal gate at a container terminal
TP	Touching points used in the sphere methods
wrt	With respect to
XT, TIT	External truck, terminal's internal truck, in a container terminal
YC, RTGC, QC	Yard crane, rubber tired gantry crane, quay crane in a container terminal

Some Technical Words

Ball center	A center used in the sphere method
Iff	If and only if
Canonical tableau	Updated tableau of a system of linear equations wrt a BV
s. to	subject to

Symbols Dealing with Sets

R^n	The n-dimensional real Euclidean vector space. The space of all vectors of the form $x = (x_1, \dots, x_n)^T$.
$\backslash$	Set difference symbol. If D, E are two sets, $D \backslash E$ is the set of all elements of D that are not in E.
$\|F\|$	Cardinality of the set F.
J, Γ_1, Γ_2	Symbols representing sets of indices.
E	Eligible set of variables to enter a basic vector in the simplex method. Sometimes used to denote a matrix. The symbols $E, \bar{E}, E^r$ are also used in Chap. 7 to denote ellipsoids.
$K, \Gamma; K^0, \Gamma^0$	Sets of feasible solutions of an LP and their interiors.
$\in$	Set inclusion symbol. $a \in D$ means that a is an element of D.
$b \notin D$	Means that b is not an element of the set D.
$\subset$	Subset symbol. $E \subset F$ means that set E is a subset of F, that is, every element of E is an element of F
$\cup$	Set union symbol.
$\cap$	Set intersection symbol.
$\emptyset$	The empty set.
ρ	Radius of a sphere or the RHS constant in the mathematical representation of an ellipsoid.
Pos cone of A	When A is a matrix, the set of all nonnegative linear combinations of column vectors of A is called its pos cone. Similarly, "pos cone of row vectors of A" refers to the set of all nonnegative linear combinations of its row vectors.

Symbols Dealing with Vectors and Matrices

$=, \geq, \leq, >, <$	Symbols for equality, greater than or equal to, less than or equal to, greater than, less than; which must hold for each component in vectors.
$\|\|x\|\|$	Euclidean norm of vector $x = (x_1, \dots, x_n)$, it is $\sqrt{x_1^2 + \dots + x_n^2}$. Euclidean distance between two vectors x, y is $\|\|x - y\|\|$.
$A = (a_{ij})$	Matrix with a_{ij} as the general element in it.
x^T, A^T	Transpose of vector x, matrix A.
A^{-1}	Inverse of the nonsingular square matrix A.

$A_{i.}$, $A_{.j}$	The ith row vector, jth column vector of matrix A.
rank(A)	Rank of a matrix A, same as the rank of its set of row vectors, or its set of column vectors.
I	Unit matrix of appropriate order.
D	A square matrix, used to denote the Hessian of a quadratic function in Chap. 9.
Diag(d)	A square diagonal matrix with the vector d along its main diagonal, and all off-diagonal entries 0.
$H(f(x))$	Hessian (matrix of second partial derivatives) of a function $f(x)$.
$\nabla f(x)$	The gradient vector of a real-valued function $f(x)$. When $f(x) = (f_1(x), \dots, f_n(x))$, this symbol denotes the Jacobian matrix of $f(x)$.
x^r, X^r	x^r denotes a point obtained in the rth iteration; X^r is a diagonal matrix with x^r along its main diagonal and all off-diagonal entries 0.
s, S	A vector, and the diagonal matrix with s along its main diagonal.
P	A pivot matrix corresponding to a pivot step; also a projection matrix; the symbols P, D are also used to denote a primal, dual pair of LPs.
η-column	The unique column in a pivot matrix different from the other unit columns.
ℓ, u, k	Bound vectors in a bounded variable LP.
e	The vector of all entries of 1 of appropriate dimension.
πA, Ax	When A is an $m \times n$ matrix, if we mention these matrix-vector products, it implies that π is a row vector in R^m and x is a column vector in R^n.

Symbols Dealing with Real Numbers

$\lvert\alpha\rvert$	Absolute value of real number α.
$n!$	n factorial.
∞	Infinity.
$\sum$	Summation symbol.
α^+, α^-	The positive and negative parts of the real number α; equal to maximum$\{0, \alpha\}$, maximum$\{0, \alpha\}$, respectively.
M	A large positive penalty parameter associated with artificial variables introduced into a model, sometimes also used to denote a matrix.

Symbols Dealing with LPs, QPs, and IPs

$x_j,\ x_{ij},\ x$	x_j is the jth decision variable in an LP or IP; x_{ij} is the decision variable associated with cell (i, j) in an assignment or a transportation problem. x denotes the vector of these decision variables.
$c_{ij};\ c_j,\ c$	The unit cost coefficient of a variable $x_{ij},\ x_j$ in an LP or IP model. c is the vector of c_{ij} or c_j.
$\pi_i,\ \pi$	Dual variable associated with the ith constraint in an LP, the vector of dual variables.
$u = (u_i),\ v = (v_j)$	Vectors of dual variables associated with rows, columns of a transportation array.
$\bar{c}_j,\ \bar{c}_{ij},\ \bar{c}$	The reduced or relative cost coefficient of variables $x_j,\ x_{ij}$ in an LP, or the transportation problem (or the associated dual slack variable); $\bar{c}$ is the vector of these relative cost coefficients.
$n,\ m$	Usually, number of variables, constraints in an LP or IP. Also, the number of sinks, sources (columns, rows) in a transportation problem.
σ	The Phase I dual basic solution.
$d = (d_j), \bar{d} = (\bar{d}_j)$	Vectors of original and updated Phase I cost vectors, respectively.
B	Usually denotes a basis for an LP in standard form.
$x_B,\ x_D$	Vectors of basic (dependent), nonbasic (independent) variables wrt a basis for an LP.
$\bar{A}_{.j},\ \bar{A}_{i.}$	The updated jth column, ith row corresponding to the original coefficient matrix A wrt a given basic vector.
(S)	In Chap. 2, denotes the active system at a given feasible solution.
$a_i,\ b_j$	In a transportation problem, these are the amounts of material available for shipment at source i, required at sink j, respectively.
$x(\lambda),\ z(\lambda),\ \lambda$	λ is a parameter denoting the value of an entering variable in the simplex method; $x(\lambda),\ z(\lambda)$ denote the resulting solution and its objective value.
θ	Usually the minimum ratio in a pivot step in the simplex algorithm for solving an LP or a transportation problem.
$\Omega_P,\ \Omega_D$	The optimum faces of the primal and dual pair of LPs.
$\mathcal{N}_1(\theta_1),\ \mathcal{N}_1(\theta_2)$	Neighborhoods of central paths used in Chap. 7.
$L_1,\ L_2,\ L_\infty$	Measures of deviation between a function and its estimated values.
K^r	The set of feasible solutions considered in the rth iteration in Sphere method 2.

$t_{\min}$, $t_{\max}$	The minimum and maximum values of the objective function in Sphere methods in Chap. 8.
c^i	The orthogonal projection of the cost vector c on the ith facetal hyperplane used in sphere methods in Chap. 8.
$\delta(x)$; $\delta[t]$	The radius of a largest inscribed sphere with center x; the center on the objective plane with objective value t in sphere methods in Chap. 8.
$B(x,\delta)$	Denotes an inscribed ball with radius δ in sphere methods in Chap. 8.
$T(x)$	The set of indices of facetal hyperplanes touching the largest inscribed ball with center x in sphere methods in Chapter 8.
$Q(x)$	A quadratic function in x.
$0-1$ variable	A variable that is constrained to take values of 0 or 1. Also called "binary variable" or "boolean variable."

Other Symbols

Superscripts, exponents	Regular letter superscripts are ordinary superscripts, for example, x^r denotes the rth point in a set of points. Superscripts in boldface letters denote exponents, for example, $2^{\mathbf{4}}$ is 16, the fourth power of 2, $n^{\mathbf{r}}$ denotes n to the power r.
$\exp(g)$	The Napieran number e (used in mathematics, = 2.71828....) raised to the power of g.
$O(n^{\mathbf{r}})$	The worst case complexity of an algorithm is said to be $O(n^{\mathbf{r}})$, if the measure of computational effort to solve a problem of size n by the algorithm is bounded above by a constant times $n^{\mathbf{r}}$.
□	Symbol indicating the end of a proof.
■	Symbol indicating the end of an example.

Symbols Dealing with Networks or Graphs

$\mathcal{N}$	The finite set of nodes in a network.
$\mathcal{A}$	The set of lines (arcs or edges) in a network.
$G = (\mathcal{N}, \mathcal{A})$	A network with node set $\mathcal{N}$ and line set $\mathcal{A}$.
(i, j)	An arc joining node i to node j.

Chapter 1
Linear Equations, Inequalities, Linear Programming: A Brief Historical Overview

This chapter, taken mostly from Murty (2006b), outlines the history of efforts that eventually led to the development of linear programming (LP) and its applications to *decision making*.

1.1 Mathematical Modeling, Algebra, Systems of Linear Equations, and Linear Algebra

One of the most fundamental ideas of the human mind, discovered more than 5,000 years ago by the Chinese, Indians, Iranians, and Babylonians, is to represent quantities that we want to determine by symbols – usually letters of the alphabet like x, y, z – then express the relationships between the quantities represented by these symbols in the form of equations, and finally, use these equations as tools to find out the true values represented by the symbols. The symbols representing the unknown quantities to be determined are nowadays called *unknowns*, or *variables*, or *decision variables*.

The process of representing the relationships between the variables through equations or other functional relationships is called *modeling* or *mathematical modeling*. The earliest mathematical models constructed were *systems of linear equations*, and soon after, the famous *elimination method* for solving them was discovered in China and India.

The Chinese text *Chiu-Chang Suanshu* (*Nine Chapters on the Mathematical Art*) composed over 2,000 years ago describes the method using a problem of determining the yield (measured in units called "tou") from three types of grains – inferior, medium, superior – given the yield data from three experiments each using a separate combination of the three types of grains. See Kangshen et al. (1999) for information on this ancient work, also a summary of this ancient Chinese text can be seen at the website: http://www-groups.dcs.st-and.ac.uk/~history/HistTopics/Nine_chapters.html).

Ancient Indian texts *Sulva Suutrah* (*Procedures Based On Ropes*) and the *Bakshali Manuscript* with origins during the same period describe the method in terms of solving systems of two (three) linear equations in two (three) variables; see

K.G. Murty, *Optimization for Decision Making: Linear and Quadratic Models*, International Series in Operations Research & Management Science 137, DOI 10.1007/978-1-4419-1291-6_1,

Joseph (1992) and also Lakshmikantham and Leela (2000) for information on these texts, and for a summary and review of this book, see http://www.tlca.com/adults/origin-math.html.

This effort culminated around 825 AD in the writing of two books by the Persian mathematician Muhammad ibn-Musa Alkhawarizmi in Arabic, which attracted international attention. The first was *Al-Maqala fi Hisab al-jabr w'almuqabilah* (*An essay on algebra and equations*). The term "al-jabr" in Arabic means "restoring" in the sense of solving an equation. In Latin translation, the title of this book became *Ludus Algebrae*, the second word in this title surviving as the modern word *algebra* for the subject, and Alkhawarizmi is regarded as the father of algebra. *Linear algebra* is the name given subsequently to the branch of algebra dealing with systems of linear equations. The word *linear* in "linear algebra" refers to the "linear combinations" in the spaces studied, and the *linearity* of "linear functions" and "linear equations" studied in the subject.

The second book, *Kitab al-Jam'a wal-Tafreeq bil Hisab al-Hindi*, appeared in a Latin translation under the title *Algoritmi de Numero Indorum*, meaning *Al-Khwarizmi Concerning the Hindu Art of Reckoning*; it was based on earlier Indian and Arabic treatises. This book survives only in its Latin translation, because all the copies of the original Arabic version have been lost or destroyed. The word *algorithm* (meaning procedures for solving algebraic systems) originated from the title of this Latin translation. Algorithms seem to have originated in the work of ancient Indian mathematicians on rules for solving linear and quadratic equations.

1.1.1 Elimination Method for Solving Linear Equations

We begin with an example application that leads to a model involving simultaneous linear equations. A steel company has four different types of scrap metal (called SM-1 to SM-4) with compositions given in the table below. They need to blend these four scrap metals into a mixture for which the composition by weight is: Al, 4.43%; Si, 3.22%; C, 3.89%; Fe, 88.46%. How should they prepare this mixture?

Compositions of available scrap metals

Type	% in type, by weight, of element			
	A	Si	C	Fe
SM-1	5	3	4	88
SM-2	7	6	5	82
SM-3	2	1	3	94
SM-4	1	2	1	96

To answer this question, we first define the decision variables, denoted by x_1, x_2, x_3, x_4, where for $j = 1$ to 4, x_j = proportion of SM-j by weight in the mixture to be prepared. Then the percentage by weight of the element Al in the mixture will be $5x_1 + 7x_2 + 2x_3 + x_4$, which is required to be 4.43. Arguing the same way for the percentage by weight in the mixture, of the elements Si, C, and Fe, we find that the decision variables x_1 to x_4 must satisfy each equation in the following system of linear equations in order to lead to the desired mixture:

$$\begin{aligned} 5x_1 + 7x_2 + 2x_3 + x_4 &= 4.43, \\ 3x_1 + 6x_2 + x_3 + 2x_4 &= 3.22, \\ 4x_1 + 5x_2 + 3x_3 + x_4 &= 3.89, \\ 88x_1 + 82x_2 + 94x_3 + 96x_4 &= 88.46, \\ x_1 + x_2 + x_3 + x_4 &= 1. \end{aligned}$$

The last equation in the system stems from the fact that the sum of the proportions of various ingredients in a blend must always be equal to 1. From the definition of the variables given above, it is clear that a solution to this system of equations makes sense for the blending application under consideration, only if all the variables in the system have nonnegative values in it. The nonnegativity restrictions on the variables are *linear inequality constraints*. They cannot be expressed in the form of linear equations, and as nobody knew how to handle linear inequalities at that time, they ignored them and considered this system of equations as the mathematical model for the problem. ■

The Gauss–Jordan (GJ) Pivot Step and the GJ (Elimination) Method

To solve a system of linear equations, each step in the elimination method uses one equation to express one variable in terms of the others, then uses that expression to eliminate that variable and that equation from the system, leading to a smaller system. The same process is repeated on the remaining system. The work in each step is organized conveniently through what is now called the *Gauss–Jordan (GJ) pivot step*.

We will illustrate this step on the following system of three linear equations in three decision variables given in the following detached coefficient table at the top. In this representation, each row in the table corresponds to an equation in the system, and the RHS is the column vector of right-hand side constants in the various equations. Normally the equality symbol for the equations is omitted.

An illustration of the GJ pivot step

Basic variable	x_1	x_2	x_3	RHS
	$\boxed{1}$	−1	−1	10
	−1	2	−2	20
	1	−2	−4	30
x_1	1	−1	−1	10
	0	1	−3	30
	0	−1	−3	20

In this step on the system given in the top table, we are eliminating the variable x_1 from the system using the equation corresponding to the first row. The column vector of the variable eliminated, x_1, is called the *pivot column*, and the row of the equation used to eliminate the variable is called the *pivot row* for the pivot step, the element in the pivot row and pivot column, known as the *pivot element*, is boxed in the above table. The pivot step converts the pivot column into the unit column with "1" entry in the pivot row and "0" entries in all the other rows by *row operations*. These row operations consist of the following:

1. For each row other than the pivot row, subtracting a suitable multiple of the pivot row from it to convert the element in this row in the pivot column, to 0.
2. At the end dividing the pivot row by the pivot element.

For example, for the GJ pivot step with the column of x_1 as the pivot column and the first row as the pivot row in the top tableau above, we need to subtract the pivot row (row 1) from row 3; add the pivot row to row 2; and as the pivot element is 1, leave the pivot row as it is. Verify from the bottom table above that these row operations convert the column of x_1 into the first unit column as required.

In the resulting table after this pivot step is carried out, the variable eliminated, x_1, is recorded as the *basic variable* in the pivot row. This row now contains an expression for x_1 as a function of the remaining variables. The other rows contain the remaining system after x_1 is eliminated, the same process is now repeated on this system.

When the method is continued on the remaining system, three things may occur:

1. All the entries in a row may become 0; this is an indication that the constraint in the corresponding row in the original system is a redundant constraint; such rows are eliminated from the tableau.
2. The coefficients of all the variables in a row may become 0, while the RHS constant remains nonzero; this indicates that the original system of equations is inconsistent, that is, it has no solution; if this occurs the method terminates.
3. If the inconsistency termination does not occur, the method terminates after performing pivot steps in all the rows; if there are no nonbasic variables at that stage, equating each basic variable to the RHS in the final tableau gives the unique solution of the system. If there are nonbasic variables, from the rows of the final tableau we get the general solution of the system in parametric form in terms of the nonbasic variables as parameters.

The elimination method remained unknown in Europe until Gauss rediscovered it at the beginning of the nineteenth century while calculating the orbit of the asteroid Ceres based on recorded observations in tracking it earlier. It was lost from view when the astronomer tracking it, Piazzi, fell ill. Gauss got the data from Piazzi, and tried to approximate the orbit of Ceres by a quadratic formula using that data. He designed the method of least squares for estimating the best values for the parameters to give the closest fit to the observed data; this gives rise to a system of linear equations to be solved. He rediscovered the elimination method to solve that system. Even though the system was quite large for hand computation, Gauss's accurate computations helped in relocating the asteroid in the skies in a few months time, and his reputation as a mathematician soared.

Europeans gave the names *Gaussian elimination method* and *Gauss–Jordan (GJ) elimination method* to two variants of the method at that time. These methods are still the leading methods in use today for solving systems of linear equations.

1.2 Review of the GJ Method for Solving Linear Equations: Revised GJ Method

The Gauss–Jordan (GJ) method for solving a system of linear equations works on the detached coefficient tableau of the system. It carries out GJ pivot steps on this tableau with each row as the pivot row, one row after the other. On each row, a pivot step is carried out at most once. The method stops when pivot steps are carried out on all the rows.

Conditions for the Existence of a Solution

First consider a single linear equation $a_1x_1 + a_2x_2 + \ldots + a_nx_n = \alpha$. This equation always has a solution if at least one of $a_1, \ldots, a_n \neq 0$; that is, when $(a_1, \ldots, a_n) \neq 0$. For example, if $a_1 \neq 0$, then $x = (\alpha/a_1, 0, \ldots, 0)^T$ is a solution of the system.

If $a = (a_1, \ldots, a_n) = 0$ and $\alpha = 0$, then this equation is a trivial equation $0 = 0$, it has no relation to the variables x, and so every x is feasible to it.

If $a = 0$ and $\alpha \neq 0$, this equation becomes the

fundamental inconsistent equation $0 = \alpha$,

where α is any nonzero number; it has no solution.

Now consider the general system of m equations in n unknowns

$$Ax = b, \tag{1.1}$$

where A, $b = (b_i)$ are $m \times n$, $m \times 1$ matrices. Let $A_{i.}$, $A_{.j}$ denote the ith row, jth column of matrix A. Then the various equations in this system are $A_{i.}x = b_i$ for $i = 1$ to m.

Theorem 1.1. Theorem of alternatives for systems of linear equations: *The system of linear equations (1.1) has no feasible solution x iff there is a linear* *combination of equations in it which becomes the fundamental inconsistent equation. That is, (1.1) has no solution iff there exists a row vector* $\pi = (\pi_1, \ldots, \pi_m)$ *such that*

$$\sum_{i=1}^{m} \pi_i A_{i.} = \pi A = 0, \tag{1.2}$$

$$\sum_{i=1}^{m} \pi_i b_i = \pi b = \alpha \neq 0.$$

The proof of this theorem comes from the GJ method itself, as will be shown later in this section. Using any solution of the alternate system (1.2), we can verify that the fundamental inconsistent equation can be obtained as the linear combination of equations in the original system (1.1), with coefficients $\pi_1, \ldots, \pi_m$; confirming that (1.1) cannot have a solution. That is why any solution π of (1.2) is known as *evidence* or *certificate of infeasibility* for (1.1).

System (1.2) is known as *the alternate system* for (1.1); it shares the same data with the original system (1.1).

Redundant Equations, Certificate of Redundancy

An equation in original system (1.1), say the ith, is said to be a redundant equation if it can be expressed as a linear combination of the others, that is, if there exists a real vector $(\pi_1, \ldots, \pi_{i-1}, \pi_{i+1}, \ldots, \pi_m)$ such that

$$A_{i.} - \sum_{t=1,\neq i}^{m} \pi_t A_{t.} = 0, \quad b_i - \sum_{t=1,\neq i}^{m} \pi_t b_t = 0.$$

Then $(-\pi_1, \ldots, -\pi_{i-1}, 1, -\pi_{i+1}, \ldots, -\pi_m)$ is known as an *evidence* or *certificate for redundancy* of the ith equation in (1.1). Such redundant equations can be eliminated from (1.1) without changing its set of feasible solutions.

In solving a system of linear equations by the GJ method, a redundant constraint will show up as a row in which all the entries including the updated RHS constant are 0.

Example 1.1. Consider the following system shown in detached coefficient form at the top of the following sequence of tableaus. We show the various tableaus obtained in solving it by the GJ method. PR and PC indicate pivot row and pivot column, respectively, in each step, and the pivot elements are boxed. "RC" indicates a "redundant constraint identified, which is eliminated from the system at this stage." After each pivot step, the entering variable in that step is recorded as the basic variable (BV) in the PR for that pivot step.

After the second pivot step, we found that the third constraint in the original system was a redundant constraint, which showed up as a row of all 0's in the current tableau. So we eliminated this constraint in all subsequent tableaus. The final basic vector obtained for the system was (x_1, x_4, x_3). There may be several different basic vectors for the system; the final one obtained under the GJ elimination method depends on the choice of entering variables in the various steps of the method.

BV	x_1	x_2	x_3	x_4	RHS	
	PC					
	[1]	1	1	1	−6	PR
	−1	1	0	−2	3	
	−2	12	5	−9	−9	
	0	4	−2	1	−8	
				PC		
x_1	1	1	1	1	−6	
	0	2	1	[−1]	−3	PR
	0	14	7	−7	−21	
	0	4	−2	1	−8	
			PC			
x_1	1	3	2	0	−9	
x_4	0	−2	−1	1	3	
	0	0	0	0	0	RC
	0	6	[−1]	0	−11	PR
x_1	1	15	0	0	−31	
x_4	0	−8	0	1	14	
x_3	0	−6	1	0	11	

The variable x_2 remained as a *nonbasic variable* (also called as *independent variable* or *free variable*). The basic solution wrt the basic vector (x_1, x_4, x_3) is $x = (x_1, x_2, x_3, x_4)^T = (-31, 0, 11, 14)^T$ obtained from the final tableau (known as the *canonical tableau* wrt present basic vector (x_1, x_4, x_3)) by setting the nonbasic variable $x_2 = 0$.

The original system has a *unique solution* iff there is no nonbasic variable left at the termination of the GJ method.

The *dimension* of the set of solutions of the system is equal to the number of nonbasic variables left at the end of the GJ method, which is 1 for this example.

From the canonical tableau, we see that the general solution of the system is $x = (x_1, x_2, x_3, x_4)^T = (-31 - 15x_2, x_2, 11 + 6x_2, 14 + 8x_2)^T$, where the free variable x_2 is a parameter that can be given any arbitrary value. ■

This version of the GJ method does not produce the evidence or certificate of redundancy when a redundant equation in the original system is identified in the method, so we do not have any way of checking whether the "0 = 0" equation appearance at that stage is genuine, or due to some errors in computation or round-off operations carried out earlier. See Chap. 1 (and Sect. 1.16 in it) in the web-book (Murty 2004) for more numerical examples of this version of the GJ method.

We will now describe an improved version of the GJ method that has the advantage of producing also the evidence whenever either a redundant equation is identified in the method or the infeasibility conclusion is reached.

1.2.1 GJ Method Using the Memory Matrix to Generate the Basis Inverse

In this version, before beginning pivot steps on the original tableau, a unit matrix I of order m, where m is the number of constraints in the system, is added by the side of the original tableau. This unit matrix is called the *memory matrix*, and its purpose is to accumulate the basis inverse; so in LP literature it is often referred to as the *basis inverse*. Here is the original tableau with the memory matrix added.

Original tableau		Memory matrix
x	RHS	
A	b	I

Now begin applying the GJ method. Remember that only the columns in the A-part associated with variables x_j are eligible to be selected as pivot columns, but all the computations are carried out on all the columns in the tableau. Suppose at some stage after some pivot steps, the current tableau is as given below.

Current tableau		Memory matrix
x	RHS	
$\bar{A}$	$\bar{b}$	$\bar{M}$

Let $A_{.j}$ and $\bar{A}_{.j}$ be the jth columns in the original-A and $\bar{A}$, respectively. Also let the entries in the ith row of the current tableau be $\bar{A}_{i.}, \bar{b}_i, \bar{M}_{i.}$. Then we will have

$$\bar{A}_{.j} = \bar{M} A_{.j}, \quad \bar{A}_{i.} = \bar{M}_{i.} A, \quad \bar{b}_i = \bar{M}_{i.} b, \quad \bar{b} = \bar{M} b. \tag{1.3}$$

So, for all $i = 1$ to m, $\bar{M}_{i.}$, the ith row of the memory matrix, gives the coefficients in an expression of $\bar{A}_{i.}$ as a linear combination of the rows in the original tableau. As $\bar{M}$ keeps track of these coefficients, it is called the memory matrix.

The equation corresponding to the ith row in the current tableau is $\bar{A}_{i.}x = \bar{b}_i$. So, if $\bar{A}_{i.} = 0$ and $\bar{b}_i = 0$, this is a redundant equation, and from the above formulas we see that $\bar{M}_{i.}$, the corresponding row in the current memory matrix, provides the evidence or certificate for this redundancy.

How to update the memory matrix when a redundant constraint is eliminated from the original system: Suppose we started with a system of m linear equations, and so the memory matrix for it is a square matrix of order m. At some stage suppose we identified the ith equation in the original system as a redundant constraint and want to eliminate it. After the elimination, the remaining system will have only $m - 1$ rows, so the memory matrix associated with it must be a square matrix of order $m - 1$. The question is: from the current memory matrix of order m, how can we get the current memory matrix for the system of remaining constraints? This is easy. When the ith constraint in the original system is identified as a redundant constraint, delete the ith row from the original tableau, also from the current tableau including the memory matrix part. Then delete the ith column also from the memory matrix part. This completes the updating of all the things for this redundant constraint deletion.

Also, if for some i we have in the current tableau $\bar{A}_{i.} = 0$ and $\bar{b}_i = \alpha \neq 0$, this row in the current tableau is the fundamental inconsistent equation, so we conclude that the original system is infeasible and terminate. Then $\bar{\pi} = \bar{M}_{i.}$ is the evidence or certificate for infeasibility of the original system. So, $\bar{\pi}$ is a solution of the alternate system (1.2).

So, this version of the GJ method has the advantage of terminating with either a solution x of the original system or a solution of the alternate system, establishing the infeasibility of the original system.

Proof of Theorem 1.1. The argument given above also provides a mathematical proof of the theorem of alternatives (Theorem 1.1) for systems of linear equations.

Example 1.2. Consider the following system with five equations in five unknowns from the left-hand part of the top tableau. For illustrative purposes, we keep redundant constraints discovered in the algorithm till the end. RC, PC, PR, and BV have the same meanings as in Example 1.1, and the pivot elements are boxed. "IC" means "inconsistent constraint, infeasibility detected."

The third constraint in the final canonical tableau represents the equation "0 = 0"; this shows that the third constraint in the original system is a redundant constraint.

From the third row of the memory matrix (also called basis inverse) in this tableau, and we see that the evidence vector for this is $(-2,-4,1,0,0)$, which implies that

> in the original system, the third constraint (which is $-2x_1+2x_2-6x_3+6x_4+2x_5=-34$) is two times constraint 1 (which is $x_1+x_2+x_3+x_4+x_5=-11$) plus four times constraint 2 (which is $-x_1-2x_3+x_4=-3$), which can be verified to be true.

The fifth constraint from the final canonical tableau is the inconsistent equation "$0=6$." From the fifth row of the basis inverse in this tableau, we see that the evidence vector for this is $\bar{\pi}=(-3,-5,0,-1,1)$. It can be verified that when you take the linear combination of equations in the original system with coefficients in $\bar{\pi}$, then you get this inconsistent equation "$0=6$". Alternately, $\bar{\pi}$ is the solution of the alternate system corresponding to the original system, which is given below (here, α turns out to be 6 for this solution $\bar{\pi}$):

	PC						Memory matrix					
BV	x_1	x_2	x_3	x_4	x_5	b						
	[1]	1	1	1	1	−11	1	0	0	0	0	PR
	−1	0	−2	1	0	−3	0	1	0	0	0	
	−2	2	−6	6	2	−34	0	0	1	0	0	
	0	3	−2	−4	−1	2	0	0	0	1	0	
	−2	6	−9	4	2	−40	0	0	0	0	1	
		PC										
x_1	1	1	1	1	1	−11	1	0	0	0	0	
	0	[1]	−1	2	1	−14	1	1	0	0	0	PR
	0	4	−4	8	4	−56	2	0	1	0	0	
	0	3	−2	−4	−1	2	0	0	0	1	0	
	0	8	−7	6	4	−62	2	0	0	0	1	
			PC									
x_1	1	0	2	−1	0	3	0	−1	0	0	0	
x_2	0	1	−1	2	1	−14	1	1	0	0	0	
	0	0	0	0	0	0	−2	−4	1	0	0	RC
	0	0	[1]	−10	−4	44	−3	−3	0	1	0	PR
	0	0	1	−10	−4	50	−6	−8	0	0	1	
x_1	1	0	0	21	10	63	6	5	0	−2	0	
x_2	0	1	0	−8	−4	−54	−2	−2	0	1	0	
	0	0	0	0	0	0	−2	−4	1	0	0	RC
x_3	0	0	1	−10	−4	44	−3	−3	0	1	0	
	0	0	0	0	0	6	−3	−5	0	−1	1	IC

$$\begin{aligned}
\pi_1 - \pi_2 - 2\pi_3 - 2\pi_5 &= 0,\\
\pi_1 + 2\pi_3 + 3\pi_4 + 6\pi_5 &= 0,\\
\pi_1 - 2\pi_2 - 6\pi_3 - 2\pi_4 - 9\pi_5 &= 0,\\
\pi_1 + \pi_2 + 6\pi_3 - 4\pi_4 + 4\pi_5 &= 0,\\
\pi_1 + 2\pi_3 - \pi_4 + 2\pi_5 &= 0,\\
-11\pi_1 - 3\pi_2 - 34\pi_3 + 2\pi_4 - 40\pi_5 &= \alpha.
\end{aligned}$$

■

Next we will discuss a computationally more efficient version of the same GJ method.

1.2.2 The Revised GJ Method with Explicit Basis Inverse

Suppose the original system that we are trying to solve is $Ax = b$, consisting of m equations in n unknowns. In many practical applications, we encounter systems in which n is much larger than m, particularly in applications involving linear programming models.

In the version of the GJ method discussed in Sect. 1.2.1, pivot computations are carried out on all the n columns of A plus the m columns of the memory matrix. Suppose after pivot steps have been carried out on some rows of the tableau, the entries in the current coefficient tableau, RHS, memory matrix are $\bar{A}$, $\bar{b} = (\bar{b}_i)$, $\bar{M}$. Then (1.3) gives us the formulae to obtain $\bar{A}_{i.}$, the ith row of $\bar{A}$ for each i; $\bar{b}_i$ for each i; and $\bar{A}_{.j}$, the jth column of $\bar{A}$, for each j, using data in $\bar{M}$ and in the original A, b.

Thus the formulae in (1.3) show that we can obtain any row or column of $\bar{A}$ as and when we need it, if we just carry out all the pivot computations in every step on the columns of the memory matrix only and update $\bar{M}$ in every step. This leads to a computationally more efficient version of the GJ method known as the *revised GJ method with explicit basis inverse*, discussed in Sect. 4.11 of Murty (2004). This is the version that is commonly used in computer implementations. This version is based on adopting a technique developed by Dantzig in the revised simplex method for linear programming, to the GJ method for solving linear equations. In this version, the current memory matrix is generally referred to as the *basis inverse*, so we will call it the *IT (inverse tableau)* and denote it by B^{-1}, instead of $\bar{M}$. The general step in this version is described next.

General step in the GJ method: Let the current inverse tableau be the following:

BV	Inverse tableau	Updated RHS
$\vdots$	B^{-1}	$\bar{b}$

Let P denote the set of rows in which pivot steps have been carried out already.

1. Select a row $i \in \{1, \dots, m\} \backslash P$ as the pivot row *(PR)* for the next pivot step.
2. For this pivot step we need PR, the updated ith row $\bar{A}_{i.}$ for the systems of equations being solved. From (1.3) we know that it is $(B^{-1})_{i.}A$, and compute it.

   If the PR, $(B^{-1})_{i.}A = 0$ and $\bar{b}_i = 0$, the ith constraint in the present original system is a redundant constraint, and in $(B^{-1})_{i.}$ we have the evidence vector for this conclusion. Eliminate this ith constraint from the original system; the ith row from the inverse tableau and the updated RHS vector, and the ith column from the inverse tableau; reduce m by 1; and look for another pivot row for the next pivot step.

   If the PR, $(B^{-1})_{i.}A = 0$, and $\bar{b}_i \neq 0$, we have in $(B^{-1})_{i.}$ evidence for the conclusion that the original system has no solution; terminate.

   If the PR, $(B^{-1})_{i.}A \neq 0$, select a nonzero entry in it as the PE (pivot element) for the next pivot step, and the variable, x_j say, containing it as the entering variable, and its column, the jth updated column $= \bar{A}_{.j} = B^{-1}A_{.j}$ (where $A_{.j}$ is the column of the entering variable x_j in the original system), as the PC (pivot column) for that pivot step. Computer programmers have developed several heuristic rules for selecting the PE from among the nonzero entries in the pivot row to keep round-off errors accumulating in digital computation under control. Put the PC by the side of the inverse tableau as below.

BV	Inverse tableau	Updated RHS	PC x_j	
$\vdots$	B^{-1}	$\bar{b}$	$\bar{A}_{.j}$	PR = i-th row

   Performing this pivot step will update the inverse tableau and the RHS vector, leading to the next inverse tableau. Now include row i in P.

3. If pivot steps have now been carried out in all the rows of the tableau, we have a solution for the original system. The basic solution for the original system wrt the present basic vector is given by setting all the nonbasic variables at 0, and the tth basic variable $=$ tth updated RHS constant for all t. Terminate.

   If there are rows in the tableau in which pivot steps have not yet been carried out, go to the next step and continue.

Example 1.3. We will now show the application of this version of the GJ method on the system solved in Example 1.2 by the regular GJ method. Remember, in this version pivot computations are carried out only on the inverse tableau and the RHS, but not on the original system. At any stage, B^{-1} denotes the inverse tableau, IT. If row i is the pivot row (PR), we will denote it by $\bar{A}_{i.} = (B^{-1})_{i.}A$. Likewise, if x_j is the entering variable (EV), its updated column, the PC, will be denoted by $\bar{A}_{.j} = B^{-1}$ (original column of x_j). RC denotes redundant constraint, and for simplicity we will not delete RCs detected. "IC" means "inconsistent constraint, infeasibility detected."

Original system							Memory matrix						
x_1	x_2	x_3	x_4	x_5	b	BV	Inverse tableau B^{-1}					PC x_1	
1	1	1	1	1	−11		1	0	0	0	0	**1**	PR
−1	0	−2	1	0	−3		0	1	0	0	0	−1	
−2	2	−6	6	2	−34		0	0	1	0	0	−2	
0	3	−2	−4	−1	2		0	0	0	1	0	0	
−2	6	−9	4	2	−40		0	0	0	0	1	−2	
Row 1 = PR above. EV x_1. Next IT on right below													
												PC x_2	
Now PR = Row 2.					−11	x_1	1	0	0	0	0	1	
$\bar{A}_{2.} = (1, 1, 0, 0, 0)A$					−14		1	1	0	0	0	**1**	PR
$= (0, 1, -1, 2, 1)$.					−56		2	0	1	0	0	4	
x_2 selected EV.					2		0	0	0	1	0	3	
PC $= B^{-1}A_{.2}$ entered.					−62		2	0	0	0	1	8	
PR = Row 3. $\bar{A}_{3.} = (-2, -4, 1, 0, 0)A = 0$. $\bar{b}_3 = 0$. RC.													
												PC x_3	
PR = Row 4. $\bar{A}_{4.} = (-3, -3$					3	x_1	0	−1	0	0	0	2	
$0, 1, 0)A = (0, 0, 1,$					−14	x_2	1	1	0	0	0	−1	
$-10, -5)$. EV $= x_3$. PC =					0		−2	−4	1	0	0	0	
$B^{-1}A_{.3}$ entered.					44		−3	−3	0	1	0	**1**	PR
					50		−6	−8	0	0	1	1	
PR = Row 5. $\bar{A}_{5.} = (-3, -5,$					63	x_1	6	5	0	−2	0		
$0, -3, 1)A = 0$. $\bar{b}_5 = 6$. IC.					−54	x_2	−2	−2	0	1	0		
Infeasible.					0		−2	−4	1	0	0		
					44	x_3	−3	−3	0	1	0		
					6		−3	−5	0	−1	1		

IT = Inverse tableau

So, the fifth equation in the updated tableau is the inconsistent equation "0 = 6," which implies that the original system has no feasible solution and the method terminates. The fifth row in the inverse tableau $\bar{\pi} = (-3, -5, 0, -1, 1)$ provides the evidence vector for this conclusion. This $\bar{\pi}$ is a solution of the alternate system corresponding to the original system. ■

1.3 Lack of a Method to Solve Linear Inequalities Until Modern Times

Even though linear equations had been conquered thousands of years ago, systems of linear inequalities remained inaccessible until modern times. The set of feasible solutions to a system of linear inequalities is called a *polyhedron* or *convex polyhedron*, and geometric properties of polyhedra were studied by the Egyptians earlier than 4000 BC while building the pyramids, and later by the Greeks, Chinese, Indians, and others.

The following theorem (Murty 2006a) relates systems of linear inequalities to systems of linear equations.

Theorem 1.2. *Consider the system of linear inequalities*

$$Ax \geq b, \tag{1.4}$$

where $A = (a_{ij})$ *is an* $m \times n$ *matrix and* $b = (b_i) \in R^m$. *So, the constraints in the system are* $A_{i.}x \geq b_i$, $i \in \{1, \dots, m\}$. *If this system has a feasible solution, then there exists a subset* $\mathbf{P} = \{p_1, \dots, p_s\} \subset \{1, \dots, m\}$ *such that every solution of the system of equations*

$$A_{i.}x = b_i, \quad i \in \mathbf{P},$$

is also a feasible solution of the original system of linear inequalities (1.4).

Proof. Let K denote the set of feasible solutions of (1.4). For any $x \in K$, the ith constraint in (1.4) is said to be *active* at x if $A_{i.}x = b_i$ and *inactive* if $A_{i.}x > b_i$.

We will now describe a procedure consisting of repetitions of a general step beginning with an initial point $x^0 \in K$.

General Step: Let $x^r \in K$ be the current point and $\mathbf{P}_r = \{i : i\text{th constraint in (1.4) is active at } x^r\}$.

Case 1: $\mathbf{P}_r = \emptyset$. In this case x^r is an interior point of K. Let $\bar{x}$ be any solution of one equation $A_{i.}x = b_i$ for some i. If $\bar{x} \in K$, define $x^{r+1} = \bar{x}$.

If $\bar{x} \notin K$, find $\bar{\lambda}$, the maximum value of λ such that $x^r + \lambda(\bar{x} - x^r) \in K$. Then $x^r + \bar{\lambda}(\bar{x} - x^r)$ must satisfy at least one of the constraints in (1.4) as an equation, define $x^{r+1} = x^r + \bar{\lambda}(\bar{x} - x^r)$.

Go back to another repetition of the general step with x^{r+1} as the current point.

Case 2: $\mathbf{P}_r \neq \emptyset$ and either x^r is the unique solution of the system of equations $\{A_{i.}x = b_i : i \in \mathbf{P}_r\}$, or $\mathbf{P}_r = \{1, \dots, m\}$. In either of these cases, $\mathbf{P} = \mathbf{P}_r$ satisfies the requirement in the theorem, terminate.

Case 3: $\mathbf{P}_r$ is a nonempty proper subset of $\{1, \dots, m\}$ and the system $\{A_{i.}x = b_i : i \in \mathbf{P}_r\}$ has alternate solutions. Let $H_r = \{x : A_{i.}x = b_i, i \in \mathbf{P}_r\}$. Let t be the dimension of H_r, and let $\{y^1, \dots, y^t\}$ be a basis for the subspace $\{A_{i.}y = 0 : i \in \mathbf{P}_r\}$.

If each of the points $y \in \{y^1, \dots, y^t\}$ satisfies $A_{i.}y = 0$ for all $i \in \{1, \dots, m\}$, then $\mathbf{P} = \mathbf{P}_r$ satisfies the requirement in the theorem, terminate.

Otherwise, let $\bar{y} \in \{y^1, \ldots, y^t, -y^1, \ldots, -y^t\}$ satisfy $A_{i.}\bar{y} < 0$ for some $i \in \{1, \ldots, m\} \backslash \mathbf{P}_r$. Find $\bar{\lambda}$, the maximum value of λ such that $x^r + \lambda\bar{y} \in K$, define $x^{r+1} = x^r + \bar{\lambda}\bar{y}$.

Go back to another repetition of the general step with x^{r+1} as the current point.

The subsets of indices generated in this procedure satisfy $\mathbf{P}_r \subset \mathbf{P}_{r+1}$ and $|\mathbf{P}_{r+1}| \geq 1 + |\mathbf{P}_r|$. So after at most m repetitions of the general step, the procedure must terminate with a subset $\mathbf{P}$ of $\{1, \ldots, m\}$ satisfying the conditions in the theorem. □

In systems of linear inequalities like (1.4) appearing in applications, typically $m \geq n$.

This theorem states that every nonempty polyhedron has a nonempty face that is an affine space. It can be used to generate a finite enumerative algorithm to find a feasible solution to a system of linear constraints containing inequalities. It involves enumeration over subsets of the inequalities in the system. For each subset do the following: eliminate all the inequality constraints in the subset from the system. If there are any inequalities in the remaining system, change them into equations. Find any solution of the resulting system of linear equations. If that solution satisfies all the constraints in the original system, done, terminate. Otherwise, repeat the same procedure with the next subset of inequalities. At the end of the enumeration, if no feasible solution of the original system has turned up, it must be infeasible.

However, if the original system has m inequality constraints, in the worst case this enumerative algorithm may have to solve $2^{\mathbf{m}}$ systems of linear equations before it either finds a feasible solution of the original system or concludes that it is infeasible. The effort required grows exponentially with the number of inequalities in the system in the worst case.

A Paradox: Many young people develop a fear of mathematics and shy away from it. But since childhood I had a fascination for mathematics because it presents so many paradoxes. Theorem 1.2 also presents an interesting *paradox.*

As you know, linear equations can be transformed into linear inequalities by replacing each equation with the opposing pair of inequalities. But there is no way a linear inequality can be transformed into linear equations. This indicates that linear inequalities are more fundamental than linear equations.

But this theorem shows that linear equations are the key to solving linear inequalities, and hence are more fundamental, this is the paradox. Again we will show later in the book that linear inequalities may play an important role for solving linear equations.

1.3.1 The Importance of Linear Inequality Constraints and Their Relation to Linear Programs

The first interest in inequalities arose from studies in mechanics, beginning in the eighteenth century. Crude examples of applications involving linear inequality models started appearing in published literature around the 1700s.

Linear programming (LP) involves optimization of a linear objective function subject to linear inequality constraints. Crude examples of LP models started appearing in published literature from about the middle of the eighteenth century. An example of an application of LP is the fertilizer maker's product mix problem discussed in Example 3.4.1 of Sect. 3.4 of Murty (2005b). It leads to the following LP model:

(1.3.1)

Maximize	$z(x) = 15x_1$	$+$	$10x_2$			Item
Subject to	$2x_1$	$+$	x_2	$\leq$	1,500	RM 1
	x_1	$+$	x_2	$\leq$	1,200	RM 2
	x_1			$\leq$	500	RM 3
	x_1	$\geq 0,$	x_2	$\geq$	0	

in which the decision variables $x_1, \ x_2$ are the tons of Hi-ph, Lo-ph fertilizers manufactured/day using three raw materials RM 1, 2, 3 for which the available supply is at most 1,500, 1,200, 500 tons/day, respectively. The limit on the supply of each of these raw materials leads to a constraint in the model, that is why these raw materials are called the "items" corresponding to those constraints in the model. The objective function to be maximized is the daily net profit from fertilizer manufacturing activities.

In this example, all the constraints on the variables are inequality constraints. In the same way, inequality constraints appear much more frequently and prominently than equality constraints in most real-world applications. In fact, we can go as far as to assert that in most applications in which a linear model is the appropriate one to use, most of the constraints are actually linear inequalities, and linear equations play only the role of a computational tool through approximations, or through results similar to Theorem 1.2. Linear equations were used to model problems mostly because an efficient method to solve them is known.

Fourier was one of the first to recognize the importance of inequalities as opposed to equations for applying mathematics. Also, he was a pioneer who observed the link between linear inequalities and linear programs in early nineteenth century.

For example, the problem of finding a feasible solution to the following system of linear inequalities (1.5) in $x_1, \ x_2$ can itself be posed as another LP for which an initial feasible solution is readily available. Formulating this problem known as a *Phase I problem* introduces one or more non-negative variables known as *artificial variables* into the model. All successful LP algorithms require an initial feasible solution at the start, so the Phase I problem can be solved using any of those algorithms, and at termination it either outputs a feasible solution of the original problem or an evidence for its infeasibility. The Phase I model for finding a feasible solution for (1.5) is (1.6), and it uses one artificial variable x_3.

$$\begin{aligned} x_1 + 2x_2 &\geq 10, \\ 2x_1 - 4x_2 &\geq 15, \\ -x_1 + 10x_2 &\geq 25, \end{aligned} \tag{1.5}$$

$$\begin{aligned} \text{Minimize} \quad & x_3 \\ \text{Subject to} \quad x_1 + 2x_2 + x_3 &\geq 10 \\ 2x_1 - 4x_2 + x_3 &\geq 15 \\ -x_1 + 10x_2 + x_3 &\geq 25 \\ x_3 &\geq 10 \end{aligned} \tag{1.6}$$

For the Phase I problem (1.6), $(x_1, x_2, x_3)^T = (0, 0, 26)^T$ is a feasible solution. In fact solving such a Phase I problem provides the most efficient approach for solving systems of linear inequalities.

Also, the duality theory of linear programming discussed in Chap. 5 shows that any linear program can be posed as a problem of solving a system of linear inequalities without any optimization. Thus, solving linear inequalities and LPs are mathematically equivalent problems. Both problems of comparable sizes can be solved with comparable efficiencies by available algorithms. So, the additional aspect of "optimization" in linear programs does not make LPs any harder either theoretically or computationally.

1.4 Fourier Elimination Method for Linear Inequalities

By 1827, Fourier generalized the elimination method to solve a system of linear inequalities. The method now known as the *Fourier* or *Fourier–Motzkin elimination method* is one of the earliest methods proposed for solving systems of linear inequalities. It consists of successive elimination of variables from the system. We will illustrate one step in this method using an example in which we will eliminate the variable x_1 from the following system.

$$\begin{aligned} x_1 - 2x_2 + x_3 &\leq 6, \\ 2x_1 + 6x_2 - 8x_3 &\leq -6, \\ -x_1 - x_2 - 2x_3 &\leq 2, \\ -2x_1 - 6x_2 + 2x_3 &\leq 2. \end{aligned}$$

x_1 appears with a positive coefficient in the first and second constraints and a negative coefficient in the third and fourth constraints. By making the coefficient of x_1 in each constraint into 1, these constraints can be expressed as

$$\begin{aligned} x_1 &\leq 6 + 2x_2 - x_3, \\ x_1 &\leq -3 - 3x_2 + 4x_3, \\ -2 - x_2 - 2x_3 &\leq x_1, \\ -1 - 3x_2 + x_3 &\leq x_1. \end{aligned}$$

The remaining system after x_1 is eliminated and is therefore

$$
\begin{aligned}
-2 - x_2 - 2x_3 &\le 6 + 2x_2 - x_3, \\
-2 - x_2 - 2x_3 &\le -3 - 3x_2 + 4x_3, \\
-1 - 3x_2 + x_3 &\le 6 + 2x_2 - x_3, \\
-1 - 3x_2 + x_3 &\le -3 - 3x_2 + 4x_3,
\end{aligned}
$$

and then $\max\{-2 - x_2 - 2x_3, -1 - 3x_2 + x_3\} \le x_1 \le \min\{6 + 2x_2 - x_3, -3 - 3x_2 + 4x_3\}$ is used to get a value for x_1 in a feasible solution when values for other variables are obtained by applying the same steps on the remaining problem successively.

However, starting with a system of m inequalities, the number of inequalities can jump to $O(m^2)$ after eliminating only one variable from the system, so this method is not practically viable except for very small problems.

1.5 History of the Simplex Method for LP

In 1827, Fourier published a geometric version of the principle behind the simplex algorithm for a linear program (vertex-to-vertex descent along the edges to an optimum, a rudimentary version of the simplex method) in the context of a specific LP in three variables (an LP model for a Chebyshev approximation problem), but did not discuss how this descent can be accomplished computationally on systems stated algebraically. In 1910, De la Vallée Poussin designed a method for the Chebyshev approximation problem, which is an algebraic and computational analogue of this Fourier's geometric version; this procedure is essentially the primal simplex method applied to that problem.

In a parallel effort, Gordan (1873), Farkas (1896), and Minkowski (1896) studied linear inequalities, and laid the foundations for the algebraic theory of polyhedra and derived necessary and sufficient conditions for a system of linear constraints, including linear inequalities to have a feasible solution.

Studying LP models for organizing and planning production (Kantorovich 1939) developed ideas of dual variables (*resolving multipliers*) and derived a dual-simplex type method for solving a general LP.

Full citations for references before 1939 mentioned so far can be seen from the list of references in Danizig (1963) or Schrijver (1986).

This work culminated in the mid-twentieth century with the development of the primal simplex method by Dantzig. This was the first complete, practically and computationally viable method for solving systems of linear inequalities. So, LP can be considered as the branch of mathematics, which is an extension of linear algebra to solve systems of linear inequalities. The development of LP is a landmark event in the history of mathematics, and its application brought our ability to solve general systems of linear constraints (including linear equations, inequalities) to a state of completion.

1.6 The Simplex Method for Solving LPs and Linear Inequalities Viewed as an Extension of the GJ Method

One of the most popular methods for solving either systems of linear constraints including linear inequality constraints, or linear programs, is the simplex method. We will discuss this method in Chap. 6. Here we will explain why this method can be viewed as the extension of the GJ method for solving systems of linear equations to these more general systems. Let $x \in R^n$ denote the column vector of decision variables.

First consider the problem of solving a system of linear constraints including inequalities. For this, the simplex method first transforms the system into a *standard form* consisting of a system of linear equations in nonnegative variables by simple transformations such as introducing nonnegative slack variables to convert inequalities into equations, and eliminating unrestricted variables using the equality constraints (see Sect. 4.1 in Chap. 4 of Murty (2005b) for a discussion of these transformations). The standard form is

$$\begin{aligned} Ax &= b, \\ x &\geq 0, \end{aligned} \tag{1.7}$$

where A is a matrix of order $m \times n$, and $b = (b_i)$ is a column vector in R^m.

1.6.1 Generating the Phase I Problem if No Feasible Solution Available for the Original Problem

Now apply the GJ method to solve the system of equations $Ax = b$ ignoring the nonnegativity restrictions on x. If this terminates with the infeasibility conclusion, clearly (1.7) is also infeasible, so terminate. Otherwise, let $x = (x_B, x_D)$ be the basic, nonbasic partition of variables obtained at the end of the GJ method, and let the final canonical tableau obtained be

BV	x_B	x_D	RHS
x_B	I	$\bar{D}$	$\bar{b}$

If $\bar{b} \geq 0$, the basic solution of (1.7) wrt the basic vector x_B, which is $(x_B, x_D) = (\bar{b}, 0)$, is a feasible solution for (1.7), and we are done.

If $\bar{b} \not\geq 0$, let r be such that $\bar{b}_r$ is the most negative element in $\bar{b}$. Now introduce a nonnegative artificial variable x_0 into the canonical tableau associated with the column vector $-e = (-1, -1, \dots, -1)^T$ of all "-1" entries, leading to the following tableau called a Phase I tableau.

BV	x_B	x_D	x_0	RHS
x_B	I	$\bar{D}$	$-e$	$\bar{b}$

In this tableau if you perform a pivot step with row r as the PR (pivot row) and x_0 as the entering variable, then you will get the canonical tableau wrt a new basic vector $x_{B'}$ containing the artificial variable x_0 as a basic variable, corresponding to which the basic solution can be verified to be ≥ 0, with the value of the artificial variable x_0 being $= -\bar{b}_r > 0$ in it.

Example 1.4. Here we provide a numerical illustration for the introduction of the artificial variable x_0. Suppose the canonical tableau obtained at the end of the GJ method on the system of equations $Ax = b$ is the following tableau with variables x_1 to x_7. The updated RHS is $\ngeq 0$, so we already introduced the artificial variable x_0 in this tableau.

BV	x_1	x_2	x_3	x_4	x_5	x_6	x_7	x_0	$\bar{b}$	
x_1	1	0	0	1	−1	2	−1	−1	−3	
x_2	0	1	0	−2	1	−1	−1	[−1]	−6	PR
x_3	0	0	1	−1	2	1	1	−1	7	
								PC		
x_1	1	−1	0	3	−2	3	0	0	3	
x_0	0	−1	0	2	−1	1	1	1	6	
x_3	0	−1	1	1	1	2	2	0	13	

The basic solution produced by the GJ method on $Ax = b$ in this problem is the basic solution corresponding to the basic vector (x_1, x_2, x_3) in which the variables x_1, x_2 are <0, so it is not feasible to the original system $Ax = b$, $x \geq 0$. The most negative variable in this solution is the basic variable x_2 in $r =$ second row. So, we introduced the artificial variable x_0 as discussed earlier, and performing a pivot step in its column with row 2 as the PR leads to a new basic vector $x_{B'} = (x_1, x_0, x_3)$ whose basic solution is a nonnegative solution to the augmented system. ■

The Phase I problem:

$$\begin{aligned} &\text{Minimize} \quad x_0 \\ &\text{subject to} \quad x_B + \bar{D}x_D - ex_0 = \bar{b} \\ &\qquad\qquad x_B, x_D, x_0 \geq 0 \end{aligned} \tag{1.8}$$

is an LP formulation for the problem of getting a feasible solution for (1.7). It is one of several different (but equivalent) ways of formulating the problem of finding a feasible solution to (1.7) as a Phase I linear program.

Solving such an LP formulation seems to be the most efficient approach for solving systems of linear constraints with inequalities. The simplex algorithm solves LPs like (1.8) starting with a nonnegative basic solution corresponding to a feasible basic vector like $x_{B'}$ by performing additional GJ pivot steps exchanging one basic variable with a nonbasic variable in each step until an optimum solution for (1.8) is obtained.

If the minimum value of x_0 in the Phase I problem (1.8) is >0, clearly the original (1.7) is infeasible. If the minimum value of x_0 in (1.8) is 0, then any optimum solution of (1.8) gives a feasible solution for (1.7), by suppressing the 0 value of x_0 from it.

If the original problem to be solved is the LP of minimizing cx subject to (1.7), then the only change in the above approach is to minimize $cx + \alpha x_0$, where α is a large positive penalty cost, instead of x_0 in (1.8). Starting with the feasible basic solution corresponding to the basic vector $x_{B'}$, the simplex method solves this LP the same way.

Starting with the canonical tableau for the system of linear equations $Ax = b$ obtained by the GJ method, the simplex method carries out additional GJ pivot steps to obtain a feasible solution for (1.7) or to solve an LP subject to (1.7). For this reason, the simplex method for LP can be considered as an extension of the GJ method for linear equations to solve systems of linear constraints including inequalities or LPs.

Linear Algebra		**Linear Programming**
Study of linear equations. Originated over 2,000 years ago.	→	Study of linear constraints including inequalities. Twentieth century extension of linear algebra.

1.7 The Importance of LP

LP has now become a dominant subject in the development of efficient computational algorithms, in the study of convex polyhedra, and in algorithms for decision making. But for a short time in the beginning, its potential was not well recognized.

Dantzig tells the story of how when he gave his first talk on LP and his simplex method for solving it, at a professional conference, Hotelling (a burly person who liked to swim in the sea, the popular story about him was that when he does, the level of the ocean raises perceptibly, see Figs. 1.1 and 1.2; my thanks to Katta Sriramamurthy and Shantisri Katta for these figures) dismissed it as unimportant

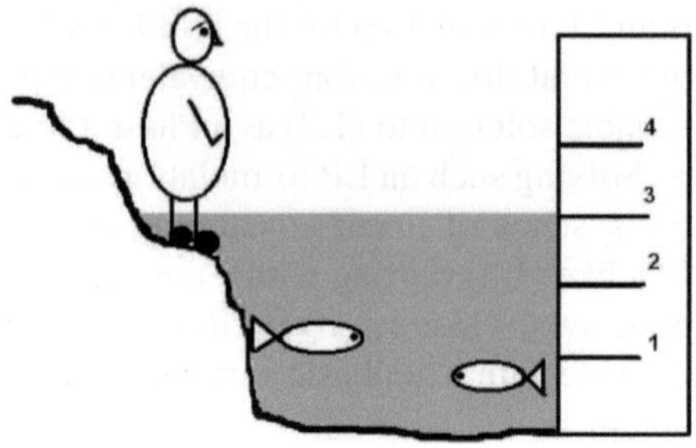

Fig. 1.1 Hotelling (a whale of a man) getting ready to swim in the ocean

Fig. 1.2 Hotelling swimming in the ocean. Watch the level of the ocean go up

since everything in the world is nonlinear. But Von Neumann came to the defense of Dantzig saying that the subject will become very important; see Page xxvii of Dantzig and Thapa (1997). The preface in this book contains an excellent account of the early history of LP from the inventor of the most successful method in OR and in the mathematical theory of polyhedra.

Von Neumann's early assessment of the importance of LP turned out to be astonishingly correct. Today, the applications of LP in almost all areas of science are so numerous, so well known, and recognized that they need no enumeration. Also, LP seems to be the basis for most of the efficient algorithms for many problems in other areas of mathematical programming. Many of the successful approaches in nonlinear programming, discrete optimization, and other branches of optimization are based on LP in their iterations. Also, with the development of duality theory and game theory (Gale 1960), LP has also assumed a central position in economics.

1.7.1 Marginal Values and Other Planning Tools that can be Derived from the LP Model

We will illustrate the very useful planning information that can be derived from an LP model for a real-world decision-making problem, using the example of the fertilizer maker's product mix problem discussed in Example 3.4.1 of Sect. 3.4 of Murty (2005b), referred to earlier in Sect. 1.3.1.

The fertilizer maker (FM) produces Hi-ph, Lo-ph fertilizers using three raw materials, RM-1, 2, 3 as inputs, whose supply is currently limited. Here is all the data on the problem.

Raw material	Input/ton of Hi-ph	Input/ton of Lo-ph	Availability (tons/day)	Cost ($/ton)
RM-1	2	1	1,500	50
RM-2	1	1	1,200	75
RM-3	1	0	500	60
Other costs[a] ($/ton)	50	40		
Market price[b]	300	175		
Net profit ($/ton made)	15	10		

[a]All manufacturing costs other than raw material costs

[b]Selling price in market of this fertilizer

So the total production cost/ton of Hi-ph = (input raw material costs) + (other production costs) $= 2 \times 50 + 1 \times 75 + 1 \times 60 + 50 = 285$\$/ton, and since its market price is \$300, production of Hi-ph leads to a net profit of $300 - 285 =$ \$ 15/ton made. The net profit from Lo-ph of \$10/ton is computed in the same way.

The market is able to absorb all the Hi-ph, Lo-ph fertilizers the company can produce, and so at present there is no limit on the production levels of these fertilizers. Defining $x_1,\ x_2$ = tons of Hi-ph, Lo-ph produced daily, the LP model for maximizing the company's daily net profit is

$$\begin{aligned}
\text{Maximize } z(x) &= 15x_1 + 10x_2 \\
\text{s. to } \quad 2x_1 + x_2 &\le b_1 = 1500 \quad \text{(RM-1 availability)} \\
x_1 + x_2 &\le b_2 = 1200 \quad \text{(RM-2 availability)} \\
x_1 \qquad &\le b_3 = \ \ 500 \quad \text{(RM-3 availability)} \\
x_1,\ x_2 &\ge 0.
\end{aligned} \tag{1.9}$$

The constraint $2x_1 + x_2 \le 1500$ requires that the feasible region of this problem should be on the side of the straight line $\{x : 2x_1 + x_2 \le 1500\}$ in Fig. 1.3. Likewise, all other constraints in (1.9) can be represented by the corresponding half-spaces in Fig. 1.3, leading to the set of feasible solutions, K of this problem as the shaded region in Fig. 1.3.

Selecting any feasible solution, $x^0 = 0$ say, we draw the objective line $\{x : z(x) = z(x^0)\}$ through it, and then move this objective line parallel to itself, increasing the RHS constant in its representation as far as possible (because in this problem we need to maximize the value of $z(x)$), while still maintaining a nonempty intersection with the feasible region. If $\hat{z}$ is the final value of the RHS constant in this process, then $\hat{z}$ is the maximum value of $z(x)$ in the problem, and any point in the intersection of $\{z(x) = \hat{z}\} \cap K$ is an optimum solution of (1.9).

Fig. 1.3 Solving the fertilizer product mix model geometrically

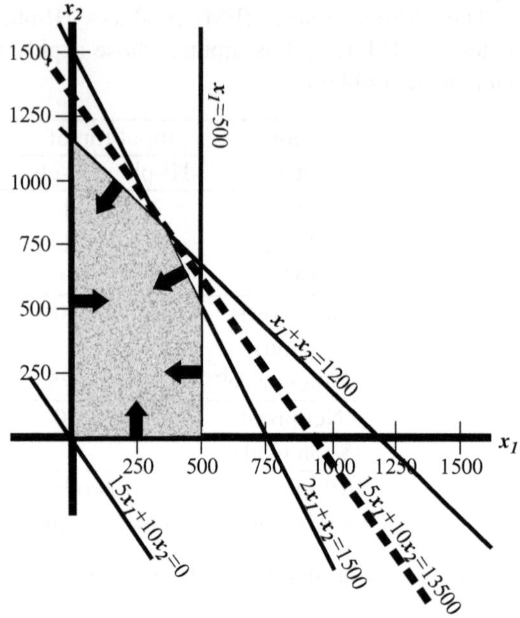

For (1.9), $\hat{z} = 13{,}500$, and the final position of the objective line is the dashed line represented by $\{x : 15x_1 + 10x_2 = 13500\}$ in Fig. 1.3. The optimum solution of (1.9) $\hat{x} = (300, 900)^T$ is unique, and it is the solution of the system:

$$\begin{aligned} 2x_1 + x_2 &= 1500, \\ x_1 + x_2 &= 1200. \end{aligned}$$

The RHS constants vector $b = (b_1, b_2, b_3)^T$ is $= (1500, 1200, 500)^T$ at present in (1.9). When the optimum solution $\hat{x}$ is implemented, the left over quantities in the daily availability of RM-1, 2, 3 are $1500 - 2\hat{x}_1 - \hat{x}_2 = 0$, $1200 - \hat{x}_1 - \hat{x}_2 = 0$, $500 - \hat{x}_1 = 200$ tons, respectively.

Thus the daily availabilities of RM-1, 2 are fully used up, while there are 200 tons of spare in the availability of RM-3 when the optimum solution $\hat{x}$ is implemented. So, to increase the net profit beyond the present maximum attainable level of \$13,500, the company has to increase the supply of either RM-1 or RM-2.

Marginal Values

Each constraint in an LP model is the material balance constraint of some item, the RHS constant in that constraint being the availability or the requirement of that item. The *marginal value* of that item (also called the marginal value corresponding to that constraint) is defined to be the rate of change in the optimum objective value of the LP per unit change in the RHS constant in the associated constraint, *while all the other data in the problem remains unchanged.*

For example, in the fertilizer product mix problem, the marginal value of RM 1 (and of the corresponding first constraint in (1.9)) is the rate of change in the maximum daily profit per unit change in the supply of RM 1 from its present value of 1,500. These rates are also called *dual variables* or the *shadow prices of the items*. These are the variables in another linear programming problem that is in *duality relationship* with the original problem. In this context, the original problem is called the *primal problem* and the other problem is called the *dual problem*. The derivation of the dual problem is discussed in Chap. 5.

So, let $b = (b_1, b_2, b_3)^T$ denote the vector of RHS constants in (1.9) and let $f(b_1, b_2, b_3)$ denote the optimum objective value in (1.9) as a function of b, while all the other data in the problem remains unchanged from their present values. Then we know that $f(1500, 1200, 500) =$ 13,500. A simple expression for the marginal value of b_1 is then

$$(f(1500 + \epsilon, 1200, 500) - f(1500, 1200, 500))/\epsilon,$$

where ϵ is either positive or negative but with a small absolute value, provided this ratio turns out to be the same for $\epsilon < 0$ or $\epsilon > 0$.

$f(1500 + \epsilon, 1200, 500)$ is the optimum objective value in the LP (1.9) with 1500 changed to $1500+\epsilon$. The set of feasible solutions of the modified problem is similar to that in Fig. 1.3 with the line corresponding to $2x_1 + x_2 = 1500$ moved parallel to itself by $|\epsilon|$ either away from 0 (if $\epsilon > 0$) or towards 0 (if $\epsilon < 0$). As long as $|\epsilon|$ is small, it is clear that the optimum solution of the modified problem is the solution of

$$\begin{aligned} 2x_1 + x_2 &= 1500 + \epsilon, \\ x_1 + x_2 &= 1200, \end{aligned}$$

which is $\hat{x}(\epsilon) = (300 + \epsilon,\ 900 - \epsilon)$ with optimum objective value of $f(1500 + \epsilon, 1200, 500) = 13500+5\epsilon = f(1500, 1200, 500)+5\epsilon$. This shows that the marginal value of b_1 (RM-1) in (1.9) is $\hat{\pi}_1 = 5$.

Similarly to compute the marginal value of b_2 (RM-2), we need to compute $f(1500, 1200 + \epsilon, 500)$, where again ϵ is a real number of small absolute value. Using the same arguments as above, we see the optimum solution of this modified problem is the solution of

$$\begin{aligned} 2x_1 + x_2 &= 1500, \\ x_1 + x_2 &= 1200 + \epsilon, \end{aligned}$$

which is $\hat{x}(\epsilon) = (300 - \epsilon, 900 + 2\epsilon)$ with optimum objective value of $f(1500, 1200 + \epsilon, 500) = 13500 + 5\epsilon = f(1500, 1200, 500) + 5\epsilon$. This shows that the marginal value of b_2 (RM-2) in (1.9) is $\hat{\pi}_2 = 5$.

To compute the marginal value of b_3 (RM-3) in (1.9), we need to compute $f(1500, 1200, 500 + \epsilon)$. In the optimum solution $\hat{x}$ of the original (1.9), we have 200 tons of RM-3 not being used in the current supply. This implies $\hat{x}$ remains optimal to the modified problem, with b_3 changed to $500 + \epsilon$ as long as $|\epsilon|$ is small. Therefore, $f(1500, 1200, 500+\epsilon) = f(1500, 1200, 500) = 13500$. Therefore, the marginal value of b_3 (RM-3) in (1.9) is $\hat{\pi}_3 = 0$. Therefore, the marginal value vector for (1.9) is $\hat{\pi} = (5, 5, 0)$.

The marginal value vector may not exist for all LPs. If $b = (b_1, \dots, b_m)^T$ is the RHS constants vector in an LP, and $f(b_1, \dots, b_m)$ is the optimum objective value function of this LP as a function of this b assuming that all the other data in the LP remains unchanged; and if the value of the ratio $(f(b_1 + \epsilon, b_2, \dots, b_m) - f(b_1, \dots, b_m))/\epsilon$ depends on whether $\epsilon > 0$ or $\epsilon < 0$, then this LP does not have marginal values. See Exercise 6.73 in Chap. 6 for an example.

In practice, we do not have to use the simple technique based on ratios of the form $(f(b_1+\epsilon, b_2, \dots, b_m) - f(b_1, \dots, b_m))/\epsilon$ to compute marginal values. In Chap. 5 we discuss techniques to check whether a general LP has the marginal values vector or not, and show that when it exists it is the optimum dual solution. We also show that when an LP is solved by any of the algorithms discussed in this book, the optimum dual solution is also obtained as a byproduct.

Interpretation of Marginal Values and an Application

In the fertilizer product mix problem (1.9), the marginal value $\hat{\pi}_1 = 5$ for RM-1 means that each unit of RM-1 available to this company at present is equivalent to 5$ in net profit. Other marginal values have a similar interpretation.

As an application of marginal values, we will show how they can be used to evaluate the profitability of producing a new fertilizer, and at what level its market price should be set so that it will be profitable in comparison with existing product lines.

Suppose this company's research lab has developed a new fertilizer with a catchy name *lushlawn*. Manufacturing lushlawn needs 3, 2, 2 tons of RM-1, 2, 3, respectively/ton and incurs other manufacturing costs of 100$/ton besides the raw material costs.

If the company decides to introduce lushlawn as a new product, at what level should they set its market price?

One ton of lushlawn needs as input a packet of (3, 2, 2) tons of RM-1, 2, 3. From the marginal value vector $\hat{\pi}$ we see that with existing product lines, this packet is equivalent to $3\hat{\pi}_1 + 2\hat{\pi}_2 + 2\hat{\pi}_3 = 25$$ of net profit.

Thus lushlawn is worth manufacturing, if it can be sold at a price that leads to a net profit of 25$/ton made. So, the breakeven market price/ton of lushlawn is $25 + (raw material costs) + (other manufacturing costs) = $25 + 3 × 50 + 2 × 75 + 2 × 60 + 100 = $435/ton. By conducting a market survey, the company can determine whether the market will accept lushlawn at a price $\geq$ this breakeven level of \$435/ton. Once this is known, the decision whether to produce lushlawn is obvious.

By providing this kind of valuable planning information, the LP model has become a highly useful decision-making tool.

1.8 Dantzig's Contributions to Linear Algebra, Convex Polyhedra, OR, Computer Science

Much has been written about Dantzig's contributions. Also, he has a personal assessment of his own contributions in Chap. 1 of his book (Dantzig 1963). As someone who started learning LP from his course at Berkeley, I will summarize here some of his contributions that are usually overlooked in other statements (for a brief account of my experiences with Dantzig see Murty (2005a)).

1.8.1 Contributions to OR

The simplex method is the first effective computational algorithm for one of the most versatile mathematical models in OR. Even though LP and also the simplex method for solving it originated much earlier than Dantzig's work, as explained in Sect. 1.5, it started becoming prominent only with Dantzing's work and OR was just beginning to develop around that time. The success of the simplex method is one of the root causes for the phenomenal development and the maturing of LP, mathematical programming in general, and OR in the second half of the twentieth century.

1.8.2 Contributions to Linear Algebra and Computer Science

Recognizing the Irrelevance of the "RREF" Concept Emphasized in Mathematics Books on Linear Algebra

Dantzig contributed important pedagogic improvements to the teaching of linear algebra. He would state all the algorithmic steps in the GJ elimination method using the fundamental tool of row operations on the detached coefficient tableau for the system with the variable corresponding to each column entered in a top row in every tableau. This makes it easier for young students to see that the essence of this method is to take linear combinations of equations in the original system to get an equivalent but simpler system from which a solution can be read out. In descriptions of the GJ method in most mathematics books on linear algebra, the variables are usually left out.

Also, these books state the termination condition in the GJ elimination method to be that of reaching the RREF (reduced row echelon form, a tableau is defined to be in RREF if it contains a full set of unit vectors in proper order at the left end). Dantzig (and of course a lot of other OR people) realized that it is not important that all unit vectors be at the left end of the tableau (they can be anywhere and can be scattered all over); also it is not important that they be in proper order from left to right. He developed the very simple *data structure* (this phrase means a strategy for storing information generated during the algorithm, and using it to improve the efficiency of that algorithm (perhaps this is the first instance of such a structure in computational algorithms)) of associating the variable corresponding to the rth unit vector in the final tableau as the rth basic variable (or basic variable in the rth row), and storing these basic variables in a column on the tableau as the algorithm progresses. This data structure makes it easier to read the solution directly from the final tableau of the GJ elimination method by making all nonbasic variables $= 0$ and the rth basic variable $=$ the rth updated RHS constant, for all r. Dantzig called this final tableau the *canonical tableau* to distinguish it from the mathematical concept of RREF. It also opened the possibility of pivot column selection strategies instead of always selecting the leftmost eligible column in this method.

Even today it is sad that in courses on linear algebra in mathematics departments, the RREF is emphasized as the output of the GJ elimination method. For a more realistic statement of the GJ method from an OR perspective see Murty (2004).

Evidence (or Certificate) of Infeasibility

As explained in Sect. 1.2, a contribution of Dantzig, the *revised simplex method*, has very important consequences to the GJ elimination method. When the GJ elimination method is executed in the revised simplex format, it produces an evidence for infeasibility automatically whenever the system of linear equations being solved has no solution.

1.8.3 Contributions to the Mathematical Study of Convex Polyhedra

Dantzig has made fundamental contributions to the mathematical study of convex polyhedra (a classical subject that has been investigated by mathematicians for more than 4,000 years) when he introduced the complete version of the primal simplex method as a computational tool.

We could only see drawings of two-dimensional polyhedra before this work. Polyhedra in higher dimensions could only be visualized through imagination. The primal simplex pivot steps are the first computational steps for actually tracing an edge (either bounded or unbounded) of a convex polyhedron. It opened a revolutionary new computational dimension in the mathematical study of convex polyhedra, and made it possible to visualize and explore higher-dimensional polyhedra through computation. At a time when research on convex polyhedra was beginning to stagnate, the simplex method has reignited the spark and enriched their mathematical study manyfold.

1.9 Interior Point Methods for LP

In the early 1980s, Narendra Karmarkar pioneered a new method for LP, an interior point method (Karmarkar 1984). Claims were made that this method would be many times faster than the simplex method for solving large-scale sparse LPs, and these claims helped focus researchers attention on it. His work attracted worldwide attention not only from operations researchers, but also from scientists in other areas.

Let me relate a personal experience. When news of his work broke out in world press, I was returning from Asia. The person sitting next to me on the flight was a petroleum geologist. When he learned that I was on the OR faculty at Michigan, he asked me excitedly "I understand that an OR scientist from India at Bell Labs made a discovery that is going to revolutionize petroleum exploration. Do you know him?!"

In talks on his algorithm that he gave at that time, Karmarker repeatedly emphasized the following points:

1. The boundary of a convex polyhedron with its faces of varying dimensions has a highly complex combinatorial structure. Any method that operates on the boundary or close to the boundary will get caught up in this combinatorial complexity, and there is a limit on improvements we can make to its efficiency.
2. Methods that operate in the central portion of the feasible region in the direction of descent of the objective function have the ability to take longer steps towards the optimum before being stopped by the boundary, and hence have the potential of being more efficient than boundary methods for larger problems.
3. From an interior point, one can move in any direction locally without violating feasibility; hence powerful methods of unconstrained optimization can be brought to bear on the problem.

Researchers saw the validity of these arguments, hence Karmarkar's talks stimulated a lot of work on these methods that stay "away" from the boundary. In the tidal wave of research that ensued, many different classes of interior point methods have been developed for LP, and extended to wider classes of problems including convex quadratic programming, monotone linear complementarity problem, and semi definite programming problems. We will discuss some popular interior point methods in a later chapter. Among them, the first is in fact the first interior point method discussed in the literature, the *primal affine scaling method* (Dikin 1967), which predates Karmarkar's work, but did not attract much attention until after Karmarkar popularized the study of interior point methods (IPMs). We will also discuss another IPM known as the *primal-dual IPM*, which is the most popular IPM for solving LPs.

1.10 Newer Methods

Many practical applications lead to large-scale LP models. All the methods discussed so far, which are based solely on matrix inversion operations, work very well if the models are very sparse, and their performance depends critically on being able to exploit the sparcity in the model to advantage. For solving large-scale models which do not fit this mold, these methods are difficult to use.

So, recent algorithmic research on LP has focused on methods that can solve LPs fast without using matrix inversion operations or using them only minimally. These methods are new, and computational experimentation with them is just beginning. We will discuss the sphere methods in this category from Murty (2006a, b), Murty and Oskoorouchi (2008a, b), Murty (2009) of Chap. 8, Murty (2008c, d) of Chap. 9 in Chaps. 8 and 9. Initial computational experiments with sphere methods for LP (summarized in Chap. 8) indicate that they have great promise.

1.11 Conclusions

The most heavily used techniques in mathematics are linear algebra tools for solving systems of linear equations whose origin goes back over 2,000 years. In the twentieth century, this has been extended into linear programming that can solve systems of linear constraints involving linear inequalities also. Now linear programming has become the most important tool for decision making.

In this book we discuss these linear programming methods, and the basic theory on which these methods are based, and also mathematical modeling techniques for intelligently modeling important decision problems in a variety of areas, as linear programs.

1.12 How to Be a Successful Decision Maker?

The aim of this book is to discuss some techniques for reaching optimum decisions in problems that can be modeled using deterministic linear and quadratic models. Successful decision making is a very complex task with many dimensions to it. Reaching an optimum decision is one aspect of it. Another important aspect not in the scope of this book is implementing the decision reached, which often requires a lot of tact. I illustrate with a story:

> "A 20-year-old lady started dating a 25-year-old man. He kept on giving her expensive gifts until one day she agreed to marry him.
>
> Two days after the wedding she realized that he had been giving her these expensive gifts mainly to trap her into marriage, but in reality he was a miser. She felt very depressed at the prospect of a possible divorce so soon after her marriage.

A month passed by during which time she got a chance to observe him closely. He was hard working, made good money, and was very nice in every respect, except that he tried to save all this money. She thought " If I can learn how to manage him, I can still have a wonderful life. Let me give it a try."

Forty years rolled by. Then her husband became sick, and on his death bed, she was serving him obediently. He said "I am going to die soon. You know very well that I love my money dearly. I want you to withdraw all my money and put it in my casket with me. I want to take it with me to my afterlife. I hope you will take the *decision* to sincerely fulfill this last request of mine."

With his hands in hers, she told him, "You have my solemn promise that your wish will be *implemented*."

Moments later he was dead. The undertaker came and the man's body was stretched out in the casket. His wife, dressed in black, and her best friend were sitting by its side. The ceremony was over, and the undertaker got ready to close the casket. Then the wife said "Wait just a moment." She went in and came out with a metal box and put it inside the casket. Then the undertaker closed it and rolled it away.

Then her friend said "I hope you were not foolish enough to put all your family's money in your husband's casket."

The loyal wife replied, "Listen, I loved my husband, and we had a long and happy married life. I made a promise that his final request would be implemented. My husband worked very hard to earn his money, and I know how much pleasure it gave him to know he would have it with him. I could not go back on my word."

The friend said, "You mean to tell me that you kept your promise?!!"

The tactful wife said, "I sure did. I got all the money together, put it in my account, and wrote him a check, and I put that cheque in the casket!"

Experience is the best teacher of "being tactful." So, I encourage all the readers to get involved in using the techniques discussed in this book in practice.

1.13 Exercises

1.1. Solve the system of equations in the scrap metal blending problem discussed in Sect. 1.1.1 using Matlab, Mathematica, Maple, or Excel, ignoring the nonnegativity restrictions on the variables. Is the solution to the system of equations unique? See if you get a solution satisfying the nonnegativity restrictions anyway.

1.2. We need a nonnegative solution to the following system of linear equations:

$$\begin{aligned} x_1 - 2x_2 + x_3 + x_4 - x_6 + 6x_7 &= 4, \\ x_1 - x_2 - 2x_3 - x_4 + x_5 + 8x_6 - 17x_7 &= -1, \\ 2x_2 + x_3 - 2x_4 - 5x_5 + 3x_6 - 4x_7 &= -3. \end{aligned}$$

Solve using any linear equation software package, ignoring the nonnegativity restrictions on variables. Does this system of equations have alternate solutions? If so, what is the dimension of the set of all solutions to this system of equations? Explain.

Did you get a solution satisfying the nonnegativity restrictions? Explain the difference in the performance of these software packages on Exercise 1.1 and this exercise. Basically, you have to explain why you were able to get a nonnegative solution for one problem using this software, and not for the other.

1.3. Solve the following systems of linear equations by hand using the GJ elimination method. Use pivot elements of 1 or -1 as far as possible to avoid the occurrence of fractions. If the system is inconsistent, obtain a solution to the alternate system in each case. How many solutions of the alternate system were you able to obtain?

$$
\begin{aligned}
(a)\qquad -x_2 + 3x_3 + 8x_4 - 7x_6 &= -3\\
7x_1 + x_2 - 2x_3 - 10x_4 + 2x_5 - 3x_6 &= -4\\
10x_1 + 2x_3 + 12x_4 + 5x_5 - 8x_6 &= 2.
\end{aligned}
$$

$$
\begin{aligned}
(b)\ -5x_1 - 2x_2 + 10x_3 + x_4 + 7x_5 + 2x_6 - 4x_7 &= 2\\
-3x_1 + 3x_2 - 2x_3 - x_4 + 2x_5 + x_6 + 3x_7 &= 3\\
-8x_1 + x_2 + 8x_3 + 9x_5 + 3x_6 - x_7 &= 6\\
x_1 + 2x_3 - 3x_5 - 4x_6 + 2x_7 &= 5\\
-7x_1 + x_2 + 10x_3 + 6x_5 - x_6 + x_7 &= 8.
\end{aligned}
$$

1.4. Solve the following system of linear equations using the GJ method.

Are there any redundant constraints in the system, and if so how many? For each of them give the certificate of redundancy.

What is the dimension of the set of solutions, S, of the system?

Give the formula for a general point in S obtained from your work.

$$
\begin{aligned}
x_1 + x_2 + x_3 + x_5 &= 11,\\
x_2 + x_3 - x_4 + x_6 + x_7 &= 5,\\
x_1 + x_3 + 2x_5 - x_7 &= 2,\\
x_1 + 2x_2 + x_4 - x_5 - x_6 + x_7 &= 24,\\
x_1 + 2x_2 + 3x_3 - 2x_4 + 2x_5 + 2x_6 + x_7 &= 12.
\end{aligned}
$$

It is required to find a solution of this system which makes $3x_1 + 4x_2 - 10x_3 + 2x_4 - 4x_5 + 6x_6$ as close to 100 as possible. Find it explaining clearly how you obtained it.

1.5. Solve the following system using the GJ method. If the system is infeasible, (1) Determine whether it can be made feasible by ignoring any one constraint from the system, (2) If the answer to (1) is "yes," which constraint can be ignored to make the system feasible, and what is a solution for the remaining system? (3) Consider the original system again. Can it be made feasible by changing the value of exactly one right-hand-side constant? If so, carry out this modification and give a feasible solution of the modified system.

$$\begin{aligned}x_1 + x_2 + x_3 - 2x_5 - 2x_7 &= -4,\\ x_1 + x_2 - x_5 - x_7 &= -1,\\ x_2 + x_3 + x_4 - x_5 + x_6 - x_7 &= -5,\\ 2x_1 + x_2 + 3x_3 - x_4 - 5x_5 - x_6 - 5x_7 &= -5.\end{aligned}$$

1.6. Apply the GJ method to the following system of linear equations:

$$\begin{aligned}x_1 + x_2 + x_7 &= 7,\\ x_2 + x_3 + x_4 - 2x_5 + 2x_6 &= 10,\\ x_1 + x_3 - x_4 + x_7 &= 9,\\ 2x_1 + x_2 + x_3 - x_4 + 2x_7 &= 20,\\ x_1 + 2x_2 + x_3 + x_4 - 2x_5 + 2x_6 + x_7 &= 12.\end{aligned}$$

If the system is infeasible, how many inconsistent constraints were you able to identify in the system? For each of them, give the certificate of inconsistency.

Now consider the original system of constraints with the original right-hand-side vector $b = (b_i) = (7, 10, 9, 20, 12)^T$. For any $i = 1$ to 5, it is possible to increase the value of b_i from its original value at a cost of c_i/unit, or decrease its value from its original value at a cost of d_i/unit, where $c = (c_i) = (9, 2, 1, 4, 3)^T$ and $d = (d_i) = (2, 1, 0, 5, 7)^T$. It is required to change the right-hand-side vector in the original system to make the system feasible. Each b_i can either be increased, decreased, or left unchanged. Formulate the problem of changing the right hand side vector of the original system to make the system feasible at minimum cost.

1.7. $\{A_{1.}, \dots, A_{r.}\}$ is a given set of row vectors in R^n. $A_{r+1.}$ is another row vector in R^n, which is not in the linear hull of $\{A_{1.}, \dots, A_{r.}\}$. Then prove that the following system of equations must have a solution $x \in R^n$.

$$\begin{aligned}A_{i.}x &= 0, \quad i = 1 \text{ to } r,\\ A_{r+1.}x &= 1.\end{aligned}$$

1.8. $f(x)$ is a given affine function of $x \in R^n$. We have a computer program by using which we can output the value of $f(x)$ for any given x, but we do not have its functional form.

As $f(x)$ is affine, we know that it can be expressed as $c_1x_1 + c_2x_2 + \dots + c_nx_n + c_0$. Explain how we can obtain the values of $c_1, \dots, c_n, c_0$ using that computer program.

1.9. A is an $m \times n$ matrix and c is a row vector in R^n. It is known that the system of linear equations $Ax = 0$, $cx = 1$ has no feasible solution x. Then show that c must be a linear combination of row vectors of the matrix A.

1.10. Solve the following system of linear equations by the GJ method.

If a feasible solution is obtained, give it. Otherwise, give a linear combination of equations in the original system that yields the fundamental inconsistent equation, explaining how you got it. If the original system is inconsistent, use the information in the updated RHS column of the final tableau to determine what changes in the original RHS column b will make the system feasible. From the current final tableau give a feasible solution of the modified system.

Original system

x_1	x_2	x_3	x_4	x_5	b
1	1	−2	1	2	3
0	−1	2	1	3	5
3	1	−2	5	12	18

1.11. The following system of linear equations is being solved by the revised GJ method with explicit basis inverse.

x_1	x_2	x_3	x_4	x_5	x_6	x_7	b
−1	0	2	−1	7	0	0	6
1	1	3	0	6	2	0	24
0	1	4	2	−5	−5	3	−2
7	6	14	7	−7	−2	6	74
0	−1	−6	0	4	0	1	−32

After some pivot steps we have the following basis inverse (also called the current memory matrix in this chapter), but the column listing the basic variables, and the updated RHS column are smeared and cannot be read correctly due to a printer problem.

	Memory matrix					
	μ_1	μ_2	μ_3	μ_4	μ_5	
	4	5	2	0	0	
	−5	−5	−2	0	0	
	−2	−2	−1	0	0	
	3	−4	−2	1	0	
	0	0	0	0	1	

Using the formulas discussed in this chapter, find (1) the basic variables associated with this inverse, (2) the basic solution, (3) evidence of redundancy and infeasibility if any, and actual expressions in terms of the equation numbers 1–5 in the original system to give proofs of these claims, (4) dimension of the set of feasible solutions, (5) rank of the coefficient matrix of the system, (6) and finally a parametric representation of the general solution of this system in terms of the independent variables.

1.12. The following system of linear equations (first tableau) has been solved by the revised GJ method with the explicit basis inverse. The second tableau given below is the basis inverse obtained at termination.

x_1	x_2	x_3	RHS
1	1	$\boxed{1}$	5
−1	3	1	6
0	4	2	11
2	3	1	10

Mem. matrix

μ_1	μ_2	μ_3	μ_4
3/2	1/6	0	−2/3
−1/2	1/6	0	1/3
−1	−1	1	0
0	−1/3	0	1/3

Using the formulas given in this chapter, find the solution of the system obtained, and explain clearly why it is unique.

1.13. A is an $m \times n$ matrix, and c is a row vector in R^n. A, c are given. Let $S = \{x : Ax = 0\}$, a subspace of R^n. We are also given that $cx = 0$ for every $x \in S$. Then prove that c must be a linear combination of row vectors of the matrix A.

1.14. $A_{1.}, \ldots, A_{m.}, c$ are $(m + 1)$ row vectors in R^n.

Express the condition for c to be in the linear hull (or subspace) of $\{A_{1.}, \ldots, A_{m.}\}$ through a system of constraints.

Write the alternate system for the system obtained above.

Very briefly mention if it is possible to solve both the original system, and the alternate system simultaneously, with the same method. Which method does this?

1.15. Let A be a given matrix of order $m \times n$ (with $n > m$) and rank r. Consider the optimization problem

$$\begin{aligned} &\text{minimize } cx \\ &\text{subject to } Ax = b \ (I) \end{aligned}$$

Assume that (I) has at least one feasible solution. Prove that this optimization problem has the property: either every feasible solution of it is optimal, or no feasible solution for it is optimal. Develop necessary and sufficient conditions for these two possible outcomes.

1.16. A company makes products P_1, P_2 using raw materials R_1, R_2, R_3. Relevant data is given below.

Item	Units/unit input for		Available
	P_1	P_2	(units/day)
R_1	4	5	3,000
R_2	2	0	1,200
R_3	1	2	900
Demand (units/day)	Unlimited	375	
Net profit ($/unit)	3	5	

Formulate the problem of finding an optimum production plan as an LP.

Solve the LP using the geometric method and find the optimum solution. Find the marginal values of the three raw materials and the demand for P_2.

If the demand for P_2 can be increased from 375 units/day by advertising locally, is it worth spending money on this advertizement?

The current prices of the resources R_1, R_2, R_3 are $2, 4, 10/unit, respectively. If additional supplies of each of these resources can be acquired, which of them has the potential for helping to increase the total daily net profit of the company? For each resource determine the breakeven price/unit at which additional supplies of it can be acquired.

Suppose the company has the opportunity to make a new product P_3. To make one unit of P_3 needs as inputs 2, 1, 2 units of resources R_1, R_2, R_3, respectively. What is the breakeven selling price of P_3 at which it becomes competitive to manufacture?

1.17. A company makes two types of discrete parts called A and B. Each part has to be cast in the casting shop, machined, and then finished. Let shops 1, 2, and 3 refer to casting, machining, and finishing shops, respectively. For $i = 1, 2, 3$, shop i has enough capacity to process either a_i units of A, or b_i units of B daily, or any combination of these two activities in proportions of these levels summing up to 1, where a_i, b_i are given below.

Shop no. i	Shop	Capacity for	
		A	B
		a_i	b_i
1	Casting	100	70
2	Machining	80	90
3	Finishing	60	110

Assume that they can sell all the castings they produce. The net profit from unit of A and B sold is $800 and 900, respectively. Ignoring the integer requirements on the units of A and B produced daily, formulate the problem of determining how many units of A and B to produce daily to maximize total net profit.

Plot the set of feasible solutions for the problem and determine the optimum solution geometrically.

Determine the marginal values of the RHS constants corresponding to the casting, machining, finishing shop capacity constraints.

If the casting shop capacity is fully used up in the optimum solution, determine how much the company could pay an outside supplier for a casting of A and B, respectively, over its production cost, and still break even.

Suddenly, the country is in the grip of an economic recession, and the company finds that it can only sell up to 55 units of A and 60 units of B daily. Discuss how this alters some of the conclusions reached earlier.

1.18. A company manufactures two types of cake mixes, A and B, using two raw materials R_1 and R_2. The following table gives the necessary data.

Raw material	Units needed to make 1 unit of		Units available
	A	B	
R_1	1	2	6,000
R_2	2	1	8,000
Net profit per unit made	7	5	
Maximum demand	3,500	2,500	

Formulate the problem of determining how many units of A and B to make, as an LP.

Solve the problem geometrically. Determine the marginal values associated with all the RHS constants in the model. Interpret them.

At this stage, how much extra profit can the company make if the supply of R_1, R_2 is increased by one unit?

A new cake mix developed by the company's kitchen needs two units of R_1 and two units of R_2 as input per unit. What is the minimum net profit that a unit of this new cake mix should make, if it were to be competitive with A, B?

1.19. A is a given matrix of order $m \times n$. The question is whether the homogeneous system $Ax = 0$ has a nonzero solution. When can you conclude that it does, directly without doing any work?

In any case, discuss an efficient method for answering this question, and find a nonzero solution when it exists, based on row operations using at most $\min\{m, n\}$ pivot steps.

References

Dantzig GB (1963) Linear programming and extensions. Princeton University Press, Princeton, NJ

Dantzig GB, Thapa MN (1997) Linear programming, vol 1. Introduction, vol 2. Theory and extensions. Springer, New York

Dikin II (1967) Iterative solution of problems of linear and quadratic programming. Sov Math Doklady 8:674–675

Farkas J (1895) Über die Anwendungen des mechanischen Princips von Fourier. Mathematische und natur wissenschaftliche Berichte aus Ungarn 12:263–281

Gale D (1960) The theory of linear economic models. McGraw-Hill, New York

Gordan P (1873) Ueber die Auflösung linearer Gleichungen mit reellen Coefficienten. Mathematische Annalen 6:23–28

Joseph GG (1992) The crest of the peacock: non-european roots of mathematics. Penguin Books, Harmondsworth, Middlesex, England

Kangshen S, Crossley JN, Lun AWC (1999) The nine chapters on the mathematical art: companion and commentary. vol xiv, p 596. Oxford University Press, Oxford (ISBN 0-19- 853936-3)

Kantorovich LV (1939) The mathematical method of production planning and organization. Manag Sci 6(4):363-422

Karmarkar N (1984) A new poolynomial-time algorithm for linear programming. Combinatorica 4:373–395

Lakshmikantham V, Leela S (2000) The origin of mathematics, p 104. University Press of America, Inc., Lanham, MD; (visit www.univpress.com for information, also a summary of this book can be seen in its book review at: http://www.tlca.com/adults/origin-math.html)

Minkowski H (1896) Geometrie der Zahlen (Erste Lieferung). Teubner, Leipzig, Germany

Murty KG (2004) Computational and algorithmic linear algebra and n-dimensional geometry. Freshman-Sophomore level linear algebra book available as a download for a requested small contribution at: http://ioe.engin.umich.edu/people/fac/ books/murty/algorithmic_linear_algebra/

Murty KG (2005a) My experiences with George Dantzig. http://www.informs.org/History/dantzig/rem_murty.htm

Murty KG (2005b) Optimization models for decision making, vol 1 (Junior level). Available as a download for a requested small contribution at: http://ioe.engin.umich.edu/people/fac/books/murty/opti_model/

Murty KG (2006a) A new practically efficient interior point method for LP. Algorithm Oper Res 1:3–19; paper can be seen at the website: http://journals.hil.unb.ca/index.php/AOR/index

Murty KG (2006b) Linear equations, inequalities, linear programs, and a new efficient algorithm. Tutorial in Operations Research, INFORMS

Schrijver A (1986) Theory of linear and integer programming. Wiley-Interscience, New York

Chapter 2
Formulation Techniques Involving Transformations of Variables

2.1 Operations Research: The Science of Better

Operations Research (OR) is the branch of science dealing with tools or techniques for decision making to optimize the performance of systems, that is, to make those systems *better*. Measures of performance, of which there may be several, are numerical criteria that gauge the quality of some aspect of system's performance, for example, annual profit or market share of a company, etc. They are of two types: (1) *profit measures:* (for these, the higher the value the better), (2) *cost measures:* (for these the lower the value the better).

OR deals with techniques for designing ways to operate the system to maximize profit measures or minimize cost measures as desired. Hence OR is the *science to make systems better*.

Linear Programming (LP) is an important branch of OR dealing with decision problems modeled as those of optimizing a linear function of decision variables subject to linear constraints that may include equality constraints, inequality constraints, and bounds in decision variables. In an LP, all decision variables are required to be continuous variables that can assume all possible values within their bounds subject to the constraints. LPs are special instances of mathematical programming. Besides LP, the subject *mathematical programming* includes network, integer, combinatorial, discrete, quadratic, and nonlinear programming.

The focus of this book is to study important aspects of LP and QP (quadratic programming) and their intelligent applications for decision making.

We refer the reader to Chap. 3 in the Junior-level book (Murty (2005b) of Chap. 1; this book can be downloaded from the website mentioned there), where decision-making problems that can be modeled directly as LPs are discussed with many illustrative examples. In this chapter we extend the range of applications of LP to include decision-making problems involving the optimization of a piecewise linear objective function subject to linear constraints. When the objective function satisfies certain properties, these problems can be transformed into LPs in terms of additional variables.

K.G. Murty, *Optimization for Decision Making: Linear and Quadratic Models*, International Series in Operations Research & Management Science 137, DOI 10.1007/978-1-4419-1291-6_2,

2.2 Differentiable Convex and Concave Functions

The concepts of convexity of functions, and of sets, are fundamental pillars in optimization theory. We already know that

> a subset $K \subset R^n$ is said to be a *convex set* if for every pair of points $x,\ y \in K$, every *convex combination* of $x,\ y$ (i.e., point of the form $\alpha x + (1-\alpha)y$ for any $0 \le \alpha \le 1$) is also in K.

A real-valued function $f(x)$ of decision variables $x = (x_1, \dots, x_n)^T \in R^n$ is said to be a *linear function* if it satisfies the following two properties that together are known as the *linearity assumptions*:

Proportionality: $f(\alpha x) = \alpha f(x)$ for all $x \in R^n,\ \alpha \in R^1$

Additivity: $f(x + y) = f(x) + f(y)$ for all $x,\ y \in R^n$

An equivalent definition is: The real-valued function $f(x)$ defined over $x \in R^n$ is a linear function, iff there exists a row vector of constants $c = (c_1, \dots, c_n)$ such that $f(x) = c_1x_1 + \ldots + c_nx_n = cx$ for all $x \in R^n$. In fact, for each $j = 1$ to n, $c_j = f(I_{.j})$, where $I_{.j}$ is the jth column vector of the unit matrix I of order n.

A real-valued function $\theta(x)$ of decision variables $x \in R^n$ is said to be an *affine function* if there exists a constant c_0 such that $\theta(x) - c_0$ is a linear function as defined earlier. Actually this constant $c_0 = \theta(0)$. Thus equivalently, $theta(x)$ is an affine function iff there exist constants $c_0,\ c_1, \dots, c_n$ such that $\theta(x) = c_0 + c_1x_1 + \ldots + c_nx_n$.

The concept of convexity of a function is defined by *Jensen's inequality* stated below; it is related to the concept of convexity of a set, but we will not discuss this relationship in this book as it is not important for the things we discuss here. A function is said to be concave if its negative is convex, but there is no corresponding concept called "concavity" for sets.

Linear and affine functions are both convex and concave; but convex and concave functions may be nonlinear. In this section, we study important properties of differentiable convex, concave functions, which may be nonlinear. A requirement is that the set on which a convex or concave function is defined must be a convex set. We will study convex, concave functions defined over R^n (or over a convex subset of it) for $n \ge 1$ in this section.

2.2.1 *Convex and Concave Functions*

A real-valued function $g(y)$ defined over some convex subset $\Gamma \subset R^n$ (Γ may be R^n itself) is said to be a *convex function* if

$$g(\alpha y^1 + (1-\alpha)y^2) \le \alpha g(y^1) + (1-\alpha)g(y^2)$$

for all $y^1,\ y^2 \in \Gamma$, and $0 \le \alpha \le 1$. This inequality defining a convex function is called *Jensen's inequality* after the Danish mathematician who introduced it.

To interpret Jensen's inequality geometrically, introduce an $(n+1)$th axis for plotting the function value. So points in this space R^{n+1} are $(y, y_{n+1})^T$, where on the y_{n+1}th axis we plot the function value $g(y)$ to get a geometric representation of the function.

The set of all points $\{(y, g(y))^T : y \in \Gamma\}$ in this space R^{n+1} is a surface, which is the *surface* or *graph of the function* $g(y)$.

The line segment $\{(\alpha y^1 + (1-\alpha)y^2, \alpha g(y^1) + (1-\alpha)g(y^2))^T : 0 \leq \alpha \leq 1\}$ joining the two points $(y^1, g(y^1))^T$, $(y^2, g(y^2))^T$ on the graph of the function is called the *chord of the function* between the points y^1, y^2 or on the one-dimensional line interval joining y^1 and y^2. If we plot the function curve and the chord on the line segment $\{\alpha y^1 + (1-\alpha)y^2 : 0 \leq \alpha \leq 1\}$, then Jensen's inequality requires that the function curve lie beneath the chord. See Fig. 2.1 where the function curve and a chord are shown for a function $\theta(\lambda)$ of one variable λ.

The real-valued function $h(y)$ defined on a convex subset $\Gamma \subset R^n$ is said to be a *concave function* if $-h(y)$ is a convex function, that is, if

$$h(\alpha y^1 + (1-\alpha)y^2) \geq \alpha h(y^1) + (1-\alpha)h(y^2)$$

for all y^1, $y^2 \in \Gamma$ and $0 \leq \alpha \leq 1$; see Fig. 2.2. For a concave function $h(y)$, the function curve always lies above every chord.

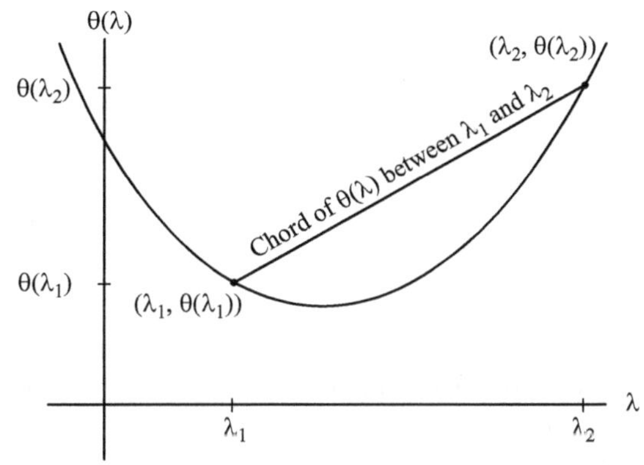

Fig. 2.1 Graph of a convex function $\theta(\lambda)$ defined on R^1 and its chord between two points λ_1 and λ_2

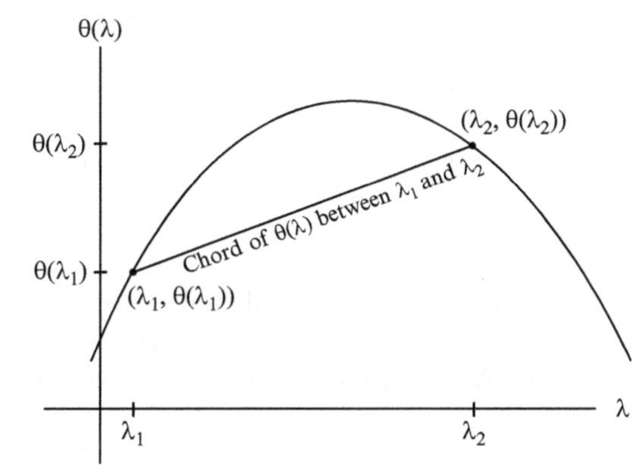

Fig. 2.2 Graph of a concave function $\theta(\lambda)$ defined on R^1 and its chord between two points λ_1 and λ_2

All linear and affine functions (i.e., functions of the form $cx + c_0$, where $c \in R^n$, $c_0 \in R^1$ are given, and $x \in R^n$ is the vector of variables) are both convex and concave.

Other examples of convex functions are $\lambda^{2\mathbf{r}}$, e^{λ} over $\lambda \in R^1$, where r is a positive integer; $-\log(\lambda)$ over $\{\lambda > 0 : \lambda \in R^1\}$; and the quadratic function $x^T Dx + cx + c_0$ over $x \in R^n$, where D is a positive semidefinite (PSD) matrix of order n (a square matrix D of order $n \times n$ is said to be a PSD (positive semidefinite) matrix iff $x^T Dx \geq 0$ for all $x \in R^n$. See Kaplan (1999); Murty (1988, 1995), or Sect. 9.1 for discussion of positive semidefiniteness of a square matrix, and the proof that this quadratic function is convex over the whole space R^n iff D is PSD).

We now derive some important properties of differentiable convex, concave functions. For this discussion, the functions may be nonlinear.

Theorem 2.1. Gradient support inequality for convex functions: *Let $g(y)$ be a real-valued differentiable function defined on R^n. Then $g(y)$ is a convex function iff*

$$g(y) \geq g(\bar{y}) + \nabla g(\bar{y})(y - \bar{y})$$

for all $y, \bar{y} \in R^n$, where $\nabla g(\bar{y}) = \left(\frac{\partial g(\bar{y})}{\partial y_1}, \ldots, \frac{\partial g(\bar{y})}{\partial y_n}\right)$ is the row vector of partial derivatives of $g(y)$ at $\bar{y}$.

Proof. Assume that $g(y)$ is convex. Let $0 < \alpha < 1$. Then $(1-\alpha)\bar{y} + \alpha y = \bar{y} + \alpha(y - \bar{y})$. So, from Jensen's inequality $g(\bar{y} + \alpha(y - \bar{y})) \leq (1-\alpha)g(\bar{y}) + \alpha g(y)$. So

$$g(y) - g(\bar{y}) \geq \frac{g(\bar{y} + \alpha(y - \bar{y})) - g(\bar{y})}{\alpha}.$$

Taking the limit as $\alpha \to 0$, by the definition of differentiability, the RHS in the above inequality tends to $\nabla g(\bar{y})(y - \bar{y})$. So we have $g(y) - g(\bar{y}) \geq \nabla g(\bar{y})(y - \bar{y})$.

Now suppose the inequality in the statement of the theorem holds for all points $\bar{y}, y \in R^n$. Let y^1, y^2 be any two points in R^n and $0 < \alpha < 1$. Taking $y = y^1$, $\bar{y} = (1-\alpha)y^1 + \alpha y^2$, we get the first inequality given below; and taking $y = y^2$, $\bar{y} = (1-\alpha)y^1 + \alpha y^2$, we get the second inequality given below.

$$g(y^1) - g((1-\alpha)y^1 + \alpha y^2) \geq \alpha(\nabla g((1-\alpha)y^1 + \alpha y^2)(y^1 - y^2),$$

$$g(y^2) - g((1-\alpha)y^1 + \alpha y^2) \geq -(1-\alpha)(\nabla g((1-\alpha)y^1 + \alpha y^2)(y^1 - y^2).$$

Multiplying the first inequality above by $(1-\alpha)$ and the second by α and adding, we get $(1-\alpha)g(y^1) + \alpha g(y^2) - g((1-\alpha)y^1 + \alpha y^2) \geq 0$, which is Jensen's inequality. As this holds for all $y^1, y^2 \in R^n$ and $0 < \alpha < 1$, $g(y)$ is convex by definition. □

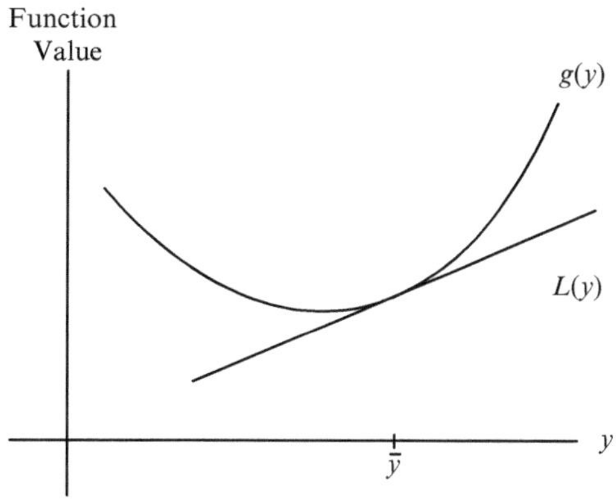

Fig. 2.3 Illustration of the gradient support inequality for a convex function

At any given point $\bar{y}$, the function $L(y) = g(\bar{y}) + \nabla g(\bar{y})(y - \bar{y})$ is an affine function of y, which is known as the *linearization* of the differentiable function $g(y)$ at the point $\bar{y}$. Theorem 2.1 shows that for a differentiable convex function $g(y)$, its linearization $L(y)$ at any point $\bar{y}$ is an underestimate for $g(y)$ at every point y; see Fig. 2.3.

The corresponding result for concave functions obtained by applying the result in Theorem 2.1 to the negative of the function is given in Theorem 2.2.

Theorem 2.2. Gradient support inequality for concave functions: *Let $h(y)$ be a real-valued differentiable function defined on R^n. Then $h(y)$ is a concave function iff*

$$h(y) \leq h(\bar{y}) + \nabla h(\bar{y})(y - \bar{y})$$

for all $y,\ \bar{y} \in R^n$, where $\nabla h(\bar{y}) = \left(\frac{\partial h(\bar{y})}{\partial y_1}, \ldots, \frac{\partial h(\bar{y})}{\partial y_n}\right)$ is the row vector of partial derivatives of $h(y)$ at $\bar{y}$. That is, the linearization of a concave function at any given point $\bar{y}$ is an overestimate of the function at every point; see Fig. 2.4.

Theorem 2.3. *Let $\theta(y)$ be a real-valued differentiable function defined on R^n. Then $\theta(y)$ is a convex [concave] function iff for all $y^1,\ y^2 \in R^n$*

$$\{\nabla\theta(y^2) - \nabla\theta(y^1)\}(y^2 - y^1) \geq 0 \qquad [\leq 0].$$

Proof. We will give the proof for the convex case, and the concave case is proved similarly.

Suppose $\theta(y)$ is convex, and let $y^1,\ y^2 \in R^n$. From Theorem 2.1 we have

$$\theta(y^2) - \theta(y^1) - \nabla\theta(y^1)(y^2 - y^1) \geq 0,$$
$$\theta(y^1) - \theta(y^2) - \nabla\theta(y^2)(y^1 - y^2) \geq 0.$$

Fig. 2.4 Illustration of the gradient support inequality for a concave function

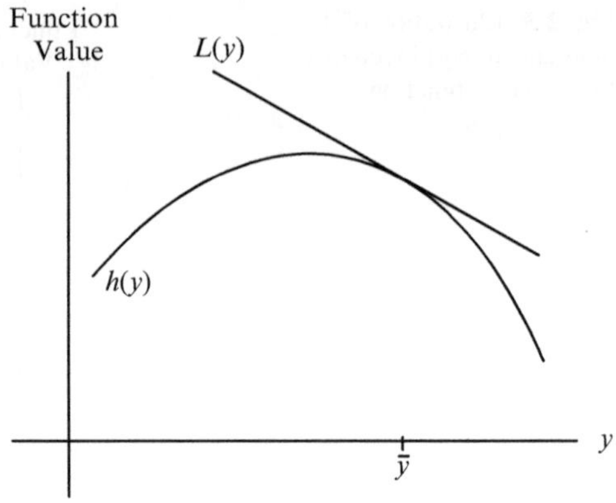

Adding these two inequalities, we get $\{\nabla\theta(y^2) - \nabla\theta(y^1)\}(y^2 - y^1) \geq 0$.

Now suppose that $\theta(y)$ satisfies the property stated in the theorem; and let $y^1,\ y^2 \in R^n$. As $\theta(y)$ is differentiable, by the mean-value theorem of calculus, we know that there exists an $0 < \bar{\alpha} < 1$ such that $\theta(y^2) - \theta(y^1) = \nabla\theta(y^1 + \bar{\alpha}(y^2 - y^1))(y^2 - y^1)$. As $\theta(y)$ satisfies the statement in the theorem, we have

$$\{\nabla\theta(y^1 + \bar{\alpha}(y^2 - y^1)) - \nabla\theta(y^1)\}\bar{\alpha}(y^2 - y^1) \geq 0 \quad \text{or}$$

$$\nabla\theta(y^1 + \bar{\alpha}(y^2 - y^1))(y^2 - y^1) \geq \nabla\theta(y^1)(y^2 - y^1).$$

But by the choice of $\bar{\alpha}$ as discussed above, the left-hand side of the last inequality is $= \theta(y^2) - \theta(y^1)$. Therefore, $\theta(y^2) - \theta(y^1) \geq \nabla\theta(y^1)(y^2 - y^1)$. Since this holds for all $y^1,\ y^2 \in R^n$, by Theorem 2.1, $\theta(y)$ is convex. □

Applying Theorem 2.3 to a function defined over R^1, we get the following result:

Result 2.1. *Let $\theta(\lambda)$ be a differentiable real-valued function of a single variable $\lambda \in R^1$. $\theta(\lambda)$ is convex [concave] iff its derivative $\frac{d\theta}{d\lambda}$ is a monotonic increasing [decreasing] function of λ.*

Hence checking whether a given differentiable function of a single variable λ is convex or concave involves checking whether its derivative is a monotonic function of λ. If the function is twice continuously differentiable, this will hold if the second derivative has the same sign for all λ. If the second derivative is ≥ 0 for all λ, the function is convex; if it is ≤ 0 for all λ, the function is concave.

Now we will discuss the generalization of Result 2.1 to functions defined on R^n for $n \geq 2$. A square matrix D of order n is said to be *positive [negative] semidefinite (PSD or [NSD])* if $x^T Dx \geq [\leq] \; 0$ for all $x \in R^n$. In Chap. 9 these concepts are defined and efficient algorithms for checking whether a given square matrix satisfies these properties are discussed.

Theorem 2.4. *Let $g(y)$ be a twice continuously differentiable real-valued function defined on R^n, and let $H(g(y)) = \left(\frac{\partial^2 g(y)}{\partial y_i \partial y_j}\right)$ denote its Hessian matrix (the $n \times n$ matrix of second partial derivatives) at y. Then $g(y)$ is convex iff $H(g(y))$ is a PSD (positive semi-definite) matrix for all y. Correspondingly, $g(y)$ is concave iff $H(g(y))$ is a NSD (negative semi-definite) matrix for all y.*

Proof. We will prove the convex case. Consider a point $\bar{y} \in R^n$.

Suppose $g(y)$ is convex. Let $\alpha > 0$ and sufficiently small. By Theorem 2.1 we have for each $x \in R^n$

$$(g(\bar{y} + \alpha x) - g(\bar{y}) - \alpha \nabla g(\bar{y}) x)/\alpha \geq 0$$

Take limit as $\alpha \to 0^+$ (through positive values of α). By the mean value theorem of calculus the left-hand side of the above inequality converges to $x^T H(g(\bar{y}))x$, and hence we have $x^T H(g(\bar{y}))x \geq 0$ for all $x \in R^n$, this is the condition for the Hessian matrix $H(g(\bar{y}))$ to be PSD.

Suppose $H(g(y))$ is PSD for all $y \in R^n$. Then by Taylor's theorem of calculus, for any $y^1, \; y^2 \in R^n$

$$g(y^2) - g(y^1) - \nabla g(y^1)(y^2 - y^1) = (y^2 - y^1)^T H(g(y^1 + \alpha(y^2 - y^1)))(y^2 - y^1)$$

for some $0 < \alpha < 1$, which is ≥ 0 since $H(g(y^1 + \alpha(y^2 - y^1)))$ is PSD. So the right-hand side of the above equation is ≥ 0 for all $y^1, \; y^2 \in R^n$; therefore $g(y)$ is convex by Theorem 2.1. □

We know that linear and affine functions are both convex and concave. Now consider the general quadratic function $f(x) = x^T Dx + cx + c_0$ in variables $x \in R^n$, its Hessian matrix $H(f(x)) = (D + D^T)/2$ is a constant matrix. Hence the quadratic function $f(x)$ is convex iff the matrix $(D + D^T)/2$ is a PSD matrix by Theorem 2.4. Checking whether a given square matrix of order n is PSD can be carried out very efficiently with an effort of at most n Gaussian pivot steps (see Kaplan (1999); Murty (1988), or Sect. 9.2 of this book, for the algorithm to use). So whether a given quadratic function is convex or not can be checked very efficiently.

For checking whether a general twice continuously differentiable nonlinear function of x outside the class of linear and quadratic functions is convex may be a hard problem, because its Hessian matrix depends on x, and the job requires checking that the Hessian matrix is a PSD matrix for every x. Fortunately, for piecewise linear (PL) functions, which we will discuss in the next section, checking whether they are convex can be carried out very efficiently even though those functions are not differentiable everywhere.

2.3 Piecewise Linear (PL) Functions

Definition: Piecewise Linear (PL) Functions: Considering real-valued continuous functions $f(x)$ defined over R^n, these are nonlinear functions that may not satisfy the linearity assumptions over the whole space R^n, but there is a partition of R^n into convex polyhedral regions, say $R^n = K_1 \cup K_2 \cup \ldots \cup K_r$ such that $f(x)$ is an affine function within each of these regions individually, that is, for each $1 \leq t \leq r$

there exist constants c_0^t, $c^t = (c_1^t, \ldots, c_n^t)$ such that $f(x) = f_t(x) = c_0^t + c^t x$ for all $x \in K_t$, and for every $S \subset \{1, \ldots, r\}$, and at every point $x \in \cap_{t \in S} K_t$, the different functions $f_t(x)$ for all $t \in S$ have the same value.

Now we give some examples of continuous PL functions defined over R^1. Denote the variable by λ.

Each convex polyhedral subset of R^1 is an interval; so a partition of R^1 into convex polyhedral subsets expresses it as a union of intervals: $[-\infty, \lambda_1] = \{\lambda : \lambda \leq \lambda_1\}$, $[\lambda_1, \lambda_2] = \{\lambda : \lambda_1 \leq \lambda \leq \lambda_2\}, \ldots, [\lambda_{r-1}, \lambda_r], [\lambda_r, \infty]$, where $\lambda_1, \ldots, \lambda_r$ are the boundary points of the various intervals, usually called the *breakpoints* in this partition.

The function $\theta(\lambda)$ is a PL function if there exists a partition of R^1 like this such that inside each interval of this partition the slope of $\theta(\lambda)$ is a constant, and its value at each breakpoint agrees with the limits of $\theta(\lambda)$ as λ approaches this breakpoint from the left, or right; that is, it should be of the form tabulated below:

Interval	Slope of $\theta(\lambda)$ in interval	Value of $\theta(\lambda)$
$\lambda \leq \lambda_1$	c_1	$c_1\lambda$
$\lambda_1 \leq \lambda \leq \lambda_2$	c_2	$\theta(\lambda_1) + c_2(\lambda - \lambda_1)$
$\lambda_2 \leq \lambda \leq \lambda_3$	c_3	$\theta(\lambda_2) + c_3(\lambda - \lambda_2)$
$\vdots$	$\vdots$	$\vdots$
$\lambda_{r-1} \leq \lambda \leq \lambda_r$	c_r	$\theta(\lambda_{r-1}) + c_r(\lambda - \lambda_{r-1})$
$\lambda \geq \lambda_r$	c_{r+1}	$\theta(\lambda_r) + c_{r+1}(\lambda - \lambda_r)$

Notice that the PL function $\theta(\lambda)$ defined in the table above is continuous, and at each of the breakpoints $\bar{\lambda} \in \{\lambda_1, \ldots, \lambda_r\}$ we verify that

$$\lim_{\epsilon \to 0^-} \theta(\bar{\lambda} + \epsilon) = \lim_{\epsilon \to 0^+} \theta(\bar{\lambda} + \epsilon) = \theta(\bar{\lambda}).$$

Here are numerical examples of continuous PL functions:

Example 2.1.

Interval	Slope of $\theta(\lambda)$ in interval	Values of $\theta(\lambda)$
$-\infty$ to 10	3	3λ
10–25	5	$30 + 5(\lambda - 10)$
25 to ∞	7	$105 + 7(\lambda - 25)$

■

Example 2.2.

Interval	Slope of $\theta(\lambda)$ in interval	Values of $\theta(\lambda)$
$-\infty$ to 100	10	10λ
100–300	5	$1{,}000 + 5(\lambda - 100)$
300–1,000	10	$2{,}000 + 10\,(\lambda - 300)$
1,000 to ∞	20	$9{,}000 + 20(\lambda - 1{,}000)$

■

Exercises

2.3.1. (1) Show that the sum of PL functions is PL. Show that a linear combination of PL functions is PL.

(2) Show that the function $\theta(\lambda) = 1/(1-\lambda)^2$ is convex on the set $-1 \le \lambda < 1$. Also, show that the function $\lambda^6 - 15\lambda^2$ is convex on the set $2 \le \lambda \le 3$.

2.3.2. Is the subset of R^2, $\{x = (x_1, x_2)^T : x_1 x_2 > 1\}$, a convex set? What about its complement?

2.3.3. Show that a real-valued function $f(x)$ of decision variables $x \in R^n$ is an affine function iff for any $x \in R^n$ the function $g(y) = f(x+y) - f(x)$ is a linear function of y.

2.3.4. Let $K_1 \cup K_2 \cup \ldots \cup K_r$ be a partition of R^n into convex polyhedral regions, and $f(x)$ a real-valued continuous function defined on R^n. Show that $f(x)$ is a PL function with this partition of R^n iff it satisfies the following properties: for each $t \in \{1, \ldots, r\}$, $x \in K_t$

(1) and all y such that $x + \alpha y \in K_t$ for some $\alpha > 0$, $f(x + \lambda y) = f(x) + \lambda((f(x + \alpha y) - f(x))/\alpha)$ for all $\lambda \ge 0$ such that $x + \lambda y \in K_t$; and

(2) for each $y^1, y^2 \in R^n$ such that $x + y^1$, $x + y^2$ are both in K_t, if $x + y^1 + y^2 \in K_t$ also, then $f(x + y^1 + y^2) = f(x) + (f(x + y^1) - f(x)) + (f(x + y^2) - f(x))$.

2.3.5. Show that the function $f(x) = (x_3^2)/(c_0 + c_1 x_1 + c_2 x_2)$ of $x \in R^3$ is a convex function on the set $\{x \in R^3 : c_0 + c_1 x_1 + c_2 x_2 > 0\}$.

2.3.1 Convexity of PL Functions of a Single Variable

We discuss convexity of PL functions next. As these functions are not differentiable at points where there slopes change, the arguments used in the previous section based on differentiability do not apply.

Result 2.2. *Let $\theta(\lambda)$ be a PL function of a single variable $\lambda \in R^1$. Let $\lambda_1, \ldots, \lambda_r$ be the various breakpoints in increasing order where its slope changes. $\theta(\lambda)$ is*

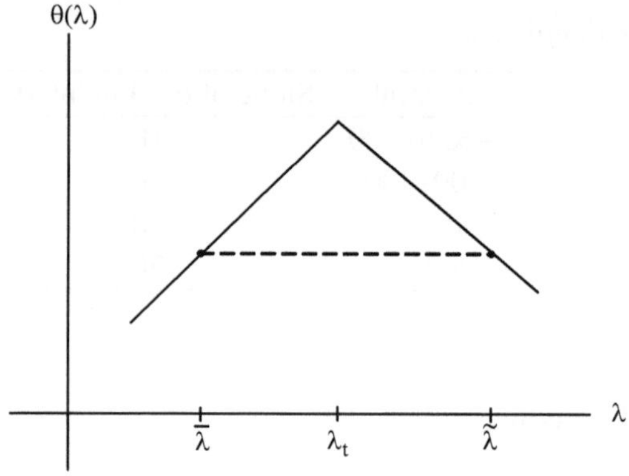

Fig. 2.5 PL function in the neighborhood of a breakpoint λ_t, where slope to the right < slope to the left

convex iff at each breakpoint λ_t its slope to the right of λ_t is strictly greater than its slope to the left of λ_t; that is, iff its slopes are monotonic increasing with the variable.

Proof. Suppose at a breakpoint λ_t, c_t = the slope of $\theta(\lambda)$ to the right of λ_t is $<c_{t-1}$ = its slope to the left of λ_t. Let $\bar{\lambda}$ be a point close to but $<\lambda_t$, where the slope of $\theta(\lambda)$ is c_{t-1}, and $\tilde{\lambda}$ is a point close to but $>\lambda_t$, where its slope is c_t. Then the graph of $\theta(\lambda)$ in the neighborhood of λ_t will be as shown by the solid line in Fig. 2.5. The chord of the function in the interval $\bar{\lambda} \leq \lambda \leq \tilde{\lambda}$ shown by the dashed line segment is below the function, violating Jensen's inequality for convex functions. So, $\theta(\lambda)$ cannot be convex.

If the slopes of the function satisfy the condition mentioned in the Result, then it can be verified that every chord lies above the function, establishing its convexity. □

The corresponding result for concave functions is: a PL function of one variable is concave iff its slope to the right of every breakpoint is less than its slope to the left of that breakpoint, that is, its slopes are monotonic decreasing with the variable. These results provide a convenient way to check whether a PL function of one variable is convex, or concave, or neither. For example, the PL function in Example 2.1 has monotonically increasing slopes, so it is convex. For the one in Example 2.2, the slope is not monotone, so it is neither convex nor concave.

2.3.2 PL Convex and Concave Functions in Several Variables

Let $f(x)$ be a PL function of variables $x = (x_1, \ldots, x_n)^T$ defined over R^n. So, there exists a partition $R^n = \cup_{t=1}^{r} K_t$, where K_t is a convex polyhedral set for all t, the interiors of $K_1, \ldots, K_r$ are mutually disjoint, and $f(x)$ is affine in each K_t; that is, we have vectors c^t and constants c_0^t such that

$$f(x) = c_0^T + c^t x \quad \text{for all } x \in K_t, t = 1 \text{ to } r. \tag{2.1}$$

Checking the convexity of $f(x)$ on R^n is not as simple as in the one-dimensional case (when $n = 1$), but the following theorem explains how it can be done.

Theorem 2.5. *Let $K_1 \cup \ldots \cup K_r$ be a partition of R^n into convex polyhedral regions, and $f(x)$ the PL function defined by the above equation (2.1). Then $f(x)$ is convex iff for each $t = 1$ to r, and for all $x \in K_t$*

$$c_0^t + c^t x = \text{Maximum}\{c_0^p + c^p x : p = 1, \ldots, r.\}$$

In effect, this says that $f(x)$ is convex iff for each $x \in R^n$

$$f(x) = \text{Maximum}\{c_0^p + c^p x : p = 1, \ldots, r.\} \tag{2.2}$$

Proof. Suppose $f(x)$ satisfies the condition (2.2) stated in the theorem. Let $x^1, x^2 \in R^n$ and $0 \leq \alpha \leq 1$. Suppose

$$f(x^1) = \text{Miximum}\{c_0^p + c^p x^1 : p = 1, \ldots, r.\} = c_0^1 + c^1 x^1, \tag{2.3}$$

$$f(x^2) = \text{Maximum}\{c_0^p + c^p x^2 : p = 1, \ldots, r.\} = c_0^2 + c^2 x^2, \tag{2.4}$$

and $f(\alpha x^1 + (1-\alpha)x^2) = \max\{c_0^p + c^p(\alpha x^1 + (1-\alpha)x^2) : p = 1, \ldots, r\}$ $= c_0^a + c^a(\alpha x^1 + (1-\alpha)x^2)$ for some a. Then

$$\begin{aligned} f(\alpha x^1 + (1-\alpha)x^2) &= \alpha(c_0^a + c^a x^1) + (1-\alpha)(c_0^a + c^a x^2), \\ &\leq \alpha(c_0^1 + c^1 x^1) + (1-\alpha)(c_0^2 + c^2 x^2) \\ &\quad \text{from (2.3), (2.4),} \\ &= \alpha f(x^1) + (1-\alpha) f(x^2). \end{aligned}$$

As this holds for all $x^1, x^2 \in R^n$ and $0 \leq \alpha \leq 1$, $f(x)$ is convex by definition.

Now suppose that $K_1 \cup \ldots \cup K_r$ is a partition of R^n into convex polyhedral regions, and $f(x)$ the PL function defined by $f(x) = c_0^t + c^t x$ for all $x \in K_t$, $t = 1$ to r, is convex. Let $\bar{x}$ be any point in R^n, suppose $\bar{x} \in K_{p_b}$. Let $x^1 \in K_1$, $x^2 \in K_2$ be any two points such that $\bar{x}$ is on the line segment L joining them, that is, $\bar{x} = \bar{\lambda}x^1 + (1 - \bar{\lambda}x^2)$ for some $0 < \bar{\lambda} < 1$. For $0 \leq \lambda \leq 1$ let $f(\lambda x^1 + (1-\lambda)x^2) = \theta(\lambda)$.

The line segment L begins in K_{p_0}, where $p_0 = 1$, and suppose it goes through $K_{p_1}, K_{p_2}, \ldots, K_{p_b}, K_{p_{b+1}}, \ldots, K_{p_s}$, where $p_s = 2$; this breaks up L into $s - 1$ intervals, each interval being the portion of L in one of the sets $K_{p_1}, \ldots, K_{p_s}$. Let the breakpoints for these intervals be $\lambda_1, \ldots, \lambda_s$ in increasing order.

So, in the interval $0 \leq \lambda \leq \lambda_1$, $\theta(\lambda) = c_0^{p_1} + c^{p_1}(\lambda x^1 + (1-\lambda)x^2) = {d_0}^{p_1} + {d_1}^{p_1}\lambda$ say. In the next interval $\lambda_1 \leq \lambda \leq \lambda_2$, $\theta(\lambda) = c_0^{p_2} + c^{p_2}(\lambda x^1 + (1-\lambda)x^2) = {d_0}^{p_2} + {d_1}^{p_2}\lambda$, etc. As $f(x)$ is continuous, $\theta(\lambda)$ is continuous, so at $\lambda = \lambda_1$, the two functions ${d_0}^{p_1} + {d_1}^{p_1}\lambda$, ${d_0}^{p_2} + {d_1}^{p_2}\lambda$ have the same value, and so on.

As $f(x)$ is convex, $\theta(\lambda)$ which is $f(x)$ on the line segment L must also be convex. So from Result 2.2 we must have ${d_1}^{p_1} < {d_1}^{p_2} < {d_1}^{p_3} < \ldots < {d_1}^{p_s}$. From this and the continuity of $\theta(\lambda)$ it can be verified that $\theta(\bar{\lambda}) = {d_0}^{p_b} + {d_1}^{p_b}\bar{\lambda} \geq {d_0}^{p} + {d_1}^{p}\bar{\lambda}$ for all $p \in \{p_1, \ldots, p_s\}$, that is,

$$f(\bar{x}) = c_0^{p_b} + c^{p_b}\bar{x} \geq c_0^{p} + c^{p}\bar{x} \quad \text{for all } p \in \{p_1, \ldots, p_s\}.$$

By varying the points x^1, x^2, the same argument leads to the conclusion that

$$f(\bar{x}) = c_0^{p_b} + c^{p_b}\bar{x} \geq c_0^{p} + c^{p}\bar{x} \quad \text{for all } p = 1 \text{ to } r.$$

Since this holds for all points $\bar{x}$, $f(x)$ satisfies (2.2). □

The function $f(x)$ defined by (2.2) is called the *pointwise supremum function* of the set of affine functions $\{{c_0}^p + c^p x : p = 1, \ldots, r\}$. Theorem 2.5 shows that a PL function defined on R^n is convex iff it is the pointwise supremum of a finite set of affine functions. In fact, in all applications where PL convex functions of two or more variables appear, they are usually seen in the form of pointwise supremum functions only. So, equations like (2.2) have become the standard way for defining PL convex functions.

In the same way, the PL function $h(x)$ defined on R^n is concave iff it is the *pointwise infimum* of a finite set of affine functions, that is, it is of the form $h(x) = \text{minimum}\{{c_0}^p + c^p x : p = 1 \text{ to } r\}$ for each $x \in R^n$.

In Fig. 2.6 we illustrate a pointwise supremum function $\theta(\lambda)$ of a single variable λ. λ is plotted on the horizontal axis, and the values of the function are plotted along the vertical axis. The function plotted is the pointwise supremum

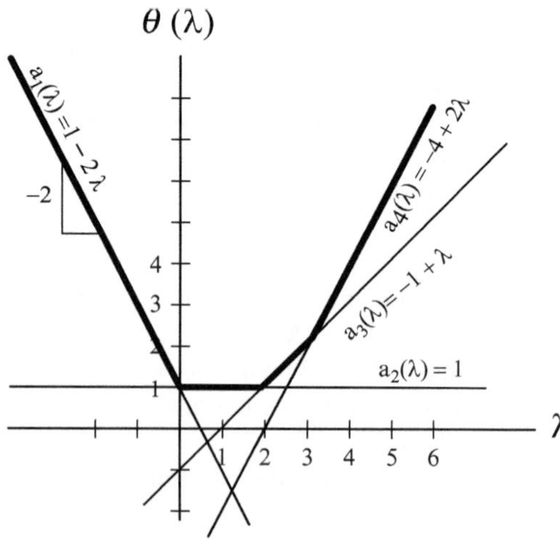

Fig. 2.6 Convexity and pointwise supremum property of a function of one variable. The various functions of which it is supremum are called $a_1(\lambda)$ to $a_4(\lambda)$

$\theta(\lambda) = \max\{a_1(\lambda) = 1 - 2\lambda,\ a_2(\lambda) = 1 + 0\lambda,\ a_3(\lambda) = -1 + \lambda,\ a_4(\lambda) = -4 + 2\lambda\}$. The graph of $\theta(\lambda)$ is plotted in the figure with thick lines. The function is:

Interval	$\theta(\lambda)$	Slope in interval
$\lambda \leq 0$	$1 - 2\lambda$	-2
$0 \leq \lambda \leq 2$	1	0
$2 \leq \lambda \leq 3$	$-1 + \lambda$	1
$\lambda \geq 3$	$-4 + 2\lambda$	2

In Fig. 2.7, we illustrate a PL concave function $h(\lambda)$ of a single variable λ, which is the pointwise infimum $h(\lambda) = \min\{a_1(\lambda) = 4 + \lambda,\ a_2(\lambda) = 3 + (1/2)\lambda,\ a_3(\lambda) = 3 - \lambda,\ a_4(\lambda) = 4 - 2\lambda\}$. The graph of $h(\lambda)$ is shown in thick lines. This function is:

Interval	$h(\lambda)$	Slope in interval
$\lambda \leq -2$	$4 + \lambda$	1
$-2 \leq \lambda \leq 0$	$3 + (1/2)\lambda$	1/2
$0 \leq \lambda \leq 1$	$3 - \lambda$	-1
$\lambda \geq 1$	$4 - 2\lambda$	-2

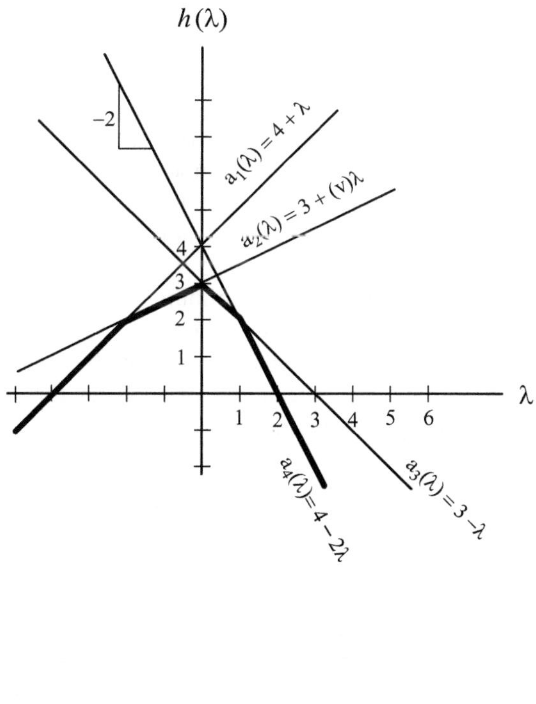

Fig. 2.7 Concavity and pointwise infimum property of a function of one variable. The various functions of which it is infimum are called $a_1(\lambda)$ to $a_4(\lambda)$

Exercises

2.3.6. Considering functions of decision variables $x = (x_1, \ldots, x_n)^T$ defined over R^n, prove that: (**1**) the sum of convex (concave) functions is convex (concave), (**2**) any positive combination of convex (concave) functions is convex (concave), (**3**) pointwise supremum of convex functions is convex, likewise pointwise infimum of concave functions is concave.

2.3.7. (**1**) Consider the function $\theta(\lambda) = |\lambda|$ of a real-valued variable λ. Draw the graph of $\theta(\lambda)$ and show that it is a PL convex function. (**2**) In the same way show that $f(x) = c|\lambda|$, where c is a constant, is PL convex if $c \geq 0$, and PL concave if $c \leq 0$. (**3**) Draw the graphs of the absolute values of affine functions $|4 + \lambda|$ and $|4 - 2\lambda|$ and show that these functions are PL convex. (**4**) For any $j = 1$ to n, show that the function $f(x) = |x_j|$ of $x = (x_1, \ldots, x_n)^T$ defined over R^n is PL convex. What are the regions of R^n within which it is linear? (**5**) Show that the function $f(x) = \sum_{j=1}^{n} c_j |x_j|$ defined over R^n is convex if $c_j \geq 0$ for all j, concave if $c_j \leq 0$ for all j. (**6**) Show that the absolute value function $f(x) = |c_0 + cx|$ of $x \in R^n$ is convex. What are the regions of R^n within which it is linear? Express this function as the pointwise supremum of a set of affine functions. (**7**) Show that the function $f(x) = \sum_{r=1}^{t} w_r |c_0^r + c^r x|$ (linear combinations of affine functions) is convex if $w_r \geq 0$ for all r, concave if $w_r \leq 0$ for all r.

2.3.8. Consider the real-valued continuous function $f(\lambda)$ of a variable λ, defined over $\lambda \geq 0$; with $f(0) = -20$; and slopes of 5, 9, 11, 8, 6, 10, respectively, in the intervals $[0, 20]$, $[20, 50]$, $[50, 60]$, $[60, 80]$, $[80, 90]$, $[90, \infty]$. Is it a convex or a concave function over $\lambda \geq 0$? If not, are there convex subsets of R^1 on which this function is convex or concave? If so, mention these and explain the reasons for the same.

2.3.9. Consider a function $\theta(x)$ defined over a convex set $\Gamma \subset R^n$. A point $\bar{x} \in \Gamma$ is said to be a **local minimum** for $\theta(x)$ over Γ if $\theta(x) \geq \theta(\bar{x})$ for all points $x \in \Gamma$ satisfying $||x - \bar{x}|| \leq \epsilon$ for some $\epsilon > 0$.

A local minimum $\bar{x}$ for $\theta(x)$ in Γ is said to be its *global minimum* in Γ if $\theta(x) \geq \theta(\bar{x})$ for all points $x \in \Gamma$. Local maximum, global maximum have corresponding definitions.

Prove that every local minimum [maximum] of $\theta(x)$ in Γ is a global minimum [maximum] if $\theta(x)$ is convex [concave]. Construct simple examples of general functions defined over R^1 which do not satisfy these properties.

Also, construct an example of a convex function that has a local maximum that is not a global maximum.

2.3.10. Show that the function $f(\lambda) = |\lambda + 1| + |\lambda - 1|$ defined on R^1 is convex, and that it has many local minima all of which are its global minima.

2.4 Optimizing PL Functions Subject to Linear Constraints

The problem of optimizing a general continuous PL function subject to linear constraints is a hard problem for which there are no known efficient algorithms. Some of these problems can be modeled as integer programs and solved by enumerative methods known for integer programs. These enumerative methods are fine for handling small-size problems, but require too much computer time as the problem size increases. However, the special problems of either:

- Minimizing a PL convex function, or equivalently
- Maximizing a PL concave function

subject to linear constraints can be transformed into LPs by introducing additional variables, and solved by efficient algorithms available for LPs. We will now discuss these transformations with several illustrative examples.

2.4.1 Minimizing a Separable PL Convex Function Subject to Linear Constraints

The negative of a concave function is convex. Maximizing a concave function is the same as minimizing its negative, which is a convex function. Using this, the techniques discussed here can also be used to solve problems in which a separable PL concave function is required to be maximized subject to linear constraints.

A real-valued function $z(x)$ of decision variables $x = (x_1, \dots, x_n)^T$ is said to be a *separable* function if it can be expressed as the sum of n different functions, each one involving only one variable, that is, has the form $z(x) = z_1(x_1) + z_2(x_2) + \dots + z_n(x_n)$. This separable function is also a PL convex function if $z_j(x_j)$ is a PL convex function for each $j = 1$ to n.

Result 2.3. *Let $\theta(\lambda)$ be the PL convex function of $\lambda \in R^1$ defined over $\lambda \geq 0$ shown in the following table:*

Interval	*Slope*	$\theta(\lambda) =$	*Interval length*
$0 = \lambda_0 \leq \lambda \leq \lambda_1$	c_1	$c_1\lambda$	λ_1
$\lambda_1 \leq \lambda \leq \lambda_2$	c_2	$\theta(\lambda_1) + c_2(\lambda - \lambda_1)$	$\lambda_2 - \lambda_1$
$\lambda_2 \leq \lambda \leq \lambda_3$	c_3	$\theta(\lambda_2) + c_3(\lambda - \lambda_2)$	$\lambda_3 - \lambda_2$
$\vdots$	$\vdots$	$\vdots$	
$\lambda_{r-1} \leq \lambda \leq \lambda_r = \infty$	c_r	$\theta(\lambda_{r-1}) + c_r(\lambda - \lambda_{r-1})$	∞

where $\lambda_1 < \lambda_2 < \ldots < \lambda_{r-1}$ and $c_1 < c_2 < ldots < c_r$ (conditions for $\theta(\lambda)$ to be convex). Then for any $\bar{\lambda} \geq 0$, $\theta(\bar{\lambda})$ is the minimum objective value in the following problem.

$$\begin{aligned} \textit{Minimize } z &= c_1\mu_1 + \ldots + c_r\mu_r \\ \textit{subject to } \quad & \mu_1 + \ldots + \mu_r = \bar{\lambda} \\ & 0 \leq \mu_t \leq \lambda_t - \lambda_{t-1} \quad t = 1, \ldots, r \end{aligned} \tag{2.5}$$

Proof. Problem (2.5) can be interpreted this way: Suppose we want to purchase exactly $\bar{\lambda}$ units of a commodity for which there are r suppliers. For $k = 1$ to r, kth supplier's rate is c_k/unit and can supply up to $\lambda_k - \lambda_{k-1}$ units only. μ_k in the problem represents the amount purchased from the kth supplier, it is ≥ 0, but is bounded above by the length of the kth interval in which the slope of $\theta(\lambda)$ is c_k. z to be minimized is the total expense to acquire the required $\bar{\lambda}$ of the commodity.

Clearly, to minimize z, we should purchase as much as possible from the cheapest supplier, and when he cannot supply any more go to the next cheapest supplier, and continue the same way until the required quantity is acquired. As the cost coefficients satisfy $c_1 < c_2 < \ldots < c_r$ by the convexity of $\theta(\lambda)$, the cheapest cost coefficient corresponds to the leftmost interval beginning with 0, the next cheapest corresponds to the next interval just to the right of it, and so on. Because of this, the optimum solution $\bar{\mu} = (\bar{\mu}_1, \ldots, \bar{\mu_r})$ of (2.5) satisfies the following special property.

Special property of optimum solution $\bar{\mu}$ of (2.5) that follows from convexity of $\theta(\lambda)$: If p is such that $\lambda_p \leq \bar{\lambda} \leq \lambda_{p+1}$, then $\bar{\mu}_t = \lambda_t - \lambda_{t-1}$, the upper bound of μ_t for all $t = 1$ to p, $\bar{\mu}_{p+1} = \bar{\lambda} - \lambda_p$, and $\bar{\mu}_t = 0$ for all $t \geq p + 2$.

This property says that in the optimum solution of (2.5) if any $\mu_k > 0$, then the value of μ_t in it must be equal to the upper bound on this variable for any $t < k$. Because of this, the optimum objective value in (2.5) is $= c_1\bar{\mu}_1 + \ldots + c_r\bar{\mu}_r \theta(\bar{\lambda})$. □

Example 2.3. – **Illustration of Result 2.3:** Consider the following PL function.

Interval	Slope in interval	$\theta(\lambda) =$	Interval length
0–10	1	λ	10
10–25	2	$10 + 2(\lambda - 10)$	15
25–30	4	$40 + 4(\lambda - 25)$	5
30–∞	6	$60 + 6(\lambda - 30)$	∞

As the slope is increasing with λ, $\theta(\lambda)$ is convex. Consider $\bar{\lambda} = 27$. We see that $\theta(27) = 48$. The LP corresponding to (2.5) for $\bar{\lambda} = 27$ in this problem is

$$\begin{aligned} \text{Minimize } z &= \mu_1 + 2\mu_2 + 4\mu_3 + 6\mu_4 \\ \text{subject to } \quad & \mu_1 + \mu_2 + \mu_3 + \mu_4 = 27 \\ & 0 \leq \mu_1 \leq 10,\ 0 \leq \mu_2 \leq 15 \\ & 0 \leq \mu_3 \leq 5,\ 0 \leq \mu_4 \end{aligned}$$

The optimum solution of this LP is obtained by increasing the values of $\mu_1, \mu_2, \mu_3, \mu_4$ one at a time from 0 in this order, moving to the next when this reaches its upper bound, until the sum of these variables reaches 27. So, the optimum solution is $\bar{\mu} = (10, 15, 2, 0)^T$ with its objective value of $\bar{\mu_1} + 2\bar{\mu_2} + 4\bar{\mu_3} + 6\bar{\mu_4} = 48 = \theta(27)$ computed earlier from the definition of this function, verifying Result 2.3 in this example. ■

If $\theta(\lambda)$ is not convex, the optimum solution of (2.5) will not satisfy the special property described in the proof of Result 2.3.

Because of this result when $\theta(\lambda)$ is PL convex, in minimizing a PL convex function in which $\theta(\lambda)$ is one of the terms, we can linearize $\theta(\lambda)$ by replacing λ by $\sum_{t=1}^{r} \mu_t$, where μ_t is a new nonnegative variable corresponding to the tth interval in the definition of $\theta(\lambda)$, bounded above by the length of this interval, and replacing $\theta(\lambda)$ by $\sum_{t=1}^{r} c_t \mu_t$.

So now consider the problem

$$\begin{aligned} \text{Minimize } \; & z(x) = z_1(x_1) + \ldots + z_n(x_n) \\ \text{subject to } \; & Ax = b \\ & x \geq 0, \end{aligned} \tag{2.6}$$

where, for each j, $z_j(x_j)$ is a PL convex function defined on $x_j \geq 0$. Suppose the various slopes for $z_j(x_j)$ are $c_j^1 < c_j^2 < \ldots c_j^{r_j}$ in that order with slopes changing at the values $d_j^1 < d_j^2 < \ldots d_j^{-1+r_j}$ for the variable x_j. Then from this discussion, the LP formulation for (2.6) involving new variables x_j^k for $k = 1$ to r_j, $j = 1$ to n is (here $\ell_j^k = d_j^k - d_j^{k-1}$ = length of the kth interval in the definition of $z_j(x_j)$)

$$\begin{aligned} \text{Minimize } \; & \sum_{j=1}^{n} \sum_{k=1}^{r_j} c_j^k x_j^k \\ \text{subject to } \; & \sum_{k=1}^{r_j} x_j^k = x_j, \quad j = 1 \text{ to } n \\ & Ax = b \\ & x \geq 0 \\ & 0 \leq x_j^k \leq \ell_j^k, \quad 1 \leq j \leq n, \; 1 \leq k \leq r_j \end{aligned} \tag{2.7}$$

Example 2.4. A company makes products P_1, P_2, P_3 using limestone (LI), electricity (EP), water (W), fuel (F), and labor (L) as inputs. Labor is measured in man hours, other inputs in suitable units. Each input is available from one or more sources.

The company has its own quarry for LI, which can supply up to 250 units/day at a cost of $20/unit. Beyond that, LI can be purchased in any amounts from an outside supplier at $50/unit.

EP is available only from the local utility. Their charges for EP are $30/unit for the first 1,000 units/day, $45/unit for up to an additional 500 units/day beyond the initial 1,000 units/day, $75/unit for amounts beyond 1,500 units/day.

Up to 800 units/day of W (water) is available from the local utility at $6/unit, beyond that they charge $7/unit of water/day.

There is a single supplier for F who can supply at most 3,000 units/day at $40/unit, beyond that there is currently no supplier for F.

From their regular workforce they have up to 640 man hours of labor/day at $10/man hour, beyond that they can get up to 160 man hours/day at $17/man hour from a pool of workers.

They can sell up to 50 units of P_1 at $3,000/unit/day in an upscale market; beyond that they can sell up to 50 more units/day of P_1 to a wholesaler at $250/unit. They can sell up to 100 units/day of P_2 at $3,500/unit. They can sell any quantity of P_3 produced at a constant rate of $4,500/unit.

Data on the inputs needed to make the various products is given in the following table. Formulate the product mix problem to maximize the net profit/day at this company.

Product	Input units/unit made				
	LI	EP	W	F	L
P_1	1/2	3	1	1	2
P_2	1	2	1/4	1	1
P_3	3/2	5	2	3	1

Maximizing the net profit is the same thing as minimizing its negative, which is = (the costs of all the inputs used/day) – (sales revenue/day). We verify that each term in this sum is a PL convex function. So, we can model this problem as an LP in terms of variables corresponding to each interval of constant slope of each of the input and output quantities.

Let LI, EP, W, F, L denote the quantities of the respective inputs used/day; and $\mathrm{P}_1,\ \mathrm{P}_2,\ \mathrm{P}_3$ denote the quantities of the respective products made and sold/day. Let $\mathrm{LI}_1,\ \mathrm{LI}_2$ denote units of limestone used daily from own quarry, outside supplier. Let $\mathrm{EP}_1,\ \mathrm{EP}_2,\ \mathrm{EP}_3$ denote units of electricity used/day at $30, 45, 75/unit, respectively. Let $\mathrm{W}_1,\ \mathrm{W}_2$ denote units of water used /day at rates of $6 and 7/unit, respectively. Let $\mathrm{L}_1,\ \mathrm{L}_2$ denote the man hours of labor used/day from regular workforce, pool, respectively. Let $\mathrm{P}_{11},\ \mathrm{P}_{12}$ denote the units of P_1 sold at the upscale market, to the wholesaler, respectively.

Then the LP model for the problem is

$$\begin{aligned}\text{Minimize } z = {} & 20\mathrm{LI}_1 + 50\mathrm{LI}_2 + 30\mathrm{EP}_1 + 45\mathrm{EP}_2 + 75\mathrm{EP}_3 + 6\mathrm{W}_1 + 7\mathrm{W}_2 + 40\mathrm{F} \\ & + 10\mathrm{L}_1 + 17\mathrm{L}_2 - 3{,}000 P_{11} - 250\mathrm{P}_{12} - 3{,}500\mathrm{P}_2 - 4{,}500\mathrm{P}_3\end{aligned}$$

subject to

$$
\begin{aligned}
(1/2)\mathrm{P}_1 + \mathrm{P}_2 + (3/2)\mathrm{P}_3 &= \mathrm{LI} \\
3\mathrm{P}_1 + 2\mathrm{P}_2 + 5\mathrm{P}_3 &= \mathrm{EP} \\
\mathrm{P}_1 + (1/4)\mathrm{P}_2 + 2\mathrm{P}_3 &= \mathrm{W} \\
\mathrm{P}_1 + \mathrm{P}_2 + 3\mathrm{P}_3 &= \mathrm{F} \\
2\mathrm{P}_1 + \mathrm{P}_2 + \mathrm{P}_3 &= \mathrm{L} \\
\mathrm{LI}_1 + \mathrm{LI}_2 = \mathrm{LI}, &\quad \mathrm{W}_1 + \mathrm{W}_2 = \mathrm{W} \\
\mathrm{EP}_1 + \mathrm{EP}_2 + \mathrm{EP}_3 &= \mathrm{EP} \\
\mathrm{L}_1 + \mathrm{L}_2 = \mathrm{L}, &\quad \mathrm{P}_{11} + \mathrm{P}_{12} = \mathrm{P}_1, \quad \text{All variables} \geq 0 \\
(\mathrm{LI}_1, \mathrm{EP}_1, \mathrm{EP}_2, \mathrm{W}_1) &\leq (250, 1{,}000, 500, 800) \\
(\mathrm{F}, \mathrm{L}_1, \mathrm{L}_2) &\leq (3{,}000, 640, 160) \\
(\mathrm{P}_{11}, \mathrm{P}_{12}, \mathrm{P}_2) &\leq (50, 50, 100).
\end{aligned}
$$

■

2.4.2 *Min-max, Max-min Problems*

As discussed earlier, a PL convex function in variables $x = (x_1, \dots, x_n)^T$ can be expressed as the pointwise maximum of a finite set of affine functions. Minimizing a function like that subject to some constraints is appropriately known as a *min-max problem*.

Similarly, a PL concave function in x can be expressed as the pointwise minimum of a finite set of affine functions. Maximizing a function like that subject to some constraints is appropriately known as a *max-min problem*. Both min-max and max-min problems can be expressed as LPs using just one additional variable, if all the constraints are linear constraints.

If the PL convex function $f(x) = \text{maximum}\{c_0^t + c^t x : t = 1, \dots, r\}$, then $-f(x) = \text{minimum}\{-c_0^t - c^t x : t = 1, \dots, r\}$ is PL concave and conversely. Using this, any min-max problem can be posed as a max-min problem and vice versa. So, it is sufficient to discuss max-min problems. Consider the max-min problem

$$
\begin{aligned}
\text{Maximize } z(x) &= \text{Minimum}\{c_0^1 + c^1 x, \dots, c_0^r + c^r x\} \\
\text{subject to } \quad Ax &= b \\
x &\geq 0.
\end{aligned}
$$

To transform this problem into an LP, introduce the new variable x_{n+1} to denote the value of the objective function $z(x)$ to be maximized. Then the equivalent LP with additional linear constraints is

$$
\begin{aligned}
\text{Maximize } \quad & x_{n+1} \\
\text{subject to } \quad & x_{n+1} \leq c_0^1 + c^1 x \\
& x_{n+1} \leq c_0^2 + c^2 x \\
& \vdots
\end{aligned}
$$

$$\begin{aligned} x_{n+1} &\le c_0^r + c^r x \\ Ax &= b \\ x &\ge 0. \end{aligned}$$

The fact that x_{n+1} is being maximized and the additional constraints together imply that if $(\bar{x},\ \bar{x}_{n+1})$ is an optimum solution of this LP model, then $\bar{x}_{n+1} = \min\{c_0^1 + c^1\bar{x}, \ldots, c_0^r + c^r\bar{x}\} = z(\bar{x})$, and that $\bar{x}_{n+1}$ is the maximum value of $z(x)$ in the original max-min problem.

Example 2.5. **Application of the Min-max Model in Worst Case Analysis:** Consider the fertilizer maker's product mix problem with decision variables $x_1,\ x_2$ (hi-ph, lo-ph fertilizers to be made daily in the next period) discussed in Sect. 1.7.1 and in Example 3.4.1 of Sect. 3.4 of Murty (2005b) of Chap. 1. This company makes hi-ph, lo-ph fertilizers using raw materials RM1, RM2, RM3 with the following data (Table 2.1):

We discussed the case where the net profit coefficients $c_1,\ c_2$ of these variables are estimated to be \$15 and 10, respectively. In reality, the prices of fertilizers are random variables that fluctuate daily. Because of unstable conditions and new agricultural research announcements, suppose that market analysts have only been able to estimate that the expected net profit coefficient vector $(c_1,\ c_2)$ is likely to be one of $\{(15, 10),\ (10, 15),\ (12, 12)\}$ without giving a single point estimate. So, here we have three possible scenarios. In scenario 1, $(c_1,\ c_2) = (15, 10)$, expected net profit $= 15x_1 + 10x_2$; in scenario 2, $(c_1,\ c_2) = (10, 15)$, expected net profit $= 10x_1 + 15x_2$; and in scenario 3, $(c_1,\ c_2) = (12, 12)$, expected net profit $= 12x_1 + 12x_2$. Suppose the raw material availability data in the problem is expected to remain unchanged. The important question is: which objective function to optimize for determining the production plan for next period.

Regardless of which of the three possible scenarios materializes, at the worst the minimum expected net profit of the company will be $p(x) = \min\{15x_1 + 10x_2,\ 10x_1 + 15x_2,\ 12x_1 + 12x_2\}$ under the production plan $x = (x_1,\ x_2)^T$. *Worst case analysis* is an approach that advocates determining the production plan to optimize this worst case net profit $p(x)$ in this situation. This leads to the max-min model:

$$\text{Maximize } p(x) = \min\{15x_1 + 10x_2,\ 10x_1 + 15x_2,\ 12x_1 + 12x_2\}$$

Table 2.1 Data for the fertilizer problem

	Tons required to make one ton of		
Item	Hi-ph	Lo-ph	Tons of item available daily
RM 1	2	1	1,500
RM 2	1	1	1,200
RM 3	1	0	500
Net profit \$/ton made	15	10	

$$\begin{aligned} \text{subject to } 2x_1 + x_2 &\le 1{,}500 \\ x_1 + x_2 &\le 1{,}200 \\ x_1 &\le 500 \\ x_1, x_2 &\ge 0. \end{aligned}$$

Its LP formulation is

$$\begin{aligned} \max \quad & p \\ \text{subject to} \quad & p \le 15x_1 + 10x_2 \\ & p \le 10x_1 + 15x_2 \\ & p \le 12x_1 + 12x_2 \\ & 2x_1 + x_2 \le 1{,}500 \\ & x_1 + x_2 \le 1{,}200 \\ & x_1 \le 500, \quad x_1, x_2 \ge 0. \end{aligned}$$

■

2.4.3 Minimizing Positive Linear Combinations of Absolute Values of Affine Functions

Let $z(x) = w_1|c_0^1 + c^1x| + \ldots + w_r|c_0^r + c^r x|$. Consider the problem:

$$\begin{aligned} \text{Minimize } & z(x) \\ \text{subject to } & Ax \ge b, \end{aligned} \tag{2.8}$$

where the weights $w_1, \ldots, w_r$ are all strictly positive. In this problem the objective function to be minimized, $z(x)$, is a PL convex function, hence this problem can be transformed into an LP. This is based on a result that helps to express the absolute value as a linear function of two additional variables, which we will discuss first.

Result 2.4. *Consider the affine function $c_0^k + c^k x$ and its value $\beta = c_0^k + c^k \bar{x}$ at some point $\bar{x} \in R^n$. Consider the following LP in two variables u, v.*

$$\begin{aligned} \textit{Minimize } & u + v \\ \textit{subject to } & u - v = \beta \\ & u, v \ge 0 \end{aligned} \tag{2.9}$$

(2.9) has a unique optimum solution $(\bar{u}, \bar{v})$, which satisfies $\bar{u}\bar{v} = 0$, and its optimum objective value $\bar{u} + \bar{v} = |\beta| = |c_0^k + c^k \bar{x}|$.

Proof. If $\beta \geq 0$, the general solution of (2.9) is $(u, v) = (\beta + \alpha, \alpha)$ for some $\alpha \geq 0$, the objective value of this solution, $\beta + 2\alpha$, assumes its minimum value when $\alpha = 0$. So in this case $(\bar{u}, \bar{v}) = (\beta, 0)$ satisfying $\bar{u}\bar{v} = 0$ and having optimum objective value of $\bar{u} + \bar{v} = \beta = |\beta|$.

If $\beta < 0$, the general solution of (2.9) is $(u, v) = (\alpha, |\beta| + \alpha)$ for some $\alpha \geq 0$, the objective value of this solution, $|\beta| + 2\alpha$, assumes its minimum value when $\alpha = 0$. So in this case $(\bar{u}, \bar{v}) = (0, |\beta|)$ satisfying $\bar{u}\bar{v} = 0$ and having optimum objective value of $\bar{u} + \bar{v} = |\beta|$.

So, the result holds in all cases. □

Example 2.6. **Illustration of Result 2.4:** Consider problem (2.9) when $\beta = -7$. The problem is

minimize $u + v$ subject to $u - v = -7$, $u, v \geq 0$.

The general solution of this problem is $(u, v) = (\alpha, 7 + \alpha)$ for $\alpha \geq 0$ with objective value $7 + 2\alpha$. So, the unique optimum solution is $(\bar{u}, \bar{v}) = (0, 7)$ and $\bar{u} + \bar{v} = 7 = |-7|$ and $\bar{u}\bar{v} = 0$. ■

In the optimum solution $(\bar{u}, \bar{v})$ of (2.9), $\bar{u}$ is usually called the *positive part of* β, and $\bar{v}$ is called the *negative part of* β. Notice that when β is negative, its negative part is actually the absolute value of β. Also, for all values of β, at least one quantity in the pair (positive part of β, negative part of β) is 0.

Commonly the positive or negative parts of β are denoted by symbols β^+, β^-, respectively. In this notation, $\beta = \beta^+ - \beta^-$ and $|\beta| = \beta^+ + \beta^-$; both β^+, β^- are ≥ 0, and satisfy $(\beta^+)(\beta^-) = 0$.

Result 2.4 helps to linearize the objective function in (2.8) by introducing two new variables for each absolute value term in it. Notice that this is only possible when all the coefficients of the absolute value terms in the objective function in (2.8) are positive. From this discussion we see that (2.8) is equivalent to the following LP with two new nonnegative variables for each $t = 1$ to r, $u_t^+ = \text{maximum}\ \{0, c_0^t + c^t x\}$, $u_t^- = -\ \text{minimum}\{0, c_0^t + c^t x\}$. u_t^+ is the *positive part* of $c_0^t + c^t x$ and u_t^- its *negative part*.

$$\begin{aligned} \text{Minimize}\ w_1[(u_1^+) + (u_1^-)] + \ldots &+ w_r[(u_r^+) + (u_r^-)] \\ \text{subject to}\ \ c_0^1 + c^1 x &= (u_1^+) - (u_1^-) \\ &\ \vdots \\ c_0^r + c^r x &= (u_r^+) - (u_r^-) \\ Ax &\geq b \\ (u_t^+),\ (u_t^-) &\geq 0, \quad t = 1, \ldots, r. \end{aligned} \tag{2.10}$$

If $(\hat{u}^+ = (\hat{u}_1^+, \ldots, \hat{u}_r^+),\ \hat{u}^- = (\hat{u}_1^-, \ldots, \hat{u}_r^-),\ \hat{x})$ is an optimum solution of (2.10), then $\hat{x}$ is an optimum solution of (2.8), and $c_0^k + c^k\hat{x} = \hat{u}_k^+ - \hat{u}_k^-$, $|c_0^k + c^k\hat{x}| = \hat{u}_k^+ + \hat{u}_k^-$; and the optimum objective values in (2.10) and (2.8) are the same.

Application of this transformation will be discussed next. This is an important model that finds many applications.

In Model (2.10), by expressing the affine function $c_0^1 + c^1x$, which may be positive or negative, as the difference $u^+ - u^-$ of two nonnegative variables; the positive part of $c_0^1 + c^1x$ denoted by $(c_0^1 + c^1x)^+ = \text{maximum}\{c_0^1 + c^1x,\ 0\}$ will be u^+, and the negative part of $c_0^1 + c^1x$ denoted by $(c_0^1 + c^1x)^- = \text{maximum}\{0,\ -(c_0^1 + c^1x)\}$ will be u^- as long as the condition $(u^+)(u^-) = 0$ holds. This condition will automatically hold as long as:

1. The coefficients of $u^+,\ u^-$ are both $\geq\ 0$ in the objective function being minimized; and
2. The column vectors of the pair of variables $u^+,\ u^-$ in the model among the constraints (not including the sign restrictions) sum to 0 (or form a linearly dependent set).

A Cautionary Note 2.1: When expressing an unrestricted variable or an affine function as a difference $u^+ - u^-$ of two nonnegative variables, and using $u^+,\ u^-$ as the positive, negative parts of that unrestricted variable or affine function, or using $u^+ + u^-$ as its absolute value, it is necessary to make sure that the condition $(u^+)(u^-) = 0$ will automatically hold at very optimum solution of the model. For this, the above two conditions must hold.

Sometimes people tend to include additional constraints involving $u^+,\ u^-$ with nonzero coefficients into the model (for examples, see Model 1 below, and Model 1 for the parameter estimation problem using the L_∞- measure of deviation in Example 2.8 below). When this is done, the Condition 2 above may be violated; this may result in the model being invalid. *So, it is better to not include additional constraints involving $u^+,\ u^-$ into the model.*

2.4.4 *Minimizing the Maximum of the Absolute Values of Several Affine Functions*

Let $z(x) = \text{Maximum}\{|c_0^1 + c^1x|, \dots, |c_0^r + c^rx|\}$. Consider the problem

$$\begin{aligned} \text{Minimize} \quad & z(x) \\ \text{subject to} \quad & Ax \geq b. \end{aligned} \tag{2.11}$$

In this problem the objective function to be minimized, $z(x)$, is the pointwise supremum of several PL convex functions, and hence is a PL convex function, hence this problem can be transformed into an LP. Combining the ideas discussed above, one LP model for this problem is Model 1 given below.

It can be verified that in this model the property $(u_t^+)(u_t^-) = 0$ for all t will hold at every optimum solution for it, so this is a valid model for the problem. But it has one disadvantage that it uses the variables $u_t^+,\ u_t^-$ representing the positive and negative parts of $c_0^t + c^tx$ in additional constraints in the model (those in the first line of constraints), with the result that the pair of column vectors of the variables

u_t^+, u_t^- among the constraints no longer form a linearly dependent set, violating Condition 2 expressed in Cautionary Note 2.1 above.

Model 1

$$\begin{aligned} &\min \quad z \\ &\text{subject to} \quad z \geq u_t^+ + u_t^-, \quad t = 1, \ldots, r \\ &c_0^1 + c^1 x = u_1^+ - u_1^- \\ &\vdots \\ &c_0^r + c^r x = u_r^+ - u_r^- \\ &Ax \geq b \\ &u_t^+, u_t^- \geq 0, \quad t = 1, \ldots, r \end{aligned} \tag{2.12}$$

It is possible to transform (2.11) into an LP model directly without introducing these u_t^+, u_t^- variables at all. This leads to a better and cleaner LP model for this problem, Model 2, with only one additional variable z.

Model 2

$$\begin{aligned} &\min \quad z \\ &\text{subject to} \quad -z \leq c_0^t + c^t x \leq z, \quad t = 1, \ldots, r \\ &Ax \geq b \\ &z \geq 0. \end{aligned} \tag{2.13}$$

The constraints specify that $z \geq |c_0^t + c^t x|$ for all t; and as z is minimized in Model 2, it guarantees that if $(\hat{z}, \hat{x})$ is an optimum solution of this Model 2, then $\hat{x}$ is an optimum solution also for (2.11), and $\hat{z}$ is the optimum objective value in (2.11).

We will now discuss important applications of these transformations in meeting multiple targets as closely as possible, and in curve fitting, and provide simple numerical examples for each.

Example 2.7. **Meeting targets as closely as possible:** Consider the fertilizer maker's product mix problem with decision variables x_1, x_2 (hi-ph, lo-ph fertilizers to be made daily in the next period) discussed in Example 3.4.1 of Sect. 3.4 of Murty (2005b) of Chap. 1 and Example 2.5 above, with net profit coefficients $(c_1, c_2) = (15, 10)$ in \$/ton of hi-ph, lo-ph fertilizers made. In these examples, we considered only maximizing one objective function, the daily net profit $= 15x_1 + 10x_2$ with the profit vector given. But in real business applications, companies have to pay attention to many other objective functions in order to survive and thrive in the market place. We will consider two others.

The second objective function that we will consider is the companies total market share, usually measured by the companies sales volume as a percentage of the sales volume of the whole market. To keep this example simple, we will measure this by the total daily sales revenue of the company. The sale prices of hi-ph, lo-ph fertilizers are \$222, \$107/ton, respectively, so this objective function is $222x_1 + 107x_2$.

The third objective function that we consider is the hi-tech market share, which is the market share of the company among hi-tech products (in this case hi-ph is the hi-tech product). This influences the public's perception of the company as a market leader. To keep this example simple, we will measure this by the daily sales revenue of the company from hi-ph sales which is \$$222x_1$.

So, here we have three different objective functions to optimize simultaneously. Problems like this are called *multiobjective optimization problems*. One commonly used technique to get a good solution in these problems is to set up a target value for each objective function (based on the companies aspirations, considering the trade-offs between the various objective functions), and to try to find a solution as close to each of the targets as possible. In our example, suppose that the targets selected for daily net profit, market share, and hi-tech market share are \$12,500, 200,000, and 70,000, respectively.

In this example, we consider the situation where the company wants to attain the target value for each objective function as closely as possible, considering both positive and negative deviations from the targets as undesirable.

When there is more than one objective function to be optimized simultaneously, decision makers may not consider all of them to be of the same importance. To account for this, it is customary to specify positive weights corresponding to the various objective functions, reflecting their importance, with the understanding that the higher the weight the more important it is to keep the deviation in the value of this objective function from its target small. So, this weight for an objective function plays the role of a penalty for unit deviation in this objective value from its target. In our example, suppose these weights for daily net profit, market share, and hi-tech market share, are 10, 6, and 8, respectively.

After these weights are given, one strategy to solve this problem is to determine the solution to implement to minimize the penalty function, which is the weighted sum of absolute deviations from the targets. This problem is (constraints on the decision variables are given in Example 2.5 above)

$$\text{Minimize penalty function} = 10|15x_1 + 10x_2 - 12{,}500|$$
$$+\ 6|222x_1 + 107x_2 - 200,000| + 8|222x_1 - 70{,}000|$$

$$\begin{aligned} \text{subject to } 2x_1 + x_2 &\le 1{,}500 \\ x_1 + x_2 &\le 1{,}200 \\ x_1 &\le 500 \\ x_1, x_2 &\ge 0. \end{aligned}$$

Linearizing this leads to the following LP:

$$\text{Minimize penalty function} = 10(u_1^+ + u_1^-) + 6(u_2^+ + u_2^-) + 8(u_3^+ + u_3^-)$$

$$\begin{aligned}
\text{subject to } 15x_1 + 10x_2 - 12{,}500 &= u_1^+ - u_1^- \\
222x_1 + 107x_2 - 200{,}000 &= u_2^+ - u_2^- \\
222x_1 - 70{,}000 &= u_3^+ - u_3^- \\
2x_1 + x_2 &\le 1{,}500 \\
x_1 + x_2 &\le 1{,}200 \\
x_1 &\le 500 \\
x_1,\ x_2,\ u_i^+,\ u_i^- &\ge 0, \quad \text{for all } i.
\end{aligned}$$

If $\hat{u}^+ = (\hat{u}_1^+,\ \hat{u}_2^+,\ \hat{u}_3^+)$, $\hat{u}^- = (\hat{u}_1^-,\ \hat{u}_2^-,\ \hat{u}_3^-)$, $\hat{x} = (\hat{x}_1,\ \hat{x}_2))$ is an optimum solution of this LP, then $\hat{x}$ is an optimum solution that minimizes the penalty function. ■

Example 2.8. **Best L_1 or L_∞ Approximations for Parameter Estimation in Curve Fitting Problems:**

A central problem in science and technological research is to determine the optimum operating conditions of processes to maximize the yield from them. Let y denote the yield from a process whose performance is influenced by n controllable factors. Let $x = (x_1, \dots, x_n)^T$ denote the vector of values of these factors, and this vector characterizes how the process is run. So, here $x = (x_1, \dots, x_n)^T$ are the independent variables whose values the decision maker can control, and the yield y is the dependent variable whose value depends on x. To model the problem of determining the optimum x mathematically, it is helpful to approximate y by a mathematical function of x, which we will denote by $y(x)$.

The data for determining the functional form of $y(x)$ is the yield at several points $x \in R^n$ in the feasible range. As there are usually errors in the measurement of yield, one makes several measurement observations of the yield at each point x used in the experiment, and takes the average of these observations as the yield value at that point. The problem of determining the functional form of $y(x)$ from such observed data is known as a **curve fitting problem.**

For a numerical example, consider the data in the following Table 2.2 obtained from experiments for the yield in a chemical reaction, as a function of the temperature t at which the reaction takes place.

The problem in this example is to determine a mathematical function $y(t)$ that fits the observed data as closely as possible.

Table 2.2 Yield at various temperatures

Temperature t	Yield, $y(t)$
−5	80
−3	92
−1	96
0	98
1	100

The commonly used strategy to solve the curve-fitting problem for the dependent variable, yield $y(x)$, in terms of independent variables, $x = (x_1, \dots, x_n)^T$, involves the following steps.

Step 1: Model function selection: Select a specific mathematical functional form $f(x, a)$ with unknown parameters say $a = (a_0, \dots, a_k)$ (these parameters are things like coefficients of various terms, exponents, etc.) that seems to offer the best fit for the yield $y(x)$.

In some cases there may be well-developed mathematical theory that specifies $f(x, a)$ directly. If that is not the case, plots of $y(x)$ against x can give an idea of suitable model functions to select.

For example, if plots indicate that $y(x)$ appears to be linear in x, then we can select the model function to be $f(x, a) = a_0 + a_1x_1 + \dots + a_nx_n$, in which the coefficients $a_0, a_1, \dots, a_n$ are the unknown parameters. This linear model function is the most commonly used one in statistical theory, and the area of this theory that deals with determining the best values for these parameters by the method of least squares is called *linear regression theory.*

If plots indicate that $y(x)$ appears to be quadratic in x, then the model function to use is $\sum_{i=1}^{n} \sum_{j=i}^{n} a_{ij}x_ix_j$ (where the coefficients a_{ij} are the parameters). Similarly, a cubic function in x may be considered as the model function if that appears more appropriate.

The linear, quadratic, cubic functions in x are special cases of the general polynomial function in x. Selecting a polynomial function in x as the model function confers a special advantage for determining the best values for the unknown parameters because this model function is *linear in these parameters.*

When the number of independent variables n is not small (i.e., ≥ 4), using a complete polynomial function in x of degree ≥ 2 as the model function leads to many unknown parameter values to be determined. That is why when such model functions are used, one normally uses the practical knowledge about the problem and the associated process to fix as many as possible of these unknown coefficients that are known to be insignificant with reasonable certainty at 0.

Polynomial functions of x of degree ≤ 3 are the most commonly used model functions for curve-fitting. Functions outside this class are only used when there is supporting theory that indicates that they are more appropriate.

Step 2: Selecting a measure of deviation: Let $f(x, a)$ be the model function selected to represent the yield, with a as the vector of parameters in it. Suppose the data available consists of r observations on the yield as in the following table.

Independent vars.	x^1	x^2	...	x^r
Observed yield	y^1	y^2	...	y^r

Then the deviations of the model function value from the observed yield at the data points $x^1, \dots, x^r$ are $f(x^1, a) - y^1, \dots, f(x^r, a) - y^r$. Some of these deviations may be ≥ 0 and some ≤ 0, but $f(x, a)$ is considered to be a good fit for the yield if all these deviations are small, that is, close to 0. In this step we have to select a single numerical measure that can check whether all these deviations are small or not.

The most celebrated and most commonly used measure of deviation is the *sum of squared deviations*, first used and developed by Carl F. Gauss, the famous nineteenth century German mathematician. He developed this measure for approximating the orbit of the asteroid Ceres with a second degree curve. This measure is also known as the L_2-measure (after the Euclidean or the L_2-metric defined as the square root of the sum of squares), and for our problem it is $L_2(a) = \sum_{k=1}^{r}(f(x^k, a) - y^k)^2$. Determining the best values of the parameters a as those that minimizes this L_2 measure $L_2(a)$ is known as the *method of least squares*.

Another measure of deviation that can be used is the L_1-measure (also known as the *rectilinear measure*); it is the sum of absolute deviations $= L_1(a) = \sum_{k=1}^{r} |f(x^k, a) - y^k|$.

A third measure of deviation that is used by some people is the L_∞-measure (also known as the *Chebyshev measure* after the Russian mathematician Tschebychev who proposed it in the nineteenth century). This measure is the maximum absolute deviation $L_\infty(a) = \max\{|f(x^k, a) - y^k| : k = 1 \text{ to } r\}$.

The L_2-measure is continuously differentiable in the parameters, but the L_1 and L_∞-measures are not (they are not differentiable at points in the parameter space where a deviation term becomes 0). That is why minimizing the L_2-measure using calculus techniques based on derivatives is easier; for this reason the method of least squares has become a very popular method for determining the best values for the unknown parameters to give the best fit to the observed data. Particularly, most of statistical theory is based on the method of least squares.

As they are not differentiable at some points, minimizing the L_1 and L_∞-measures may be difficult in general. However, when the model function $f(x, a)$ is linear in the parameter vector a (this is the case when $f(x, a)$ is a polynomial in x), then determining a to minimize the L_1 or L_∞-measures can be transformed into LPs and solved very efficiently. That is why parameter estimation to minimize the L_1 or L_∞-measures is becoming increasingly popular when $f(x, a)$ is linear in a.

The parameter vector that minimizes the L_2-measure is always unique, but the problem of minimizing L_1 or L_∞-measures usually have alternate optima. There are some other differences among the L_2, L_1, L_∞-measures worth noting. Many people do not like to use the L_∞-measure for parameter estimation, because it determines the parameter values to minimize the deviations of extreme measurements (which are often labeled as "outliers" in statistical literature), totally ignoring all other observations. Both L_1, L_2-measures give equal weight to all the observations.

The L_2-measure would be the preferred measure to use when $f(x, a)$ is not linear in the parameter vector a, because it is differentiable everywhere. When $f(x, a)$ is linear in a, the choice between L_2, L_1-measures of deviation to use for parameter estimation is a matter for individual judgement and the availability of suitable software for carrying out the computations required.

Step 3: Parameter estimation: Solve the problem of determining $\bar{a}$ that minimizes the measure of deviation selected.

The optimum solutions for the problems of minimizing $L_2(a)$, $L_1(a)$, $L_\infty(a)$ may be different. Let $\bar{a}$ denote the optimum a-vector that minimizes whichever measure of deviation has been selected for determining the best a-vector. The optimum objective value in this problem is known as the *residue*. If the residue is "small, " $f(x, \bar{a})$ is accepted as the functional form for $y(x)$.

If the residue is "large, " it is an indication that $f(x, a)$ is not the appropriate functional form for the yield $y(x)$. In this case go back to Step 1 to select a better model function for the yield, and repeat this whole process with it.

Finally, the question of how to judge whether the residue is "small" or "large". Statistical theory provides some tests of significance for this judgement when using the method of least squares. These are developed under the assumption that the observed yield follows a normal distribution. But, in general, the answer to this question depends mostly on personal judgement.

When $f(x,\ a)$ is linear in a, a necessary and sufficient condition for optimality for the problem of minimizing $L_2(a)$ is $\frac{\partial L_2(a)}{\partial a} = 0$. This is a system of linear equations in a, which can be solved for determining the optimum solution $\bar{a}$.

The problems of minimizing $L_1(a)$ are $L_\infty(a)$ when $f(x,\ a)$ linear in a can be transformed into an LP. We will show how to do this using the example of yield in the chemical reaction as a function of the temperature t of the reaction; data for which is given in Table 2.2 above.

Estimates of the Parameter Vector *a* that Minimize $L_2(a)$: Suppose plots indicate that the yield in this chemical reaction, as a function of the reaction temperature, $y(t)$ can be approximated closely by a quadratic function of t. So we take the model function to be $f(t,\ a) = a_0 + a_1 t + a_2 t^2$, where $a = (a_0,\ a_1,\ a_2)$ is the parameter vector to be estimated.

So, $f(-5,\ a) = a_0 - 5a_1 + 25a_2$, hence the deviation between $f(t, a)$ and $y(t)$ at $t = -5$ is $a_0 - 5a_1 + 25a_2 - 80$. Continuing this way, we see that

$$L_2(a) = (a_0 - 5a_1 + 25a_2 - 80)^2 + (a_0 - 3a_1 + 9a_2 - 92)^2 + (a_0 - a_1 + a_2 - 96)^2$$
$$+ (a_0 - 98)^2 + (a_0 + a_1 + a_2 - 100)^2,$$
$$L_1(a) = |(a_0 - 5a_1 + 25a_2 - 80)| + |(a_0 - 3a_1 + 9a_2 - 92)| + |(a_0 - a_1 + a_2 - 96)|$$
$$+ |(a_0 - 98)| + |(a_0 + a_1 + a_2 - 100)|,$$
$$L_\infty\{(a) = \max\{|(a_0 - 5a_1 + 25a_2 - 80)|, |(a_0 - 3a_1 + 9a_2 - 92)|$$
$$|(a_0 - a_1 + a_2 - 96)|, |(a_0 - 98)|, |(a_0 + a_1 + a_2 - 100)|\}.$$

So, the method of least squares involves finding a that minimizes $L_2(a)$. The necessary and sufficient optimality conditions for this are $\frac{\partial L_2(a)}{\partial a} = 0$, which are

$$5a_0 - 8a_1 + 36a_2 = 466,$$
$$-8a_0 + 36a_1 - 152a_2 = -672,$$
$$36a_0 - 152a_1 + 708a_2 = 3{,}024.$$

It can be verified that this has the unique solution of $\bar{a} = (\bar{a}_0, \bar{a}_1, \bar{a}_2) = (98.6141, 1.1770, -0.4904)$. So the fit obtained by the method of least squares is $f(t, \bar{a}) = 98.6141 + 1.1770t - 0.4904t^2$, with a residue of 3.7527, in L_2-measure units.

Estimates of the Parameter Vector *a* that Minimize $L_1(a)$: The problem of minimizing $L_1(a)$ is the following LP:

$$
\begin{aligned}
\text{Minimize } & \sum_{i=1}^{5}(u_i^+ + u_i^-) \\
\text{subject to } (a_0 - 5a_1 + 25a_2 - 80) &= u_1^+ - u_1^- \\
(a_0 - 3a_1 + 9a_2 - 92) &= u_2^+ - u_2^- \\
(a_0 - a_1 + a_2 - 96) &= u_3^+ - u_3^- \\
(a_0 - 98) &= u_4^+ - u_4^- \\
(a_0 + a_1 + a_2 - 100) &= u_5^+ - u_5^- \\
u_i^+, u_i^- &\geq 0, \text{ for all } i.
\end{aligned}
$$

One of the optimum solutions of this problem is $\bar{a} = (\bar{a}_0, \bar{a}_1, \bar{a}_2) = (98.3333, 2, -0.3333,)$; the fit given by this solution is $f(t, \bar{a}) = 98.3333 + 2t - 0.3333t^2$, with a residue of 3, in L_1-measure units.

Estimates of the Parameter Vector *a* that Minimize $L_\infty(a)$: One LP model discussed earlier for the problem of minimizing $L_\infty(a)$ is the following:

Model 1:

$$
\begin{aligned}
& \text{Minimize } z \\
& \text{subject to } z \geq (u_i^+ + u_i^-) \text{ for all } i \\
(a_0 - 5a_1 + 25a_2 - 80) &= u_1^+ - u_1^- \\
(a_0 - 3a_1 + 9a_2 - 92) &= u_2^+ - u_2^- \\
(a_0 - a_1 + a_2 - 96) &= u_3^+ - u_3^- \\
(a_0 - 98) &= u_4^+ - u_4^- \\
(a_0 + a_1 + a_2 - 100) &= u_5^+ - u_5^- \\
u_i^+, u_i^- &\geq 0, \text{ for all } i.
\end{aligned}
$$

One of the optimum solutions of this model is $\hat{a} = (\hat{a}_0, \hat{a}_1, \hat{a}_2) = (98.5, 1, -0.5)$, so the fit given by this solution is $f(t, \hat{a}) = 98.5 + t - 0.5t^2$, with a residue of 1, in L_∞-measure units. The corresponding values of positive and negative parts of the deviations in this optimum solution are $\hat{u}^+ = (1, 0, 1, 0.5, 1)$ and $\hat{u}^- = (0, 1, 0, 0, 0)$, and it can be verified that this optimum solution satisfies $(u_t^+)(u_t^-) = 0$ for all t.

Even though this Model 1 is a perfectly valid LP model for the problem of minimizing the L_∞-measure of deviation, it has the disadvantage of using the variables u_t^+, u_t^- representing the positive and negative parts of deviations in additional constraints in the model, as explained earlier.

A more direct model for the problem of minimizing $L_\infty(a)$ is the following Model 2 given below. As explained earlier, Model 2 is the better model to use for minimizing $L_\infty(a)$. One of the optimum solutions for this model is the same $\hat{a}$ that was given as the optimum solutions for Model 1, so it leads to the same fit $f(t, \hat{a})$ as described under Model 1.

Model 2:

$$
\begin{aligned}
& \text{Minimize } z \\
\text{subject to} \quad & -z \leq (a_0 - 5a_1 + 25a_2 - 80) \leq z \\
& -z \leq (a_0 - 3a_1 + 9a_2 - 92) \leq z \\
& -z \leq (a_0 - a_1 + a_2 - 96) \leq z \\
& -z \leq (a_0 - 98) \leq z \\
& -z \leq (a_0 + a_1 + a_2 - 100) \leq z \\
& z \geq 0
\end{aligned}
$$

All three methods, the L_2, L_1, L_∞ methods lead to reasonably good fits for the yield in this chemical reaction, so any one of these fits can be used as the functional form for yield when the reaction temperature is in the range used under this experiment. ■

2.4.5 *Minimizing Positive Combinations of Excesses/Shortages*

In many systems, the decision makers usually set up target values for one or more linear functions of the decision variables whose values characterize the way the system operates. Suppose the decision variables are $x = (x_1, \dots, x_n)^T$ and a linear function $\sum a_j x_j$ has a target value of b.

Targets may be set up for many such linear functions. If each of these desired targets is included as a constraint in the model, that model may not have a feasible solution either because there are too many constraints in it, or because some target constraints conflict with the others. That is why in these situations one does not normally require that the target values be met exactly. Instead, each linear function with a target value is allowed to take any value, and a solution that minimizes a penalty function for deviations from the targets is selected for implementation.

For the linear function $\sum a_j x_j$ with target value b, the **excess** at the solution point x (or the *positive part of the deviation* $(\sum a_j x_j - b)$) denoted by $(\sum a_j x_j - b)^+$ and the **shortage** at x (the *negative part of the deviation* $(\sum a_j x_j - b)$) denoted by $(\sum a_j x_j - b)^-$ are defined to be

$$\text{if}\left(\sum a_j x_j - b\right) \geq 0, \text{ excess}\left(\sum a_j x_j - b\right)^+ = \left(\sum a_j x_j - b\right),$$
$$\text{shortage}\left(\sum a_j x_j - b\right)^- = 0$$

$$\text{if}\sum a_j x_j - b) \leq 0, \text{excess}\left(\sum a_j x_j - b\right)^+ = 0,$$
$$\text{shortage}\left(\sum a_j x_j - b\right)^- = |\left(\sum a_j x_j - b\right)|.$$

Therefore, both excess and shortage are always ≥ 0, and the penalty term corresponding to this target will be $\alpha\left(\sum a_j x_j - b\right)^+ + \beta\left(\sum a_j x_j - b\right)^-$, where $\alpha,\ \beta \geq 0$ are, respectively, the penalties per unit excess, shortage ($\alpha,\ \beta$ may not be equal, in fact one of them may be positive and the other 0) set by the decision makers.

The *penalty function* = sum of the penalty terms corresponding to all the targets, by minimizing it subject to the essential constraints on the decision variables, we can expect to get a compromise solution to the problem. If it makes the deviations from some of the targets too large, the corresponding penalty coefficients can be increased and the modified problem solved again. After a few iterations like this, one usually gets a reasonable solution for the problem.

The minimum value of the penalty function is ≥ 0, and it will be 0 iff there is a feasible solution meeting all the targets. When there is no feasible solution meeting all the targets, the deviations from some targets will always be nonzero; minimizing the penalty function in this case seeks a balance among the various deviations from the targets, that is, it seeks a good compromise solution.

By expressing the deviation $(ax - b)$, which may be positive or negative, as the difference $u^+ - u^-$ of two nonnegative variables, the excess $(ax - b)^+$ defined above will be u^+ and the shortage $(ax - b)^-$ defined above will be u^- as long as the condition $(u^+)(u^-) = 0$ holds. For this, remember the precautions expressed in the Cautionary Note 2.1 given above.

Example 2.9. We provide an example in the context of a simple transportation problem. Suppose a company makes a product at two plants P_i, $i = 1, 2$. At plant P_i, a_i (in tons) and g_i (in \$/ton) are the production capacity and production cost during regular time working hours; and b_i (in tons) and h_i (in \$/ton) are the production capacity and production cost during overtime working hours.

The company has dealers in three markets, M_j, $j = 1, 2, 3$ selling the product. The selling price in different markets is different. In market M_j, the estimated demand is d_j (in tons), and up to this demand of d_j tons can be sold at the selling price of p_j (in \$/ton), beyond which the market is saturated. However, in each market j, there are wholesalers who are willing to buy any excess over the demand at the price of s_j (in \$/ton).

The cost coefficient c_{ij} (in \$/ton) is the unit transportation cost for shipping the product from plant i to market j. All this data is given in the following table.

	c_{ij} for $j =$			a_i	b_i	g_i	h_i
	1	2	3				
$i = 1$	11	8	2	900	300	100	130
2	7	5	4	500	200	120	160
d_j	400	500	200				
p_j	150	140	135				
s_j	135	137	130				

We want to formulate the problem of finding the best production, shipping plan to maximize net profit (=sales revenue − production costs), as an LP. There is no requirement that the amount shipped to any of the markets should equal or exceed the demand at it, in fact any amount of the available product can be shipped to any of the markets. Clearly the decision variables in this problem are

x_{ij} = tons shipped from P_i to M_j; $i = 1, 2$; $j = 1, 2, 3$

y_i = tons produced in P_i, $i = 1, 2$

y_{i1}, y_{i2} = tons of regular, overtime production at P_i, $i = 1, 2$.

The essential constraints in this problem are the production capacity constraints, these cannot be violated. They are

$$\begin{aligned} x_{11} + x_{12} + x_{13} &= y_1 = y_{11} + y_{12} \\ x_{21} + x_{22} + x_{23} &= y_2 = y_{21} + y_{22} \\ 0 \le y_{i1} \le a_i, &\quad 0 \le y_{i2} \le b_i \\ \text{for } i &= 1, 2 \end{aligned} \tag{2.14}$$

From the production costs, we see that the slope of the production cost function at each plant is monotonic increasing, hence it is PL convex and its negative is PL concave. So, this negative production cost that appears as a term in the overall objective function to be maximized can be expressed as $-(g_1 y_{11} + h_1 y_{12} + g_2 y_{21} + h_2 y_{22})$.

The demand d_j at market j is like a target value to ship to that market, but the actual amount sent there can be anything. For each unit of excess sent over the demand, there is a drop in the sales revenue of $(p_j - s_j)$/unit. So the total sales revenue can be expressed as $(\sum_{i=1}^{2} x_{ij})p_j - (\sum_{i=1}^{2} x_{ij} - d_j)^+(p_j - s_j)$. So, our problem is

$$\text{Maximize } \sum_{j=1}^{3} \left[\left(\sum_{i=1}^{2} x_{ij} \right) p_j - \left(\sum_{i=1}^{2} x_{ij} - d_j \right)^+ (p_j - s_j) \right]$$

$$-(g_1 y_{11} + h_1 y_{12} + g_2 y_{21} + h_2 y_{22}) - \sum_{i=1}^{2} \sum_{j=1}^{3} c_{ij} x_{ij}$$

subject to the constraints (2.14). Putting it in minimization form and linearizing, it is

$$\text{Minimize } (g_1 y_{11} + h_1 y_{12} + g_2 y_{21} + h_2 y_{22}) + \sum_{i=1}^{2}\sum_{j=1}^{3} c_{ij} x_{ij}$$

$$-\sum_{j=1}^{3}\left[\left(\sum_{i=1}^{2} x_{ij}\right) p_j - u_j^{+}(p_j - s_j)\right]$$

Subject to (2.14), and

$$\sum_{i=1}^{2} x_{ij} = d_j + u_j^{+} - u_j^{-} \quad \text{for all } j$$

$$u_j^{+},\ u_j^{-} \geq 0 \quad \text{for all } j.$$

■

2.5 Multiobjective LP Models

So far we discussed only problems in which there is a single well-defined objective function specified to be optimized. In most real-world decision-making problems there are usually several objective functions to be optimized simultaneously. In many of these problems, the objective functions conflict with one another; that is, moving in a direction that improves the value of one objective function often makes the value of some other objective function worse. See (Charnes and Cooper (1977), Hwang and Masud (1979), Keeney and Raiffa (1976), Sawaragi et al. (1985), Steuer (1986)), for a discussion of multiobjective optimization.

When dealing with such a conflicting set of objective functions, even developing a concept of *optimality* that every one can agree on has turned out to be very difficult.

With the result there is no universally accepted concept of optimality in multiobjective optimization.

Hence, all practical methods for handling multiobjective problems focus on finding some type of a *compromise solution.*

Let $x = (x_1, \ldots, x_n)^T$ denote the vector of decision variables. Let $z_1(x), \ldots, z_k(x)$ denote the k objective functions to be optimized simultaneously. If any one of them is to be maximized, replace it by its negative, so all the objective functions are to be minimized. Then this multiobjective LP is of the form

$$\begin{aligned} \text{Minimize } & z_1(x), \ldots, z_k(x) \text{ simultaneously} \\ \text{subject to } & Ax = b \\ & Dx \geq d \\ & x \geq 0. \end{aligned} \tag{2.15}$$

It is possible that each objective function is measured in its own special units. A feasible solution $\bar{x}$ to the problem is said to be a *pareto optimal solution* (various other names used for the same concept are: *vector minimum, nondominated solution, equilibrium solution, efficient solution*, etc.) to (2.15) if there exists no other feasible solution x that is better than $\bar{x}$ for every objective function and strictly better for at least one objective function; that is, if there exists no feasible solution x satisfying

$$\begin{aligned} z_r(x) &\leq z_r(\bar{x}) \quad \text{for all } r = 1 \text{ to } k \text{ ; and} \\ z_r(x) &< z_r(\bar{x}) \quad \text{for at least one } r. \end{aligned}$$

A feasible solution that is not a nondominated solution is called a *dominated solution* to the problem. Clearly, a dominated solution is never a desirable solution to implement, because there are other solutions better than it for every objective function. So for a feasible solution to be a candidate to be considered for (2.15), it must be a nondominated solution only.

Nobel Prize in This Area: The mathematical theory of nondominated solutions is very highly developed. John Nash was awarded the *1994 Nobel Prize* in economics for proving the existence of nondominated solutions for certain types of multiobjective problems, and a highly popular Hollywood movie "A Beautiful Life" has been made based on his life.

Very efficient algorithms have been developed for enumerating the set of all nondominated solutions to multiobjective LPs; this set is commonly known as the **efficient frontier**. However, typically there are far too many nondominated solutions to multiobjective LPs, and so far no one has been able to develop a concept for the best among them, or an efficient way to select an acceptable one. So, much of the highly developed mathematical theory on nondominated solutions remains unused in practice.

Example 2.10. Consider a multiobjective LP in which two objective functions $z(x) = (z_1(x), z_2(x))$ are required to be minimized simultaneously. Suppose $\bar{x}$ with objective values $z(\bar{x}) = (100, 200)$ and $\hat{x}$ with $z(\hat{x}) = (150, 180)$ are two nondominated feasible solutions for this problem. The solution $\bar{x}$ is a better solution than $\hat{x}$ for objective function $z_1(x)$, but $\hat{x}$ is better than $\bar{x}$ for $z_2(x)$. In this pair, improvement in the value of $z_1(x)$ comes at the expense of deterioration in the value of $z_2(x)$, and it is not clear which solution is better among these two. ∎

The question can be resolved if we can get some quantitative *compromise (or tradeoff)* information between the two objectives; that is, how many units of $z_2(x)$ are the decision makers willing to sacrifice to improve the value of $z_1(x)$ by one unit? Unfortunately, such compromise information is not available in multiobjective problems; that is what makes them hard to solve.

As another illustration, consider a problem in which two objective functions z_1, z_2 are required to be minimized simultaneously. If $\bar{x}$ is a feasible solution to the problem with values $\bar{z}_1$, $\bar{z}_2$ for the two objective functions, we represent $\bar{x}$ by the point $(\bar{z}_1, \bar{z}_2)$ in the z_1, z_2-plane. In Fig. 2.8, we mark the points in the z_1, z_2-plane corresponding to feasible solutions of the problem. They form the dotted region in

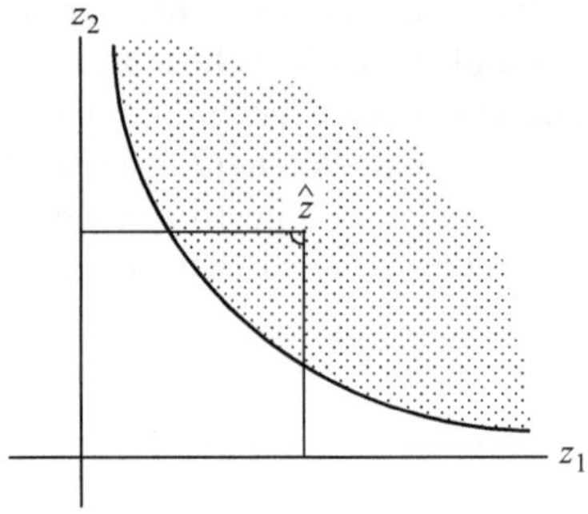

Fig. 2.8 Dotted region consists of points in the objective plane corresponding to feasible solutions. The point $\hat{z}$ does not correspond to a pareto optimum point, since points in the cone region marked by the angle sign, which are in the dotted area, correspond to strictly superior feasible solutions on one or both objective functions. The thick boundary curve corresponds to the efficient frontier

the z_1, z_2-plane in Fig. 2.8. A feasible solution corresponding to a point such as $\hat{z} = (\hat{z}_1, \hat{z}_2)$ in the interior of the shaded region is not a pareto optimum, as feasible solutions corresponding to points in the shaded region satisfying $z_1 \leq \hat{z}_1$, $z_2 \leq \hat{z}_2$ are strictly better for one or both of the objective functions. So, for this problem, pareto optimum solutions are those corresponding to points on the thick boundary curve in Fig. 2.8, and there are an infinite number of them.

Feasible solutions in the efficient frontier for this problem correspond to points on the thick boundary curve in Fig. 2.8. As points representing solutions trace out this efficient frontier, if there are gains in the value of one objective function, there will be losses in the value of the other.

The reader should not be fooled by the word *optimum* in the phrase *pareto optimum.* In a multiobjective model, a pareto optimum does not have the nice optimality properties that we have seen in single objective models. Remember that a pareto optimum point is just a feasible solution with the property that any move from it, if it leads to a gain in the value of one objective function, it also leads to a loss in the value of another objective function. Usually there are many such points, and it is hard to determine which efficient solution is better unless we have some idea of how much one unit decrease in the value of z_2 is worth in terms of units of z_1.

2.5.1 *Practical Approaches for Handling Multiobjective LPs in Current Use*

As pointed out earlier, if complete compromise (or exchange, or tradeoff) information between unit values of the various objective functions is available, it will make it much easier to handle the multiobjective problem. Considering (2.15), suppose the decision makers determine that $c_1 (=1)$ units of $z_1(x)$ (in whatever units this objective function is measured in) is equivalent to (or has the same merit or value as)

c_2 units of $z_2(x)$ (in its own units), that is equivalent to c_3 units of $z_3(x), \ldots,$ which is also equivalent to c_k units of $z_k(x)$. This vector $c = (c_1, \ldots, c_k)$ gives complete compromise or exchange information between the various objective functions in this problem, and so can be called the *exchange vector*. As $z_r(x)$, c_r are in the same units, the quantity $(1/c_r)z_r(x)$ is a dimensionless quantity, and we can form the sum $\sum_{r=1}^{k}(1/c_r)z_r(x)$ and use it as a single objective function that measures the value of the solution vector x. Hence, given the exchange vector c, the multiobjective problem (2.15) is equivalent to the single objective problem of minimizing $\sum_{r=1}^{k}(1/c_r)z_r(x)$ subject to the constraints in (2.15).

Unfortunately, in many real-world applications, this exchange vector is not available. Usually there may be several decision makers interested in the solution of this multiobjective problem, and each one may have a different opinion of what the value of the exchange coefficient c_r should be for each r. So, there is no universal agreement on the exchange vector, and the challenge is to obtain a satisfactory solution of the multiobjective problem, without explicitly using any exchange vector.

Even though the practical approaches in use for handling multiobjective problems do not mention exchange vectors directly, they get it indirectly using different wording that the various decision makers find easier to answer.

2.5.2 *Weighted Average Technique*

This technique uses the tradeoff information in the form of what are called *weights* measuring the relative importance of the various objective functions, and these weights can be interpreted also as cost coefficients attached to unit values of the various objective functions. The process of generating these weights will be easier if all the objective functions are transformed and measured in common units, say money units, scores, etc.

Let $w = (w_1, \ldots, w_k)$ be the vector of weights given. From the discussion above, forming the sum $\sum_{r=1}^{k} w_r z_r(x)$ makes sense, and this technique takes the solution of the multiobjective LP (2.15) to be an optimum solution of the single objective LP:

$$\begin{aligned} \text{Minimize } z(x) = \textstyle\sum_{r=1}^{k} w_r z_r(x) & \\ \text{subject to } \quad Ax &= b \\ Dx &\geq d \\ x &\geq 0. \end{aligned} \tag{2.16}$$

It can be shown that if all $w_r > 0$, then every optimum solution of (2.16) is a nondominated solution for (2.15). So, this type of optimizing a positive weighted combination of all the objective functions is commonly used to generate a nondominated solution for the problem. But the solution obtained depends critically on the choice of the weights $w_1, \ldots, w_k$ used in combining the original objective functions $z_1(x), \ldots, z_k(x)$ into the composite objective function $z(x)$ in (2.16).

There may be several decision makers who have a stake in determining the optimum solution to be selected for implementation. They may not all agree on the choice of the weight vector to be used. It usually takes a lot of planning, discussion, and negotiations, and many compromises, before a weight vector that everyone can agree upon is arrived at. For this negotiation process, it is often helpful to solve (2.16) with a variety of weight vectors and review the optimum solutions that come up, before selecting one of them for implementation.

Example 2.11. Consider the fertilizer problem discussed in Example 2.7, in which the constraints on the decision variables x_1, x_2 = tons of hi-ph, lo-ph fertilizer made daily are

$$2x_1 + x_2 \leq 1{,}500, \;\; x_1 + x_2 \leq 1{,}200$$
$$x_1 \leq 500, \;\; x_1, x_2 \geq 0.$$

For hi-ph, lo-ph, the selling prices are \$222, 107/ton, respectively; and the net profit coefficients are \$15, 10, respectively. The important objectives all to be maximized are net profit $z_1(x) = 15x_1 + 10x_2$, total sales revenue (used as a measure of market share); $z_2(x) = 222x_1 + 107x_2$, sales revenue from hi-ph sales (used as a measure of hi-tech market share); $z_3(x) = 222x_1$; all measured in units of Dollar. The multiobjective problem is to maximize $z_1(x), z_2(x), z_3(x)$ simultaneously, subject to the constraints on x_1, x_2 given above.

Suppose the decision makers have decided that the weights for the objective functions $z_1(x), z_2(x), z_3(x)$ (measuring their relative importance) are 0.5, 0.25, 0.25, respectively. Then we take a compromise solution for this multiobjective problem to be an optimum solution of the single objective function LP:

$$\text{Maximize} \;\; 0.5(15x_1 + 10x_2) + 0.25(222x_1 + 107x_2) + 0.25(222x_1)$$

$$\text{subject to} \;\; 2x_1 + x_2 \leq 1{,}500, \;\; x_1 + x_2 \leq 1{,}200$$
$$x_1 \leq 500, \;\; x_1, x_2 \geq 0.$$

■

2.5.3 *The Goal Programming Approach*

The goal programming approach is perhaps the most popular method used for handling multiobjective problems in practice. It has the added conveniences that different objective functions can be measured in different units, and that it is not necessary to have all the objective functions in the same (either maximization or minimization) form. This method developed by A. Charnes has nice features that appeal to the intuition of business people; that is why it is the common method in usage. Several other references on goal programming are given at the end of this chapter (Charnes and Cooper (1977), Hwang and Masud (1979), Keeney and Raiffa (1976), Sawaragi et al. (1985), Schniederjans (1995), Sponk (1981), Steuer 1986).

The most appealing feature of this method is that instead of trying to optimize each objective function, the decision maker is asked to specify a **goal** or **target value** that realistically is the most desirable value for that function (the name of the method comes from this feature). Considering (2.15), we will denote the goal selected for $z_r(x)$ by g_r for $r = 1$ to k. The decision makers are also required to specify a unit penalty coefficient $\alpha_r \geq 0$ for each unit the value of $z_r(x)$ is in excess of the goal g_r, and a unit penalty coefficient $\beta_r \geq 0$ for each unit the value of $z_r(x)$ is short of the goal g_r. These penalty coefficients play the role of exchange or tradeoff coefficients between the various objective functions discussed earlier in this method. In terms of this goal setting, the objective functions are divided into three types:

Type 1: Those for which the higher the value the better: Each of these objective functions should really be maximized; for each of them the goal is like a minimum acceptable value for it. Objective values $\geq$ the goal are the most desirable; those below the goal are to be avoided as far as possible, and are penalized with positive penalties. So, for objective functions $z_r(x)$ of this type, $\alpha_r = 0$ and $\beta_r > 0$.

Type 2: Those for which the lower the value the better: These objective functions should be minimized, for them the goal is like a maximum acceptable value. Objective values $\leq$ goal are desirable, those $>$ the goal are penalized. So for $z_r(x)$ of this type, $\alpha_r > 0$ and $\beta_r = 0$.

Type 3: Those for which the preferred value is the goal: For these objective functions their goal is the most desirable value, and both deviations above or below the goal are penalized, So, for objective functions $z_r(x)$ of this type, both $\alpha_r > 0$ and $\beta_r > 0$.

At any feasible solution x, for $r = 1$ to k, we express the deviation in the rth objective function from its goal, $z_r(x) - g_r$, as a difference of two nonnegative variables

$$z_r(x) - g_r = u_r^+ - u_r^-, \quad u_r^+, u_r^- \geq 0,$$

where u_r^+, u_r^- are the **positive** and **negative parts of the deviation** $z_r(x) - g_r$ as explained earlier. That is, $u_r^+ = \text{maximum}\{0, z_r(x) - g_r\}$ and $u_r^- = \text{maximum}\ \{0, -(z_r(x) - g_r)\}$.

Given this information, the goal programming approach takes the solution of the multiobjective problem (2.15) to be a feasible solution that minimizes the penalty function $\sum_{r=1}^{k}(\alpha_r u_r^+ + \beta_r u_r^-)$. So, it takes the solution for (2.15) to be an optimum solution of the single objective LP.

$$\begin{aligned}
&\text{Minimize} \quad \sum_{r=1}^{k}(\alpha_r u_r^+ + \beta_r u_r^-)\\
&\text{subject to}\\
&z_r(x) - g_r = u_r^+ - u_r^-, \quad r = 1 \text{ to } k\\
&Ax = b, \quad Dx \geq d\\
&u_r^+,\ u_r^- \geq 0, \quad r = 1 \text{ to } k.
\end{aligned} \tag{2.17}$$

As all α_r and $\beta_r \geq 0$, and from the manner in which the values for α_r, β_r are selected, an optimum solution of this problem will try to meet the targets set for each objective function or deviate from them in the desired direction as far as possible. If the optimum solution obtained for this problem is not considered satisfactory for (2.15), the search for a better solution can be continued using this same single objective LP model with revised goals, or penalty coefficients, or both.

It can be shown that this goal programming approach is equivalent to the positive linear combination approach when all the objective functions $z_r(x)$ are linear.

Example 2.12. Consider the multiobjective problem of the fertilizer manufacturer discussed in Examples 2.7 and 2.11. Suppose the first objective function $z_1(x) =$ net daily profit = \$$(15x_1 + 10x_2)$ with a goal of \$13,000 is a Type 1 objective coefficient with penalty coefficients for excess, shortage of $\alpha_1 = 0, \beta_1 = 0.5$.

Suppose the second objective function $z_2(x) =$ the market share, now measured by the daily fertilizer tonnage sold $= (x_1 + x_2)$ tons, with a goal $g_2 = 1{,}150$ tons, is also a Type 1 objective function with penalty coefficients for excess, shortage of $\alpha_2 = 0, \beta_2 = 0.3$.

Suppose the third objective function, $z_3(x) =$ hi-tech market share now measured by daily hi-ph tonnage sold $= x_1$ tons, is a Type 3 objective function with a goal $g_3 = 400$ tons, with penalty coefficients for excess, shortage of $\alpha_3 = 0.2, \beta_3 = 0.2$.

With this data, the goal programming model for this problem is

$$\text{Minimize } 0.5u_1^- + 0.3u_2^- + 0.2(u_3^- + u_3^+)$$

subject to

$$\begin{array}{llllll}
15x_1 & +10x_2 & +u_1^- - u_1^+ & & & = 13{,}000 \\
x_1 & +x_2 & & +u_2^- - u_2^+ & & = 1{,}150 \\
x_1 & & & & +u_3^- - u_3^+ & = 400 \\
2x_1 & +x_2 & & & & \leq 1{,}500 \\
x_1 & +x_2 & & & & \leq 1{,}200 \\
x_1 & & & & & \leq 500
\end{array}$$

$$x_1,\ x_2,\ u_1^-, u_1^+, u_2^-,\ u_2^+, u_3^-,\ u_3^+ \geq 0$$

An optimum solution of this problem is:

$$\hat{x} = (\hat{x}_1,\ \hat{x}_2)^T = (350,\ 800)^T.$$

The solution $\hat{x}$ attains the goals set for net daily profit, and total fertilizer tonnage sold daily, but falls short of the goal on the hi-ph tonnage sold daily by 50 tons. The vector $\hat{x}$ is the solution for this multiobjective problem obtained by goal programming, with the goals and penalty coefficients given above. ■

2.6 Exercises

2.1. Hiring workers for a new project: A contractor is working on a project expected to last an year divided into six periods. He needs workers for this project, who can be hired in two different ways:

Steady workers: These people are hired at the beginning of the year for the whole year, they get paid at the rate of c_1\$/period for each of the six periods of the year ($c_1 = 20{,}000$)

Casual workers: These people are hired at the beginning of each period for that period only. As the availability of people seeking employment in the various periods varies, the rate of pay per casual worker in period t is expected to be c_{2t}\$ in period t for $t = 1$–6 ($c_{2t} =$ 13,000, 17,000, 23,000, 25,000, 21,000, 15,000, respectively, for $t = 1$–6).

The company estimates that they will need d_t workers in period t for $t = 1$–6 ($d_t = 15, 20, 23, 17, 25, 28$, respectively, for $t = 1$–6).

In any period t, if the number of workers on the payroll in that period exceeds d_t, then the excess workers will have to be paid their contracted salary, but they will essentially be idle in that period. To enforce responsible hiring, the Head Office charges the contractor an amount of \$2,500 penalty/idle worker/period.

Formulate the problem of determining the optimal hiring policy to meet worker requirements to minimize the total cost + penalty, as a linear program, ignoring the integer restrictions on the number of workers hired.

2.2. Blend for foundry sand: A steel foundry uses a lot of sand for preparing molds for castings. They have five different suppliers for sand denoted by S_1 to S_5, each has an upper limit that they can supply per week. Sand is classified into four different particle sizes; the supply from each supplier has a different particle size composition. For the castings the company produces, the ideal particle size composition of sand is known, and the demand for sand at the company is 800 tons/week. Data for the following items is tabulated below.

f_{ij} = fraction by weight of particle size j in sand supplied by S_i, $i = 1$–5, $j = 1$ to 4
U_i = upper limit on tons of sand that S_i can supply per week, $i = 1$–5
g_j = fraction by weight of particle size j in the ideal sand for the company, $j = 1$–4,

This data is tabulated below.

Supplier i	f_{ij} For $j =$				U_i
	1	2	3	4	
S_1	0.10	0.40	0.30	0.20	100
S_2	0.20	0.10	0.50	0.20	250
S_3	0.30	0.58	0.02	0.10	50
S_4	0.40	0.15	0.30	0.15	500
S_5	0.60	0.17	0.18	0.05	200
g_j	0.45	0.30	0.15	0.10	demand = 800

Clearly each supplier's sand deviates from the desired ideal sand. So, the company has decided to blend the sands from the various suppliers into a mixture and use it.

Write down the decision variables needed to model the problem of finding an optimal blend that matches the ideal particle size composition as closely as possible. Develop a measure of deviation of the prepared mixture's particle size composition from the ideal, and formulate the problem of minimizing this measure of deviation as a linear program to find the best blend.

2.3. Investment fund allocation among different shopping malls: A company wants to invest their funds among different shopping malls in the Atlanta retailing system. Their ultimate goal is of course to maximize the expected return from their investment. They believe that they can achieve this by making sure that the sales volumes in the shopping malls in which they invest remain high. Their research has shown that a major factor affecting the sales volume in a shopping mall is its *patronization rate*, which is defined as the share of the *shopping trips in the area* that it attracts, and this can be estimated through observations.

The company has investigated 15 shopping malls in the Atlanta area. Of these, they have decided to consider only shopping malls that attract 7% or more of the total shopping trips in the Atlanta area for investment. That narrowed the list to only five shopping malls.

Two other major factors affecting the sales volume at a shopping mall are the *average income level* (annual income per member of household) of families in the neighborhood of the mall and the *number of major tenants* in the mall. The following table gives the data on these major factors for the five selected malls. In this table, PPR_j = predicted patronage rate of mall j, as a percentage of the total of the five malls; AI_j = average annual income level per household member in the neighborhood of mall j in Dollar; MT_j = number of major tenants in mall j, for j =1–5.

Mall j	PPR_j	AI_j	MT_j
1. Cumberland	20.5	19,700	4
2. Greenbriar	16.3	10,500	2
3. Lenox Square	25.3	18,800	3
4. Northlake	22.2	16,700	3
5. West End	15.7	8,200	1

The decision variables to be determined are the fractions x_j of their total investment to allocate to mall j, for j = 1–5. They want to determine these decision variables to meet the following three goals as closely as possible (all these goals have the same level of importance).

Goal 1: The x_j's should be proportional to the PPR_j's as closely as possible.

Goal 2: The x_j's should be proportional to the AI_j's as closely as possible.

Goal 3: The x_j's should be proportional to the MT_j's as closely as possible.

Formulate the problem of determining the decision variables to meet all these goals as closely as possible, using goal programming (Khorramshshgol and Okoruwa 1994).

2.4. Finding an optimum mix of water from different underground wells for people in the Gaza Strip: WHO has set up standards of 250 mg ℓ^{-1} for chlorides and 50 mg ℓ^{-1} for nitrates as upper bounds for these chemicals in drinking water. The Gaza Strip is a narrow strip along the Mediterranean sea, with a population of around 1.1 million people, which has serious water quality problems not only for agricultural and industrial use, but also for drinking. Ground water is the only significant source of water they have. Because of waste water contamination, uncontrolled use of agricultural fertilizer and pesticides and industrial pollutants, which directly penetrate to the groundwater reservoir through pores in the rocks, and indirectly by decomposition, salts in some wells are at high levels. For example, level of nitrates in the water in some wells exceeds 300 mg ℓ^{-1} (six times WHO upper limit), and the level of chlorides may exceed WHO limit by four to five times. Drinking water with nitrate level over 150 mg ℓ^{-1} poses an extreme risk of blue baby syndrome in infants and carcinogenic effects in adults. Similarly, high levels of chlorides in water make it unacceptable for drinking due to the salinity it causes, and also causes high blood pressure in those who drink it.

No.	Well name	Capacity Q ($m^3\,h^{-1}$)	Chlorides mg ℓ^{-1}	Nitrates mg ℓ^{-1}
1	Sheikh Radwan 1	180	273	135
2	Sheikh Radwan 1A	180	245	140
3	Sheikh Radwan 3	150	1,015	135
4	Sheikh Radwan 4	180	1,085	90
5	Sheikh Radwan 7	180	553	175
6	Sheikh Radwan 7A	180	485	115
7	Sheikh Radwan 8	150	133	80
8	Sheikh Radwan 9	190	133	90
9	Sheikh Radwan 10	190	90	60
10	Sheikh Radwan 11	190	110	80
11	Sheikh Radwan 12	180	110	45
12	Sheikh Radwan 13	180	460	200
13	Sheikh Radwan 15	180	110	45
14	Sheikh Radwan 16	180	110	45
15	Sheikh Ejleen 1	150	810	110
16	Sheikh Ejleen 2	120	440	80
17	Saffa well 1	200	600	225
18	Saffa well 2	150	400	225
19	Saffa well 3	100	740	215
20	Saffa well 4	180	590	110

This question deals with achieving an optimum strategy for mixing the water from different wells to produce drinking water for the people in Gaza, which meets WHO limits for nitrates and chlorides. There are 20 municipal wells as sources for the water; data on the capacity of each well, nitrate, chloride level in the water from each well are given above.

Develop a mathematical model to determine how much water from each of the wells should be drawn (subject to the capacity constraint) and mixed to produce drinking water at the rate of $450\,\text{m}^3\,\text{h}^{-1}$, satisfying the WHO limits on nitrates and chlorides as closely as possible (Agha 2006).

2.5. Locations for new water reservoirs: A small country has a more or less rectangular shape, suppose it is represented by a rectangle in the nonnegative orthant of the x, y-Cartesian plane with $0 \leq x \leq 10$ and $0 \leq y \leq 40$.

The country's water supply comes from six underground wells located at points $(3.21, 39.41), (3.94, 38.70), (5.09, 34.43), (6.05, 34.50), (6.33, 35.82), (6.64, 36.48)$, respectively.

The country has decided to set up three different holding reservoirs of equal capacity to hold the water pumped from the wells. The total quantity pumped into each of the reservoirs to be set up will be 1.4×10^6 m^3/year. The amounts pumped from the six wells will be 23,546, 311,112, 317,543, 1,182,600, 1,182,600, 1,182,600 m^3 per year, respectively.

There are three population centers in the country, each has a municipal water reservoir for supplying water to people in that center. These municipal water reservoirs are located at points (2.23, 10.11), (4.50, 20.22), (7.96, 30.21), respectively.

Because of the topography of the country and the way water distribution pipes are laid, distance between any two points $a = (a_1, a_2)$ and $b = (b_1, b_2)$ for water distribution purposes can be measured using the rectilinear distance between a and b, which is $= |a_1 - b_1| + |a_2 - b_2|$.

Formulate the problem of finding the locations (x_i, y_i), $i = 1$–3 of the three proposed reservoirs to minimize the sum of the rectilinear distances of each of the proposed reservoirs to each of the supply wells and the municipal reservoirs, as an LP.

2.6. How to manage a prime raw material supply curtailment: Coastal States Chemicals and Fertilizers (CSCF) located between Baton Rouge and New Orleans in Louisiana makes eight different products using NG (natural gas) as a prime raw material. Following table provides relevant information on the production of these products at CSCF.

Product	Prod. capacity (tons day^{-1})	Rate[a]	NG input (10^3FT^3 ton^{-1})	Net profit (\$ ton^{-1})
Phosphoric acid	400	80	5.5	60
Urea	250	80	7.0	80
Ammonium phosphate	300	90	8.0	90
Ammonium nitrate	300	100	10.0	100
Chlorine	800	60	15.0	50
Caustic soda	1,000	60	16.0	50
Vinyl chloride manomer	500	60	12.0	65
Hydroflouric acid	400	80	11.0	70

[a]Present production rate as a % of capacity.

Cajan Pipeline Co., the main supplier of NG, informed CSCF that they may have to curtail their supply of NG due to shortages in the availability of NG.

It is required to develop contingency plans to determine the new production rates for the various products if there is a (a) 20% and (b) 40% NG supply curtailment. Model the problem for developing these plans optimally as an LP model. Then discuss how to find the impact that natural gas shortages have on company profits (Iverstine and Kinard 1977).

2.7. Financial management at NFP (not-for-profit) organizations: NFPs are usually funded through grants from many different sources, with each individual funding source putting specific restrictions on how and when their grant money can be spent. For example, certain funding sources dictate what maximum and/or minimum percent of a specific employee's salary can be charged to their grant depending on the work the employee does for the NFP agency. Some sources impose a requirement that funds used from their grant be matched with funds from grants from other specified types of sources.

Here we consider the problem of assigning employee costs and the other costs of the organization to the available grants subject to all the constraints. These costs are classified into three categories: *direct*, *common*, and *indirect costs*.

Direct costs are those that can be traced specifically to a particular program of the NFP (such as the salary of a nurse practitioner working full time in a community clinic run by the NFP. The salary of a medical assistant who works in two departments is also a direct cost split across these two departments according to the proportion of time devoted to each).

Common costs are those nonadministrative costs that are shared across the agency and not easily traced to a particular program; examples are rent, utilities, office supplies, etc.

Indirect costs are management and general administration expenses, like employee benefits, agency management related expenses, etc.

An employee with an FTE (full-time-equivalent) of 1 (i.e., a full-time employee) is either dedicated to one specific program or splits his/her work hours across multiple programs. Managerial employees typically spend a portion of their work hours on specific programs (referred to as *program-related FTE*) and reminder on agency management (*management-related FTE*), which will be treated as indirect cost to the agency.

Consider an NFP with four employees working on two programs according to the following data.

Name	FTE				$/month	
	On prog.		Related to		Salary	Benefits
	1	2	Prog.	Mgt.		
John	1	0	1	0	5,000	450
Jill	0.30	0.45	0.75	0	4,000	300
Dick	0	0	0	1	9,000	900
Sue	0	0.50	0.50	0.50	6,000	550
Total	1.3	0.95	2.25	1.50	24,000	2,200

The total monthly common costs for the agency are $5,000.

Of the total monthly FTE of 3.75, management-related FTEs are 1.5, or 40% of overall FTEs are dedicated to management. So, 40% of this agency's monthly common costs of $5,000, that is, $2,000, are management related and treated as agency's indirect costs. The remaining 60% of monthly common costs, that is, $3,000, are program related and treated under direct costs of the agency.

The total monthly indirect costs for this agency comprise of all benefits, managerial salaries, management related common costs, that is, $\$2{,}200 + 9{,}000 + 3{,}000 + 2{,}000 = \$16{,}200$. The total monthly direct costs for this agency consist of all program related employee salaries and program related common costs $= \$5{,}000 + 4{,}000 + 3{,}000 + 3{,}000 = \$15{,}000$.

For the next month, this agency has funding from three separate grants. The amounts of these grants and restrictions on how these amounts can be spent are summarized below:

For Grant 1: Amount available for the month is $14,000. Restrictions are (1) between 20% and 40% of grant money should be devoted to Program 1 direct costs, (2) between 30% and 60% of grant money should be devoted to Program 2 direct costs.

For Grant 2: Amount available for the month is $7,000. Restriction is (1) No more than 40% of grant money to be spent for indirect costs.

For Grant 3: Amount available for the month is $10,000. Restriction is (1) At least 40% of grant money to be spent for direct costs.

For all the grants: Restriction is: Some overspending for the month above the available amount is allowed, but as far as possible the overspending amount should be kept within 2% of the available amount.

It is required to determine what portion of each employee's salary and benefits, and what portion of the common costs will be charged to each of the grants for the coming month. Formulate the problem of determining these things as a goal programming problem to satisfy each of the restrictions on spending mentioned above as closely as possible (Mehrotra et al. 2006).

2.8. Advertising media selection problem: There is a lot of interest in media selection for effective advertising. In this problem we consider selection from nine promotional media well known for their objective specific effects. We consider four different objectives, each measured in terms of the number of consumers induced to move from a lower hierarchical stage to a higher level of cognitive development by an advertising event. The four objectives are the following:

1. z_1 = **Awareness:** This objective measures the number of new consumers who become aware of the manufacturer's product or brand as a result of a promotional effort. This objective is most likely to be influenced by network radio and TV jingle slogan campaigns.
2. z_2 = **Knowledge:** This measures the number of consumers who acquire product-or-brand related knowledge as a result of the promotional event. This objective is most likely influenced by noncompetitive descriptive copy promoting the merits of the advertised product.
3. z_3 = **Preference:** This measures the number of consumers who acquire a preference for the product or brand of interest. This criterion is influenced by competitive copy favorably comparing the advertiser's brand to competitor's brands.
4. z_4 = **Purchase:** This measures the number of consumers who actually purchase the product sold. Best influenced by retail store copy in the local media.

The various promotional media considered are the following:

1. Slogan/jingle campaign on network radio
2. Slogan/jingle campaign on network TV
3. Descriptive copy in national magazine
4. Descriptive copy on network TV
5. Competitive copy in national magazine
6. Competitive copy on network TV
7. Retail store copy on spot radio
8: Retail store copy on spot TV
9: Retail store copy in top 100 newspapers

Let x_j denote the number of campaigns to be held on media j, for $j = 1$ to 9; these are the decision variables whose values need to be determined.

Even though all x_j are integer variables, we will treat them as continuous variables in this problem; for implementation, the noninteger values in the solution will be rounded to nearest integer values.

Each of the above objective function values depend on the decision vector $x = (x_j)$. For $i = 1$ to 4, z_i can be estimated by the linear function $100(\sum_{j=1}^{9} c_{ij}x_j)$, where data on the c_{ij} coefficients, desired upper bounds u_j on the variables x_j, $d_j =$

cost in Dollar per single campaign of promotional media j, and the total budget are given below (following table gives money figures in \$100 units). The c_{ij} values are estimated through data obtained from marketing surveys.

c_{ij} for	Promotional media j								
	1	2	3	4	5	6	7	8	9
$i = 1$	150	500	100	100	100	100	50	50	50
2	10	50	200	350	150	250	10	10	10
3	0	0	150	250	200	350	10	10	10
4	0	10	15	15	20	50	100	200	300
u_j	26	26	12	12	12	12	6	6	6
d_j	69	757	780	1,515	1,100	1,515	100	950	3,500

Budget = 35,000. All money figures in \$100 units.

Notice that each of the four objective functions are measured in the same units (number of consumers), so combining them into some type of a composite objective function is perfectly valid.

The mediaplanner of the company wants to use the advertizing budget available most effectively. Formulate this as a multiobjective optimization problem, and discuss a good approach for solving it (Steuer and Oliver 1976).

2.9. Time transportation problems with multiple objectives: Bharat Coking Coal in the state of Bihar, India, gets coal from six different mines, and uses it to make coking coals at six coking plants, each coking plant making a different grade of coking coal.

For $i = 1$ to 6: p_i denotes the percent of sulfur in the coal from mine i, $i = 1$ to 6; and a_i is the maximum units of coal that mine i can ship per period. For $j = 1$ to 6: U_j denotes the maximum allowable percent of sulfur in the input coal at coking plant j for $j = 1$ to 6, and b_j denotes the units of coal required at plant j per period. For $i, j = 1$ to 6, let x_{ij} denote the units of coal shipped from mine i to plant j in a period, and t^1_{ij} the units of time that coal shipped from mine i takes to reach plant j. This data is given below.

Mine i =	t^1_{ij} data for plant j =						a_i	p_i
	1	2	3	4	5	6		
1	4	4	3	2	10	4	8	0.4
2	8	2	4	8	10	4	15	0.8
3	1	3	8	2	8	6	10	0.6
4	8	3	4	8	1	8	6	1.4
5	4	2	4	3	8	4	4	1.6
6	5	3	5	4	9	5	7	0.4
b_j	7	10	9	4	1	3		
U_j	0.7	0.5	0.7	0.6	0.5	0.8		

(1) The *time minimizing transportation problem* is the problem of minimizing the time t_1, by which shipments from all the mines reach their respective destinations, assuming that all the shipments leave each mine at time point 0 (this type of objective function is encountered commonly in transportation of perishable goods, delivery of emergency supplies, etc.). A single objective transportation problem with this type of objective function is known in the literature as the *bottleneck transportation model.*

In this numerical example we have a bottleneck transportation problem with the additional constraint that the sulfur content of coal delivered at plant j has to be $\leq U_j$ for $j = 1$ to 6.

(a) Formulate this problem.

(b) Notice that the objective function in this model is not linear. Discuss an efficient approach for solving this model (Hint: This can be modeled as a bounded variable LP with variable upper bounds on the variables.).

(2) Now consider the same coal shipping example with multiple time objective functions to be minimized simultaneously, instead of the single time objective function considered earlier. For instance, in this example, let t_{ij}^3 be the units of time that a truck going from mine i to plant j takes to reach the highway after leaving mine i (after leaving mine i, the truck goes through the congested neighborhood of mine i for some distance before reaching the highway taking it to plant j. The time taken for this portion of travel called "congestion time" is t_{ij}^3). Also, let t_{ij}^2 denote the time that the truck going from mine i to plant j takes for travel between two checkposts on the way. The multiobjective version consider the three time objective functions; those given by the cost data matrices (t_{ij}^1), (t_{ij}^2), (t_{ij}^3) measuring the total travel time, the travel time for travel between two checkposts on the way on each route; and the third measuring the congestion time that each shipment takes to begin highway transit to its destination after leaving the mine where it originates. So, the three objective functions to be minimized are $t_1 = \text{maximum}\{t_{ij}^1 : (i, j) \text{ such that } x_{ij} > 0\}$, $t_2 = \text{maximum}\{t_{ij}^2 : (i, j) \text{ such that } x_{ij} > 0\}$, and $t_3 = \text{maximum}\{t_{ij}^3 : (i, j) \text{ such that } x_{ij} > 0\}$.

Discuss an approach how this multiobjective bottleneck transportation model can be solved (Singh and Saxena 2003).

2.10. Selecting investment opportunities: There are five projects in which money can be invested, covering a 4-year horizon. Details of these projects are given below. In this description, *payout* refers to either the return on investment or the principal given back.

Project A accepts an investment between \$10,000 and \$500,000 at the beginning of year 1. This investment is for a 2-year period. The project pays out 30cents/\$ invested at the end of the 1st year and \$1/\$ invested at the end of the second.

Project B is identical to Project A, except that it is available at the beginning of year 2.

Project C is a first year investment available at the beginning of year 1, which pays out \$1.10/\$ invested at the beginning of year 2. It accepts investments of \$20,000 or over.

Project D is a 3-year investment available at the beginning of year 1, which pays out \$1.75/\$ invested at the beginning of year 4.

Project E becomes available at the beginning of year 3 and will pay out \$1.20/\$ invested at the beginning of year 4. It accepts an investment of at most \$750,000.

Payout received from any of these projects in years 2 and 3 may be reinvested in others, which are available for investment at that time. In addition, short-term (1 year) bank accounts yielding 6% interest are available for any money not invested in the projects in the given year.

Suppose we start year 1 with \$1 million of our money to put into a mix of these opportunities, but no more there after, although we will reinvest payouts. The planning horizon considered therefore is beginning of year 1 to beginning of year 4. All cash received at the beginning of year 4 will be withdrawn.

(1) Develop a model to find an investment plan over the planning horizon, which will maximize the total money from the investments by the beginning of year 4
(2) Develop a model to find an investment that maximizes the NPV (net present value) of all the payout money, assuming a discount rate of 10%/year, and that payouts received are not reinvested but cashed out.

(Source: "Case 12-3B, "Mitchell Enterprises," from Vatter et al. 1978).

2.11. Production planning at a rubber company: Rubicon is a rubber company that used to manufacture a variety of rubber products including tires for forklift trucks and small tractors in the 1960s. At this time the company finds it advantageous to take short lead-time contracts to make small runs of regular automobile snow tires for a large distributor of auto replacement tires. These tires, bearing the distributor's trademark and made to the distributor's specifications, utilize surplus capacity at Rubicon. The contract with the distributor calls for a staged delivery schedule of the two types of snow tires (nylon, fiberglass) over the three summer months as indicated below.

Delivery schedule

Date	No. tires to be delivered	
	Nylon	Fiberglass
31 June	4,000	1,000
30 July	8,000	5,000
31 August	3,000	5,000
Total	15,000	11,000
Price/tire	\$7	\$9

Only two types of machines, the Wheeling and Regal machines, can be used in molding tires of the sort covered by the contract. These machines are fully booked until the first of June. After that date, unused capacity would be available spasmodically between other contracts as given below.

Prod. hours available		
Month	Wheeling machines	Regal machines
June	700	1,500
July	300	400
August	1,000	300

The two types of molding machines are similar except that the Wheeling machines are somewhat faster for making both types of tires than the older Regal machines. Here is the data (costs are for the 1960s).

Machine	Prod. hr/tire for type		Mc. running cost/hr
	Nylon	Fiberglass	
Wheeling	0.15	0.12	\$10.07
Regal	0.16	0.14	\$9.75
Material costs/tire	\$3.10	\$3.90	
FPS[a]costs/tire	\$0.23	\$0.23	

[a]Finishing, packaging, shipping

There is a small storage area adjacent to the production shop where up to one month's production could be kept until they are delivered to the distributor or to a warehouse where tires can be stored from one month to the next at a cost of \$0.10/tire/month. Shipping is scheduled three days prior to the end of the month for delivery before the last day of the month.

(1) Formulate the problem of minimizing the cost of producing and storing the tires to meet this contract on time. Find an optimum solution of your model using any of the available LP software systems.
(2) An additional Wheeling machine was due to arrive at the end of August. For a \$200 fee, its arrival can be expedited to a month earlier. This early arrival would make available 172 additional hours of Wheeling machine time in August for this snow tire work.
Discuss how the model for the problem will change if this expediting is carried out, solve the new model, and decide whether the new machine's arrival should be expedited.
(3) From the optimum solutions, determine a tentative schedule for the maintenance department, indicating when the yearly maintenance check on the various machines could be performed.
(4) The snow tire distributor has found that sales of fiberglass tires had been very good the previous year, so they indicated that they may ask Rubicon to increase their supply of this type of tires. Explain the strategy that Rubicon should adopt if this request comes through (Hint: This needs knowledge of sensitivity analysis discussed later in Chap. 6).

(Source: From Vatter et al. 1978).

2.12. Manpower planning at a mutual life insurance company: At a MLIC (Mutual Life Insurance Company), life insurance is sold through full-time agents of the company. This exercise refers to the manpower planning effort carried out in the company in 1962 for the 1963–1975 horizon. The most critical element in long-range planning is identifying the company's personnel requirements, as this is the variable most likely to affect the sales of new life insurance.

A newly hired agents have relatively low productivity (in terms of new life insurance volume generated per year) in the beginning, but their productivity improves with their years of service at the company. Also, there is considerable turnover among the agents, many leave the company after working for some years for other jobs (e.g., out of 100 agents hired in a year, on average only 22 remain with the company 4 years later).

From past records at the company, agent survival rates and agent productivity rates have been estimated as a function of length of service, and this information in terms of P_r, S_r, T_r is given below, where for $r = 1, \cdots$

P_r = The probability that an agent who is in the rth year of service at the company surviving this year (i.e., continuing in service to go on to the $(r+1)$th year of service)

S_r = Average production rate (i.e., sales of new life insurance commissioned in that year in units of \$1,000) of a person in the rth year of service at the company and surviving into the next year of service

T_r = Average production rate (i.e., sales of new life insurance commissioned in that year in units of \$1,000) of a person in the rth year of service at the company and terminating his service at the company by the end of that year.

r = year of	Money unit = \$1,000		
service	P_r	S_r	T_r
1st	0.74	126	42
2nd	0.55	336	61
3rd	0.71	366	62
4th	0.77	404	84
5th	0.79	417	75
6th or higher	0.91	474	116

For $r \geq 6$, P_r, S_r, T_r remain the same.

Using this information, answer the following questions.

(1) For $r = 1$ to 6 and higher, what is the probability that a person hired at the beginning of first year will survive at the company r years and goes on into the $(r+1)$th year of service?

(2) For $r = 1$ to 6 and higher, what is the probability that a person hired at the beginning of first year will survive at the company r years and terminate his service at the company during the rth year of service?

(3) For a person hired at the beginning of first year, define for $r = 1, \ldots, PC_r =$ the expected sales volume (in units of \$1,000) that this person can be expected to generate during the rth year. Find this PC_r for $r = 1, \cdots$.

(4) The company has set up sales goals for each of the years 1970–1975. Let these be denoted by G_i for i = 1970–1975. Goals for sales volume in several consecutive years in future are set in order to ensure some continuity in sales growth. The company requires that the actual estimated sales in those years must be greater than or equal to these goals.

During the 1962 calender year, the company hired 430 new agents. For i = 1963–1975, let x_i = the number of new agents to be hired during the calender year i. The company's objective is to be a growth company, but at the same time does not want to grow so fast as to become unmanageable. To realize these, they want that for each i, x_i should be between x_{i-1} to $(1.5)x_{i-1}$.

Ignoring the integer restrictions on the variables x_i, set up a linear programming model to determine (x_i : for i = 1963–1975) that minimizes the total number of new agents hired between 1963 and 1975 subject to all the constraints mentioned above.

For (G_i : i = 1970–1975) = (351, 416, 482, 548, 615, 682) and (636, 712, 793, 874, 958, 1,042), respectively, solve the model using some LP software, and compare the two optimum solutions.

(Source: Davis and Webster 1968).

2.13. Forecasting expected demand for paint: Past data of a paint manufacturing company that distributes and sells

Year (t)	y	ℓ	x	p
1	337.2	6.666	14.50	200
2	404.2	5.338	15.75	202
3	402.1	4.321	16.78	206.6
4	452.0	6.117	17.44	208.8
5	431.0	5.559	19.77	212.9
6	582.0	7.920	23.76	216.0
7	596.6	5.816	31.61	218.2
8	620.8	6.113	32.17	221.7
9	513.6	4.258	35.09	224.5
10	606.9	5.591	36.42	227.9
11	628.0	6.675	36.58	231.0
12	602.7	5.543	37.14	234.0
13	656.7	6.693	41.30	238.0
14	778.5	7.638	45.62	242.0
15	827.6	7.752	47.38	247.0

its paints nationwide, over a 15-year period data on ℓ = home improvement loans granted in the year in the nation as a whole in units of \$1 billion, p = population of the country in millions, x = index of building construction started in the nation as a whole, are given above. Formulate the problem of determining y = its total yearly sales (in \$1 million units), as a function of some of these variables on which y may depend.

2.14. Estimating labor costs for preparing different menu items: A catering company prepares and serves specialized meals for groups who hire their services to celebrate some occasion or other with a meal party. All the meals include a main course (there are four choices for this: M_1, chicken; M_2, meat; M_3, fish; and M_4, pasta/vegetarian), an appetizer (there are three choices for this: A_1, Fruit cocktail; A_2, Nut and date cocktail; and A_3, Melon and Prosciutto), a dessert (there are three choices under this: D_1, A piece of pie; D_2, A cake or pastry; and D_3, Strawberries with ice cream), an optional salad, and optional liqueurs and other after dinner drinks. Of course, all meals include serving of wine or champagne during the main course, and coffee/tea after the main course as standard features.

Contract number	Labor cost/ person in $	Choices in agreed menu				
		Main course	Appetizer	Dessert	Salad	Liqueurs
1	6.65	M_1	A_1	D_2	1	0
2	7.25	M_1	A_1	D_1	1	1
3	6.15	M_4	A_1	D_2	0	0
4	6.25	M_4	A_1	D_2	0	1
5	6.95	M_1	A_1	D_1	1	0
6	5.75	M_4	A_1	D_2	0	0
7	7.50	M_3	A_1	D_3	1	1
8	7.60	M_3	A_1	D_3	1	0
9	7.75	M_3	A_1	D_3	1	0
10	5.70	M_4	A_1	D_2	0	0
11	6.95	M_1	A_1	D_1	1	0
12	6.85	M_4	A_1	D_2	0	1
13	7.20	M_1	A_1	D_3	1	1
14	7.35	M_2	A_1	D_3	1	0
15	7.80	M_3	A_2	D_3	0	1
16	8.15	M_2	A_2	D_1	1	0
17	9.15	M_3	A_2	D_3	1	1
18	7.50	M_2	A_2	D_1	0	1
19	6.50	M_4	A_2	D_2	0	0
20	8.50	M_3	A_2	D_3	1	0
21	7.50	M_1	A_2	D_2	0	0
22	8.70	M_2	A_2	D_3	1	1
23	9.00	M_3	A_2	D_3	1	1
24	7.80	M_3	A_2	D_1	1	0
25	7.50	M_1	A_3	D_1	1	0
26	6.00	M_4	A_3	D_2	0	0
27	8.50	M_3	A_3	D_1	1	1
28	7.80	M_2	A_3	D_1	1	1
29	7.25	M_1	A_3	D_2	0	1
30	6.20	M_1	A_3	D_2	0	0
31	5.75	M_4	A_3	D_2	0	0
32	7.10	M_1	A_3	D_1	1	0
33	7.00	M_2	A_3	D_2	0	1
34	7.50	M_3	A_3	D_1	1	1
35	6.15	M_4	A_3	D_2	0	0

1 = yes, 0 = No

The catering company provides from its own store the necessary food and other ingredients for preparing the meals; but hires workers from a sister company for preparing, assembling, and serving the meals. The sister company charges our catering company for the man hours of labor provided to prepare the meals according to the agreed upon menu, serving them, and handling the event. Data on these labor costs from 35 different meal contracts (with 100–150 guests each) is in the table given above.

Develop an LP model to determine the labor costs associated with each choice available under the main course, appetizer, dessert, salad, and liqueurs.

2.15. Estimating the sales potential of a new store at a proposed location: A department store chain wants to develop an estimate of sales for a store as a function of the demographic and other relevant information of the site where it is located.

No.	SV	SA	Data related to trading zone of store								
			P	FS	B	H	MFI	HG	HO	HV	RP
1	192	122.5	643	3.2	1.4	1.2	105	53	11	28	94
2	236	111.7	743	3.5	23.9	22.4	82	53	7	24	70
3	156	123.4	872	3.4	11.1	0.9	135	64	53	30	146
4	133	106.0	720	3.1	5.3	1.3	124	52	19	29	141
5	102	116.9	491	3.6	26.0	1.3	126	60	63	27	127
6	86	85.0	286	4.0	4.4	0.4	132	51	86	25	154
7	108	90.3	159	3.9	7.7	1.0	141	56	67	36	147
8	74	74.9	190	3.7	8.4	1.1	125	55	61	25	123
9	149	122.8	530	3.4	6.3	1.3	127	59	59	30	124
10	242	120.9	1,113	3.3	14.4	7.8	106	54	14	32	90
11	148	120.3	133	3.5	5.1	0.7	148	59	51	47	136
12	148	64.4	337	3.4	6.7	0.4	154	61	48	44	147
13	117	112.6	309	3.7	6.4	1.6	120	55	52	27	110
14	112	93.8	230	3.6	11.2	0.3	139	51	72	30	131
15	47	49.1	244	4.0	5.9	2.9	123	54	82	23	154
16	110	118.1	280	3.6	9.8	0.4	126	51	65	25	134
17	97	104.5	217	4.2	1.4	1.0	141	44	91	30	154
18	80	110.1	518	3.7	8.3	0.6	139	47	77	28	150
19	90	105.0	202	3.9	4.1	0.3	157	50	90	32	161
20	73	77.2	180	3.9	2.8	0.6	153	55	89	33	138
21	88	85.7	166	4.0	0.6	0.7	128	49	85	27	158
22	132	100.0	647	3.7	28.9	1.1	126	51	70	26	130
23	84	116.1	210	3.6	7.1	3.4	123	53	56	30	111
24	81	74.1	422	3.7	6.6	0.4	147	66	81	30	155
25	134	113.2	607	3.4	18.4	1.9	120	54	39	28	123

Each department store has its *trading zone*, that is, the zone from which it draws its traffic. For a given size of store, all else being equal, the more people in the trading zone, the higher the sales volume at the store. For a typical suburban store, 75% of all customers come from within a 5–10 mile radius; that is why this area is defined as the store's trading zone. The characteristics of the trading zone like its population (**P** in units of 1,000 people), the average family size (**FS**), the median family income (**MFI** in units of \$100), the selling area in the store (**SA** in 1,000 ft^2), the % of hard goods in the store (**HG**), the percent of home owners (**HO**), the median home value (**HV** in units of \$1,000), the median rent paid/month by those living in rented quarters (**RP** in \$) the percent of blacks (**B**), and hispanics (**H**) in the population in the trading zone are all expected to influence the sales volume in the store. It is possible to get some of this information using computerized census tract data, but the rest has to be collected.

The chain already has 25 stores in the NY area. Data on the annual sales volume (**SV** in units of \$100,000) at these stores and on the values of these explanatory factors for the sites where they are located is given below (all these data is 1970s data, so the \$ refers to 1970s US \$). The first column gives the store number.

Develop an LP model to express the expected annual sales at a store in terms of the explanatory variables on which data is provided, that best fits the data.

Solve the model using an available LP software package and find the best fit functional form for the expected annual sales.

There are two potential new sites where the chain has the opportunity to open new stores. These sites are in densely populated areas with good access and little competition, and as there are not enough stores near them, it is believed that residents of those areas are spending significant amounts outside of their zone. Here is the data on the characteristics of the two sites:

Site 1: (P, FS, B, H, MFI, HO, HV, RP) = (955, 3.7, 40.0, 10.8, 84, 10, 23, 80). From the land area available, the selling area in the store set up can be 125,000 ft^2, which can be increased to 146,000 ft^2 if necessary.

Site 2: (P, FS, B, H, MFI, HO, HV, RP) = (431, 3.5, 13.8, 6.6, 94, 11, 18, 83), and the store can have a selling area of 120,000 ft^2.

Use the functional form found above to predict the expected sales volume in a store is set up at sites 1, 2. Which among these sites should get priority for building the next store?

2.16. (a) Developing a formula for new car sales volume at US franchised dealers by quarter: The following table gives data on the total volume of sales of new cars at US franchised dealers all over US by quarter from 1958 to 1974.

Quarter	Sales*	Quarter	Sales	Quarter	Sales	Quarter	Sales
1958, 1st	1069	1962, 2nd	1887	1966, 3rd	1744	1970, 4th	1490
2nd	1142	3rd	1360	4th	2218	1971, 1st	1977
3rd	927	4th	1946	1967, 1st	1744	2nd	2281
4th	1151	1963, 1st	1721	2nd	2235	3rd	1654
1959, 1st	1323	2nd	2056	3rd	1692	4th	2676
2nd	1613	3rd	1498	4th	1897	1972, 1st	2078
3rd	1289	4th	2059	1968, 1st	2021	2nd	2537
4th	1263	1964, 1st	1844	2nd	2321	3rd	2165
1960, 1st	1515	2nd	2171	3rd	1934	4th	2541
2nd	1739	3rd	1753	4th	2349	1973, 1st	2473
3rd	1330	4th	1895	1969, 1st	2030	2nd	2742
4th	1559	1965, 1st	2196	2nd	2347	3rd	2247
1961, 1st	1212	2nd	2381	3rd	1927	4th	2208
2nd	1550	3rd	1822	4th	2161	1974, 1st	1771
3rd	1180	4th	2366	1970, 1st	2207	2nd	2166
4th	1624	1966, 1st	2207	2nd	2189	3rd	1949
1962, 1st	1559	2nd	2209	3rd	1654	4th	1562

There is a long-term growth trend in sales due to growth in the population and increases in standard of living. It is believed that the *trend factor* is of the form $(1069)a^{\mathbf{t}}$, where a is a parameter to be estimated from the data and t is the number of the quarter, counting with 1958, 1st quarter as the one corresponding to $t = 0$.

New car sales are also affected by the season of the year. Let s_1 to s_4 denote the seasonal factors associated with the seasons corresponding to the first to the fourth quarters of the year.

Another important thing affecting new car sales is the fluctuation in general business activity caused by the so-called *odd year–even year* business cycle. Let these *cyclical factors* corresponding to the odd, even years be denoted by c_1, c_2 respectively. So, our model is

Total new car sales at all franchised dealers in US for a quarter = (the trend factor for the quarter)(seasonal factor for the quarter)(cyclical factor for the year).

Give an LP formulation for the problem of determining the values of various parameters in this model, which gives the best fit for the data.

(b) Developing a formula for monthly champagne sales in France: Historical data on the sales of champagne in France (in millions of bottles) by month for the last six years is given in the following table. French champagne sales have a strong seasonal pattern, but also exhibit a steady growth trend with time.

Using plots of data as necessary develop an appropriate model for (1) French champagne sales in month j of year t, in terms of t, j, and (2) the annual sales in year t, in terms of t. Formulate the problem of determining the values for the unknown parameters in the model to give the best fit to the data given, as an LP.

Year, month	Sales*	Year, month	Sales	Year, month	Sales	Year, month	Sales
(1, 1)	2.82	(2, 7)	3.03	(4, 1)	5.38	(5, 7)	3.97
2)	2.67	8)	1.76	2)	3.09	8)	1.72
3)	2.76	9)	3.60	3)	3.72	9)	5.05
4)	2.72	10)	4.47	4)	4.51	10)	6.92
5)	2.95	11)	6.84	5)	4.52	11)	9.86
6)	3.04	12)	8.36	6)	4.54	12)	11.33
7)	2.28	(3, 1)	3.11	7)	3.66	(6, 1)	4.02
8)	2.21	2)	3.01	8)	1.64	2)	3.96
9)	2.92	3)	4.05	9)	4.74	3)	4.51
10)	4.31	4)	3.52	10)	5.43	4)	4.28
11)	5.77	5)	3.94	11)	8.31	5)	4.97
12)	7.31	6)	3.99	12)	10.65	6)	4.68
(2, 1)	2.54	7)	3.26	(5, 1)	3.63	7)	3.52
2)	2.48	8)	1.58	2)	4.29	8)	1.82
3)	3.03	9)	3.53	3)	4.15	9)	5.22
4)	3.27	10)	5.21	4)	4.12	10)	6.87
5)	3.78	11)	7.62	5)	4.65	11)	10.80
6)	3.23	12)	9.25	6)	4.75	12)	13.92

2.17. Goal programming for project selection: IML (Indian Mines Ltd.) is the largest coal-producing company in India, producing various grades of coal and contributing more than 90% of the country's production. Being a public sector company, several of the mines operated by it are losing money. The company has decided to face the challenges of (a) increasing total annual production, (b) reducing the average cost of production/ton, and (c) making their mines profitable, or at least reducing their losses from present levels to some reasonable levels.

In this pursuit, the company has decided to invest in two types of mine projects: (1) reconstruction mine projects to enhance production capacity of some existing, operating mines, and (2) opening up new mines. They have made a list of eight different projects of which three are reconstruction projects, and five are new mine projects to invest in. Following table provides data on these projects.

In this table: No. = project number, IN = total investment (in million rupees) needed to complete the project (however, we assume that the company may decide to invest any fraction of this total in each project, more details on this given below); PP = for an existing mine this is the present annual production level at the mine in million tons; FP = future estimated annual production in million tons if project receives full investment; PC = for an existing mine, this is the present total annual operating cost in million rupees; FC = future estimated total annual operating cost of the mine if project receives full investment; PPR = for an existing mine, this is the annual profit/loss (+ or −) in million rupees; FPR = future estimated annual profit/loss in million rupees of the mine if it receives full investment; PM = for an existing mine, this is the present manpower in persons; FM = future estimated manpower at the mine in persons, if it receives full investment.

No.	IN	Production		Cost		Profit/loss		Manpower	
		PP	FP	PC	FC	PPR	FPR	PM	FM
				Existing mines reconstruction					
1	106.2	0.225	0.315	115.4	127.2	−40.0	−18.9	1792	1615
2	118.0	0.27	0.45	160.6	220.4	−66.2	−40.3	3042	3042
3	130.6	0.18	0.45	72.6	106.0	−30.3	48.8	293	417
				New Mines					
4	708.7		0.6		265.5		−45.5		2196
5	94.7		0.195		50.1		−83.8		958
6	694.8		0.563		241.7		22.7		1643
7	196.2		0.293		86.4		−8.7		728
8	123.1		0.24		75.5		−1.04		237

For simplicity assume that on each project j, the company may decide to invest a fraction x_j of the total investment needed for it, where $0 \leq x_j \leq 1$ for all j and $x = (x_j)$. Make the simplifying linearity assumption that the effect of x for an existing mine j will be: values after the decision is implemented, of: annual production, annual operating cost, annual profit/loss, manpower, will be: (its present value)$(1 - x_j) + x_j$(future estimated value mentioned in the tableau); and for a new mine j these values will be: x_j (future estimated value mentioned in the tableau).

A team of managers at the company has selected the following as important targets or goals for the investment decision: (1) if possible, keep total investment in all projects to within 1,100 million rupees, (2) try to achieve an average production cost of 350 rupees/ton of coal produced at these mines, (3) try to keep the total loss at these project sites to within 60 million rupees as far as possible, and (4) keep the total manpower deployment at these project sites $\leq$ the desired upper limit of 9,000. Also, after consulting several executives and experts, it has been determined that the weights measuring the achievements of various goals should be 0.298 for the capital investment goal, 0.252 for the production cost goal, 0.241 for the profit/loss goal, and 0.209 for the manpower goal.

The company would like to impose a constraint that the total estimated annual production at these sites should reach or exceed 2 million tons.

(1) Formulate the problem as a linear goal programming problem.
(2) Discuss what changes should be made in the model if it is required that each project should either be not taken up at all; or if it is taken up, it should be completed (i.e., partial funding of projects is not allowed) (Mukherjee and Bera 1995).

2.18. A bicriterion assignment problem: Let K denote the set of feasible solutions of the usual $n \times n$ assignment problem represented by the system of constraints: $\{x = (x_{ij}) : \sum_{i=1}^{n} x_{ij} = 1 \text{ for all } j = 1 \text{ to } n; \sum_{j=1}^{n} x_{ij} = 1 \text{ for all } i = 1 \text{ to } n\}$.

In the usual cost-minimizing linear assignment problem (LAP), the decision variable x_{ij} takes the value 1 [0] if job i is assigned [not assigned] to machine j for execution; c_{ij} represents the job completion cost of carrying out job i on machine j, and the objective function to minimize is $z_1(x) = \sum_{i=1}^{n} \sum_{j=1}^{n} c_{ij} x_{ij}$.

All extreme point solutions of K (called **assignments**) make all variables x_{ij} take only values 0 or 1. That is why all algorithms for LAP try to obtain extreme point optimum solutions, thereby meeting the 0-1 requirements on all the variables without imposing these requirements explicitly.

Let t_{ij} denote the time (in hours, say) needed to complete job i on machine j. Then the time taken to complete all jobs if assignment $\bar{x} = (\bar{x}_{ij})$ is implemented is $\bar{t} = \max\{t_{ij} : (i, j) \text{ such that } \bar{x}_{ij} = 1\}$, assuming that all the jobs are started simultaneously.

Normally when work is going on, the company needs to have some supervisors to supervise the work. Suppose the cost of providing the supervisory force is F\$/h. Both the job completion cost and the supervisory cost need to be minimized, and so this leads to a bicriterion assignment problem: minimize $z_1(x)$ and $z_2(x) = F \max\{t_{ij} : x_{ij} = 1\}$ subject to $x \in K$ and $x_{ij} \in \{0, 1\}$ for all i, j.

(1) Prove that this bicriterion assignment problem has at most n^2 efficient solutions.
(2) Prove that an optimum solution for the single objective problem of minimizing $z_1(x) + z_2(x)$ over $x \in K$ is attained at an efficient solution of the bicriterion assignment problem.

Clearly, $z_2(x)$ is not a linear function of x. Consider the following algorithm for the problem of minimizing $Q(x) = z_1(x) + z_2(x)$ (not a linear function) over $x \in K$. Let ℓ be a lower bound for $z_2(x)$ over $x \in K$. For example, you can arrange t_{ij}s in increasing order in a list, and take ℓ to be the nth element from the bottom in this ordered list. Let $c^0 = (c^0_{ij} = c_{ij})$.

Find an optimum assignment, x^0, for the LAP of minimizing $\sum_{i=1}^{n} \sum_{j=1}^{n} c^0_{ij} x_{ij}$. Let $z^0_1 = z_1(x^0), z^0_2 = z_2(x^0),\ Q^0 = z^0_1 + z^0_2$. Define $x^* = x^0,\ Q^* = Q^0$. Set $k = 1$ and go to Step 1.

General Step k**:** Consists of two substeps.

Substep 1: Define for all $i,\ j, c^k_{ij} = c_{ij}$ if $t_{ij} < z^{k-1}_2$, or $= \infty$ if $t_{ij} \geq z^{k-1}_2$.

Substep 2: Find an optimum assignment x^k minimizing $\sum_{i=1}^{n} \sum_{j=1}^{n} c^k_{ij} x_{ij}$. Let $z^k_1 = z_1(x^k),\ z^k_2 = z_2(x^k),\ Q^k = z^k_1 + z^k_2$.

If $z^k_1 \geq Q^* - \ell$, then x^* is an optimum solution of the problem and Q^* is the optimum objective value, terminate.

If $Q^k < Q^*$, update x^* to x^k, Q^* to Q^k, and go to the next step by adding 1 to k.

In this algorithm show that z^k_1 is nondecreasing, while z^k_2 is strictly decreasing. Using this show that the algorithm obtains an optimum solution minimizing $Q(x)$ over K. When

$$c = \begin{pmatrix} 6 & 3 & 5 & 8 & 10 & 6 \\ 6 & 4 & 6 & 5 & 9 & 8 \\ 11 & 7 & 4 & 8 & 3 & 2 \\ 9 & 10 & 8 & 6 & 10 & 4 \\ 4 & 6 & 7 & 9 & 8 & 7 \\ 3 & 5 & 11 & 10 & 12 & 8 \end{pmatrix}, \quad (t_{ij}) = \begin{pmatrix} 4 & 20 & 9 & 3 & 8 & 9 \\ 6 & 18 & 8 & 7 & 17 & 8 \\ 2 & 8 & 20 & 7 & 15 & 7 \\ 12 & 13 & 14 & 6 & 9 & 10 \\ 9 & 8 & 7 & 14 & 5 & 9 \\ 17 & 13 & 3 & 4 & 13 & 7 \end{pmatrix},$$

find an optimum assignment minimizing $Q(x)$ (Geetha and Nair 1993; Bakshi et al. 1979; Berman et al. 1990).

2.19. Planning diets for diabetic patients: An important part of managing diabetes is very careful diet planning and sticking to it conscientiously. Diabetic patients have to observe recommended dietary allowances on carbohydrates, fats, and calories; and also make sure that their daily diet consists of other nutrients such as vitamin B_6, iron, etc., within prescribed limits.

Nutrient	Average units/serving of food group						RDA
	MI	VEG	FT	BR	ME	OI	
CL	118	20	54	79	65	61	1800
PR	9.3	1.4	0.7	2.6	8.2	0.8	46
CH	13	5	13	15	0.6	0.6	$\leq$180
F	3.1	0.2	0.3	0.8	3.5	6.2	78
CA	334	44	15	14	7	6.7	800
FE	0.15	0.75	0.54	0.79	0.78	0.15	18
VC	2.3	22.84	22.53	0	0	0	45
VB6	90.2	91.09	58.04	67.06	102	17.78	2000
VB12	1	0	0	0	1.3	0.01	3
VE	0.05	1.51	0.78	0.38	0.13	2.45	12
MG	28.8	21.6	9.9	19.87	8.9	7.71	300
ZN	1.2	0.4	0.07	0.47	2.2	0.13	15

The usual approach to menu planning is a manual one based on *exchange lists* of food groups, but most patients find this tedious and difficult to use. This exercise deals with menu planning considering both client food preferences and recommended dietary allowances, using an interactive goal programming approach.

In their exchange lists for meal planning, the ADA (American Diabetes Association) forms various foods into six food groups: MI (milk), VEG (vegetables), FT (fruits), BR (breads, this group also includes crackers, cereal, starchy vegetables), ME (meats), OI (oils). For each food group, 1 serving is defined to be the following quantity: MI (1 cup or 245 g), VEG (1 cup cooked vegetables or 80 g), FT (averaged value of 100 g), BR (1 slice or 25 g for bread and/or crackers, 56 g for cereal, or 118 g for starchy vegetables), ME (1 oz or 30 g of lean meat), OI (1 teaspoon or 14 g of oil, butter, or margarine).

For each sex and age group of the patient, ADA has provided RDA (recommended dietary allowance) for various nutrients. Among all the nutrients, experimentation has indicated that meeting the RDAs for a selected list of 12 generally produced a diet sufficient also in the others. These 12 nutrients are CL (calories), PR (proteins, measured in grams), CH (carbohydrates, measured in grams), F (fat, measured in grams), CA (calcium, measured in milligrams), FE (iron, measured in micrograms), VC (vitamin C, measured in milligrams), VB6 (vitamin B_6, measured in milligrams), VB12 (vitamin B_{12}, measured in micrograms), VE (vitamin E, measured in IU), MG (magnesium, measured in milligrams), and ZN (zinc, measured in milligrams).

The patient is a young adult female. For her sex and age group, the RDA for various nutrients are given in the last column of the table given below. The table also gives the average value of the nutrient content/serving of each food group, obtained using the Ohio State University. Nutrient Data Bank. She has specified the minimum and maximum number of servings of each food group that she would like to include in her daily diet. These bounds are MI (2–5 servings), VEG (1–5 servings), FT (1–5 servings), BR (3–6 servings), and OI (2–5 servings). Also, the ADA has recommended that no more than 35% of calories in the diet should be derived from the MI and OI groups. All these define the constraints on the decision variables in the model for the problem.

The RDA number for each of the 12 nutrients is the goal for the content of that nutrient in the daily diet. The aim is to find a diet in which the contents of the nutrients CL, CH, and F are less than the corresponding goals as far as possible, while the contents of the other nutrients are as close to the goal as possible. Take the weight (a measure of the importance attached to that goal achievement) for goals on CL, CH, and F as 5 each; the weight for each of the vitamin goals as 4; and the weight for all the other goals as 3. Formulate the problem of determining the number of servings of each of the food groups to include in the daily diet, to meet the goals subject to the constraints mentioned above (Rugg et al. 1983).

2.20. Job allocation to operators at a book typesetting company: An agency typesets books for publishers. At present they are starting work on m different books, and n operators are available to work on typesetting them. The operators work independently, each at his own speed, but all of them begin work on the portions assigned to them at time point 0 and continue until it is finished. The agency can split each book into any number of fractions, so that different fractions can be performed by different operators.

If x_{ij} denotes the fraction of the ith book allotted to the jth operator, from past experience they can estimate that it will take that operator $\gamma_{ij} x_{ij}$ units of time, where the $m \times n$ matrix $\Gamma = (\gamma_{ij})$ is given. The total working time of an operator will be the sum of the times needed to perform all activity fractions assigned to him/her.

Formulate the problem of determining the fractions of each job to be allocated to each operator, so as to minimize the clock time by which all the jobs are completed, as an LP. Give this formulation for the numerical example in which $m = 5$, $n = 3$, and

$$\Gamma = \begin{pmatrix} 13 & 17 & 7 \\ 37 & 23 & 11 \\ 19 & 2 & 31 \\ 5 & 3 & 29 \\ 1 & 1 & 1 \end{pmatrix}.$$

(Andreatta et al. 1993).

2.21. Tele-marketing: A market research center (MRC) needs to collect data on three types of products. They hire hourly employees to call potential customers and gather this information. During a certain hourly period, the number of calls to be made regarding the three types of products is given in the following table. Here expected call duration is in minutes (min).

Product type	1	2	3
No. calls to make	90	150	270
Expected call duration	2.3	3.4	1.9

(1) There are six grades of employees with different salaries (wages/hour given in dollars) for making these calls. Each employee can handle calls on a subset of products depending on their expertise and past experience. Here is the relevant data.

	Call types they handle					
Grade →	1	2	3	4	5	6
Type 1	*	–	*	–	*	*
2	–	*	*	*	–	*
3	–	–	–	*	*	*
Wage/hour	10	8	25	20	22	30

"*" indicates grade can handle this type

Let x_j = number of employees of grade j employed during the hour for making these calls. Taking x_j as a continuous variable, model the problem of getting the job done at the smallest cost.

(2) Now assume that they have a 3 h period to complete the above job, but a limit of only 14 employees who can work in the office at any point of time. They have the opportunity to hire the employees for a period of 1 h, or 2 h, or 3 h for this work. Those hired for 1 h are to be paid at the hourly rate given above. Those hired for a 2 h period are paid at the above rate for the first hour, and 90% of the above rate for the second hour. Those hired for the 3 h period are to be paid for the first two hours as mentioned in previous sentence, but only 80% of the hourly rate given in the above table for the third hour.

Discuss the model to find the best plan to hire people to complete the job at minimal cost in this situation. (Problem given by Vincent F. Yu.)

2.22. Minimizing losses from attacks on bank armored vehicles: In large cities, each bank usually has several branches scattered in different areas of the city. On each working day, the need arises for the bank to transfer different quantities of money between its branches in the city. Banks use armored vehicles for these money transfers. While these armored vehicles are traveling on the roads, they are sometimes attacked by robbers to loot the money inside them.

Designing safe routes for their armored vehicles is a serious problem that banks face. The chance of an attack on these bank vehicles is usually very low, but it increases in hard economic times as unemployment in the city goes up. When an attack occurs, if the attackers are successful, the bank stands to loose all the money inside the vehicle at that time. One strategy that banks can use to minimize the expected loss from these attacks is to put an upper bound, u, on the amount of money that can be carried inside an armored vehicle at any time.

As the upper bound u decreases, the expected total daily mileage, m, of all the armored vehicles of the bank in the city increases as these vehicles may have to make more trips to carry out the needed money transfers. So, let $m(u)$ denote this expected total daily mileage of the banks armored vehicles as a function of u.

Using simulation, the functional form of $m(u)$ has been determined to be the following PL function. Here, $u,\ m(u)$ are given in coded money, distance units.

Range of u	$m(u)$
40–50	$110 - 3(u - 50)$
35–40	$140 - 5(u - 40)$
30–35	$165 - 10(u - 35)$

As the expected daily mileage increases, their public exposure increases, and this is expected to increase the chance of an attack on them. Let $p(m)$ denote the probability of an attack on one of these armored vehicles while on the road in a day, as a function of m. Using data on the past records of these occurrences and other information, the Statistics Division of the bank came up with the following estimates of $p(m)$.

Range of m	Estimate of $p(m)$
$110 - 155$	$0.0018 + 0.0005((m - 110)/45)$
$155 - 215$	$0.0023 + 0.0007((m - 155)/60)$

From past data, we know that once an attack occurs on one of their armored vehicles on the road, the probability that the attack will be repelled with no loss of money inside is 0.25 and the probability that all the money inside will be lost is 0.75.

Using this information determine the optimum value of u, explaining clearly how it is obtained.

(1) With fuel prices going up, extra mileage increases the cost of armored vehicle operations too. Discuss how the model will change if it is required to minimize the sum of expected losses from attacks on armored vehicles plus the expected cost of armored vehicle operations to carry out the needed money transfers.

2.23. Referring to Sect. 2.5, prove that if $\bar{x}$ is an optimum solution of (2.16) minimizing a positive weighted combination of all the objective functions in a multi-objective minimization problem (2.15), then $\bar{x}$ is a nondominated solution for (2.15).

2.24. Concave function minimization: Consider the problem: Minimize $\theta(x)$ subject to $Ax = b,\ x \geq 0$, where A is a matrix of order $m \times n$ and $\theta(x)$ is a concave function. If this problem has an optimum solution, show that it must have an optimum solution which is an extreme point of the set of all feasible solutions.

2.25. Minimizing scrap in aluminium foil slitting: At an aluminium foil mill, the input is in the form of coils of aluminium foil in standard widths (SW), which they slit into foil of different widths for a variety of end users. Customers order foil specifying the width, gauge, and surface finish required. The mill selects an appropriate coil from stock and slits it into the desired width. On a particular day they have standard widths A to G in stock, and need to fill customer widths (CW) I to V. Following table gives the weight of scrap in pounds, which will result from slitting the customer orders of each width on the leftmost column of the table, from the SW on the top line (an entry of "." indicates that the SW is not suitable for that CW). We need to determine which SW to use in slitting each CW to minimize total scrap generated for filling all customer orders. Can this problem be modeled as an LP? Formulate this problem. (Lanzenauer 1975).

CW\SW	A	B	C	D	E	F	G
I	61	19	69	4	46	26	45
II	15	.	44	52	66	95	27
III	94	55	.	85	65	67	.
IV	42	48	11	62	13	.	.
V	23	.	.	58	.	.	.

2.26. We need to find a vector $x = (x_1, x_2)^T$ satisfying each of the following equations as closely as possible: $(2x_1+4x_2, x_1+2x_2, 4x_1+5x_2, 4x_1+3x_2, 3x_1+4x_2, 6x_1+3x_2)^T = (240, 130, 320, 170, 60, 45)^T$. Formulate this as an LP and solve it.

2.27. We need to find a vector $x = (x_1, x_2, x_3)^T \geq 0$ satisfying the following constraints as closely as possible. Formulate this as an LP.

$$
\begin{aligned}
x_1 + x_2 + x_3 &= 100,\\
x_1 + x_2 - x_3 &\geq 60,\\
-x_1 + x_2 + x_3 &\geq 70,\\
-x_1 + x_2 - x_3 &\leq 20.
\end{aligned}
$$

2.28. Yearly production plan at a brewery: The demand from retailers for the brewed product at a brewery is highly seasonal. For the months of January to December for the planning year, the demand in units (1 unit = 1,000 barrels) is estimated to be $(5, 4, 5, 7, 10, 12, 14, 13, 11, 8, 10, 7)^T$.

Each month of the planning year, the brewery can operate at four production levels; one shift/day, two shifts/day, or each with overtime. Following table provides the production data for each of these levels. Money is measured in units, which we will abbreviate as *mu*.

Prod. level	Prod. capacity (units/month)	Labor cost mu/month
One shift	5	3000
One shift + overtime	7	4000
Two shifts	9	6000
Two shifts + overtime	12	8000

Increasing from one shift level in a month to two shift level the next month incurs a cost of 1,500 mu; while decreasing from two shift level to one shift level incurs a cost of 1,000 mu.

They can hold brewed product between the brewhouse and the bottleshop in refrigerated tanks. Every unit inventory held in these tanks at the end of a production month to the next month costs 300 mu.

Shortage in a month incurs a shortage cost of 500 mu/unit short; also this shortage has to be made up in succeeding months in order to meet total demand. January begins with two units of brewed product in inventory, and would be operating at the one shift production level. Formulate the problem of developing a minimum cost production plan for the planning year that would result in a closing inventory at the end of the year of two units. (From Lanzenauer (1975)).

2.29. Toy-store problem: A toy-store chain has several stores in the midwest. For the coming X-mas season, they need to place orders with their overseas suppliers before the end of May for delivery in time for the X-mas sales period.

As unsold toys at the end of the X-mas season do not contribute much to the profit of the company, they base their order quantities quite close to the expected sales volume. From experience over the years they observed that the X-mas sales volume has a high positive correlation with the DJA = Dow Jones average (a measure of the economic status of the region prior to the sales period), and a high negative correlation with the percent unemployment rate in the region. Following table gives data on the DJA during the months of February, March, April (these are independent variables x_1, x_2, x_3), the percent unemployment in the region during this period (independent variable x_4), and the toy sales volume in the region in millions of dollars during the X-mas sales season (dependent variable y) between 1990 and 2001.

From the above discussion it is reasonable to assume that the expected value of y can be approximated by a function $a_0 + a_1x_1 + a_2x_2 + a_3x_3 + a_4x_4$, where the parameters satisfy $0 \leq a_1 \leq a_2 \leq a_3$ and $a_4 \leq 0$. Write the LP formulation of the problem of finding parameter values that give the closest fit to data by the L_1 measure of deviation.

Year	x_1	x_2	x_3	x_4	y
2001	10,690	10,185	10,306	4.3	59
2000	10,533	10,525	10,798	4.0	60
1999	9,356	9,550	10,307	4.3	54
1998	8,307	8,664	8,940	4.6	47
1997	6,828	6,727	6,800	5.2	36
1996	5,435	5,527	5,579	5.5	28
1995	3,927	4,080	4,239	5.5	20
1994	3,898	3,723	3,634	6.5	17
1993	3,344	3,405	3,434	7.1	14
1992	3,247	3,253	3,294	7.4	13
1991	2,798	2,903	2,895	6.6	11
1990	2,607	2,665	2,673	5.2	10

2.30. Oil Refinery Optimization: A refinery has a distillation capacity of 100,000 barrels of crude/day in its fractionator. Here crude oil is basically heated, and as the temperature increases, different products called DN (distillation naphtha), DHO (distillation heating oil), DGO (distillation gas oil), and P (pitch) are given off in vapor form and are collected at various levels. The refinery gets crude oil from three different countries, these are called crudes 1, 2, 3. All the crudes and the various products are measured by volume in barrels. The output statistics from the distillation of each of the available crudes are tabulated below.

Distillation output	Yield (barrels/barrel) from distillation of		
	Crude 1	Crude 2	Crude 3
DN	0.19	0.16	0.02
DHO	0.27	0.32	0.24
DGO	0.38	0.27	0.26
P	0.05	0.13	0.39
Price ($/barrel)	23.25	22.00	20.50
Available (barrels/day)	60,000	90,000	80,000

It can be verified that the total volume of outputs from the distillation of one barrel of crude is <1. The loss is due to evaporation and unusable heavy residuals.

DHO can be sold directly as heating oil. DGO can be sold directly as diesel fuel. Sale prices of these products are given below.

DHO and DGO can also be processed further in a catalytic cracker. The catalytic cracker can either process a maximum of 100,000 barrels/day of DHO, or a maximum of 50,000 barrels/day of DGO, or a combination of these in proportion of these levels adding up to 1. Also, when processing DHO, the catalytic cracker can be run either at a normal level or at a high severity level. The high severity level helps to convert more of the DHO into naptha as seen from the table below. In processing DGO, the catalytic cracker is run at normal level only and never on high severity level.

Catalytic cracker outputs	Output (barrels/barrel of feed)		
	DHO normal level	DHO high severity level	DGO normal level
Catalytic naptha (CN)	0.18	0.32	0.48
Catalytic heating oil (CHO)	0.80	0.69	0.52
Pitch (P)	0.11	0.10	0. 15

The cracking process converts the feed into products whose density is smaller than that of the feed, that is why the volume of outputs from this process is greater than the feed volume.

The pitch (from fractionator and catalytic cracker) can be combined with CHO (two parts of CHO to 17 parts of pitch) and sold as heavy fuel oil. DN and CN can be combined (20 parts of DN with 17 parts or greater of CN) and sold as gasoline. The quality of the gasoline improves with the proportion of CN in this blend. The following table gives the selling prices (all money figures in this exercise are in 1995 US $) and demand for the various final products.

Final product	Selling price ($/barrel)	Daily demand
Gasoline	43.25	Up to 40,000 barrels
Heating oil	39.75	Up to 40,000 barrels
Diesel fuel	39.00	Any amount
Heavy fuel oil	30.00	Any amount

The processing cost on the fractionator is estimated to be $0.60/barrel of crude processed. On the catalytic cracker, the processing costs are $1.50/barrel of fresh feed at the high severity level and $0.95/barrel of fresh feed at the normal level. Formulate the problem of determining how much of each final product to produce daily in order to maximize daily net profit, as an LP.

2.31. Consider the following optimization problem in n variables $x_1, \ldots, x_n$. In this problem, $a,\ h$ are positive data elements.

$$\begin{aligned} \text{Minimize} \quad & z(x) = cx \\ \text{subject to} \quad & Dx \geq d \\ & \sum_{j=1}^{n} |x_j - a| \leq h \\ & x \geq 0. \end{aligned}$$

Notice that there is one constraint here involving absolute value terms. Is it possible to transform this problem into a linear program? Explain clearly. Discuss how one can handle this problem.

2.32. A UM Dental School statistician has conducted research on the condition of the teeth of 8-year-old kids in Michigan and estimated that the average number of decayed teeth in the mouth of an 8-year-old Michigan kid is 1.8.

Given this data, it is required to find the minimum and maximum possible values for the fraction of 8-year-old Michigan kids who have two or more decayed teeth in their mouths. Give LP formulations for the problems of finding this minimum and maximum (two separate problems).

Also comment on why an LP model is appropriate for these problems.

2.33. Optimum assignment of students to schools: A school district consists of r neighborhoods, s schools, and in each of these schools g is the number of grades or levels (like fourth grade, fifth grade, etc.) for students. Here are the data elements: k_{ju} = capacity for the number of students of grade u in school j; s_{iu} = number of students in neighborhood i studying in grade u in a particular school year; and d_{ij} = minutes that a student from neighborhood i has to spend in school bus daily to get to school j, for $j = 1$ to s, $u = 1$ to g, and $i = 1$ to r.

Treating the number of students as a continuous variable, formulate the problem of assigning students in this school district to schools to minimize the total number of minutes spent in the school bus daily by all the students as a linear program. Also, discuss whether this problem can be formulated as a transportation problem.

2.34. Recruiting workers for a new plant: A paper company is setting up a new plant in a city for which they need to recruit the following numbers of workers: 2,000 nonprofessional workers and 800 professional workers.

There are four categories of workers: men (nonminority), women (nonminority), men (minority), women (minority). The estimated average cost of recruiting per worker in each category is given below:

Cost of recruiting/worker

Category	Nonprofessional	Professional
Minority (men or women)	\$740	\$1560
Women (nonminority)	850	1450
Men (nonminority)	570	1290

The company has established the following goals in order of priority:

Goal 1: The percentage of women (minority or nonminority) among the workers should be at least 30% as far as possible.
Goal 2: The percentage of minorities (men or women) among the workers should be at least 20% as far as possible.
Goal 3: The recruiting cost should be at most \$2 million as far as possible.

It is required to find how many workers to recruit in each category in order to meet the goals as closely as possible. Ignoring the integer nature of the variables, formulate this problem as a linear program using a goal programming approach.

If you need to make any additional assumptions, please state them very clearly.

2.35. Short-term cash flow management: A department store has revenues coming from their operations, and bills to pay out, at the end of each month. These quantities for a 6-month period are given below.

Month	On last day of month	
	Revenue from operations	Bills to pay
July	$1,000	5,000
Aug	2,000	5,000
Sep	2,000	6,000
Oct	4,000	2,000
Nov	7,000	2,000
Dec	9,000	1,000

Clearly, they have a short-term cash flow problem at the beginning of the period, as they have no other income to pay the bills other than the revenue. They have two available sources to borrow money.

Bank: The bank can loan a maximum of $7,000. The loan can be taken on 31 July, and it should be paid back together with 9% interest on 31 Dec (early payback does not reduce the interest amount).

S&L: The S&L can loan any amount of money for a period of one month, any number of times. The loan amount can be taken on the last day of any month, and it should be paid back together with 3% interest on the last day of following month.

Formulate the problem of determining how the store can minimize the cost of paying their bills on time each month in the period, as a linear program.

2.36. Optimizing currency exchange transactions: A company holds 1.2 billion Japanese yen, 10.5 billion Indonesian rupiahs, and 28 million Malaysian ringgits. Here are the exchange rates and transaction costs. JY = Japanese yen, IR = Indonesian rupiah, MR = Malaysian ringgit, UD = US$, CD = Canadian$, EE = European euro, EP = English pound, MP = Mexican peso.

From	Exchange rate to							
	JY	IR	MR	UD	CD	EE	EP	MP
JY	1	50	0.04	0.008	0.01	0.0064	0.0048	0.0768
IR		1	0.0008	0.00016	0.0002	0.000128	0.000096	0.001536
MR			1	0.2	0.25	0.16	0.12	1.92
UD				1	1.25	0.8	0.6	9.6
CD					1	0.64	0.48	7.68
EE						1	0.75	12
EP							1	16
MP								1

From	Transaction cost %							
	JY	IR	MR	UD	CD	EE	EP	MP
JY		0.5	0.5	0.4	0.4	0.4	0.25	0.5
IR			0.7	0.5	0.3	0.3	0.75	0.75
MR				0.7	0.7	0.4	0.45	0.5
UD					0.15	0.2	0.2	0.2
CD						0.2	0.1	0.1
EE							0.05	0.5
EP								0.5
MP								

(1) The portion below the main diagonal in the first table is left blank because each entry in it can be obtained from the corresponding entry above the diagonal (every currency considered can be converted into every other currency). The second table is symmetric around the main diagonal, that is why only the portion above the main diagonal is given. The transaction cost for converting one currency, say c_1 units of currency 1, into any other currency is $(c_1/100)\times$ (entry for this currency pair in the second table) in units of the currency being converted (i.e., here currency 1). Formulate the problem of finding the most cost effective method to convert these holdings into US$, as a min cost flow problem.

(2) Suppose there are transaction limits for converting yen, rupiah, and ringgits (only these currencies, no limits for converting other currencies) as shown in the following table (unit = equivalent of million US$). Then find the most cost effective way of converting as much of these currency holdings into US$ as possible.

From	Transaction Limits							
	JY	IR	MR	UD	CD	EE	EP	MP
JY	–	5	5	2	2	2	4	8
IR	5	–	2	1	2	1	3	2
MR	3	5	–	2	3	3	2	1

2.37. Blending of residual fuel oil: Crude oil is separated into many different products such as naphtha (used for making gasoline), diesel oil, etc. The most expensive products produced from crude oil are these higher volatile fractions. After all these highly profitable fractions are removed, one of the fractions that remains at the bottom is heavy residual fuel oil; its market price is quite low compared to the above products. The sulfur content in heavy fuel oil is usually high, and burning it requires special equipment; therefore, this fraction is mostly used as fuel in ships (i.e., ocean-going vessels) under the name "*bunker fuel*." The quality of heavy fuel oil depends on the type of crude oil from which it is produced, and measured by three important characteristics: SG (specific gravity = weight/volume), SC (sulfur content, percent by weight), VBI (viscosity blend index, which blends linearly when quantities are measured in volume units like gallons). To meet the specifications on these characteristics, bunker fuel sold to customers is usually a mixture of various fractions.

Suppose we have three different fractions available to blend into bunker fuel: AF (asphalt flux), CO (clarified oil), and K (kerosene). Data on them is given below.

Fraction	Value of characteristic			Price
	SG	SC	VBI	\$/gallon
AF	0.98	2.39	0.966	1.86
CO	0.91	2.20	0.740	1.20
K	1.20	0.20	0.347	2.40
Specs. in	≤ 1.06	≤ 2	≤ 0.808	
Bunker fuel	≥ 1		≥ 0.738	

Bunker fuel can be sold at \$1.80/gallon. Select appropriate units for measuring the three constituent fractions, and assuming that the blending cost is negligible, formulate an LP model for determining an optimum blend for bunker fuel.

2.38. Gasoline blending: A gasoline blending company has six types of raw gasolines available with data given below.

Type	Oc. R. (octane rating)	Available b/day	Cost \$/gallon
T1	72	2,500	2.40
T2	81	3,800	2.55
T3	86	4,300	2.75
T4	89	5,500	2.85
T5	96	1,500	3.10
T6	99	1,000	3.20

1 barrel = 42 gallons.

They blend these raw gasolines into two grades of fuel: regular and premium. Minimum specifications on Oc. R. for regular is 85 and for premium it is 95. They need to produce a minimum of 9,000 b of regular/day and between 2,000 and 4,000 b/day of premium. The selling prices of these grades are \$3.05/g and \$3.20/g for regular and premium, respectively. Formulate an LP to determine an optimum blending plan for this company (Dantzing and Thappa (1997) of Chap. 1).

2.39. A production planning problem: Over the next 5-week horizon, a company has to produce and deliver to its customer a special commodity according to the following schedule: 200 units each at the end of weeks 1 and 2; then 300, 400, 500 units, respectively, at the end of weeks 3, 4, and 5.

The production process requires workers who received special training. The training of a new worker, carried out by a trained worker, takes 1 week. They begin week 1 with 10 trained workers on hand. Each trained worker can train up to five new workers during a week if assigned to training during that week, or produce up to 10 units of the product if assigned to production during that week. Only workers trained at the company are used for production or training of new workers.

Shortages in delivery cost \$100/unit short /week short, until delivered. By the end of the fifth week all required deliveries must be completed. A unit of product produced ahead of schedule must be stored at the company until its delivery time, this storage cost is \$20/unit/week.

Salaries of workers are \$15,000/week if used for production, \$20,000/week if used to train new workers. Treating the number of workers used for the various tasks each week as continuous variables, formulate the problem of determining an optimum production/training schedule as an LP model (Dantzig (1963) of Chap. 1).

2.40. Minimizing the cost of production level changes: A company has a process that produces a product P. The production rate of this process measured in tons/month can be changed, but this change costs money.

The company's policy is to change the production rate, if necessary, only at the beginning of each month. During any month, the process keeps on producing at the same rate as it was set at the beginning of that month.

To increase the production rate from existing level costs $c_1 = \$100$ per ton/month increase. To decrease production rate from existing level costs $c_2 = \$50$ per ton/month decrease. When the production rate is changed, the amount of increase or decrease in the rate has to be ≤ 100 tons.

The first month's production rate can be set at any nonnegative level; there is no cost for this.

Production in a month can either be used for meeting the demand in that month or put in storage for meeting the demand later on. The storage room has a capacity for storing up to $s = 1{,}000$ tons and collects storage charges at the end of each month on the total amount in storage then, at the rate of $c_3 = \$10$/ton. They have 100 tons in storage at the beginning of month 1.

Demand data (d_i = demand (in tons) in period i: $i = 1$ to 6) is (100, 500, 800, 600, 300, 500), and demand in each month has to be met exactly. At the end of month 6 they would like to have at least 250 tons in storage.

The actual production cost/ton is the same in all the months so the company needs to determine the production, storage plan in months 1–6 to minimize the sum of storage and production level changing costs while satisfying all the constraints. Formulate this as an LP and justify your formulation carefully.

2.41. $x = (x_1, \dots, x_n)^T \in R^n$, where $n \geq 4$ is the vector of decision variables in the optimization problem

$$\begin{aligned} \text{Minimize} \quad & \theta(x) \\ \text{subject to} \quad & Ax = b \\ \text{and} \quad & x \geq 0, \end{aligned}$$

where A is a given $m \times n$ matrix and $\theta(x) = \text{Maximum}\{x_1, 5x_2, 10x_3\} - \text{Minimum}\{x_1, 5x_2, 10x_3\}$.

Is it possible to transform this problem into a linear program? Why? If it is, give that transformation and a clear explanation why it will work.

2.42. Six period production planning problem: A company divides the year into six planning periods for planning purposes. Here is the relevant data for one important product of this company for the coming year.

Period j	Total Prod. cost (\$/ton) $= c_j$	Prod. capacity (tons) k_j
1	20	15,000
2	25	20,000
3	30	22,000
4	40	30,000
5	35	27,000
6	50	25,000

There is unlimited demand for this product. The company has two main customers for it who have very different purchasing strategies.

Customer 1 wants to sign a purchasing deal for the whole year with the company, for any amount of the product, at $p_1 =$ 85\$/ton; but he stipulates a phased delivery: If q_1 tons is the quantity the company agrees to sell him, he wants 10% of this quantity delivered at the end of each of periods 1, 2, 5, 6, and 30% of this quantity delivered at the end of each of periods 3 and 4.

Customer 2 will purchase any quantity of the product available at the end of each period, but only at the rate $p_2 =$ 65\$/ton.

Product manufactured during a period has to be sold by the end of that period or disposed off, and it cannot be stored into the next period.

Formulate the problem of maximizing the net profit of the company during the year as an LP

2.43. Speciality chemicals manufacturing: A speciality chemicals company makes three chemicals denoted by C_1, C_2, C_3; each in a separate division. These are rare chemicals used for highly specialized applications, every one who uses them always use all three of them in combination in proportions: C_1:C_2:C_3 = 1:2:4 (i.e., each unit of C_1 is always used in combination with 2 units of C_2 and 4 units of C_3, etc.). Also as they are very expensive chemicals, whenever a customer buys any of them, they always buy all of them in measured quantities in the above proportion.

Hence the customer demand for these chemicals can be measured unambiguously by the demand for any one of them (since the demand for the other two can be obtained from that one using the above proportions). Four expensive raw materials RM_1 to RM_4 are used in making these chemicals. Here is the relevant data for a period.

Raw material	Units of input to make 1 unit C_1	C_2	C_3	Max units available
RM_1	2	0	1	3,000
RM_2	4	3	0	7,000
RM_3	0	3	1	5,000
RM_4	0	1	4	6,000
Labor man hours.	3	2	4	

They can allocate up to 10,000 man hours of labor during regular time for the production of these chemicals, and up to another 3,000 man hours during overtime.

(1) Select suitable units for measuring customer demand for these chemicals, and formulate an LP model for determining the maximum customer demand that the company can meet in this period.
(2) The prices of raw materials RM_1 to RM_4 in money units/unit are 250, 175, 300, 400. The labor needed is highly skilled labor, regular time cost/man hour is 5, and overtime cost is 7.5 money units.

Assuming that the maximum customer demand that can be met (in units selected in (1)) is d, formulate an LP model for meeting this demand exactly, at minimum cost.

2.44. Milk procurement: Organic Valley (OV) is a company that buys organic raw milk from individual farms in the milk-producing area of Wisconsin, divided into Regions 1 and 2. The farms vary in size, but all are small to medium. The composition of raw milk in terms of butter content and its "separation properties" tend to be more or less the same among the farms in each region, but the price/liter of raw milk varies from farm to farm. So, OV has classified the farms in each region into two classes, depending on the average price/liter of raw milk. Here is the data:

The average butter content of raw milk from Regions 1 and 2 is 40, 35 g/liter, respectively. The two classes in Region 1 are 1.1 and 1.2 and those in Region 2 are 2.1 and 2.2. In class 1.1 and 1.2, the average price (cents)/liter is 45 and 50, and the maximum availability in liters/day is 2,000, unlimited, respectively. In class 2.1 and 2.2, the average price (cents)/liter is 40 and 47, and the maximum availability in liters/day is 3,000 and unlimited, respectively.

In procuring milk each day, depending on the quantity to be purchased from each region, OV allocates that quantity among the two classes in that region so as to minimize the total cost.

For each region $i = 1, 2$, give the expression $f_i(x_i)$ for the price of buying x_i liters of raw milk from region i, explaining the reason for it.

Raw milk may be used in the blending process as it is or it may be passed through a separator before blending. Each liter from Region 1 when separated yields 0.5 L with butter 20 g/L, and another 0.5 L with butter 60 g/L at a cost of 3 cents/L. Similarly each liter from Region 2 when separated yields 0.5 L with butter 15 g/L, and another 0.5 L with butter 55 g/L at a cost of 3 cents/L.

The blending process blends the various milks into two products (grades of milk called cream and regular) with the following specifications.

Product	Min. required butter content (g L^{-1})	Mkt. price (cents L^{-1})	Demand (L day^{-1})
Cream	52	150	1,000
Regular	22	80	8,000

Any leftover unsold milk is sold to an ice cream plant for 38 cents/L as long as the butter content in it is $\geq$15 g/L.

Formulate the problem of maximizing OV's net daily profit as an LP.

2.45. Milk blending: We have eight different grades of milk available with different butter contents. For $i = 1$ to 8, the butter content of grade i is 15, 20, 40 45, 47, 50, 55, 60 g/L, respectively, and x_i liters is the quantity of this grade of milk available.

We have a customer to whom milk can be sold, but he will only buy milk in which the butter content is $\geq$42 g/L.

We have a blender that can blend any combination of several grades of milk. From the grades of milk that we have, it is required to determine the maximum quantity of milk satisfying the customer's specification on butter content that we can produce using this blender. Formulate this problem as an LP.

2.46. Finding an interior point of a convex polyhedron: Let $K = \{x : Ax \geq b\}$. An $x \in K$ is called an interior point of K if it satisfies $Ax > b$, boundary point of K otherwise. Formulate the problem of finding an interior point of K, or establishing that K has no interior point, as an LP.

On the other hand, if the goal is to find a boundary point of K, how can one find it?

2.47. Optimizing bets at a horse race: A person has β\$ to bet on n horses competing in a race. The race course has the following pay-off policy: If x_i\$ is bet on the ith horse, the pay-off from this bet is $\alpha_{0i} x_i$ if that horse comes first in the race, $\alpha_{1i} x_i$ if it comes second in that race, 0 otherwise.

Formulate the problem of determining how much to bet on each horse in the race to maximize the minimum net gain, irrespective of which of the n horses come in first or second in the race.

2.48. Optimal revision of estimates to satisfy monotonicity: A sequence of parameters $(a_1, \ldots, a_5)$, which are known to be monotonic increasing, is being estimated from data subject to random fluctuations. The estimation process estimates each a_i independently from the others. The estimates obtained are denoted by $(\bar{a}_1, \ldots, \bar{a}_5)$, because of random fluctuations this sequence may not be monotonic increasing with i. It is required to revise the estimate $\bar{a}_i$ into x_i, for $i = 1$ to 5, making the smallest possible changes, so that the revised sequence $(x_1, \ldots, x_5)$ is monotonic increasing. Formulate this as an LP.

2.49. Revising property tax rates: A city currently collects property taxes within its boundary; this is the major source of its revenue. Having incurred budget deficits for several years, the city has appointed a committee to revise the property tax rates, and also suggest other taxes that the city can levy to balance the budget.

Besides property taxes (PT), the committee is investigating the possibility of including food and drug taxes (FDT), general sales tax (ST), and a gasoline tax (GT).

The tax base of property values is \$550 million, and PT rate is an annual percentage of this value. The annual food and drug sales volume and general sales volume are \$35 million and \$55 million, respectively; again, the FDT and ST rates are percentages of these volumes. Annual gasoline sales volume within the city is estimated at 8 million gallons; the GT rate will be in the form of cents/gallon sale.

The committee wants to determine the PT, FDT, ST, GT rates so as to achieve the following goals as far as possible:

G1: Total tax revenue should be $\geq$ 16 million

G2: FDT should be $\leq$ 10% of total tax revenue

G3: ST should be equal to 20% of all taxes collected

G4: GT should be $\leq$ 3 cents/gallon

G5: GT should be $\geq$ 12% of the total tax revenue.

The committee has decided that missing **G1** (per million) \$, **G4** (per 1 cent/gallon), and **G2, G3,** or **G5** (per 1%) should all carry the same penalty.

Formulate the problem of deciding the PT, FDT, ST, GT rates so as to minimize the total penalty as an LP model.

2.50. A property of convex functions: Let $f(x)$ be a real-valued function in $x \in R^n$.

Prove that if $f(x)$ is a convex function and α is any real number, then $\{x : f(x) \leq \alpha\}$ = the set of all points x satisfying $f(x) \leq \alpha$ must be a convex set.

2.51. Arranging financing for construction work: The construction of a bridge starts on 1 Sept of year 1, and is expected to be completed on 31 Aug of year 5. For this, the city has agreed to make the following payments to the contractor.

Date	Amount to be paid (\$million)
1 September, year 1	12
1 September, year 2	14
1 September, year 3	18
1 September, year 4	15

The city will raise this money by selling bonds on each of the above dates. Let x_j denote the \$amount of bonds to be sold on the jth of these dates, $j = 1$ to 4.

All bonds accumulate simple interest only on the principal amount. Four year bonds, three year bonds, two year bonds, one year bonds pay, respectively, 10%, 9%, 8%, 7% simple interest yearly. All bond amounts plus accrued interest will have to be paid to bondholders in a simple lump sum payment on 1 September of year 4.

If the money collected in a bond is not used to pay the contractor on the day the bond is sold, the city immediately invests it in a short-term (exactly one year) time deposit, which pays the principal +6% interest paid exactly one year later. So, on each of the dates in above tableau, contractors are paid out of money collected from bond sales on that day + any money coming from repayment of short-term deposit made 1 year before, and any remaining amount is invested in short term deposits right away.

The Federal government agreed to pay the city $59 million, the total construction cost of bridge, on 1 September of year 4. The city will use this money towards lump sum payments to be made to bondholders on that day.

It is required to determine optimum values of x_j's and short term deposits to be made in each year of project horizon so as to minimize total interest money to be paid to bond holders on 1 September of year 4 when they get lump sum payments. Formulate this as an LP.

2.52. Managing a two-dam reservoir system: The state has two dams with reservoirs denoted by R_1, R_2, respectively, on a river. Water for the river comes mainly from melting snows on mountains before R_1. Estimated availability of water at R_1 is 294 MAF (million acrefeet)/year.

At R_1, the state has agreed to supply $\geq$24 MAF/year to a town there. Remaining water at R_1 flows down to R_2 losing 20% to evaporation along the way. At R_2 state releases some water to farmers there for irrigation, and the rest passes through a hydroelectric generator and then flows further downstream.

The town at R_1 pays $0.5/AF for water supplied. Farmers pay $0.2/AF for irrigation water released to them. Water passing through hydroelectric generators earns $0.8/AF.

The state has a target of 100 MAF/year for releasing to farmers for irrigation, and would be happy if this target can be exceeded. The penalty for downfall in irrigation water release below target is 10 penalty units/MAF downfall.

The state has a target of $144 million for total income/year from this system, and would be happy if this target can be exceeded. Penalty for downfall in total income below target is 4 penalty units per $million downfall.

Give a goal programming model for allocating available water to the three uses in this two-objective optimization problem.

2.53. $f(x)$ is a convex function defined in the space of $x = (x_1, x_2)^T$, satisfying following property:

Property: Inside the circle with $x^* = (3,2)^T$ as center and radius 1, x^* is the unique minimum point for $f(x)$, that is, for all points $\bar{x}$ inside the circle, $\bar{x} \neq x^*$, we have $f(\bar{x}) > f(x^*)$.

Take any point x outside this circle. Using Jensen's inequality, prove that it is not possible for x to satisfy $f(x) < f(x^*)$.

2.54. Managing Tunis Water Resources System: The Tunis Water Resources System (TWRS) comprises seven reservoirs interconnected by water conveyance structures (WCS) (a canal and three pipelines). Each reservoir provides water for a number of demand centers, and a demand center may receive water from more than one reservoir.

The reservoirs (all with first initial "R") are RJ (Reservoir Joumine, or Joumine), RBM (Ben Metir), RK (Kasseb), RBH (Bou Heurtma), RM (Mellegue), RSS (Sidi Salem), and RS (Siliana). Data on their capacities and the expected incremental inflows from precipitation in its catchment area in the month of May are given below:

Reservoir	RJ	RBM	RK	RBH	RM	RSS	RS
Capacity ($10^6 m^3$)	120	44	72	103	89	510	62
Expected May inflow ($10^6 m^3$)	2.9	0.9	1.7	3	15.3	19.6	2.5

There are demand centers all identified with "D" as the initial letter in their abbreviations, with demand for water in the month of May as given below.

Demand center	Expected May demand ($10^6 m^3$)
DIMA	1.2
DBLI	0.9
DJE	2
DIBH	10.7
DINE	0.3
DIAEA	9.3
DISI	4.3
DIMSC	13.1
DIBV	1.8
DTO	2.6
DTU	5

The following Fig. 2.9 shows the WCS with its canals (in solid lines), pipelines (in dashed lines), with their monthly flow capacities in units of 10^6 m^3/month, and directions of flow indicated by arrows.

It is required to determine how much water should be supplied in the month of May to each demand center through the various canals and pipelines to meet all the demands, such that the amount of water in each reservoir at the end of May is as close as possible to the amount in it at the beginning of May. Formulate this as an LP. (Data from Nandalal and Bogardi 2007.)

2.55. Maximizing returns from short-term investments: In the automobile industry, companies make improvements in their car models and start selling the improved versions of the cars at the beginning of each year. These new versions

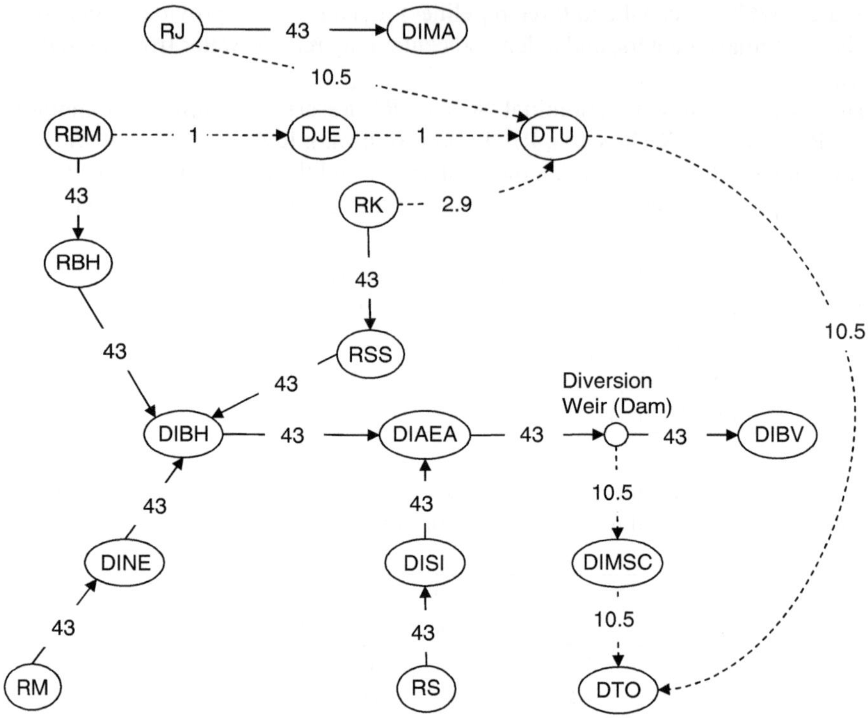

Fig. 2.9 This figure shows the WCS with its canals (*in solid lines*), pipelines (*in dashed lines*), with their monthly flow capacities in units of 10^6 m^3/month, and directions of flow indicated by *arrows*. Each draws water from a reservoir node, and as it passes through each demand node satisfying the demand there, the quantity of water it carries keeps decreasing. For the sake of this problem, we ignore evaporation losses

enjoy high sale volumes during the first half of the year, after which the sale volumes tend to level off during the second half of the year. That is why income to these companies in the form of sale revenues from dealers tends to be higher in the first half of the year than during the second half.

By the middle of the year, quality problems with the versions of car models being sold become well known, and companies then set up new design teams to work on fixing up these problems in the new versions of these car models to be released in the market at the beginning of next year. That is why design changing expenses for these companies tend to be higher in the second half of the year than during the first half.

Hence, the income for these companies tends to be higher than their expenses during the first half of the year and vice versa during the second half. These companies normally operate on annual accounting, and their profits for the year are declared only at the end of the year. So, to maximize their annual profits, these companies normally invest their surplus income over expenses in the first half of the year, in short term investment opportunities, to be cashed as needed or by the end of the year.

We provide the data for a company that divides the year into four periods of three months each. Data on the expected income and expenses in each period is given below. There are two investment opportunities, I_1 and I_2 for short-term investments. In each of these opportunities, any amounts can be invested for periods ranging from one, two, or three periods (can assume that all transactions take place only on the last day of each period) with fractional returns given in a table below (i.e., if x is the amount invested, r is the fractional return for a duration of k periods, and this investment will pay out $x(1 + r)$ at the end of k periods after investing).

Period	Expected income	Expected expenses
1	200	100
2	180	110
3	150	180
4	170	250
Annual total	700	640

Returns from short-term investments		
Duration	Fractional returns in	
(no. periods)	I_1	I_2
1	0.02	0.01
2	0.05	0.06
3	0.08	0.09

Remember that all investments during the year will be cashed by the end of the year. Also, the maximum invested over the year in any investment opportunity cannot exceed 150 money units. The only money that the company invests in these investments is any money on hand (from income in that period either from dealers or from earlier investments that are cashed). Formulate the problem of determining how much the company can invest in each investment opportunity, so as to maximize the money left over in the companies hands by the end of the year.

2.56. Cake mix blending: A premium-quality cake mix is a blend of six different powdered ingredients called I_1 to I_6.

There are three weighing machines (WM-1 to WM-3) for measuring the quantities of the ingredients. Each weighing machine has one or more sizeable hoppers attached to it, each hopper holding one of the ingredients.

The following table indicates the ingredients contained in the hoppers attached to each of the weighing machines (* mark indicates that one hopper of the corresponding weighing machine contains the ingredient, blank entry indicates that none of the hoppers on this weighing machine contains that ingredient).

	I_1	I_2	I_3	I_4	I_5	I_6
WM-1	*		*	*		
WM-2		*		*	*	
WM-3	*	*			*	*

An ingredient can be weighed on a weighing machine only if one of its hoppers holds that ingredient. For example, on WM-1 we can weigh ingredients I_1, I_3, I_4, but none of the others. Also, I_1 can be weighed on WM-1 or WM-3 or both of these, but not on WM-2.

On each weighing machine, the ingredients are emptied from one of its hoppers at a time, each hopper emptying for a preset duration depending on the quantity of that ingredient to be measured on that weighing machine into the scale of that weighing machine. After all the hoppers finish delivering their ingredients, the contents of the scale are conveyed to a bin where they will be blended into the batch.

The weighing machines work simultaneously and independently in the plant, and the whole operation is computer controlled. As soon as the scales on all the three weighing machines deliver their contents to the bin, the batch is complete, weighing stops, the scales are cleaned, and preparations take place for the next batch.

Some ingredients flow faster than the others from the hoppers.

g_i = time in seconds for 1 kg of ingredient I_i to flow from the hoppers onto the scales of a weighing machine, for $i = 1$ to 6.

Here is the data:

Ingredient	I_1	I_2	I_3	I_4	I_5	I_6
kg needed for a batch	130	140	101	290	29	21
g_i	1	1.2	1.3	1.1	0.8	0.9

The capacity of the scales on each of the three weighing machines is 400 kg.

The problem is to determine how much quantity of each ingredient should be weighed on each weighing machine in order to minimize the clock time for weighing a batch.

Give a formulation of this problem as a linear program. Define all your decision variables very clearly, and write the objective function and all the constraints carefully. Justify your model.

2.57. In R^2, the *rectilinear distance* between two points $a = (a_1, a_2)$ and $b = (b_1, b_2)$ is defined to be $d(a, b) = |a_1 - b_1| + |a_2 - b_2|$.

We are given four points in R^2. These are $a^1 = (-10, 20)$, $a^2 = (60, 30)$, $a^3 = (40, 10)$, $a^4 = (80, 60)$. The *weights* associated with these points are $(w_1, w_2, w_3, w_4) = (0.2, 0.1, 0.3, 0.4)$, respectively.

It is required to find a point $x \in R^2$ to minimize $z(x) = \max\{w_t d(x, a^t) : t = 1$ to $4\}$. Give a linear programming formulation of this problem.

2.58. Operating a hydroelectric power system: An agency operates two water reservoirs on a river system. Each of these reservoirs has a hydroelectric power generator, and all the water released from this reservoir flows through this generator and generates power. AF = acrefeet, KAF = kiloacrefeet or 1,000 AF are units for measuring the volume of water. KWH = kilowatthour is a unit for measuring electric power.

The amount of power generated when an AF of water is released from the reservoir depends nonlinearly on the *head* which is determined by the amount of water in the reservoir at the time of release; however, to model the problem as a linear program, we will assume that it is a constant = the average amount given in the following table.

For operational decisions, an year is divided into six periods (or seasons) of two months each. For simplicity, assume that each period is an interval of time with uniform characteristics, that is, the inflows during this period from precipitation occur at a constant rate.

During each period, the agency is required to release an amount of water $\geq$ a specified minimum amount from each reservoir for downstream uses. All water released from reservoirs flows through the power generators and hence produces electricity. During each period that water is released at a constant rate so that the total release for the period equals the planned amount.

At any instant of time, if reservoir is at its full capacity, additional inflowing water spills over the dam. The spilled water does not generate any electric power. As an example, if water stored in Reservoir 1 with its capacity of 4,000 KAF at the beginning of a period is 3,200 KAF, the inflow during this period into this reservoir is 1,400 KAF, and the water released during this period from this reservoir is 200 KAF, then $3200 + 1400 - 200 - 4000 = 400 KAF$ will be the spilled water from this reservoir during this period.

Here is the data on the two reservoirs:

Reservoir	Capacity	Power generated by releasing 1 AF[a]	Water stored at year beginning
1	4000 KAF	310 KWH	1800 KAF
2	6000 KAF	420 KWH	2500 KAF

[a]Average amount per AF as mentioned above

Power sales: They have an industrial customer (IC) who buys power on an annual basis and pays \$70/1,000 KWH. The IC stipulates that if the annual KWH of power sold to them is denoted by e_0, then $p_i e_0$ KWH of it must be delivered in period i, for $i = 1$ to 6; where the p_i are given in the following table. In each period, any power not sent to the IC is sold in that period itself to the regional power grid at a price of \$50/1,000 KWH.

Period	Inflow (KAF) into reservoir		Min release, KAF from reservoir		p_i
i	1	2	1	2	
1	550	2620	200	310	0.1
2	1,470	2300	200	580	0.15
3	990	1230	200	990	0.15
4	150	730	200	1,500	0.3
5	30	410	200	560	0.2
6	160	500	200	400	0.1

At the end of the year, the amount of water in storage in each reservoir must be the same as that at the beginning of the year.

It is required to determine how much water to release from each reservoir in each period, the annual amount of power to sell IC, and the amount of power to sell the regional power grid in each period, so as to maximize the total annual return from power sales subject to all the above constraints. Formulate this as an LP, but do not solve numerically.

2.59. Consider the LP: Minimize some objective function, $z = cx$ subject to the following constraints in standard form.

x_1	x_2	x_3	x_4	x_5	x_6	x_7	
1	0	0	0	3	1	−1	13
0	1	0	0	7	−2	2	29
0	0	1	0	−2	0	0	45
0	0	0	1	8	1	−1	83

$x_j \geq 0$ for all j

Assuming that the problem has an optimum solution, prove that there must be an optimum solution of the problem in which at least one of the two variables x_6 or x_7 is 0.

2.60. Consider a company making five different car models indexed by $i = 1$ to 5, in the luxury car segment. In the market there are 20 other car models indexed by $i = 6$ to 25, made by other manufacturers in this segment.

The sales volume Q_i during a quarter (of the year) of any of these car models depends on the sale price P_i of this model and the sale prices of other car models in this segment. For a particular quarter, we are given the following data:

For $i = 1$ to 5, S_i (= unsold inventory in the dealer lots at the beginning of the quarter) and D_i^0 (projected demand at current price levels) of car model i

C = total production capacity of car models $i = 1$ to 5 at the company's plant in this quarter

For $i = 1$ to 5, P_i^0 = price of car model i at the end of the previous quarter

For $i = 6$ to 25, P_i (= best estimate of the price of car model i of the competitor for this quarter),

a_{ij} for $i = 1$ to 5 and $j = 1$ to 25 is the rate of change in the expected demand of car model i in this quarter per unit change in P_j

c_i for $i = 1$ to 5 is the total production cost/unit of the ith car model.

Since these car models are competing with each other in the marketplace, change in the selling price of any one of them affects the demand for the other car models too in this quarter; the a_{ij} coefficients measure these effects, and these coefficients are estimated from market surveys on customers buying new cars. So, what we are assuming is that the expected demand in the quarter for the ith car model for $i = 1$ to 5 is $D_i = D_i^0 - \sum_{j=1}^{25} a_{ij} p_j$, where P_6 to P_{25} are given data and P_1 to P_5 are the prices set for car models $i = 1$ to 5, respectively, for this quarter, these are decision variables the company can choose (so, for $i = 1$ to 5, $P_i - P_i^0$ is the change in prices of these models from last quarter to this quarter through rebates or price increases).

Let I_i denote the production level of the ith car model in this quarter by the company for $i = 1$ to 5. Then the amount $S_i + I_i$ is the volume available for sale of this model during this quarter. Assuming that the sales and production volumes of each car model are continuous variables, formulate the problem of determining P_i, I_i to maximize the company's net profit during the quarter (problem formulated by Robert Bordley).

2.61. A is a matrix of order $m \times n$ of full row rank, and S is the subspace of R^n, $S = \{x \in R^n : Ax = 0\}$.

For any pair of points $p = (p_j)$, $q = (q_j) \in R^n$, the L_1 **distance** between them, denoted by $||p - q||_1$, is defined to be $\sum_{j=1}^{n} |p_j - q_j|$.

Given a point $y \in R^n$, $y \notin S$, the L_1 projection of y onto the subspace S is defined as an optimum solution of the problem of minimizing $||y - x||_1$ over $x \in S$. Formulate the problem of finding an L_1 projection of a given point $y \in R^n$, $y \notin S$ onto the subspace S as a linear programming problem.

Is this L_1 projection of y onto S a unique point? Illustrate with a numerical example (from Brooks and Dula 2008).

2.62. Consider the problem discussed in Exercise 2.61. If S is a hyperplane in R^n containing the origin, that is, $S = \{x : a_1x_1 + \ldots + a_nx_n = 0\}$, where $(a_1, \ldots, a_n) \neq 0$ and $y \in R^n, y \notin S$; derive the set of all points in S which are at the minimum L_1-distance to y (J.H. Dula).

2.63. Check whether the real-valued function $f(x_1, x_2) = -x_1 + (x_1 + x_2) e^{\{-(x_1+x_2)\}}$ is a convex function (R. Saigal).

2.64. $w = (w_1, \ldots, w_n) > 0$ is a given vector of positive weights. For $y = (y_1, \ldots, y_n)^T$ let

$$L(y) = \text{Maximum}\{w_i y_i : i = 1, \ldots, n\}$$

$a \in R^n$ is given, and we are required to find $x = (x_1, \ldots, x_n)^T$ to minimize $L(a - x)$ subject to $x_j - x_{j+1} \leq 0$ for all $j = 1$ to $n - 1$. Formulate this problem as an LP.

How does the formulation change if we have additional constraints: $0 \leq x_j \leq 1$ for all j? (V.A. Ubhaya).

2.65. Let $\lambda \in R^1$ and $\epsilon > 0$ is a positive constant. Define $\theta(\lambda) = \epsilon \log\{1 + e^{[g(\lambda)/\epsilon]} + 1\}$, where $g(\lambda) = a + b\lambda$, an affine function. Show that $\theta(\lambda)$ is a convex function of λ.

References

Agha SR (2006) 'Use of goal programming and integer programming for water quality management- A case study of Gaza Strip. Eur J Oper Res 174:1991–1998

Andreatta G, Filipi C, Romanin-Jacur G (1993) The linear balancing flow problem. Eur J OR 64:68–82

Bakshi HC, Bhatia HL, Puri MC (1979) Trade-off relations in assignment problems. Cahiers du Centre d'Etudes de Receherche Operationnelle 21:367–374

Berman O, Einav D, Handler G (1990) The constrained bottleneck problem in networks. Oper Res 38:178–181

Brooks PR, Dula JH (2008) Note on the L_1 projection onto a subspace. Virginia Commonwealth University

Charnes A, Cooper WW (1977) Goal programming and multiple objective optimization, Part I. Eur J Oper Res 1(1):39–54

Davis KR, Webster FE Jr (1968) Sales force management-text and cases. The Ronald Press

Geetha S, Nair KPK (1993) A variation of the assignment problem. Eur J Oper Res 68:422–426

Hwang CL, Masud AS (1979) Multiple objective decision making methods and applications: A state of the art survey. Springer, New York

Ignizio JP (1976) Goal programming and extensions. Lexington Books, Lexington, MA

Iverstine and Kinard (1977) Cases in production and operations management. Charles E. Merill Publishing Co., Columbus, Ohio

Kaplan W (1999) Maxima and minima with applications. Wiley, New York

Keeney RL, Raiffa H (1976) Decisions with multiple objectives. Wiley, New York

Khorramshshgol R, Okoruwa AA (1994) A goal programming approach to investment decisions: A case study of fund allocation among different shopping malls. Eur J Oper Res 73:17–22

Lanzenauer CHV (1975) Cases in OR. Holden-Day, San Francisco

Lee SM (1972) Goal programming for decision analysis. Auerbach, Philadelphia

Mehrotra V, Ozluk O, Salem CL (2006) An application of ORMS to not-for-profit financial management: Optimal assignment of operating costs to restricted funding sources. Working Paper, College of Business, San Francisco State University

Mukherjee K, Bera A (1995) Application of goal programming in project selection decision: A case study from the Indian coal mining industry. Eur J Oper Res 82:18–25

Murty KG (1983) Linear programming. Wiley, New York

Murty KG (1988) Linear complementarity, linear and nonlinear programming. Helderman Verlag, Berlin; Can be accessed on the web from Murty's webpage at: http://www-personal.engin.umich.edu/ murty/

Murty KG (1995) Operations research: Deterministic optimization models. Prentice Hall, Englewood Cliffs, NJ

Nandalal KDW, Bogardi JJ (2007) Dynamic programming based operation of reservoirs, applicability and limits. Cambridge University Press, UNESCO

Rugg MM, White GP, Endres JM (1983) Using goal programming to improve the calculation of diabetic diets. Computers OR 1(4):365–373

Sawaragi Y, Nakayama YH, Tanino T (1985) Theory of multiobjective optimization, (vol. 176. Mathematics in science and engineering). Academic, Orlando, FL, ISBN 0126203709

Schniederjans MJ (1995) Goal programming : Methodology and applications. Kluwer, Boston

Singh P, Saxena PK (2003) The multiple objective time transportation problem with additional restrictions. Eur J Oper Res 14(6):460–476

Sponk J (1981) Interactive multiple goal programming: Applications to financial management. Martinus Nijhoff, Boston

Steuer RE (1986) Multiple criteria optimization: Theory, computations, and application. Wiley, New York

Steuer RE, Oliver RL (1976) An application of multiple objective linear programming to media selection. Omega 4(4):455–462

Vatter PA, Bradley SP, Frey SC Jr, Jackson BB (1978) Quantitative methods in management: Text and cases. Richard B. Irwin Inc., Homewood, IL

Zeleny M (1974) Linear multiobjective programming. Lecture notes in economics and mathematical systems, vol. 123. Springer Verlag, Berlin

Chapter 3
Intelligent Modeling Essential to Get Good Results

3.1 The Need for Intelligent Modeling in Real World Decision Making

Operations Research/Management Science (OR/MS) theory has developed efficient algorithms for solving some single-objective optimization models that are highly structured.

In real-world applications, decision problems tend to have many complex features, uncertainty influencing many important aspects, and several other complications. Constructing a mathematical model to find an optimum decision in such problems is often a very difficult task that requires a lot of skill. The trouble is that none of the models discussed in OR theory may fit perfectly the problem you need to solve. As Wolfram (2002) suggests "··· the idea of describing behavior in terms of mathematical equations works well where the behavior is fairly simple. It almost inevitably fails whenever the behavior is more complex."

So none of the standard techniques discussed in OR literature may apply directly to the real problem. In this case the model has to be an approximate representation of the real problem. It has to be constructed to satisfy the aim that the optimum solution found for the model will be a good solution for the real one. Hence, there is a wide gulf between real-world problems and mathematical models for which efficient algorithms have been developed in theory. To bridge this gulf and get good results, it is essential to model real-world problems intelligently. Heuristic modeling techniques, approximations, relaxations, hierarchical modeling techniques with substitute objective functions for each stage (Murty and Djang 1999), and heuristic algorithms serve as a bridge between the two sides of this gulf (Fig. 3.1) (Murty et al. (2005b) of Chapter 1). This theoretical justification is also supported by Ackoff (1996), Dantzig (1990), Geoffrion (1976), and Simon (1987).

In this mode of application, the methods developed in ORMS theory are not the main methods for solving complex real-world problems, but become valuable tools for designing intelligent approaches to handle them. We illustrate this process with three case studies on work done for developing decision support systems in three different real-world decision-making problems.

K.G. Murty, *Optimization for Decision Making: Linear and Quadratic Models*, International Series in Operations Research & Management Science 137, DOI 10.1007/978-1-4419-1291-6_3,

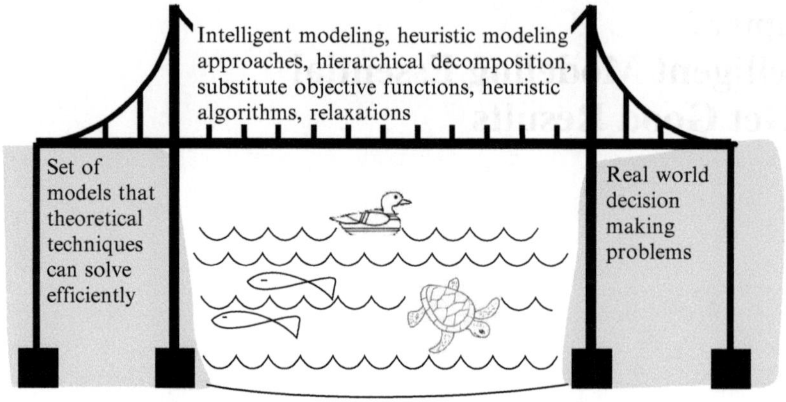

Fig. 3.1 "Mathematical models for which we have theoretically efficient algorithms" and "real-world problems we encounter in practice" are like the two banks of a very large river with a wide gulf between them. "Intelligent modeling, heuristic modeling approaches, hierarchical decomposition, substitute objective functions, heuristic algorithms, relaxations" serve as a bridge for this gulf

3.2 Case Studies Illustrating the Need for Intelligent Modeling

3.2.1 Case Study 1: Application in a Container Shipping Terminal

Most of the cargo transported in ocean-going vessels around the world today can be classified into two types:

- Bulk shipping of huge quantities of commodities such as crude oil, coal, ore, grain, etc., which are shipped using specialized vessels called bulk carriers
- Containerized shipping in which a variety of goods are packed into standard-size steel containers that are shipped on vessels

In this section, we focus on containerized shipping. A *container terminal* (or *terminal* in short) in a port is the place where ocean-going container vessels dock on berths, and unload inbound (import) containers (empty or filled with cargo), and pick up outbound (export) containers. The terminals have *storage yards* (*SYs*) for the temporary storage of these containers (Fig. 3.2).

Containers, Storage Blocks, and Yard Cranes

Containers are steel boxes of dimensions (all measurements are in feet) $20 \times 8 \times 8.5$ or $20 \times 8 \times 9.5$ (called 20-ft containers), or $40 \times 8 \times 8.5$ or $40 \times 8 \times 9.5$ (called 40-ft containers), or specialized slightly larger-size boxes (e.g., refrigerated containers for cargo that must be kept at specified cold temperatures during transit).

Fig. 3.2 Containers in temporary storage at the storage yard (SY). At water's edge, see quay cranes (QC) unloading, loading containers from, into vessels. In SY, see yard cranes (YC) that store and retrieve containers from storage

In measuring terminal throughput and vessel capacity, etc., a unified unit, *TEU* (20-ft equivalent unit), is commonly used, with each 40-ft or larger container being counted as 2 TEUs.

The storage yard (SY) in a terminal is usually divided into rectangular-shaped regions called *storage blocks* or *blocks*. A typical block has seven rows (or lanes) of spaces, six of which are used for storing containers in **stacks** or **columns**, and the seventh reserved for truck passing. Each row typically consists of over twenty 20-ft container stacks stored lengthwise end to end. For storing a 40-ft container stack, two 20-ft stack spaces are used.

In each stack, containers are stored one on top of another. The placing of a container in a stack, or its retrieval from the stack, is carried out by huge cranes called yard cranes (*YCs*). The most commonly used yard cranes are *rubber tired gantry cranes* (*RTGCs*) that move on rubber tires. The RTGC stands on two rows of tires and spans the seven rows of spaces of the block between the tires (Fig. 3.3). The bridge (top arm) of the RTGC has a spreader (container picking unit) that can travel across the width of the block between rows one to seven. The RTGC can move on its tires along the length of the block. With these two motions, the RTGC can position its spreader to pick up or place down a container in any stack of the block or on top of a truck in the truck passing row.

The height of an RTGC determines the height of each stack (i.e., the number of containers that can be stored vertically in a stack). Older models of RTGCs are five-level-high RTGCs. This model can store only four containers in a stack; the

Fig. 3.3 Six rows of stored containers and seventh (*rightmost*) lane for truck passing in a storage block. You can see the RTGC lowering a retrieved container from storage onto a truck

fifth level is needed for container movement across the width of the block. Newer models are six-level-high RTGCs. They can store five containers in a stack and use the sixth level for container movement.

The unloading of containers from a vessel, or the loading of containers into a vessel, is carried out by huge cranes called *quay cranes* (*QCs*) (Fig. 3.4).

Vessels bring import containers that are unloaded at the docks, and brought to the SY by the terminal's internal trucks (TIT) for temporary storage until a customer's *external truck* (*XT*) comes to pick it up. At that time, the RTGC retrieves the import container from storage, puts it on the XT, which then leaves through the terminal gate (TG).

Customers send export containers to the terminal on their XTs which enter through the TG and go to a block where the RTGC removes that container and puts it in storage. When the vessel into which it is to be loaded arrives on the dock, the RTGC retrieves the export container from storage and puts it on a TIT, which takes it to the dock for loading into the vessel by the QC.

Shipping lines rate container terminals largely based on *vessel turnaround time* (average time the terminal takes to unload and load a docked vessel) and the related measure *quay crane rate (QC rate)* (average number of containers moved by a quay-crane per hour). Technically, the QC rate can be as high as 40, but at most terminals it hovers around 20–25. There are only about 40 major shipping lines in the world, and they all take very seriously their decision about which container terminals to

Fig. 3.4 A docked vessel and QCs unloading inbound containers from it and loading outbound containers into it. Also see a *TIT* (terminal's internal truck) that transports containers between the berth and the SY

patronize. With several new ports and container terminals opening up all over the world, container terminals face growing competition and have to provide premium service through higher productivity to stay in business.

The roads inside the terminal get congested with container truck traffic. Traffic flow in a container terminal is akin to the circulation of blood in a human being – life depends on it. As TITs get stuck in traffic, QC rate goes down as QCs have to wait for TITs to take away import containers unloaded or to bring export containers to load. So, one of the most serious problems in terminal operations is routing the trucks optimally to minimize congestion. This and several other problems in developing a decision support system for various daily operations in the terminal to maximize its productivity and equipment utilization are discussed in Murty et al. (2005a, b); Zhang et al. (2003); Murty (2007); Petering and Murty (2009).

The policy for allocating storage spaces to arriving containers directly influences the routing of container trucks and is therefore an important factor in controlling congestion. Developing a policy for allocating storage spaces to arriving containers optimally is a large and complex problem at busy container terminals, because it involves a dynamic terminal system that may have several thousands of trucks passing through the terminal gate daily; several vessels arriving and leaving each day; several thousand TEUs of constantly accessed storage space; stochastic arrivals and departures of containers; and many complex objectives to be optimized simultaneously.

Important features of the terminal's container storage and retrieval workload that must be taken into account are (1) the distribution of work over time is variable because of uncertainties in weather conditions, road, and sea traffic conditions, and other factors; (2) it is impossible to control the order in which containers arrive or their times of arrival; (3) work (i.e., storage) must be carried out at arrival time without delay.

Here we discuss the mathematical modeling of the problems of allocating storage spaces to arriving containers and of routing the trucks inside the terminal optimally to minimize congestion.

The Planning Period and Input Data for Decision Making

Container terminals work three 8-h shifts around the clock, every day. It is convenient to make the planning horizon for decision making equal to the time interval of a shift or less.

There are two important considerations for selecting the appropriate length of the planning period. One is that it should be possible to estimate the total workload in the period with reasonable accuracy. Another is that this total workload should be distributed as uniformly as possible over time during this period. Operators have found that a 4-h period (i.e., half of a shift) comes closest to meeting both these considerations adequately. With shorter time periods, the accuracy of the estimate of total workloads is poor. With longer time periods, the distribution of the workload over time tends to be highly uneven. Hence, terminals have found it convenient to organize their work with a 4-h period, half-a-shift, as the planning period for decision making. A day is divided into six planning periods: 00:00–4:00, 4:00–8:00, 8:00–12:00, 12:00–16:00, 16:00–20:00, 20:00–24:00 h.

Our main focus is to route trucks optimally to minimize congestion. The policy used for allocating storage spaces to arriving containers directly influences the routing of container trucks, and is therefore the most important factor in controlling congestion.

The First Model Considered

When we started working on this problem, we searched the literature for previous work on this problem. Even though none of the published papers at that time considered congestion as the objective function to optimize, several publications discussed integer programming models for storage space allocation to arriving containers. The storage spaces allotted to arriving containers determines the routes that TITs take to carry these containers from the berth to the SY, and thus directly influence congestion on the road inside the terminal.

The decision variables for the models discussed in those papers were typically binary variables of the form $x_{ijk\ell}$, which takes the value 1 if the ith container coming from the jth QC position on a vessel is stored in the kth stack of block ℓ, and value of 0 otherwise.

Typically, there are a large number of container storage positions in the yard, some of which are already occupied by containers in storage at the beginning of the planning period. Assigning a specific open storage position for each of the containers expected to arrive in a period leads to a huge integer program that takes a long time to solve. The set of occupied storage positions in the yard changes every minute, and it is very difficult (almost impossible) to control this change. Before the integer programming model for storage space allocation can be solved, the data in the model will change significantly, making it impossible to implement the output from the model. Because of this we realized that these traditional integer programming models for storage space allocation are not only impractical, but also inappropriate.

Choice of Decision Variables to Model Storage Space Allocation

There are a large number of container storage positions in the yard, some of which are already occupied by containers in storage at any point of time, and determining a specific open storage position for each newly arriving container in the arrival stream leads to a huge mathematical model that will be impractical to solve. So, we have decided to model the problem as a two-stage problem. Stage 1 determines only the block in which each arriving container will be stored. Once the container arrives in the block where it is to be stored, Stage 2 allocates an open position in the block at the time of the container's arrival, using a simple online algorithm, for storing it. This leads to the second model that we considered for modeling the problem.

The Second Model Considered

Here we discuss the model for Stage 1, that of determining a block where each container arriving for storage in the planning period will be stored. It is based on a commonly used approach discussed in the literature for this problem; it uses a *batch-processing strategy*. It considers all the containers expected to arrive and leave at each node (berth or the terminal gate) during the planning period as a batch, and models the problem of handling these batches at the various nodes optimally over the planning period to minimize congestion.

The first step in developing this model with the goal of controlling the congestion on the road system is to develop a measure for congestion. In city traffic with roads of widely varying capacities and vehicles of many different types flowing on them, it is very hard to characterize congestion by a single measure that every one can agree on. But in the small road system inside the terminal, all roads have the same capacity and the only vehicles on them are container trucks, and so it is far simpler to develop measures of congestion on it.

For this we represent the terminal road system by a directed network $G = (\mathcal{N}, \mathcal{A})$, where $\mathcal{N}$ = set of nodes (each block, berth unloading/loading position, road intersection; the TG (terminal gate) is a node in it) and $\mathcal{A}$ = set of arcs (each left or right road segment joining a pair of nodes is an arc in it).

Each pair

- (Berth unloading/loading position, block) for import containers unloaded from vessels at this position, going to the block for storage
- (Block, berth unloading/loading position) for export containers stored and retrieved from this block, going to this position for being loaded into a hatch of a docked vessel
- (TG, block) for export containers being brought into terminal by customers on their truck, sent for storage in this block until the vessel into which they will be loaded arrives at the terminal
- (Block, TG) for import containers stored in this block, retrieved from storage and placed on the customer's truck, being taken out through the TG

is an origin–destination pair for trucks that have to go from the origin to the destination on it. These trucks constitute a separate commodity that flows on G. Let T denote the number of these various (origin, destination) pairs; it is the number of different commodities (container trucks with this (origin, destination)) flowing on the arcs in G.

Suppose the planning period is period k. Planning for this period k is carried out in the previous period, period $k-1$. By period $k-1$, the terminal has a reasonable idea of which vessels will be docked in each of their berths during period k, the expected number of import containers that will be unloaded from each hatch of each docked vessel and sent for storage in the SY during period k, the expected number of export containers that will be retrieved from storage in each block of the SY and sent for loading into each hatch of each docked vessel in period k, the expected number of import containers that will be retrieved from storage in each block of the SY (these containers will be picked up by customer trucks and leave the terminal through the TG), and the expected number of export containers that will be brought into the terminal for storage in the SY until their vessel arrives. All this information constitutes the data for this model, which is a *multicommodity flow model*. For example, here is an illustrative sample of data for a planning period:

- Data at the TG: 400 export containers to go from the TG to some block in SY for storage.
- Data at a block (B_1) in SY: 40 export containers to be retrieved from storage and dispatched to loading position 1 on berth 4, 10 export containers similarly to loading position 2 in berth 8, 20 stored import containers to be retrieved and sent out through the TG.
- Data at a position on a berth (Berth 1): 80 import containers to be unloaded and sent for storage in some block.

The decision variables in this multicommodity flow model are

f_{ij}^r = units (i.e., number of trucks) of commodity r passing through arc (i,j) in the planning period, for $r = 1$ to t, $(i,j) \in \mathcal{A}$,

$$f_{ij} = \sum_{r=1}^{T} f_{ij}^r \text{ for } (i,j) \in \mathcal{A}.$$

The vector $f = (f_{ij}^r)$ is the flow vector of various commodities on G in the planning period. Given f, a measure of congestion on G in this planning period is $\theta = \max\{f_{ij} : (i, j) \in \mathcal{A}\}$.

Also, as distributing traffic on all arcs in G as much as possible leads to the best possible traffic position, another measure of congestion is $\theta - \mu$, where $\mu = \min\{f_{ij} : (i, j) \in \mathcal{A}\}$.

Considering all the containers expected to arrive and leave at each node (TG, or berth position, or block) during the planning period as a batch, the multicommodity flow model formulates the problem of handling each of these batches at the various nodes optimally to minimize any one of the measures of congestion discussed above. In that process it determines the allocation of blocks for storing each container expected to arrive for storage during the planning period.

We do not give the details of all the constraints in this multicommodity flow model, but point out that it may be a large-scale LP with thousands of constraints for container terminals at busy ports. However, commercial LP software systems are able to solve such large-scale models reasonably fast, and it turns out that this model can be solved to optimality in a few minutes of computer time. From the output, we get the storage block allocation for each arriving new container, and the route to be followed by the truck carrying it.

The output from this model turned out to be difficult to implement, as drivers resented being told which routes to take. Also, as the model is solely based on estimated total workload during the entire planning period, we realized that the output from it works well only if that workload (number of containers handled/minute) is distributed more or less uniformly over time during the planning period.

In reality at the terminal where we did this project, the number of containers handled/hour during the planning period varied from 50 to 400. This model did not take this wide fluctuation in workload over time at the terminal into account at all. So the output from this model turned out to be poor.

The Third Model Considered

Actually, the arriving containers at a terminal form a continuous unending stream. Handling this stream using discretization techniques (i.e., those like the second model above, which handle all the containers expected to arrive during the planning period as a batch) produces poor solutions because the rate of arrivals in the stream fluctuates widely over time.

We therefore had to develop dynamic, real-time approaches to allocate storage blocks, and routing trucks to distribute the traffic uniformly among the various road segments continuously over time to alleviate congestion.

At that time we did not know how to develop a dynamic plan, but observing the terminal operations over some time gave a clue. Let

> the *fill-ratio* in block i at time t which is $= f_i(t) =$ (number of containers in storage in block i at time t)/(total number of storage spaces for containers in block i)

We observed that the fill-ratio in a block is highly positively correlated with the truck traffic in and around that block (i.e., the number of containers moving in and out of this block/minute). So, maintaining the fill-ratios in all the blocks nearly equal will ensure that the volume of traffic in the neighborhood of all the blocks will be nearly equal. This in turn ensures equal distribution of traffic on the terminal roads, and hence minimizes congestion. This leads to a *substitute objective function technique* (illustrated in Murty and Djang (1999)) for controlling congestion. For the planning period k, it divides the storage space assignment decision into three stages:

Stage 1: Determine Container Quota Numbers for Blocks: The *container quota for block i* for the planning period is defined to be the number of containers arriving at the terminal for storage during this planning period, which will be dispatched for storage to block i, a decision variable that we denote by x_i.

In the model for Stage 1 to determine x_i, there should be a constraint that $\sum_i x_i$ should be equal to the total number of new containers expected to arrive for storage during the planning period, to make sure that every one of them can be allotted to some block for storage within its container quota. The algorithm that we developed for Stage 1 is discussed later.

Stage 1 determines only the container quota numbers for the blocks, not the identities of containers that will be stored in each block. The identities will be determined by the dispatching policy discussed in Stage 2.

Stage 2: Dispatching Policy for Container Trucks at the TG and at the Berths: This is actually a policy that specifies to which block in the storage yard each truck carrying a container arriving for storage in the terminal will be dispatched.

To implement the policy that we developed for making this decision, the terminal has to monitor continuously over time: $w_i(t)$ = the number of trucks waiting in block i to be served by the yard cranes there at time point t. As part of our work on this project, the terminal where we did this work developed systems to monitor $w_i(t)$ continuously over time for each block i. They also developed a *dispatching cell* that has the responsibility of dispatching each truck in the arriving stream to a block, and this cell gets this $w_i(t)$ information continuously over time for each block i.

As time passes during the planning period, the dispatching cell also keeps track of how many containers in the arriving stream have already been sent to each block for storage; when this number becomes equal to the container quota number for that block, they will not send any more containers for storage to that block during the planning period. For each block i, let

> $x_i^R(t) = x_i -$ (number of new containers sent to block i for storage up to time t in the planning period) = remaining container quota number for block i at time t in the planning period.

Regardless of how we determine the container quota numbers x_i, if we send a consecutive sequence of arriving container trucks to the same block in a short time interval, we will create congestion at that block. To avoid this possibility, we must ensure that the yard crane in that block has enough time to unload a truck we send there before we send another.

Hence this policy dispatches each truck arriving (at the terminal gate and at each berth) at time point t in the period, to a block i satisfying $w_i(t) = \text{Min}\{w_j(t) : j$ satisfying $x_j^R(t) > 0\}$, that is, a block with a remaining positive quota that has the smallest number of trucks waiting in it.

Stage 3: Storage-Position Assignment Policy Within the Block: The storage-position-assignment policy assigns storage positions to containers arriving at a block to minimize reshuffling (i.e., the average number unproductive moves of the YC to retrieve a stored container). This is another important objective function to optimize in terminals to maximize the utilization of YCs. For this, we developed a heuristic online algorithm, which is working well. As our aim here is to discuss the work done to minimize congestion on the roads, we will not describe this policy in detail, but refer the interested reader to Murty et al. (2005a, b).

Algorithm Developed for Stage 1 to Determine Container Quota Numbers for Blocks: The Fill-Ratio Equalization Policy

The decision variable x_i, the container quota number for block i for the planning period k, is the number of containers arriving in this period to be dispatched for storage to block i. The data needed for this decision are the following:

- a_i = the number of stored containers that will remain in block i at the end of this period if no additional containers are sent for storage there during this period (this is = (the number of stored containers expected to be in this block at the beginning of this planning period − the number expected to be retrieved from storage during this planning period); this a_i is easily estimated from the data available in the previous period $k - 1$).
- N = the number of new containers expected to arrive at the terminal in this period for storage.
- B, A = the total number of blocks in the storage yard, the number of storage positions in each block.

The fill-ratio equalization policy determines the decision variables x_i for this period to make sure that the fill ratios in all the blocks are as nearly equal as possible at one time point during the period, namely the end of the period.

The fill ratio in the whole yard at the end of this period will be $F = (N + \sum_i a_i)/(AB)$. If the fill ratios in all the blocks at the end of this period are all equal, they will all be equal to F. Thus, this policy determines x_i to guarantee that the fill ratio in each block will be as close to F as possible, by the least sum of absolute deviations measure, by the end of this period. To determine x_i, this leads to the following LP model that has very special structure, with only $B + 1$ (=100 at the terminal where we did this work) constraints.

$$\text{Minimize} \sum_{i=1}^{B} (u_i^+ + u_i^-)$$

$$\text{subject to} \sum_{i=1}^{B} x_i = N$$

$$a_i + x_i - A \times F = u_i^+ - u_i^-$$

$$x_i,\ u_i^+,\ u_i^- \geq 0.$$

Also, using the very special structure of this problem, we developed the following simple combinatorial scheme to obtain its optimum solution.

Step 1: Rearrange blocks so that a_i increases with i.
Step 2: Determine x_i in increasing order of i using
$x_1 = \text{Minimum}\{N,\ A \times F - a_1\}$, and for $i \geq 2$
$x_i = \text{Minimum}\{\text{Maximum}\{0, A \times F - a_i\},\ N - \sum_{r=1}^{i-1} x_r\}$.

Numerical Example: Suppose there are $B = 9$ blocks in the terminal, each with $A = 600$ storage spaces. Suppose we expect N = 1,040 containers to arrive for storage during the planning period. Data on a_i, the previously stored containers remaining in storage at the end of this planning period in block i, already in increasing order of i, are given in the following table.

i	a_i	x_i	$a_i + x_i$
1	100	300	400
2	120	280	400
3	150	250	400
4	300	100	400
5	325	75	400
6	350	35	385
7	375	0	375
8	400	0	400
9	450	0	450
Total	2570	1040	

The fill-ratio over the whole yard at the end of this period will be $F = ((\sum_i a_i) + 1040)/(9 \times 600) = (2570 + 1040)/(5400) \approx 0.67$, and so the average number of containers in storage/block at this point of time will be $A \times F \approx 400$. So, the

LP model for determining the container quota numbers for blocks, the x_i's in this planning period, is

$$\begin{aligned} \text{Minimize} \quad & \sum_{i=1}^{9} (u_i^+ + u_i^-) \\ \text{subject to} \quad & \sum_{i=1}^{9} x_i = 1040 \\ & a_i + x_i - 400 = u_i^+ - u_i^- \quad \text{for all } i \\ & x_i,\ u_i^+,\ u_i^- \geq 0 \quad \text{for all } i. \end{aligned}$$

Our combinatorial scheme determines x_i in the order $i = 1, 2, \ldots$ to bring $a_i + x_i$ to 400 until all 1,040 new containers are allotted to blocks for storage. Under this policy, $a_i + x_i$ will be the number of stored containers in block i at the end of this period. From the above table, we can verify that the values of all $a_i + x_i$ are nearly equal in this numerical example.

This policy determines only the quota number of containers for each block, not which containers will be stored in each block. That is determined by the dispatching policy discussed earlier.

The policy of determining the container quota numbers x_i to maintain fill-ratios of all blocks nearly equal and the dynamic dispatching policy for arriving container trucks ensure that container truck traffic at each point of time is distributed evenly on all portions of the terminal road system, and thus makes sure that the traffic situation is the best that can be for the prevailing workload in the terminal. To implement these policies, the company had to develop equipment for continuously monitoring how many trucks are waiting to be served by the RTGCs in each block and relaying this information to the people directing container truck traffic. But it was very beneficial as these policies turned out to be highly effective and helped reduce congestion, and consequently reduced the truck turnaround time by over 20%.

In this example two standard approaches for modeling the problem lead to large-scale models that gave very poor results. But the third approach tried to minimize congestion by controlling the variation in the fill-ratios of blocks coupled with a dynamic truck dispatching policy; it led to a small model that was easy to solve. So, it can be viewed as a *substitute objective function technique* (one that controls the original objective function by optimizing a highly correlated substitute objective function that is much easier to control), and it has been highly effective. *This example shows that to get good results in real-world applications, it is necessary to model the problems intelligently.*

(optimization techniques) + (information technology) + (intelligent modeling) → makes a powerful combination for solving practical problems.

3.2.2 *Case Study 2: Application in a Bus Rental Company*

Faced with the imminent retirement of two senior employees who used to make decisions on bus allocations to customers' requests manually, a bus rental company in Seoul, South Korea, asked us to develop a decision-making support system (DMSS) to help the young fresh graduate employee who will be taking over this job from them.

Practice has shown that allocation and routing decisions made manually by human operators with long experience are usually nearly optimal, and it is very hard to beat those decisions using a computerized DMSS. Therefore, the company asked us to design an i-DMSS (intelligent DMSS) that can help the new decision maker to reach decisions comparable in quality to those made by the retiring pair of senior decision makers.

In this section we discuss this decision problem, its context, the models we used to solve it, the algorithms we used in the i-DMSS to solve these models in Murty and Kim (2006), and how this i-DMSS is used to make the decisions daily. It produces solutions 10–20% more economical than the manual decisions .

The Problem Description

We discuss the process of analyzing, modeling, and developing an i-DMSS to solve the bus allocation problem. The application involves solving the same type of a problem daily with new data for each day. This is a typical problem that arises at bus rental companies in major cities all over the world, even though the constraints and other features of the problem may vary from company to company.

This company rents buses (with drivers provided) to customer groups who request them. The group size varies considerably. To serve customer needs economically, the company rents two-size buses; a small 15-seat bus for small groups (they have five of these buses), and a larger 45-seat bus for larger groups (they have 20 of these buses).

Each customer request (called *a job*) completely specifies the route that the group wants to take, the starting location and starting time, any intermediate stops in the job, and the ending location and the ending time. As the customer completely specifies the route in their job, there is no routing to be done by the company. We denote

n = number of jobs for which buses are to be allotted on a day,

$\underline{t}_i$, $\bar{t}_i$ = starting and ending clock times of job i, $i = 1$ to n,

p_i, q_i = starting and ending locations of job i, $i = 1$ to n,

L = set consisting of the two bus depots, and all the distinct sites among the starting and ending locations of the n jobs,

d_{ij} = distance from i to j measured in expected travel time in minutes, for $i, j \in L$, $i \neq j$,

$D = (d_{ij})$ = the $|L| \times |L|$-matrix *distance data matrix* for the problem for that day.

The difference between the job ending and starting times, $\bar{t}_i - \underline{t}_i$, is called *the duration of job i*. This varies between 0.5 and 20 h, but more than 75% of the jobs this company receives have durations ≤ 5 h.

The company gets up to 100 requests each day. Jobs are classified into *large-group jobs* and *small-group jobs* depending on the corresponding customer group size. Large-group jobs need a 45-seat bus, while a small-group job can be serviced by either a 15-seat bus or a 45-seat bus. While servicing a job, the bus and its driver should be at the disposal of the customer corresponding to that job, that is, two jobs cannot be combined into a bus at the same time. The company's charges for each customer depend on the size of the bus they need, the duration of their whole job, and the total mileage on that job.

The company keeps its buses at two depots in different locations. At each depot they have a staff of drivers for the jobs served by buses from that depot. On days when their own buses are not adequate, the company itself rents buses from other vendors. All the data about jobs to be served on a day is available at the company by the day before. They finalize all the bus and driver allocations for each day by the evening of the day before, so that the drivers can take the buses from their depots to the starting locations of the first jobs, on time. The drivers bring the buses back to their depots from the ending locations of their last jobs.

As several jobs are of short duration, for economical operation the company likes to pack as many jobs as possible, one after the other, into each bus's daily work schedule known as its *work-sequence* for the day. Suppose a bus handles jobs numbered $i_1, \cdots, i_h$ in that order on a day, then its work-sequence for that day is the sequence of jobs $(i_1, \cdots, i_h)$. Then for each $g = 1$ to $h - 1$ after completing job i_g at clock time $\bar{t}_g$ at location q_g, the driver has to drive that bus to the starting location p_{g+1} of the next job i_{g+1} before its starting time $\underline{t}_{g+1}$. So, for $(i_1, \cdots, i_h)$ to be a work-sequence, the condition $\underline{t}_{g+1} - \bar{t}_g \geq$ (driving time from q_g to p_{g+1}) must hold for all $g = 1$ to $h - 1$. In this case, the drives from q_g to p_{g+1} for $g = 1$ to $h - 1$ of this bus are called *empty load drives* on this work-sequence.

During an empty load drive, the company is incurring the cost of keeping the bus running (fuel etc. + driver's wages) on its own without any customer paying for it.

All the buses start at their depot and return to their depot after their last job in their work-sequence for the day. So for this bus, the quantity $[\bar{t}_h - \underline{t}_1$ + (driving time from depot to p_1) + (driving time from q_h to depot)], measured in hours, represents the time in hours the driver of that bus worked that day, and this quantity is called the *duration of that work-sequence* (note that this depends on the depot of the bus to which this work-sequence is assigned).

The driver's wages for a day are proportional to the number of hours she/he has worked. For this reason drivers have a strong desire to maximize the number of hours that they work, but driving fatigue can lead to serious accidents; that is why the company likes to keep the duration of work-sequences to less than 12 h. Work-sequences of duration over 12 h are called *long-duration work-sequences*. However, there are some single jobs that are themselves of duration over 24 h, and these long-duration jobs are quite lucrative to the company. These jobs usually have many intermediate stops of considerable length during which the driver has nothing to do but wait, so she/he can rest, take a nap, and thus refresh her/himself. Long-duration jobs and work-sequences with such intermediate rest periods do not contribute to fatigue, and hence should be considered differently from other long-duration work-sequences involving fatigue causing continuous driving. So, the company has decided to allow such long-duration work-sequences, but set it as a goal to keep the percentage of these long-duration work-sequences to 50 as far as possible so that they can alternate a long-duration work-sequence allotted to a driver one day with a short-duration work-sequence the next day.

The problem is a multiobjective problem. The most important objective is to minimize *OBJ1* = the total number of buses used to handle the jobs = the total number of work-sequences into which the jobs are partitioned. This also involves minimizing the expenses on renting other vendors buses used to handle the jobs.

Customers pay for all the travel within the jobs, but the company does not collect any money for the travel from the depot to the starting location of the first job in the work-sequence, and back from the ending location of the last job in the work-sequence to the depot, and the travel from the ending location of a job to the starting location of the next job within a work-sequence. The second most important objective to minimize *OBJ2* = the total cost of these empty load drives of all the buses.

The third most important objective stated as a goal is to keep *OBJ3* = the percentage of long-duration work-sequences below 50 as far as possible.

The company had two full-time employees who were doing these allocations over a long period of time manually using a map of the various locations involved each day. With the imminent retirement of these experienced decision makers, the company asked us to develop a computerized DMSS to help the new young person who will replace these two old timers.

Models in the Literature for This Problem

This problem arises at all bus rental companies in major cities all over the world with minor variations in the constraints. Also, many other allocation problems in different areas of applications have features similar to this problem. So, all these problems can be modeled using the same type of mathematical model. Several published papers in the literature suggest using a 0–1 integer programming model with 0–1 decision variables of the form

$$x_{ijkl} = \begin{cases} 1, \text{ if the } i\text{th job is included in the } j\text{th worksequence,} \\ \quad \text{for which a bus of size } k \text{ from depot } \ell \text{ is allocated} \\ 0, \text{ otherwise} \end{cases}$$

for $i = 1$ to n; $k = 1, 2$; $\ell = 1, 2, 3$ ($\ell = 1, 2$ denotes the two depots of the company, and $\ell = 3$ denotes "outside vendors" for buses); and $j = 1$ to t, where t is an upper bound on the number of work sequences on this day.

Some other papers in the literature modeled these problems with different binary integer variables of the form

$$x_{jgkl} = \begin{cases} 1, \text{ if jobs } j,\ g \text{ are both allocated to the same bus of} \\ \quad \text{size } k, \text{ from depot } \ell \\ 0, \text{ otherwise} \end{cases}$$

for $j,\ g = 1$ to n; $k = 1, 2$; $\ell = 1, 2, 3$.

With either set of decision variables, either to model OBJ3 (upper bound of 50 on the percent of long-duration work-sequences) as a constraint in the model or to include a penalty term for violating this desired upper bound is very hard. So, for simplicity, these publications assume that there should be no long-duration work-sequences at all, as it is easy to model this constraint through linear constraints.

Even with all these simplifications, the model ends up having a very large number of constraints.

Also, with decision variables having four subscripts like the above, the number of binary variables in the model becomes too large. When n is of the order of 100 as in our application, the resulting 0–1 integer programming model with large number of variables and constraints needs many days of computer time to solve even if we have supercomputers to solve them. Our client company is a small company operating with low margins of profit, and has only ordinary PCs available for this work, and has shown no interest in acquiring software systems for large integer programs for this work.

Overall Approach Used for Developing the i-DMSS

So, we applied hierarchical decomposition to break up the problem into several stages that are simpler to model and solve. We found that it becomes easier to handle OBJ2 if it is split into two parts as OBJ2.1 + OBJ2.2, where

- OBJ2.1 = cost of empty load drives in-between consecutive jobs on all the work-sequences used,
- OBJ2.2 = cost of empty load drives from the depot to the starting location of the first job in the work-sequence, and from the ending location of the last job in the work-sequence to the depot,

for all the buses used. It can be verified that OBJ2.1 is uniquely determined by the set of work-sequences into which the jobs are partitioned and does not depend at all on the depot from which a bus is allotted to each of the work-sequences adopted. Likewise, OBJ2.2 mainly depends on the depots from which buses are allotted to work-sequences and not so much on how the work-sequences are formed.

We found that the overall model becomes a lot simpler if the two parts OBJ2.1 and OBJ2.2 are optimized in separate stages of our approach. We will optimize OBJ1 and OBJ2.1 together in the stage that determines the initial set of work-sequences. Minimizing OBJ2.2 is taken as the objective for determining from which depot to allot a bus to each work-sequence after the set of work-sequences to adopt is finalized.

Drivers are paid on an hourly basis for the duration of the work-sequences that they operate. So, drivers are usually unhappy if they are assigned to work-sequences whose duration is too small on a day. So, the company would also like to make sure that most of the work-sequences adopted have durations $\geq$ some threshold lower bound; currently this lower bound $\delta = 6$ h.

Decomposition for Two Types of Buses

In this problem, we have large-group jobs and small-group jobs; 15-seat buses that can serve small-group jobs only and 45-seat buses that can serve all the jobs. On some days when there are a large number of small-group jobs, the company may experience a shortage of 15-seat buses to handle all of them. On such days, instead of renting some extra 15-seat buses from outside vendors, the company has found it to be much more economical to assign some of its own 45-seat buses to small-group jobs. Because of this, we use the following hierarchical procedure for handling the allocation of two types of buses to the various jobs:

Phase 1: First consider only the small-group jobs for which the 15-seat bus is suitable. Apply the procedure to partition this subset of jobs into work-sequences. Find the total working time associated with each of the work-sequences. Select a threshold value, say δ hours (currently $\delta = 6$), as a lower bound for a day's working time. For all work-sequences associated with a working time $\delta \geq 6$, assign 15-seat buses to the extent they are available, using the model discussed below, with the only sources for the buses as the two depots of the company.

Phase 2: The jobs on all the work-sequences associated with working time $\delta < 6$ and all the other work-sequences for which 15-seat buses may not have been allocated in Phase 1 are combined with the set of large-group jobs. We then apply the procedure to allocate 45-seat buses for this set of jobs.

The Model to Partition the Set of Jobs into Work-sequences

How to Minimize OBJ1? Minimizing OBJ1 requires that we partition the set of n jobs into the smallest possible number of work-sequences. There is an efficient network model for this problem. For $i = 1$ to n, represent job i by node i in a directed network.

Job i is called a long-duration job if its duration $\bar{t}_i - \underline{t}_i$ is ≥ 12 h, nodes corresponding to such jobs are left as isolated nodes in the network without any arcs incident at them. These jobs are already too long, and we will not consider combining them with any other jobs in a work-sequence for a bus.

If neither of nodes i, j is a long-duration job, include an arc from node i to node j if a bus can handle job j after handling job i (the condition for including this arc is (driving time from q_i to p_j) + $\bar{t}_i \leq \underline{t}_j$). Let the resulting network be $G = (N, A)$, where N = set of nodes in G and A = set of arcs. G is called the *job precedence acyclic network* for the problem.

By making small perturbations in the starting times of the jobs if necessary, we number the jobs serially in increasing order of their starting times. Then it is clear that if $i > j$, then $(i, j) \notin A$. This implies that the numbering of the nodes in G is an acyclic numbering (i.e., all arcs go from a node to another node with higher number). Hence G is an acyclic network, and it is the job precedence acyclic network for this set of jobs.

We define a simple chain in G to be either a set containing a single node or a sequence of more than one node, $i_1, \ldots, i_h$ satisfying the condition that $(i_{g-1}, i_g) \in A$ for $g = 2$ to h. Thus it corresponds to the usual notion of a simple chain in network terminology (Ford and Fulkerson (1962); Murty (1992)) when there are two or more nodes in it. However, a single node by itself is also considered as a simple chain (it has no arcs) in this context. Therefore, every work-sequence corresponds to a simple chain in G, and vice versa.

As each work-sequence corresponds to a simple chain in the acyclic network and vice versa, the problem of partitioning the set of jobs into the smallest number of work-sequences is the same as that of finding a minimum cardinality *simple chain cover* for all the nodes in G, which is known in network programming literature as *Dilworth's minimal chain decomposition problem.*

An efficient algorithm for it based on the maximum cardinality bipartite matching algorithm has been developed by Fulkerson (1956) and discussed in Ford and Fulkerson (1962); see also Murty (1992). The algorithm involves finding a maximum cardinality matching in the bipartite network $B = (N_1, N_2; A_1)$ with node set $N_1 = \{R_1, \ldots, R_n\}$, $N_2 = \{C_1, \ldots, C_n\}$, and edge set $A_1 = \{(R_i, C_j) : (i, j)$ is an arc in A in $G\}$.

Note 3.1. The maximum cardinality matching problem in the bipartite network B is just an *assignment problem* with a 0–1 cost matrix (d'_{ij}), where $d'_{ij} = 0$ if (i, j) is an arc in B and $d'_{ij} = 1$ otherwise. If $\bar{x} = (\bar{x}_{ij})$ is the optimum assignment for this problem, then the maximum cardinality matching in B is the set of arcs $\{(i, j) : \bar{x}_{ij} = 1$ and $d'_{ij} = 0\}$, and its cardinality is the number of arcs in it.

Suppose the cardinality of a maximum cardinality matching in B is r. Then it is shown that the minimum number of simple chains needed to cover all the nodes in G is $s = n - r$; that is, in our problem at least s work-sequences or buses are needed to cover all the jobs. From any maximum cardinality matching M in B, an easy procedure is available for deriving a set of simple chains in G to cover all the nodes in G (see Ford and Fulkerson (1962); Murty (1992)). This procedure consists of obtaining the set of arcs $\{(i, j) : (R_i, C_j)$ is an edge in the matching M in $B\}$, it is the set of arcs in a node disjoint collection of simple chains in G, this collection of simple chains is a minimum cardinality simple chain cover for the nodes of G. Find it. The sequence of jobs corresponding to nodes in the order in which they appear on each of these simple chains is a work-sequence for a bus, and hence each of these simple chains can also be interpreted as a bus route.

Numerical Example: Job (or node) i is called a predecessor (or ancestor) of job j if j can be handled by a bus after i, then we include the arc (i, j) in the job predecessor acyclic network G defined above. In addition, node i is called an immediate predecessor of j if it is a predecessor of j, and there is no other predecessor k of j for which i is also a predecessor.

Typically, G will have too many arcs. So, in this $n = 9$ node example we show only a subnetwork $\bar{G}$ of G, with an arc (i, j) in it only if i is an immediate predecessor of j in Fig. 3.5. With this, j can be handled by a bus after i if there is a simple chain (or directed path) from i to j in $\bar{G}$.

So, after job 1, a bus can take up any of the jobs 3, 4, 6, 7, 8, or 9 in this example. The bipartite network to apply Fulkerson's algorithm on this example is given in Fig. 3.6.

M is a maximum cardinality matching in B with six edges. So, a minimal chain decomposition of G in this example has $9 - 6 = 3$ chains, and the arcs on these three chains are $\{(1,3),\ (3,7),\ (2,4),\ (4,6),\ (6,8),\ (5,9)\}$. Hence the three chains are $\{1, 3, 7\}, \{2, 4, 6, 8\}, \{5, 9\}$ with nodes appearing in the order listed, and these are the work-sequences in a partition of the nine jobs in this example into the smallest number of work-sequences.

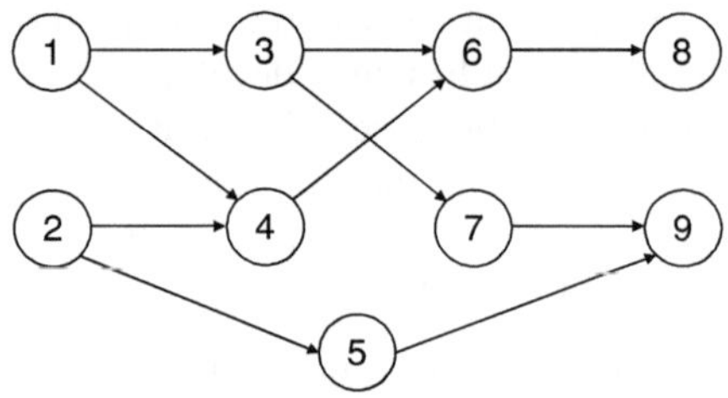

Fig. 3.5 The subnetwork for the example

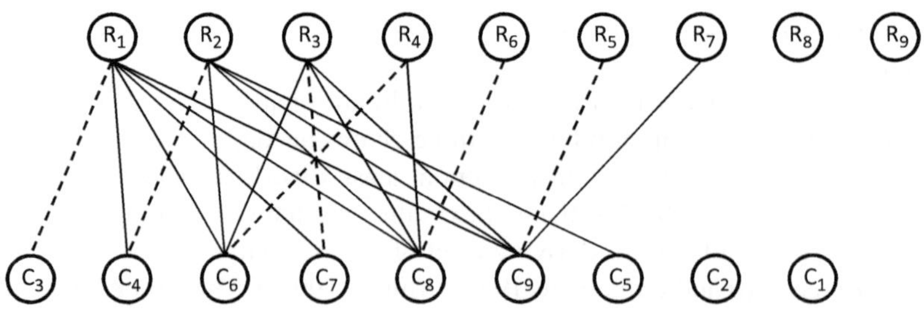

Fig. 3.6 The bipartite network B for this example. A maximum cardinality matching in it $M = \{(R_1, C_3), (R_3, C_7), (R_2, C_4), (R_4, C_6), (R_6, C_8), (R_5, C_9)\}$ is marked with dotted edges

This algorithm provides an efficient approach when OBJ1 is the only objective to consider, ignoring the constraints on work-sequence durations implied by OBJ3 and the lower bounds on them.

Initial Set of Work-sequences: The strategy solves the following:

Problem 3.1. Find $s =$ minimum number of work-sequences into which the set of jobs $\{1, \ldots, n\}$ can be partitioned as discussed earlier. There may be several such minimal partitions. Among all of them find the one that has minimum value for OBJ2.1.

The arc (i, j) in the network represents the opportunity of a bus servicing job j after servicing job i. If this happens, then that bus travels from location q_i after finishing the servicing of job i at clock time $\bar{t}_i$ to location p_j to start the servicing of job j at clock time $\underline{t}_j$. Then in the time interval between $\bar{t}_i$ and $\underline{t}_j$ of length $\underline{t}_j - \bar{t}_i$ hours representing the empty load drive corresponding to this arc (i, j), the bus and the driver are working but the company gets no profit from it. This time has been estimated to cost at the rate of \$40/h. Hence we take the empty load drive cost of arc (i, j) in G to be \$ $c_{ij} = 40(\underline{t}_j - \bar{t}_i)$; this is the cost coefficient of arc (i, j) in G and the corresponding arc (R_i, C_j) in B for OBJ2.1.

The maximum cardinality matching problem in B usually has many alternate optimum solutions, and any one of them can be used to get a minimum cardinality simple chain cover for the nodes in G. So, to partition the set of jobs into work-sequences minimizing OBJ1 and OBJ2.1 simultaneously, we need to find a minimum cost maximum cardinality matching in the bipartite network B with (c_{ij}) as the vector of arc cost coefficients. As B is bipartite, this can be found very efficiently using the min cost max flow algorithm.

In fact, this problem is the following minor variation of the assignment problem:

$$\begin{aligned}
\text{Minimize } & \sum [c_{ij} x_{ij} : \text{ over arcs } (i, j) \text{ in } B] \\
\text{s. to } & \sum_{i=1}^{n} x_{ij} = 1 \text{ for all } j = 1 \text{ to } n \\
& \sum_{j=1}^{n} x_{ij} = 1 \text{ for all } i = 1 \text{ to } n \\
& \sum [x_{ij} : \text{ over arcs } (i, j) \text{ in } B] = r \\
& x_{ij} \geq 0 \text{ for all } i, j
\end{aligned}$$

and we need an integer solution for this LP. The special structure of this problem guarantees that every extreme point optimum of this LP (discussed later in Chaps. 4–6) will be integer. If $\bar{x}$ is an extreme point optimum of this LP, then the matching in B corresponding to it, defined by $\{(i, j) : \bar{x}_{ij} = 1$, and (i, j) is an arc in $B\}$, is a minimum cost maximum cardinality matching in B.

If $\bar{M}$ is a min cost max cardinality matching in B, the set of arcs $\{(i, j) : (R_i, C_j) \in \bar{M}\}$ is a node disjoint collection of simple chains in G. The collection of work-sequences corresponding to these simple chains minimizes OBJ2.1 among all partitions of the set of jobs into the smallest possible number of work-sequences. This set of work-sequences is a solution for Problem 1; it is the partition of the jobs into an initial set of work-sequences to be considered in the heuristic approach for handling OBJ3 next.

How to Handle OBJ3? The partition of jobs $\{1, \ldots, n\}$ into a set of work-sequences obtained above considered only optimizing OBJ1 and OBJ2.1 in this priority order, but totally ignored OBJ3. Now we will discuss how to modify this partition taking OBJ3 into account. Define for a work-sequence its *total working time* = the difference between the ending time of the last job and the starting time of the first job on the work-sequence expressed in hours.

The duration of a work-sequence defined earlier depends on the depot from which the bus for this work-sequence is allotted, and it is equal to its total working time plus the driving time from and to the depot. Giving an allowance of 1 h for driving from and to the depot, we call work-sequences for which the total working time is greater than or equal to 11 h (=safety time − 1) as *long work-sequences*, and these are the work-sequences most likely to violate the maximum duration constraint.

These long work-sequences are of two types. The first type is the single job work-sequences, which consist of just one long-duration job. The second type is multiple job work-sequences that are long. Typically, less than 10% customer service requests are long-duration jobs. The company likes these because they generate higher fees, and they try to assign them with equal frequency among all their drivers. Long-duration jobs almost always contain nondriving rest periods during which the driver can take a nap and get refreshed. For this reason, single-job long-work-sequences are never considered a problem.

OBJ3 is a goal requiring that the percentage of long-duration work-sequences should be ≤ 50 as far as possible. When this goal is violated, some of the multiple job long-duration work-sequences are modified using the following heuristic strategies:

> On each of these long multiple-job work-sequences, the longest arc (i.e., an arc (i, j) on this work-sequence having maximum $(\underline{t}_j - \bar{t}_i)$ value) is deleted from the network G, and the algorithm discussed above applied on the remaining network. Almost always this produces a new collection of work-sequences that satisfies the goal on the percentage of long work-sequences while increasing the number of work-sequences only slightly. Otherwise, this process is repeated.

If the current list of work-sequences consists of some short-duration work-sequences (i.e., those whose durations are $\leq$ lower bound specified for it), they may be suitable to be allotted to drivers who were allotted long-duration work-sequences the previous day. If any other short-duration work-sequences remain in the list, they are handled using the procedure similar to the above.

Clearly the optimum values of OBJ1 and OBJ2.1 obtained in the first run of the algorithm are lower bounds for the minimum values of these respective objective functions in the specified priority order for partitioning the set of jobs considered into work-sequences while satisfying the goal in OBJ3. We use these lower bounds to compare the quality of the final solutions obtained.

The Model to Allocate a Bus to Each Work-sequence

After finalizing the partition of the set of jobs into work-sequences, we turn to the problem of assigning buses to these work-sequences. These buses can come either from depot 1, 2, or outside vendors. The assignment of buses to work-sequences will be carried out so as to minimize OBJ2.2 (the cost of empty load drives from the depot to the starting location of the first job, and from the ending location of last job to the depot, for all the company's own buses used) + the rental cost of buses from outside vendors that are used, which includes besides OBJ2.2 the rental cost and the cost of such empty load drives of buses rented from outside vendors. Let

- $p =$ number of work-sequences in the final set,
- $c_t,\ d_t =$ cost of the empty load drive at the beginning and at the end of the tth work-sequence if a bus is assigned to it from depot 1 and 2, respectively,
- $e_t =$ cost of renting a bus for the tth work-sequence from an outside vendor,
- $N_1,\ N_2 =$ number of buses available at depot 1 and 2, respectively.

As e_t is typically much larger than c_t or d_t, the number of buses rented from outsider vendors will be $(p - N_1 - N_2)^+ = \text{maximum}\ \{0,\ p - N_1 - N_2\}$.

We have three sources for buses, sources 1, 2, and 3 (these are depot 1, 2, and outside vendors, respectively), with availability of buses equal to $N_1,\ N_2,\ (p - N_1 - N_2)^+$, respectively. Each work-sequence requires exactly one bus. Clearly the problem of assigning buses to work-sequence can be modeled as a $3 \times p$ transportation problem, with the $(3 \times p)$ cost matrix whose tth column is $(c_t,\ d_t,\ e_t)^T$ for $t = 1$ to p.

Numerical Results Results obtained by using the i-DMSS on requested job data at the company over a 5-day period from the past are shown in the following table. The first column "A" is the day number. The second column "B" gives the number of small-group jobs on the day. The third and fourth columns, "**1, 2.1**," give values of OBJ1 and OBJ2.1, respectively, in the set of initial work-sequences obtained for small-group jobs. They are the lower bounds for these objectives for small-group jobs. The fifth, sixth, and seventh columns, "**1, 2.1, 2.2**," give the final values for OBJ1, OBJ2.1, and OBJ2.2, respectively, for small-group jobs. The eighth column "B" gives the number of remaining jobs for which 45-seat buses are to be allotted. The ninth and tenth columns give the values for OBJ1 and OBJ2.1 in the initial set of work-sequences obtained for these jobs. They are the lower bounds for OBJ1 and OBJ2.1, respectively, for handling these jobs. The eleventh, twelfth, and thirteenth columns give the final values for OBJ1, OBJ2.1, and OBJ2.2.

Numerical Results

A	Small-group jobs						Large-group jobs					
	B	Omit*		Meet*			B	Omit*		Meet*		
		1	**2.1**	**1**	**2.1**	**2.2**		**1**	**2.1**	**1**	**2.1**	**2.2**
1	14	5	3070	5	3070	2170	54	22	8100	23	9520	7700
2	13	5	2650	5	2650	2480	43	17	6490	17	8810	7540
3	12	5	2320	5	2320	1980	44	18	9980	19	11800	7480
4	14	5	3600	5	3600	2200	51	21	9150	21	9630	8400
5	15	4	2300	4	2300	1800	55	23	9570	24	9810	8330

* Refers to results – omitting OBJ3, meeting OBJ3 – respectively.

In the table, the final values for OBJ1 and OBJ2.1 for the small-group jobs are the same as the lower bound for these respective objective functions for this subset of jobs. To meet OBJ3, for large-group jobs, the value of OBJ1 increased by at most one over its lower bound on three of the five days, while OBJ2.1 increased about 9.5% on an average over its lower bound. This shows that the final solutions obtained by the i-DMSS had objective values quite close to the lower bounds for these objective functions.

We found that the results obtained by using the i-DMSS on this data are between 10% and 20% more economical than the manual decisions made by the senior employees for those days.

Conclusions

We described how we used relaxations, hierarchical decomposition, and heuristics to model and analyze the complex problem of allotting buses at a bus rental company, and presented the numerical results obtained with this approach. The new person responsible to make the allocation decisions at the company uses the DMSS in an interactive manner to make all the decisions in about a couple of hours every evening.

3.2.3 *Case Study 3: Allocating Gates to Flights at an International Airport*

The problem of assigning gates to flights of various types (arrival, departure, connection, and intermediate parking flights) is an important decision problem in daily operations at major airports all over the world. Strong competition between airlines and increasing demand of passengers for more comfort have made the measure of quality of these decisions at an airport an important performance index of airport management. That is why mathematical modeling of this problem and the application of OR (operations research) methods to solve those models have been studied widely in OR literature.

The dynamic operational environment in modern busy airports, increasing numbers of flights and volumes of traffic, uncertainty (random deviations in data elements like arrival, departure times from flight time tables and schedules), its multiobjective nature, and its combinatorial complexity make the flight-gate allocation a very interesting decision problem both from a theoretical and a practical point of view.

Responsibility for gate allocations to flights rests with different agencies at different airports. At some airports gate allocation decisions are made by the airport management themselves for all their customer airlines. At others, some airlines lease gates from the airport on long term contracts. Then those airlines make gate allocation decisions for their flights themselves.

Typically international airports have features quite distinct from those of regional airports. The common characteristics of busy international airports all over the world are that they usually serve a large number of different airlines; they normally serve a large number of flights spreading over most of the 24-h day; they have to accommodate planes of various types and sizes; and a considerable percentage of their flights are long-haul flights coming from long distances. These features and the fact that international airports are much bigger and have much higher volumes of traffic compared to regional or domestic airports make the problem of assigning gates to flights at an international airport somewhat harder in practice than that at a regional airport.

In this section we discuss the process of developing a DSS (decision support system) and appropriate mathematical models and algorithms to use for making gate allocation decisions at an international airport from Yu et al. (2009). Normally international airports have both cargo and passenger flights, but in this section we will only consider gate allocation decisions for passenger flights.

Flights use different types of planes (large-wingspan planes, short-wingspan planes, planes for long-haul flights, short-haul flights, etc.) depending on the expected passenger volume on the flight, length of the flight, and several other considerations. As mentioned earlier, airport gates are also classified into different types depending on their size, location in the airport (those in the central portion of the airport, remote gates, etc.); these gate characteristics determine their desirability to airlines for their flights.

So, for each flight for which a gate is to be assigned, we need the set of all the gates to which it can be assigned. This set of gates eligible to be assigned to a flight is classified into first (most preferred by airline), second (second most preferred), and third (least preferred) categories in order of their preference by the airlines. Sometimes at some airports there may also be a fourth category of gates in decreasing order of preference. This data is available and it will be used in constructing the objective function to optimize, for determining the allocation of gates to flights.

The material in this section is based on the gate allocation project we carried out at the Taoyuan International Airport (code name TPE) serving Taiwan. For details, see our paper Yu et al. (2009). As an example, at TPE there are 78 gates (A1–A9, B1–B9, C1–C10, D1–D10, 501–525, 601–615) serving 27 different types of flight planes ranging from the small CL-604 to the large Boeing 777 and Airbus 380. Among the 78 gates, those in the A, B, C, D sections are equipped with passenger

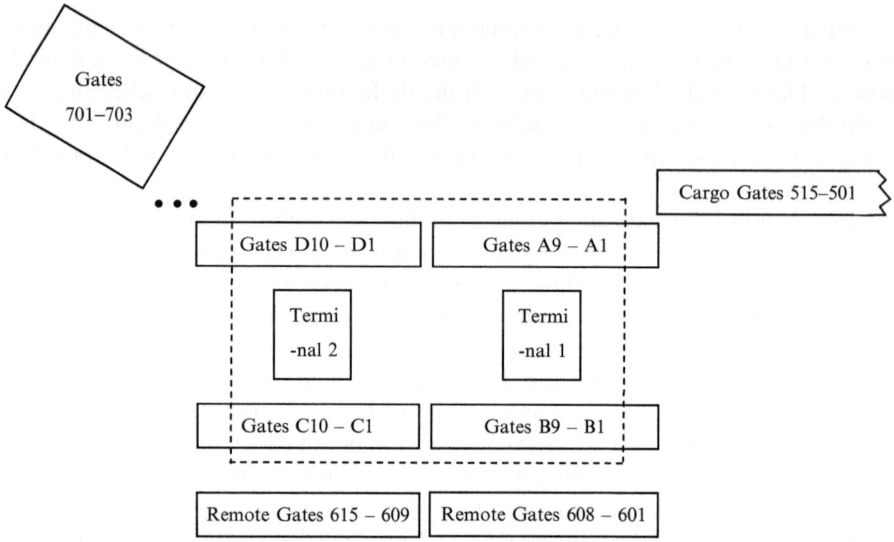

Note: Gates within the dash-line rectangle are considered as the central gates.

Fig. 3.7 Gate Layout of the Airport TPE

bridges (also called *jettys*) for passengers to get out of the plane and walk inside the terminal; 40 are apron gates (501–525, 601–615) (i.e., these are open air gates without a passenger bridge, i.e., passengers have to either walk or be taken by bus to and from inside the airport to this gate position) as shown in Fig. 3.7. In addition, there are also three emergency gates 701–703 (this designation means that these gates are too far away from the terminals and are to be used only when no other gates are available), which are also apron gates. "Central gates" refers to gates considered to be in the central portion of the airport, which has an operating Skytrain line between Terminals 1 and 2, with a train running back and forth every 2–5 min. Using this, if necessary, passengers can get from any of these central gates to any other within 10–15 min; see Fig. 3.7.

At TPE, even though the classification of gates into first, second, and third categories mentioned above for their flights differs from airline to airline, it tends to be very similar. Airlines like to have their flights use gates in the terminal in which most of their activities take place. So, in general, for most airlines the first-category gates are those among the A, B, C, D sections in the terminal in which they have their operations. Remote gates 601–615 are in the second-category for their flights. Most airlines tend to place Cargo gates (numbers 501–525) in the third-category. Emergency gates numbered 701–703 may be considered in the fourth-category for almost all the airlines.

Data Elements with Uncertainties in Them

The other input data for gate assignment decisions are flight arrival and departure times. This data is subject to considerable uncertainty. As the difference between the actual landing time of a flight and its scheduled landing time is a random variable, gate allocations for flights on a day cannot be finalized based on information available the previous day about their landing times. That is why on day $t-1$ the airports make a tentative gate allocation plan for all the flights on the next day t based on the information about their landing and departure times available on day $t - 1$, and on day t they revise these tentative gate allocations as needed by changes in flight landing and departure times. Normally on day $t-1$, by 2:00 PM all the tentative landing and departure time information for all the flights on day t becomes available. Using this information, the airport prepares the tentative gate allocation plan for day t by 10:00 PM on day $t - 1$.

The Various Objectives to be Optimized

There are various costs associated with gate assignments which need to be optimized simultaneously.

As the gate allocation problem is a multiobjective problem, we solve it by combining the various objectives into a single penalty function, and then determining the gate allocations to minimize the combined penalty function. Here we will discuss the various objective functions considered, and the penalty coefficients corresponding to them that can be used in a DSS developed for gate allocations.

An Objective Commonly Considered in the Literature: In published literature on the gate allocation problem in OR journals, the most frequently used objective is minimization of the walking distance of all the passengers inside the airport. This is also used as the most important goal in the design of airport terminals. This appealing objective is easily motivated and clearly understood, but it leads to very difficult models that can hardly be solved.

A disturbing fact that has surfaced in the last few years is the large number of airline passengers feeling various degrees of uneasiness and actual sickness (some even suffering severe health consequences) from sitting without any physical activity for considerable periods of time on medium to long airline flights. In view of this, we feel that it is inappropriate to place a great emphasis on minimizing the total walking distance of all airline passengers inside airports. On the contrary, maybe encouragement should be provided for passengers to walk around when they have the time. Moreover, this objective does not measure any real cost, and is not high on either the airport's or any airline's list of important objective functions to be optimized. Also, the availability at many international airports nowadays, of rapid transit (also called Skytrain and other names) service between various terminals in the airport, or clusters or centers of gates separated by some distance and walking belts inside each terminal for covering long distances makes this objective function even less important. Airline managers whom we talked to tell us that even though this objective function is emphasized heavily in OR literature, their current practice of assigning gates to flights automatically takes care of this objective function, because in it, gates that are far away from the central part of the airport are assigned to a flight only when there is no gate in the central part available at the time of its landing.

Compared to this objective function of the total walking distance of all the passengers, the other objective functions like OBJ1 and 2 discussed below are real costs that are considered high-priority objectives by both the airport and all the airlines. So, we will not consider this objective function in our model.

However, there is a class of passengers – transfer passengers – who have only a limited time (like an hour or so), who are greatly inconvenienced by having to walk a long distance between their arrival and destination gates. One objective that we will consider is minimizing $OBJ =$ the total walking distance that transfer passengers with a limited time to walk between their arrival and departure gates.

The number of passengers on each flight and the destination of each of them in the airport (either the exit or the gate assigned to some other flights) fluctuate randomly and widely from day to day. So, to evaluate this objective function reasonably accurately, we need data on the number of passengers transferring between every pair of gates, and hence modeling the problem of minimizing this objective function needs binary variables with four subscripts, and consequently leads to a large 0–1 integer programming model that is hard to solve.

So, we will handle this OBJ indirectly. One way of achieving this objective is to make sure that arriving flights carrying more than a certain number of transfer passengers are assigned to gates in the *central part of the terminal*, that is, gates more or less equidistant from gates in all corners of the terminal (this guarantees that wherever the departure gate may be, transfer passengers on those flights have to travel only a small distance to reach them).

We can identify the set of those centrally located gates in the terminal. For an arriving flight j with more than the prescribed number of transfer passengers and an eligible gate i outside this central set include a penalty term corresponding to this objective function in the penalty cost coefficient c_{ij} corresponding to this assignment of gate i to flight j. Using this, we can incorporate this OBJ into OBJ1 discussed below.

In the project at TPE discussed in Yu et al. (2009), the central part of the airport consists of the A, B, C, D sections shown in Fig. 3.7. The first category gates mentioned above are all gates in the central part of TPE; the second category gates are at some distance to the central part; and the third category gates are much farther away. So, the penalty terms corresponding to the allocation of second and third category gates to such flights in OBJ1 will reflect the impact of this objective OBJ in the penalty function.

OBJ1: This objective measures the costs or penalties associated with the gate assigned to a flight.

Emergency and cargo gates (these terms are used normally for gates some distance away from the central portion of the airport; typically these are apron gates without passenger bridges) may only be used during periods of heavy flight traffic when regular passenger gates (i.e., those with jetties or passenger bridges) are not available for assignment, and only for certain types of flights for which such gates fit the flights. If such a gate i is eligible to be assigned to a flight j, then OBJ1 measures the penalty term corresponding to this assignment. Also, such an assignment usually results in charges to airlines for coaches and sometimes towing charges, etc. These are real costs that the airlines and the airport want to minimize. So, whenever towing and coach charges are incurred in the assignment of an eligible gate i to a flight j, we include these costs (scaled appropriately) in the penalty coefficient c^1_{ij} corresponding to this assignment.

Even when a regular gate i is assigned to a flight j, there may be a penalty corresponding to this assignment depending on whether gate i is in the central part of the airport are not, if flight j is carrying a lot of transfer passengers, and the preference category (first, second, or third as mentioned earlier) to which gate i belongs for flight j. All these penalty terms corresponding to various gate assignments to flights are to be determined by the decision makers.

OBJ 2: This objective measures the cost associated with the time the plane spends circling around the airport, as well as waiting on the ground after landing before beginning to taxi to the gate.

The plane is asked to circle the airport when there is no gate to receive it if it lands right away, and even the taxiway is too full for it to wait after landing. In the past, this type of "necessity for the plane circling around the airport" used to occur sometimes, but nowadays at most airports around the world, this has become an extremely rare event. So, we ignore this in our mathematical model.

After landing, the plane will be asked to wait on the taxiway if the gate i to which it is assigned is either not free momentarily or the path from the landing point on the runway to gate i is blocked momentarily by some obstruction.

So in OBJ2, we will consider only the costs and penalties for planes having to wait on taxiways before beginning taxiing to the gate. If a flight lands at 8:00 PM say, but the gate allocated to it is not going to be free until 8:15 PM, and taxiing needs only 5 min, then the plane will be asked to wait on the taxiway for 10 min before beginning to taxi to the gate. Such events occur during peak times of the day at most busy airports. They cause inconvenience to the passengers and the flight crew on the plane.

We can measure the penalty for such an event by $c^2_j t_{ij}$, where

- t_{ij} = time in minutes that flight j plane has to wait on the taxiway after landing before beginning to taxi to gate i if it is allotted to gate i,
- c^2_j = the penalty/minute waiting on taxiway for flight j.

The penalty coefficient c^2_j may depend on the plane size, average passenger load in flight j, etc. Suitable values for c^2_j have to be determined by airport management or the concerned decision makers.

OBJ3: On each planning day this objective plays a role only in updating the tentative gate allocations made the day before for this day. Most airports want to make as few changes as possible in the tentative gate allocations made already. However, some airports may not consider this objective important.

If this is considered important we can handle this objective by trying to minimize the number of changes made in the tentative gate allocations. While updating the gate allocations on the planning day, we will include a penalty term c^3_{ij} for choosing gate i for flight j in the final allocations, where $c^3_{ij} = 0$, if gate i is the tentative gate allotted to flight j; 1 otherwise.

Outputs Needed

Typically most airports have the following goals for the gate allocation effort.

- Each day by about 2:00 PM or so the airport has all the information on the scheduled landing times for all the flights next day. Using this data, prepare by 10:00 PM, a tentative gate allocation plan for all the flights next day.
- For each arriving and departing flight on the planning day, the gate allocation for it should be finalized about 2 h before its actual arrival and departure based on the latest information available about it.

The Planning Scheme to be Used

Flights arrive and land, and depart continuously over time. So, arriving and departing flights form a continuous stream, and before a flight arrives or departs, we need to make a decision about its gate allocation.

Suppose a flight A arrives at 8:00 PM, and a gate, 1 say, is allotted to it. Then gate 1 will be occupied by this flight in the period 8:00–10:00 PM say, and is unavailable for allocation to other flights arriving in this interval. Thus, the allocation of a particular gate to a flight limits the choice of gates to flights arriving and departing later. Consequently, in the above example, the allocation of gate 1 to flight A at 8:00 PM may lead to undesirable allocations to other flights arriving between 8:00 and 10:00 PM.

For this reason, most of the literature on the gate allocation problem insists on making the gate allocation decisions for all the flights in a day simultaneously using a large mathematical model covering the whole day. Because of this, researchers claim that their model outputs the global optimum gate allocation plan for the whole day, without getting trapped in suboptimal plans over shorter intervals of time.

We need to point out that there is a serious and fundamental flaw in this argument.

Most airports are committed to treating all their customer airlines equally, and do not give special privilege or preference to any particular airline. This implies that "first arrived, first assigned" policy should be adhered to strictly; similar policy also holds for departing flights. This means that each flight should be allotted to the best gate available for allocation at the time of its arrival/departure, irrespective of how it affects the availability of gates to flights arriving/departing after this flight.

For example, suppose two gates 1, 2 are the only ones available between 5 and 5:30 PM; of these 1 is a first-category gate and 2 is a third-category gate. Suppose flight A lands at 5 PM, and flight B lands at 5:15 PM. By this policy, we must assign gate 1 to flight A. It would violate the "first arrived, first assigned" policy to assign gate 2 to flight A for the sake of assigning the first-category gate 1 to flight B arriving later than flight A.

The "first arrived, first assigned" policy is one that all the airports in the world claim to adhere to. In our conversations with airport officials, we were told that they have to follow this policy to maintain good business relations with all their customer airlines. This policy implies that each flight should be allotted the best gate for it available at the time of its arrival/departure, and hence ideally it is best to determine gate allocations by an online algorithm that makes real-time decisions for each flight based on availability of gates at the time of its arrival/departure.

However, with lots of flights arriving/departing in short durations, and the necessity to announce gate allocations at least 2 h before the actual arrival/departure based on the best information available at that time, and the desire to keep a large percentage of these allocation decisions unchanged as far as possible, it is very difficult to make gate allocations totally online for each flight just in time for its arrival/departure. So, we adopt the following practical strategy that is close in spirit to online decision making, and yet is easy to implement in practice. We select a short planning interval (e.g., a 15 min or 30 min interval), and determine the best gate assignments for all flights arriving/departing in this interval, at gates that will be available for assignment at some point of time in this interval, minimizing the penalty function discussed earlier, using a simple static mathematical model. If the optimum solution obtained violates the "first arrived, first assigned" policy for some flights, then it is easy to modify that solution (using swapping and other manual moves) into one that satisfies that property, as the planning interval is short and at the time of decision making all the necessary data for this interval is known accurately.

For this reason we develop the following planning scheme for making gate allocation decisions. In contrast, models in the literature for gate allocation totally ignore the "first arrived, first assigned" policy.

Selection of the Planning Interval

We divide the day into short planning intervals for gate allocation decisions. Decisions are made for the intervals in chronological order, and decisions made for an interval are taken as fixed in making decisions for future intervals.

In the spirit of keeping close to online decision making, we find that taking the planning interval length as 30 min is convenient and works well. So, we describe the mathematical model in terms of 30-min planning intervals (interval length can be changed from 30 min to similar short duration as appropriate).

When interval k is the planning interval, gate allocation solutions for flights arriving in time intervals $\leq k-1$ are fully known, and that information can be used to simplify many of the gate assignment constraints in the model for planning interval k.

For example, the assignment of a large aircraft to a particular gate may imply that adjacent gates can only accept aircraft of a certain size, or are even completely blocked. So, if gate 1 is going to be used by a large aircraft flight in time interval $k-1$, and that plane will continue to stay at that gate for some time during interval k, then adjacent gates of gate 1 can simply be made ineligible for allocation to flights using planes of nonacceptable size during planning interval k.

Thus, the choice of our short-duration planning interval allows us to both avoid the effects of uncertainty in data elements, and also makes it possible to solve the problem using a simpler mathematical model that is easier to solve. Also, it is easier to make simple modifications in the output allocation manually for implementation.

Peak and Off-Peak Periods

We divide the day into 48 intervals of 30-min each, and number them serially 1 to 48 with 1 representing the 12:00–12:30 AM interval. Then we can plot the average number of flights arriving/departing in each half-hour interval of the day based on past data at the airport. From this we can see what the peak periods are for flights arriving/departing at this airport. In fact, different working days may have different peak periods. Identifying these peak periods provides very useful information for managing gate allocations at the airport while keeping all customer airlines happy. For solving the gate allocation problem, the peak periods offer most challenge; in off-peak periods the problem is relatively easy to solve. During peak times, the FOOs (flight operations officers of the airport responsible for gate allocations) may have to persuade airlines to shorten the gate occupancy times for their flights in order to accommodate more flights, and most airlines are usually happy to cooperate.

In Fig. 3.8 we show the plot of the average number of flights arriving/departing in each half-hour interval of the day based on June 2007 data at TPE. From this we see that there are two peak periods of the day for number of flights arriving/departing at TPE, one in the period 7:00–10:00 AM and another in the period 3:00–5:00 PM. Corresponding charts for the data in various months of 2007 confirm the same observation. However, this pattern is somewhat different for different days of the week (Sunday to Saturday).

How is the Gate Occupancy Time of a Flight Determined

Most airports use fixed blocks of time for flights based on the size of the plane normally used for the flight (like a 1.5-h block for flights that normally use large wingspan planes, a 1-h block for those that use medium wingspan planes, etc.).

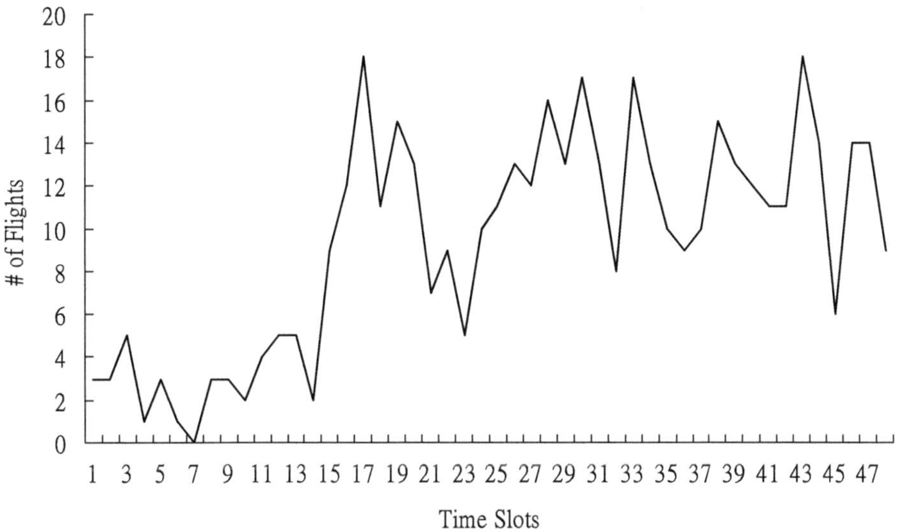

Fig. 3.8 This figure shows the typical averages of the total number of flights arriving, departing in each 30-min interval of a day. Time intervals are numbered serially 1–48 (1 is the time slot from midnight to 12:30 AM) are shown on the horizontal axis. On the vertical axis we plot the number of flights arriving and departing in the interval

TPE normally allocates 60-min gate time to arrival flights (i.e., these are arriving flights for which this airport is the final destination, 90-min gate time to departure flights (these are flights which originate at this airport) for flight preparation, passenger disembarking, embarking, flight clean up, ground service to the plane, etc. For transit flights (these are flights that arrive and make a brief stop at this airport, and then depart for another destination), the allocated gate time is 150-min, the sum of gate times of an arrival flight and a departure flight. However, these times may vary depending on the size of the aircraft and the time of the day. During peak times, the FOOs may shorten the gate time to accommodate more flights. On average, a 30-min buffer time is scheduled between two flights using the same gate to take into account the uncertainty in flights' actual arrival/departure time. This time may also be cut short during peak hours, as the FOOs try to persuade flights occupying gates to finish up their work as soon as possible. The FOOs usually have the full cooperation of all the airlines in this effort during peak periods.

Strategy Used for Making Gate Allocation Decisions for a Planning Day

We will discuss below a **procedure** that gives the mathematical model for making gate allocation decisions in a single 30-min planning interval assuming that the arrival and departure times for all the flights arriving and departing in that interval are known exactly, and discusses how to solve it. Here we will discuss how to use that procedure to generate the outputs needed for the planning day.

To Make the Tentative Gate Allocation Plan for All the Flights on the Planning Day: This plan has to be prepared by 10:00 PM of the day before the planning day. The data used for making this tentative plan are the scheduled arrival, departure times for all the flights on the planning day, which becomes available by 2:00 PM of the day before the planning day.

The planning day consists of $k = 1$ to 48 planning intervals of 30-min each. The allocation decisions in these intervals are made in chronological order one after the other, starting with the first planning interval (00:00–00:30 h), using the procedure described below. So, at the time of making decisions for the kth interval, the gate allocations for flights in the $(k - 1)$th interval would have been finalized and become available.

To Update and Make Final Gate Allocation Decisions for a Planning Interval on the Planning Day: Consider the kth planning interval. Gate allocation decisions for flights arriving, departing in this interval are finalized 2 h before the beginning of this interval.

Nowadays flight arrival and departure information is being continuously updated, and this real-time information is delivered continuously to all airport organizations who use it. About 2.5 h before the beginning of the planning interval, the kth, the arrival/departure times for flights in the planning interval are known reasonably precisely. Also by this time gate allocations for flights in the $(k - 1)$th interval would have been finalized and are known. Gate allocation decisions for the planning interval are finalized using all this data with the procedure discussed below.

It is possible that some last-minute changes occur in the arrival/departure times of flights in the planning interval, after the gate allocation plan for this interval is finalized. Nowadays, such changes are rare and only few in number. Any necessary changes in gate allocations to accommodate these last minute changes in arrival, departure times are carried out by the gate allocation officers manually.

Procedure for Gate Assignments to Flights in a Planning Interval

Consider the kth planning interval on the planning day.

There may be some flights expected to arrive towards the end of the $(k - 1)$th interval for which gates have not been assigned in the planning work for that interval. These flights will also be considered for gate assignment in this kth planning interval.

Let

- J, n : n is the number of flights that need to have a gate assigned in this planning interval k. This includes flights that depart or land at some point of time in this planning interval, and flights that are expected to land before, but have not been assigned to a gate in the previous interval. J denotes the set of these flights, and the index j is used to denote a general flight in J.

- i : is the index used to denote a gate in the airport (includes all gates, remote and emergency gates also if they can be used by some flights during heavy peak times) that are expected to be available for assignment to flights in this interval. If a flight is going to be occupied for the entire kth planning interval by a flight assignment made in earlier periods, then it is not even considered in this model.
- x_{ij} : is the decision variable defined for gate i and flight $j \in J$; this variable takes the value 1 if flight j is assigned to gate i in this planning interval, or 0 otherwise.

If a gate i is not suitable to assign to flight j for whatever reason (e.g., if flight j uses a large plane and gate i is not of a size appropriate for it, etc.), then it is made ineligible for assignment to flight j, and the corresponding variable x_{ij} is not even considered in the model for planning interval k.

Similarly, several of the gate assignment constraints can be taken care of by this ineligibility classification. For example, as mentioned in the previous section, if gate i is adjacent to a gate occupied by a large plane, and that plane will be there for some time during interval k, then it is made ineligible for all flights $j \in J$ with planes of unacceptable size during planning interval k, and the corresponding variables x_{ij} do not appear in the gate assignment model for this interval.

Let

- $G_j =$ set of gates i that are eligible to be assigned to flight $j \in J$ in planning interval k
- $I = \cup_{j \in J} G_j =$ set of gates i that are eligible to be assigned to at least one flight $j \in J$ in this planning interval
- $F_i =$ set of flights $j \in J$ for which gate $i \in I$ is eligible to be assigned in this planning interval

We will combine the various objectives into a single penalty function to be minimized to determine an appropriate compromise between the various objectives, while assuring some of the hard constraints in gate assignment.

Let

- c_{ij}: the combined positive penalty coefficient associated with the decision variable x_{ij}. It is the sum of positive penalty coefficients associated with x_{ij} corresponding to the various objectives (these are determined by the decision makers based on trade-offs between the various objectives).

When considering gate allocations in planing interval k, flights that are expected to arrive in time interval $k-1$, but have not been assigned to gates then, should be given preference. Also, the airport may consider giving preference to certain flights that arrive in planning period k itself.

So, partition the set of flights J that need gate assignments in this planning interval k into $J_1 \cup J_2$, where

- $J_1 =$ subset of the flights in J that have to be given first preference for gate assignments

- J_2 = the remaining flights in J.

We first determine the maximum number r of flights to which gates can be assigned in this planning period k, subject to the constraints mentioned in the previous sections. It leads to the following *transportation model.*

There is a constraint corresponding to each flight that needs a gate assigned in this planning interval k, the one corresponding to each flight $j \in J_1$ is an equality constraint specifying that each of these flights must be assigned one gate for itself (because we are required to give these flights first preference for gate allocation); the one corresponding to each flight $j \in J_2$ is an inequality constraint specifying that this flight needs one gate for itself if an eligible one for it can be found.

There is one inequality constraint corresponding to each gate that becomes free at some point of time in this planning interval k and can receive a flight from that time onwards, specifying that this gate can accommodate at most one flight for which it is eligible to be assigned.

The objective in this model is to maximize the total number of eligible flight-gate assignments that can be made in this planning interval. The model is

$$
\begin{aligned}
r = \quad & \text{Maximum value of} \quad \sum_{j \in J} \sum_{i \in G_j} x_{ij} \\
\text{Subject to} \quad & \sum_{i \in G_j} x_{ij} = 1 \quad \text{for each } j \in J_1 \\
& \sum_{i \in G_j} x_{ij} \leq 1 \quad \text{for each } j \in J_2 \\
& \sum_{j \in F_i} x_{ij} \leq 1 \quad \text{for each gate } i \in I \\
& x_{ij} \geq 0 \quad \text{for each } j \in J,\ i \in G_j.
\end{aligned}
$$

If this model turns out to be infeasible, it is an indication that there are not enough eligible gates available in this planning interval to even assign to all the flights $j \in J_1$. Then, the airport authorities can modify and relax some of the eligibility requirements for gate assignments if possible, or modify the set J_1 as appropriate, and solve the model with revised information.

When this model is feasible, the maximum objective value r in it gives the maximum number of flights among those needing gates in this planning interval, for which gates eligible for them can be assigned in this interval. If $r < n = |J|$, the remaining $n - r$ flights in J for which eligible gates cannot be assigned in this planning interval will have to be transferred to the next interval for gate assignments.

The above model only provides the maximum number of gate assignments that can be made in the planning interval k. It does not try to find optimal gate assignments. When the above model is feasible, the optimum gate allocations are found by solving another mathematical model, which is also a transportation (or network flow) model. It tries to minimize the composite penalty function constructed above, subject to the same constraints as in the above model, and the additional constraint that the total number of flight-gate assignments should equal the maximum possible number r found above. It is

$$\begin{aligned}
\text{Minimize} \quad & \sum_{j \in J} \sum_{i \in G_j} c_{ij} x_{ij} \\
\text{Subject to} \quad & \sum_{i \in G_j} x_{ij} = 1 \quad \text{for each } j \in J_1 \\
& \sum_{i \in G_j} x_{ij} \le 1 \quad \text{for each } j \in J_2 \\
& \sum_{j \in F_i} x_{ij} \le 1 \quad \text{for each gate } i \in I \\
& \sum_{j \in J} \sum_{i \in G_j} x_{ij} = r \\
& x_{ij} \ge 0 \quad \text{for each } j \in J,\ i \in G_j.
\end{aligned}$$

Transportation (network flow) models are easy to solve. Typically, we can expect to have at most 100–200 flights to deal with in any planning interval at busy international airports. For problems of this size, either of the above models will require at most 0.5 s of a common PC time to solve, using software programs available today. An optimum solution $\bar{x} = (\bar{x}_{ij})$ obtained for the second model provides an optimum gate allocation through the interpretation that

flight $j \in J$ is assigned to gate i in the optimum gate allocation $\bar{x}$ if $\bar{x}_{ij} = s1$.

There may be other constraints that have not been included in this model. If so, they can be added to the model. Or the gate assignment team may use their expert judgment to modify an optimum solution of this model into another that can be implemented.

Summary

We described the common features of the problem of allocating gates to passenger flights at an international airport, and how many of the models for it in published literature seem to focus on optimizing an objective function that the actual decision makers do not consider important. This actually makes the models for this problem in published literature very complex and difficult to solve. We outline a scheme for developing a DSS for this problem that leads to a much simpler model that is easy to implement in practice. In this approach, we applied the mathematical model in two stages. In the first stage, we just maximized the number of flights to which gates have been allocated, and in the second stage we optimized the actual allocation of gates to these flights in terms of the other objective functions (OBJ 1, 2, 3) discussed in this section. All the airport officials that we talked to felt that top priority should be given to maximize the number of flights that are gated, this is the reason for the two-stage approach.

3.3 Murty's Three Commandments for Successful Decision Making

From the work in these case studies and other similar projects, I noticed that the following are important for successful decision making:

(a) Looking at the problem from all possible angles and not just in one way
(b) Constructing an intelligent mathematical model to analyze the problem and to solve it
(c) Being very tactful in selling the optimum solution obtained from the model and in implementing it.

3.4 Exercises

3.1. Allocating shelfspace to items in a grocery store: This exercise deals with the question of determining the amount of shelfspace to allocate in a grocery store to various items sold by the store, in order to maximize the expected sales volume (or the expected profit) of the store.

To model this problem, first we need to measure shelfspace in some units, say units of 1 ft length. Let $S =$ total shelfspace units available to allocate.

Items may be small (such as a screw) or reasonable size such as a 2-kg box of breakfast cereal. A small item such as a screw will never be displayed on the shelf individually; they are normally displayed in a container of reasonable size that holds many screws, maybe even has several compartments for screws of various sizes. So, we will assume that items are measured in units of comparable size. In fact, we will measure each item in the same units of shelfspace needed to display it. So, for $i =$ 1 to n; units of item $i =$ the amount of item i that one shelfspace unit will hold (if a single copy of an item takes more than 1 shelfspace unit to display, maybe the unit of item i will be a fraction <1 of a single copy of the item).

Select a unit of time, such as a day or a week that is convenient to work with. Let x_i denote the units of shelfspace allotted to item i. These are the decision variables, the optimum values of which have to be determined, and $x = (x_i)$.

For $i = 1$ to n, let $d_i =$ expected demand (or sales) for item i in the store in units/unit time. We can expect that d_i increases as the exposure of the store's customers to this item i increases. Exposure of customers to this item increases as the number of units of it on display on the shelves increases. So, we can expect that d_i increases as x_i increases. So, we represent this demand by $d_i(x_i)$ as a function of x_i; its slope is expected to be positive.

Suppose that the shelfspace allotted to item i is fully stocked at time point 0, that is, at that point of time x_i units of item i are on display. As customers pick up units of item i to purchase, the number on display will keep decreasing until this shelfspace is restocked again. The expected time between restockings of this item can be derived from the distribution of demand, but for this exercise assume that it can be approximated by $x_i/d_i(x_i)$. Restocking actually involves two tasks:

Task 1: Checking periodically the stock level of this item on the shelf to see if it has reached a safety stock level and needs to be replenished.

Task 2: Replenishing the stock level of this item to x_i, which requires going into the warehouse, bringing in those copies of the item, and arranging them on the shelf.

So, $\ell_i(x_i)$ = labor in man hours needed for tasks 1 and 2 to maintain stock levels of this item on the shelf, has an inverse relationship with x_i. As x_i increases, $\ell_i(x_i)$ decreases and vice versa.

Let c_i = cost in money units/man hour of labor to manage stock levels. Let p_i = profit in money units/unit of item i sold. Then the net profit to the company is the profit from sales − cost of labor to manage stock levels on the shelves. For i = 1 to n, let LB_i and UB_i denote lower and upper bounds for x_i specified by store management. Also, experience may indicate some relationships between the x_i's. For example, everyone who buys coffee powder, or semia, may also buy an equal quantity of sugar; this then leads to the constraint $x_1 \geq x_2 + x_3$, where items 1, 2, 3 are sugar, coffee powder, semia, respectively.

Discuss how the functional forms for $d_i(x_i)$ and $\ell_i(x_i)$ can be determined experimentally. Also, assuming that both $d_i(x_i)$, $\ell_i(x_i)$ are linear functions of x_i, formulate the problem of determining an optimum x to maximize expected net profit to the store.

References

Ackoff R (1996) The future of operational research is past. J Oper Res Soc 30(2):93–104

Dantzig GB (1990) The diet problem. Interfaces 20(4):43–47

Ford LR, Fulkerson DR (1962) Flows in networks. Princeton University Press, Princeton, NJ

Fulkerson DR (1956) Note on Dilworth's decomposition theorem for partially ordered sets. Proc Am Math Soc 7:701–702

Geoffrion AM (1976) The purpose of mathematical programming is insight, not numbers. Interfaces 7(1):81–93

Murty KG (1992) Network programming. Prentice Hall, Englewood Cliffs, NJ

Murty KG (2007) Yard crane pools and optimum layouts of container terminals. J Ind Syst Eng 1(3):190–199; can be downloaded from website: http://www.jise.info/issues/volume1no3/14.pdf

Murty KG, Djang PA (1999) The U.S. Army National Guard's Mobile Training Simulators location and routing Problem. Oper Res 47(2):175–182

Murty KG, Kim WJ (2006) An i-DMSS based on bipartite matching and heuristics for rental bus allocation, In: Gupta J, Mora M, Forgionne G (eds) Intelligent decision making support systems. Springer Verlag, Berlin

Murty KG, liu J, Wan Y-W, Lin R (2005a) A DSS (decision support system) for operations in a container terminal. Decision Support Syst 39(3):309–332

Murty KG, Wan Y-W, Liu J, Tseng MM, Leung E, Lai KK, Chiu HWC (2005b) Hongkong International Terminals gains elastic capacity using a data-intensive decision-support system. Interfaces 35(1):61–75

Petering MEH, Murty KG (2009) Effect of block length and yard crane deployment systems on overall performance at a seaport container transshipment terminal. Comput Oper Res 36:1711-1725

Simon H (1987) Two heads are better than one: the collaboration between AI and OR. Interfaces 17:8–15

Wolfram (2002) New kind of sciences. Wolfram Media, United Kingdom

Yu VF, Murty KG, Wan Y-W, Dann J, Lee R (2009) Developing a DSS for allocating gates to flights at an international airport. Int J Decision Support Syst Technol 1(1):46–68

Zhang C, Liu J, Wan Y-W, Murty KG, Linn R (2003) Storage space allocation in container terminals. Transport Res B Methodol 37(10):883–903

Chapter 4
Polyhedral Geometry

4.1 Hyperplanes, Half-Spaces, and Convex Polyhedra

A *hyperplane* in R^n is the set of feasible solutions of a single linear equation. Let $H = \{x : a_1x_1 + \cdots + a_nx_n = a_0\}$, where the coefficient vector $(a_1, \ldots, a_n) \neq 0$, be a hyperplane in R^n.

Only when $n = 2$ (i.e., in R^2 only) every hyperplane is a straight line, and vice versa. In Fig. 4.1, we show the hyperplane (straight line in R^2) corresponding to the equation $x_1 + x_2 = 1$.

When $n \geq 3$, hyperplanes are not straight lines. Figure 4.2 shows a portion of the hyperplane corresponding to the equation $x_1 + x_2 + x_3 = 1$ in R^3.

For any n, each hyperplane $H = \{x : a_1x_1 + \cdots + a_nx_n = a_0\}$ in R^n divides the space into two *half-spaces* H^+ and H^-, where

$$H^+ = \{x : a_1x_1 + \cdots + a_nx_n \geq a_0\},$$
$$H^- = \{x : a_1x_1 + \cdots + a_nx_n \leq a_0\}.$$

The half-spaces H^+ and H^- are closed half-spaces with $H^+ \cap H^- = H$ and $H^+ \cup H^- = R^n$. Thus, each linear inequality constraint corresponds to a half-space, and vice versa. In Fig. 4.1, we show the half-spaces defined by the hyperplane $\{x : x_1 + x_2 - 1\}$ in R^2.

4.1.1 Expressing a Linear Equation as a Pair of Inequalities

As $H^+ \cap H^- = H$, each hyperplane is the intersection of the two half-spaces defined by it. Equivalently, in any system of constraints, an equation, $a_1x_1 + \cdots + a_nx_n = a_0$, can be replaced by the pair of inequalities $a_1x_1 + \cdots + a_nx_n \geq a_0$ and $a_1x_1 + \cdots + a_nx_n \leq a_0$.

K.G. Murty, *Optimization for Decision Making: Linear and Quadratic Models*, International Series in Operations Research & Management Science 137,
DOI 10.1007/978-1-4419-1291-6_4,

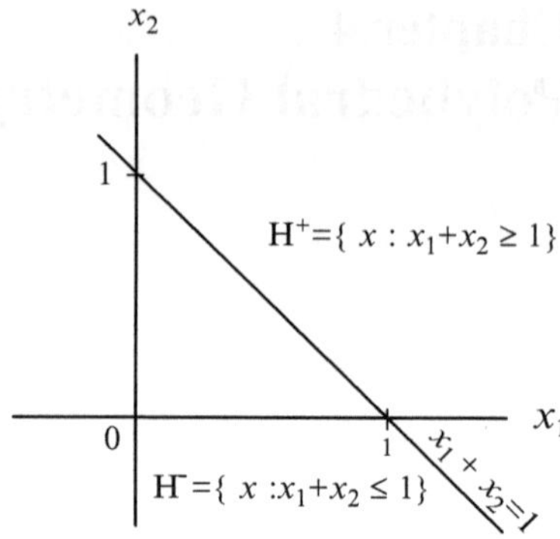

Fig. 4.1 A hyperplane in R^2 (in R^2 it is a straight line). You can also see the two half-spaces into which it divides the space

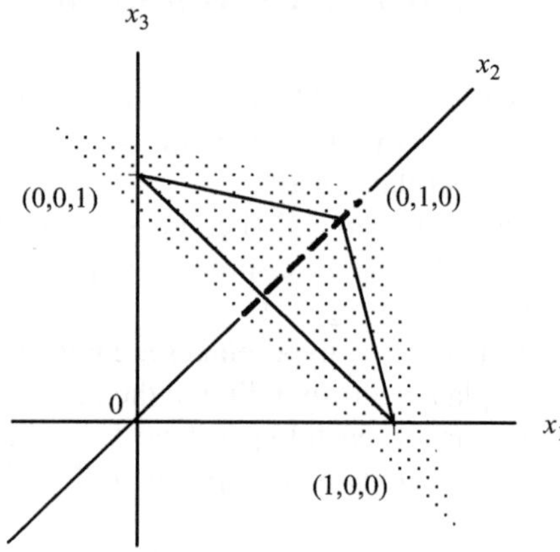

Fig. 4.2 Hyperplane represented by $x_1 + x_2 + x_3 = 1$ in R^3, part of which is shaded

Convex Polyhedron

The intersection of a finite number of half-spaces is a geometric object known as a *convex polyhedron*. So every convex polyhedron is the set of feasible solutions of a system consisting of a finite number of linear inequalities. As every linear equation is equivalent to a pair of linear inequalities, we can define a convex polyhedron to be the set of feasible solutions of a system of linear constraints containing a finite number of linear inequalities and/or linear equations.

Geometric properties of convex polyhedra are being studied for a long time. The Egyptian Pharaohs studied convex polyhedra over 5,000 years ago for designing and constructing the Great Pyramids.

The set of feasible solutions of any linear programming problem is a convex polyhedron. Because of the development of LP (linear programming) in the mid-twentieth century, the study of convex polyhedra has gained added momentum. The advent of LP has added a computational dimension to the study of convex polyhedra and their boundary structure.

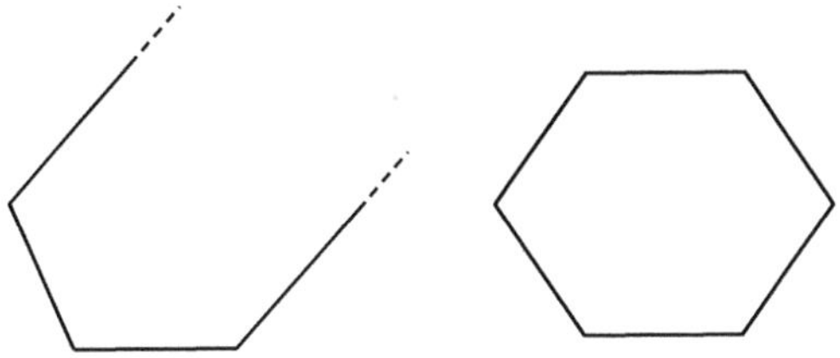

Fig. 4.3 Unbounded (*left*) and bounded (*right*) convex polyhedra in R^2

The boundary of a convex polyhedron is divided into its various boundary faces of various dimensions and has a highly intricate combinatorial structure. The set of optimum solutions of any linear program is a boundary face of its feasible solution set. Also, the simplex method – the main method used for solving LPs in the third quarter of the twentieth century, which continues to be one of the most popular methods today – operates only on the one-dimensional faces of the set of feasible solutions. This has resulted in a very intensive study of the boundary structure of convex polyhedra since the mid-twentieth century.

In this chapter we will discuss important results on the geometry of convex polyhedra that are essential to the study of LP and the simplex method for solving it. For a more detailed and complete treatment of the theory of convex polyhedra, the reader is referred to the mathematical books (Grünbaum 1967; Ziegler 1994).

Convex Polytopes

A convex polyhedron may be unbounded or bounded. In Fig. 4.3, on the left is an unbounded convex polyhedron, and on the right is a bounded convex polyhedron in R^2. A bounded convex polyhedron is known as a *convex polytope*.

4.1.2 Straight Lines, Half-Lines, and Their Directions

The most convenient and commonly used representation of a straight line is the parametric representation involving a single parameter λ say, as

$$L = \{x(\lambda) = (x_j(\lambda)) = (a_j + \lambda b_j) : \lambda \text{ takes all real values}\}$$

where $a = (a_1, \dots, a_n)^T$, $b = (b_1, \dots, b_n)^T$ are given vectors such that $b \neq 0$. For example, the straight line in Fig. 4.1 in R^2 joining the points $(1, 0)^T$ and $(0, 1)^T$ can be represented as $\{x(\lambda) = (1, 0)^T + \lambda(-1, 1)^T : \lambda$ takes all real values$\}$.

For any positive integer n, in R^n, a straight line is uniquely characterized by any two distinct points on it. For example, if a, c are two distinct vectors in R^n (i.e., $a \neq c$), the straight line through them is $\{x(\lambda) = a + \lambda(a - c)$: λ takes all real values$\}$.

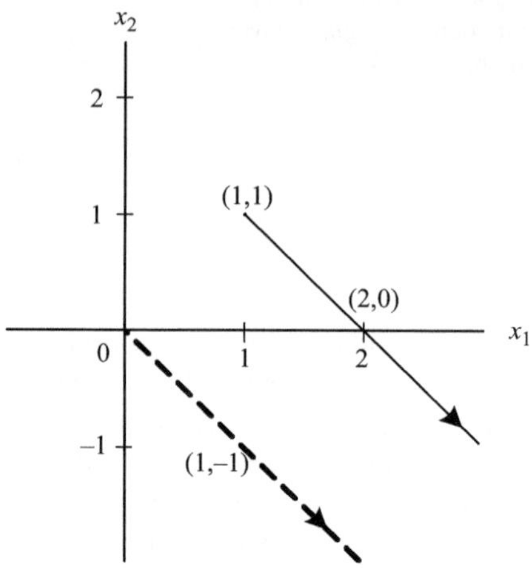

Fig. 4.4 The *solid line* in the figure is the half-line $\{(1,1)+\lambda(1,-1) : \lambda \geq 0\}$. The point $(2,0)$ on it is obtained by moving from $(1,1)$ a step length of 1 in the direction of $(1,-1)$. The *dashed line* in the figure is the ray of $(1,-1)$, which is parallel to the above half-line

In the parametric representation of a straight line, if we let the parameter λ take only nonnegative values (instead of all real values), we get a *half-line*, in this case the coefficient vector of the parameter λ in the representation is called the *direction* of that half-line.

So, the half-line starting at a point a in R^n in the direction b in R^n has the parametric representation $L^+ = \{x : a + \lambda b : \lambda \geq 0\}$. The point $a + \lambda b$ in L^+ is said to be obtained by moving from the point a, a step length of λ in the direction b. Figure 4.4 illustrates a half-line in R^2.

Any half-line beginning at 0 is also called a *ray*. So, the half-line beginning at 0 in the direction of $b \neq 0$, that is, $\{\lambda b : \lambda \geq 0\}$ is the ray of b.

4.1.3 Convex Combinations, Line Segments

Given a finite set of points $\Gamma = \{x^1, \dots, x^r\}$ in R^n, any point of the form $\alpha_1 x^1 + \cdots + \alpha_r x^r$, where the coefficients satisfy $\alpha_1 \geq 0, \dots, \alpha_r \geq 0$, $\alpha_1 + \cdots + \alpha_r = 1$, is known as a *convex combination* of points in Γ. The set of all such convex combinations is known as the *convex hull* of Γ.

If $r = |\Gamma| = 2$, its convex hull is known as the *line segment* joining the two points in Γ. In this case $\Gamma = \{x^1, x^2\}$, a point on the line segment joining x^1, x^2 can be represented using a single parameter as $\alpha x^1 + (1-\alpha)x^2$, where $0 \leq \alpha \leq 1$. As α increases from 0 to 1, this point moves along the line segment joining the two points in Γ from x^2 to x^1. Figure 4.5 shows the convex hull of six points in R^2, a line segment, and a pyramid in R^3, which is the convex hull of five points.

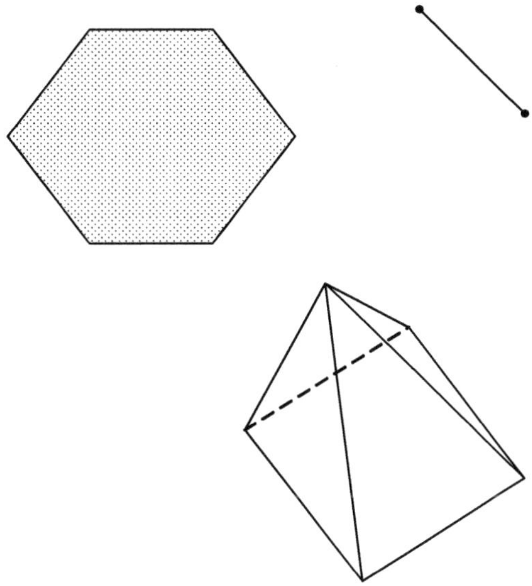

Fig. 4.5 Some convex hulls in R^1, R^2, and R^3

4.2 Tight (Active)/Slack (Inactive) Constraints at a Feasible Solution $\bar{x}$

Let K be the set of feasible solutions of the general system of linear constraints

$$A_{i.}x \begin{cases} = b_i & \text{for } i = 1 \text{ to } p \\ \geq b_i & \text{for } i = p+1 \text{ to } m \end{cases} \tag{4.1}$$

consisting of inequalities and maybe some equality constraints. Let $\bar{x} \in K$, that is, a feasible solution of (4.1). *Tight* (or *active*) constraints at $\bar{x}$ in (4.1) are all the constraints in (4.1) that are satisfied as equations at $\bar{x}$. Hence

All equality constraints in (4.1)	Always active (tight) constraints at every feasible solution $\bar{x}$
All inequality constraints in (4.1) satisfied as equations at $\bar{x}$	Active (tight) constraints at the feasible solution $\bar{x}$
All inequality constraints in (4.1) satisfied as strict inequalities at $\bar{x}$	Inactive (slack) constraints at the feasible solution $\bar{x}$.

Example 4.1. Consider the feasible solution $\bar{x} = (-1, 10, 0, 5, 10)^T$ to the following system of constraints

$$\begin{aligned} x_1 + x_2 + x_3 + x_4 + x_5 &= 24 \\ x_1 - x_2 + x_3 - x_4 &\geq -16 \\ -x_2 - x_3 + x_4 + x_5 &\leq 8 \end{aligned}$$

$$-x_1 + x_2 + x_3 - 2x_4 + 2x_5 \leq 31$$
$$-x_1 + x_2 + x_3 - 2x_4 + 2x_5 \geq 21$$
$$-2 \leq x_1 \leq 10, \;\; 0 \leq x_2, x_3 \leq 10$$
$$x_5 \text{ unrestricted}, \;\; x_4 \geq 0.$$

In this system there is one equality constraint, and 11 inequality constraints (each bound on a variable is also an inequality constraint).

The constraints that $\bar{x}$ satisfies as equations are the first (equality constraint); second, fifth, upper bound on x_2, lower bound on x_3 (all inequality constraints); these are the constraints that are active at the feasible solution $\bar{x}$.

The other constraints in the system, the third, fourths, lower bound on x_2, upper bound on x_3, lower bound on x_4 (all inequality constraints) are inactive at $\bar{x}$. ■

Example 4.2. Here we will give a two-dimensional example so we can also illustrate the concepts of active/inactive constraints at a point geometrically. Remember that in R^2 every linear equation corresponds to a straight line (hyperplane in R^2), and a point $\bar{x}$ satisfies that equation iff it is on the corresponding straight line.

A linear inequality represents the half-space on one side of the corresponding straight line, this constraint is active at $\bar{x}$ iff $\bar{x}$ lies on that straight line, inactive if it lies in the interior of the half-space. Now consider the following system of constraints in two variables x_1, x_2. Consider the points $x^1 = (1,1)^T$, $x^2 = (1,2)^T$, $x^3 = (2,2)^T$, all of which are feasible solutions of this system. Figure 4.6 shows the feasible region of this system.

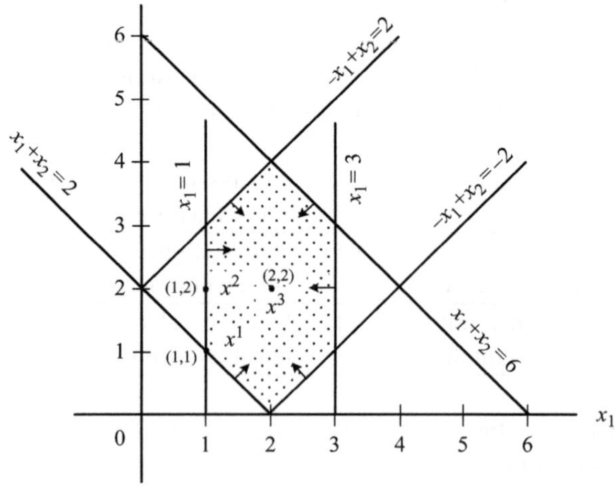

Fig. 4.6 Feasible region is *shaded*. The *thick dot* in the interior of the feasible region is $x^3 = (2,2)^T$

$$\begin{aligned} x_1 + x_2 &\geq 2 \\ -x_1 + x_2 &\leq 2 \\ x_1 + x_2 &\leq 6 \\ -x_1 + x_2 &\geq -2 \\ x_1 &\geq 1 \\ x_1 &\leq 3 \\ x_2 &\geq 0. \end{aligned} \tag{4.2}$$

The solution x^1 satisfies only constraints 1, 5 in (4.2) as equations, hence these are the tight (active) constraints at x^1; the other constraints in (4.2) satisfied as strict inequalities at x^1 are inactive or slack at x^1.

The solution x^2 satisfies only constraint 5 in (4.2) as an equation; all the other constraints are satisfied as strict inequalities by x^2. So, only the 5th constraint in (4.2) is active at x^2, all other constraints are inactive or slack at x^2.

The solution x^3 satisfies all the constraints in (4.2) as strict inequalities. Hence there are no active constraints at x^3 in (4.2); all are inactive. Clearly x^3 is an interior point of the set of feasible solutions of (4.2), as we can see clearly from Fig. 4.6. ■

4.2.1 What is the Importance of Classifying the Constraints in a System as Active/Inactive at a Feasible Solution?

When at a feasible solution $\bar{x}$, a fundamental strategy in many optimization algorithms is to look for a direction y to move, that will lead to improvements in objective value while maintaining feasibility. If a constraint is active at $\bar{x}$, then unless the direction of movement y is selected with special care, this constraint may be violated even for moves of small step lengths, however small. On the other hand, when a constraint is inactive, it will continue to remain satisfied when we move in any direction from $\bar{x}$, as long as the step length is sufficiently small.

That is why in selecting the direction y to move at $\bar{x}$, the algorithms ignore the inactive constraints, and only make sure that none of the active constraints will be violated by the move. This points out the importance of classifying and identifying the active constraints at the current feasible solution.

Let $\bar{x} = (\bar{x}_1, \dots, \bar{x}_n)^T$ be a feasible solution for a system of linear constraints and suppose

$$a_{11}x_1 + \dots + a_{1n}x_n = a_{01}$$

is an equality constraint in the system. As $\bar{x}$ is a feasible solution of the system, we have $a_{11}\bar{x}_1 + \dots + a_{1n}\bar{x}_n = a_{01}$, and this constraint is active at $\bar{x}$. If we move in the direction y from $\bar{x}$ a positive step length λ, we get the point $\bar{x} + \lambda y$. For this point to satisfy this equality constraint, we need $\sum_{j=1}^{n} a_{1j}(\bar{x}_j + \lambda y_j) = a_{01}$, and since $a_{11}\bar{x}_1 + \dots + a_{1n}\bar{x}_n = a_{01}$ we need $a_{11}y_1 + \dots + a_{1n}y_n = 0$, that is, the direction y has to satisfy the *homogeneous equation* obtained by setting the RHS constant in every equality constraint in the system to 0 in order to retain feasibility after moving in the direction y from $\bar{x}$.

Now consider an inequality constraint in the system:

$$a_{21}x_1 + \cdots + a_{2n}x_n \geq a_{02}.$$

We need $\sum_{j=1}^{n} a_{2j}(\bar{x}_j + \lambda y_j) \geq a_{02}$ for $\bar{x} + \lambda y$ to satisfy this constraint. So, if this constraint is inactive at $\bar{x}$, that is, $\sum_{j=1}^{n} a_{2j}\bar{x}_j > a_{02}$, then $\bar{x} + \lambda y$ will continue satisfying this constraint for all directions y as long as λ is sufficiently small.

If this constraint is active at $\bar{x}$, i.e., $\sum_{j=1}^{n} a_{2j}\bar{x}_j = a_{02}$, then this constraint will continue to be satisfied by $\bar{x} + \lambda y$ for $\lambda \geq 0$ iff $\sum_{j=1}^{n} a_{2j}y_j \geq 0$, that is, the direction y satisfies the *homogeneous inequality* obtained by setting the RHS constant in this constraint to 0.

4.3 Subspaces, Affine Spaces, Convex Polyhedra; Binding, Nonbinding, Redundant Inequalities; Minimal Representations

A *subspace* of R^n is the set of feasible solutions of a system of homogeneous linear equations, that is, a system of the form $Ax = 0$ with all RHS constants 0.

An *affine space* of R^n is the set of feasible solutions of a general system of linear equations, that is, a set of the form $\Gamma = \{x \in R^n : Ax = b\}$, where A is a matrix of order $m \times n$ and b is a column vector in R^m. If $\bar{x} \in \Gamma$, the transformation of variables $y = x - \bar{x}$, called *translating $\bar{x}$ to the origin*, translates Γ into the subspace $\{y : Ay = 0\}$ in the space of the new variables y.

As defined in Sect. 4.1, a *convex polyhedron* is the set of feasible solutions K of a system of linear constraints consisting of a finite number of linear equations and inequalities of the form

$$\begin{aligned} Ax &= b, \\ Dx &\geq d, \end{aligned} \tag{4.3}$$

where the inequality constraints include all the bound restrictions on individual variables in the system. Suppose A is of order $m \times n$ and D is of order $p \times n$.

An inequality constraint in (4.3), say the first $D_{1.}x \geq d_1$, is said to be a *redundant inequality constraint*, if its removal from the system (4.3) does not change its set of feasible solutions; that is, iff the minimum value of $D_{1.}x$ subject to the other constraints in (4.3) without this constraint is itself $\geq d_1$.

So, whether a given inequality constraint in the system is redundant or not can be determined by solving a single LP. As an example in the system of three inequality constraints in variables $(x_1, x_2)^T$: $x_1 \geq 4$, $x_2 \geq 6$, $x_1 + x_2 \geq 5$; you can clearly verify that the third constraint $x_1 + x_2 \geq 5$ is redundant.

So, all the redundant inequalities in the system (4.3) can theoretically be identified by solving at most p LPs, and deleted from the system without changing its set of feasible solutions. For the sake of discussion in this section, let us assume that this has been done, but for notational convenience we will continue to denote the system by the same symbols. Under this assumption, all inequality constraints in (4.3) are now nonredundant.

Now an inequality constraint in the system (4.3) – say the first, $D_{1.}x \geq d_1$, – is said to be a *binding [nonbinding] inequality constraint* if β = maximum value of $D_{1.}x$ subject to (4.3); satisfies $\beta = d_1$ [$\beta > d_1$].

As an example, in the system of 3 inequality constraints in variables $(x_1, x_2)^T$; $x_1 \geq 4, x_2 \geq 6, -(x_1 + x_2) \geq -10$, you can clearly verify that all the constraints are binding, that is, they hold as equations at every feasible solution.

If any inequality constraint in system (4.3) is binding, it can be changed into an equality constraint without changing the set of feasible solutions of the system. Thus, theoretically, with an effort of solving at most $2p$ LPs, we can identify and eliminate all redundant inequality constraints, and also identify all binding inequality constraints among the remaining and convert them into equality constraints; all this without changing the set of feasible solutions of the system. After this effort is completed, any remaining inequality constraints in the system are nonbinding. There are two possibilities at this stage:

1. There are no inequality constraints left in the remaining system (i.e., all inequality constraints in the original system are either redundant or binding); in this case the convex polyhedron K is itself an affine space.
2. There are ≥ 1 inequality constraints in the final system, in this case the convex polyhedron K is not an affine space.

Affine spaces are the focus of discussion in linear algebra textbooks (e.g., Murty (2004) of Chap. 1). In the sequel *we will study convex polyhedra that are not affine spaces*, that is, in the system of constraints representing them, there is at least one nonbinding inequality constraint.

In the system of constraints representing K at this stage, among the system of equality constraints there may be redundant equality constraints, and these can be identified by performing GJ (Gauss-Jordan) pivot steps on this system of equations, and eliminated from the system without changing the set of feasible solutions.

The final system of constraints representing K after all these operations are carried out contains no redundant equality constraints and no redundant or binding inequality constraints; it is known as a *minimal representation* for the convex polyhedron K. For notational convenience, we continue to denote the minimal representation of K by (4.3) itself using the same symbols, where A, D are matrices of orders $m \times n$, $p \times n$, respectively.

Then the affine space $\{x : Ax = b\}$ of dimension $n - m$ is the smallest dimensional affine space containing K, and the dimension of K itself is $n - m$. This affine space is a space in which K has full dimension.

In the minimal representation for K, each equality constraint can be used to eliminate one variable from the system. For example, if the equality constraint is $A_{i.}x = a_{i1}x_1 + \cdots + a_{in}x_n = b_i$, and if $a_{i1} \neq 0$, then from this equation we have

$$x_1 = (b_i - a_{i2}x_2 - \cdots - a_{in}x_n)/a_{i1};$$

this expression can be substituted for x_1 in all the remaining constraints and x_1 eliminated from them; the expression for x_1 above can be stored at another location to determine the value of x_1 in the solution from the values of other variables in it determined from the remaining system, and now the equality constraint $A_{i.}x = b_i$ can also be eliminated from the system. In effect, now the remaining system in the remaining variables $(x_2, \ldots, x_n)^T$ is a representation of K in the space of these variables.

The same process can now be repeated on the remaining system with an equality constraint in it, and the process continued the same way. As there are m nonredundant equality constraints at the start of this process, in the end we will be left with a system in $(n - m)$ variables consisting of inequality constraints only. This is a space in which K has full dimension, and its representation in this space consists of inequality constraints only.

Now consider the general LP model: minimize cx subject to a general system of constraints consisting of equations and inequalities. In LP models obtained from practical applications, the system may contain redundant constraints and binding inequality constraints. If an algorithm for LP requires a minimal representation for the set of feasible solutions as the input, that algorithm will not be practical because identifying each redundant or binding inequality constraint in the original LP model by the process discussed above itself requires solving additional LPs. The algorithms that we will discuss for LP are able to solve the general LP model itself directly, using the constraints in that model.

Exercises

4.3.1. The set of feasible solutions of a system of homogeneous linear inequalities $\{x : Ax \geq 0\}$ is called a *convex polyhedral cone*. Consider the convex polyhedron $K = \{x : Ax \geq b\}$. If there exists a point $\bar{x} \in K$ satisfying all the constraints in the representation of K as equations (i.e., all these constraints are active at $\bar{x}$), show that K is a translate of a convex polyhedral cone.

4.4 The Interior and the Boundary of a Convex Polyhedron

Consider a convex polyhedron K in the space in which it has full dimension, say R^n. Then from Sect. 4.3 we know that K can be represented by a system of linear inequalities in $x = (x_1, \ldots, x_n)^T$. In the mathematical study of convex polyhedra, a point $\bar{x} \in K$ is said to be

- A *boundary point* of K iff there is at least one active constraint at $\bar{x}$
- An *interior point* of K if there are no active constraints at $\bar{x}$,

in the system of constraints representing K. The *boundary* of K is the set of all boundary points and its *interior* is the set of all its interior points. The boundary of K is divided into subsets called *faces* of K, properties of which are discussed in detail in the next section. They have a very complex and fascinating mathematical structure that is being studied for over 2,000 years.

Some of the properties of faces of K are exploited in algorithms such as the simplex method for LP. The rest of this chapter is mainly concerned with studying some of the most important properties in the facial structure of K. The active constraints at a boundary point $\bar{x} \in K$ play a major role in determining what type of faces of K contain $\bar{x}$.

4.5 Supporting Hyperplanes, Faces of a Convex Polyhedron, Optimum Face for an LP

4.5.1 Supporting Hyperplanes

Consider a general system of linear constraints in variables $x = (x_1, \dots, x_n)^T \in R^n$ in which the inequality constraints include all the bound constraints on individual variables.

$$\begin{aligned} Dx &= d, \\ Fx &\geq g. \end{aligned} \tag{4.4}$$

Let K denote the set of feasible solutions of (4.4). A *supporting hyperplane* for K is a hyperplane H in R^n satisfying: K is completely contained on one side of H and $H \cap K \neq \emptyset$. Two figures in Fig. 4.7 show examples of supporting hyperplanes (straight lines in R^2) for convex polyhedra in R^2.

In Fig. 4.8, in the figure on the left K is in one side of the hyperplane H, but its intersection with K is $\emptyset$. In the figure on the right the hyperplane H has a nonempty intersection with K, but K is in both sides of H. So, in both the figures in Fig. 4.8, H is not a supporting hyperplane of K.

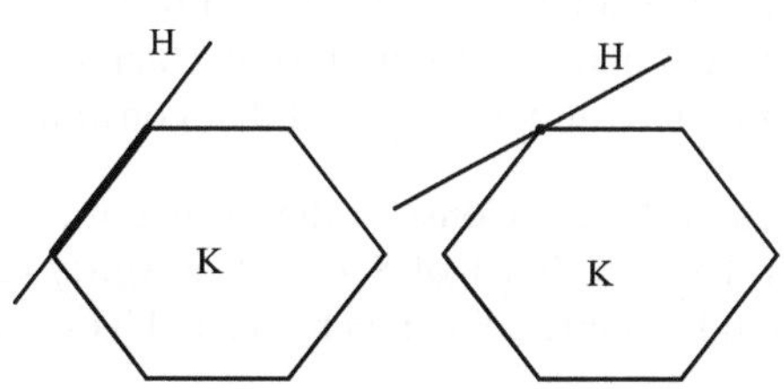

Fig. 4.7 In both the figures here, H is a supporting hyperplane of K

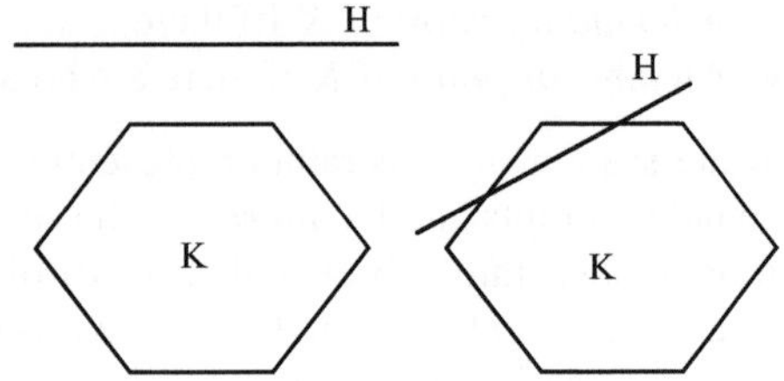

Fig. 4.8 In both these figures, H is not a supporting hyperplane of K

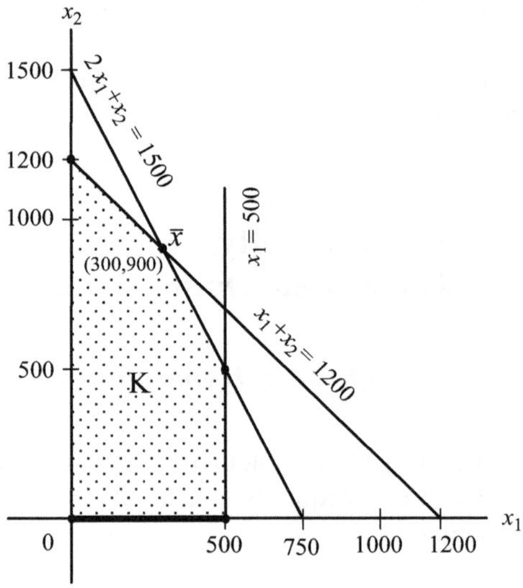

Fig. 4.9 The set of feasible solutions, K, of the fertilizer problem. $\{\bar{x}\}$ is a face of K of dimension 0, and is called an extreme point of K. The thick line segment is a face of K of dimension 1, and is called an edge of K

4.5.2 *Faces of a Convex Polyhedron*

A *face* of a convex polyhedron K is either the empty set, or K itself, or the intersection $H \cap K$ of K with a supporting hyperplane H. Faces of K, other than K and $\emptyset$, are called *proper faces*.

As K is strictly on one side of the supporting hyperplane H, if $H \cap K \neq \emptyset$, it can be seen that $H \cap K$ must be the set of feasible solutions of a system obtained from (4.4) by making a subset of the inequality constraints in it into equations and leaving the others unchanged.

So an alternate definition is: Given the convex polyhedron which is the set of feasible solutions of (4.4), a proper face of it is the set of feasible solutions of a system obtained from (4.4) by changing a subset of inequality constraints in it into equations and leaving the other constraints unchanged.

Example 4.3. Consider the system of constraints in the fertilizer problem discussed in Example 3.4.1 of Sect. 3.4 in Murty (2005b) of Chap. 1 and also in Sect. 1.7.1, and Example 2.5 in this book, and its set of feasible solutions K shown in Fig. 4.9.

$$\begin{aligned} 2x_1 + x_2 &\le 1500, \\ x_1 + x_2 &\le 1200, \\ x_1 &\le 500, \\ x_1, x_2 &\ge 0. \end{aligned}$$

Consider the system obtained by changing the first two inequality constraints into equations:

$$\begin{aligned} 2x_1 + x_2 &= 1500, \\ x_1 + x_2 &= 1200, \\ x_1 &\le 500, \\ x_1, x_2 &\ge 0. \end{aligned}$$

The first two equations in this system have the unique solution $\bar{x} = (300, 900)^T$, which also satisfies the remaining constraints. So, the only solution of this system is $\bar{x}$, and it constitutes a face of K by itself. A single point that by itself is a face of a convex polyhedron is called an *extreme point* of the convex polyhedron. Each extreme point of a convex polyhedron is a face of dimension 0.

Next consider the system obtained by changing only inequality, $x_2 \ge 0$, into an equation and leaving the others unchanged. This system is

$$\begin{aligned} 2x_1 + x_2 &\le 1500, \\ x_1 + x_2 &\le 1200, \\ x_1 &\le 500, \\ x_1 &\ge 0, \\ x_2 &= 0. \end{aligned}$$

Clearly, the set of feasible solutions of this system is the line segment joining the points $0 = (0,0)^T$ and $(500,0)^T$ marked with a thick line in Fig. 4.9. This line segment is a face of K of dimension 1, such a face of a convex polyhedron is called an *edge*. ■

The following theorem explains the role of faces in linear programming.

Theorem 4.1. *The set of optimum solutions of a linear program is a face of its set of feasible solutions.*

Proof. The general LP can always be put in the following form that we consider:

$$\begin{aligned} \text{Minimize } \; z(x) &= cx \\ \text{subject to } \; Dx &= d \\ Fx &\ge g. \end{aligned} \tag{4.5}$$

Let K denote the set of feasible solutions of this LP.

If $K = \emptyset$ or if $K \neq \emptyset$ but (4.5) has no optimum solution (this happens if $z(x)$ is unbounded below on K), then the set of optimum solutions of this LP is $\emptyset$, and by definition this is a face of K. So, the theorem holds in these cases.

If $c = 0$, then every point in K has the same value for $z(x)$, hence the set of optimum solutions of this LP is K itself, and by definition this is a face of K, so the theorem holds in this case also.

So, we assume that $c \neq 0$, and that (4.5) has an optimum solution. Let $\bar{x} \in K$ be an optimum solution of (4.5) and $H = \{x : cx = c\bar{x}\}$. So, $H \cap K \neq \emptyset$ as it contains $\bar{x}$, and it is the set of optimum solutions of (4.5). Also, as $\bar{x}$ minimizes cx on K, we have $cx \geq c\bar{x}$ for all $x \in K$, so K is in the half-space $\{x : cx \geq c\bar{x}\}$, which is one side of the hyperplane H. Therefore, H is a supporting hyperplane for K; and hence $H \cap K$, the set of optimum solutions of (4.5), is a face of K. So, the theorem holds in this case too. □

For this reason, the set of optimum solutions of an LP is usually referred to as its *optimum face*.

In the following sections we will discuss zero- and one-dimensional faces, which are the main tools used by the simplex method to solve an LP.

4.6 Zero-Dimensional Faces, or Extreme Points, or Basic Feasible Solutions (BFSs)

Let K be the convex polyhedron, which is the set of feasible solutions of

$$\begin{aligned} Dx &= d, \\ Fx &\geq g, \end{aligned} \tag{4.6}$$

where any bound restrictions on individual variables are already included among the inequality constraints in (4.6). There are several equivalent definitions for zero-dimensional faces of K, and some of them call them by different names. We will discuss these various definitions now.

Definition 4.1. Geometric Definition: A point $\bar{x} \in K$ is said to be an *extreme point* (also called *corner point*, or *vertex*, or *zero-dimensional face*) of K if it cannot be expressed as a convex combination of two other distinct points in K; that is, if $x^1, x^2 \in K$, and $\bar{x} = \alpha x^1 + (1 - \alpha)x^2$ for some $0 < \alpha < 1$ implies that $x^1 = x^2 = \bar{x}$.

Example 4.4. See the point $\bar{x}$ in the feasible solution set for the fertilizer problem in Fig. 4.9. Verify that it satisfies the property stated in Definition 4.1, and hence is an extreme point of this convex polyhedron. Verify that the other extreme points of this convex polyhedron are $(0, 0)^T$, $(500, 0)^T$, $(0, 1200)^T$, all marked with dots in Fig. 4.9. ■

This geometric definition does not provide an efficient criterion to check whether a given point in a higher-dimensional convex polyhedron is an extreme point. For this we provide an algebraic definition next, in which extreme points are called *basic feasible solutions (BFSs)*.

Definition 4.2. Algebraic definition: BFS for a system of linear constraints: Consider the general system of linear constraints (4.6) and let $\bar{x}$ be a feasible solution for it. Let the *active system* at $\bar{x}$ be the system of active constraints at $\bar{x}$ treated as a system of equations, denote it by (S). $\bar{x}$ is said to be a BFS for (4.6) iff it is the unique solution for this active system at $\bar{x}$ (i.e., for (S)), or equivalently, iff the set of column vectors of variables in this active system is linearly independent.

Definition: Nondegenerate, Degenerate BFSs: When $\bar{x}$ is a BFS of (4.6), it is said to be a *nondegenerate BFS* if the active system (S) at $\bar{x}$ is a square system (i.e., number of equations in it = number of variables in it), and *degenerate* if the number of equations in this active system is > number of variables in it.

Note 4.1. The classification into nondegenerate, degenerate solutions is defined only for basic feasible solutions; it does not apply to feasible solutions that are not BFSs.

Example 4.5. Consider the system of constraints in the fertilizer problem discussed in Example 4.3, which we reproduce below.

$$\begin{aligned} 2x_1 + x_2 &\leq 1500, \\ x_1 + x_2 &\leq 1200, \\ x_1 &\leq 500, \\ x_1, x_2 &\geq 0, \end{aligned}$$

and the feasible solution $\bar{x} = (300, 900)^T$ for it. The first two constraints are the only active constraints at $\bar{x}$, and hence the active system at $\bar{x}$ is

$$\begin{aligned} 2x_1 + x_2 &= 1500, \\ x_1 + x_2 &= 1200. \end{aligned}$$

The set of column vectors of variables in this active system $\{(2,1)^T, (1,1)^T\}$ is linearly independent, and hence $\bar{x}$ is the unique solution for it. So $\bar{x}$ is a BFS for this system. Also, as the active system is a square system, $\bar{x}$ is a nondegenerate BFS for this system.

In the same way, verify that all the solutions $(500, 500)^T$, $(500, 0)^T$, $(0, 0)^T$, $(0, 200)^T$ are all nondegenerate BFSs of this system.

Now consider the feasible solution $\tilde{x} = (200, 1000)^T$ of this system. There is only one active constraint, the second, at $\tilde{x}$. Hence the active system at $\tilde{x}$ is

$$x_1 + x_2 = 1200.$$

This active system has many solutions, so $\tilde{x}$ is not a BFS of this system. ■

Example 4.6. When the system of constraints is in the form

$$Ax = b, \qquad (4.7)$$
$$x \geq 0,$$

where A is an $m \times n$ matrix (in this system all the variables are nonnegative variables, and all other constraints are equality constraints), it is said to be in *standard form*. Here is a numerical example given in detached coefficient tableau form.

x_1	x_2	x_3	x_4	x_5	
1	1	4	12	2	16
0	1	1	3	−4	4

$x_j \geq 0$ for all j

The coefficient matrix for this system A is a 2×5 matrix, and the first two constraints are equations. Consider the feasible solution $x^1 = (12, 4, 0, 0, 0)^T$ for this system. The active system at x^1 is

x_1	x_2	x_3	x_4	x_5	
1	1	4	12	2	16
0	1	1	3	−4	4
0	0	1	0	0	0
0	0	0	1	0	0
0	0	0	0	1	0

As x^1 is a solution of this active system, it is the unique solution iff the column vectors of the variables in this system is a linearly independent set. As the bottom three rows of the coefficient matrix in this system are three different unit vectors, this holds iff the column vectors in the remaining system after striking off these three unit vectors and the columns of the "1" entries in them, that is, $\Gamma = \{(1,0)^T, (1,1)^T\}$, are linearly independent. Notice that Γ is the set of column vectors in A of the original system, of variables x_1, x_2, which are strictly positive in x^1. It is linearly independent. So, x^1 is the unique solution of this active system; hence x^1 is a BFS of this system in standard form, and as the active system is a square system, x^1 is a nondegenerate BFS in this example.

Now, consider the feasible solution $x^2 = (0, 0, 1, 1, 0)^T$ for this system. The positive variables in this system are x_3, x_4. The active system at x^2 is

x_1	x_2	x_3	x_4	x_5	
1	1	4	12	2	16
0	1	1	3	−4	4
1	0	0	0	0	0
0	1	0	0	0	0
0	0	0	0	1	0

Again as the bottom three rows of the coefficient matrix in this system are three different unit vectors, we see that the set of column vectors of variables in this active system is linearly independent iff $\{(4,1)^T,\ (12,3)^T\}$ is. It is not. So, x^2 is not the unique solution for this active system, and hence is not a BFS of this system. ■

We will now show that the two definitions, the geometric definition of an extreme point and the algebraic definition of a BFS, are equivalent.

Theorem 4.2. *Let K be the set of feasible solutions of the general system of linear constraints*

$$\begin{aligned} Dx &= d, \\ Fx &\geq g. \end{aligned} \tag{4.8}$$

Then every extreme point of K by the Geometric Definition 4.1, is a BFS of (4.8) by the Algebraic Definition 4.2, and vice versa.

Proof. Let $\bar{x}$ be a BFS of (4.8), that is, it satisfies the conditions for being a BFS as stated in Definition 4.2. Among the inequality constraints in $Fx \geq g$, put all of them that $\bar{x}$ satisfies as equations together, suppose they form $F_1 x \geq g^1$, and let the remaining ones that $\bar{x}$ satisfies as strict inequalities form $F_2 x \geq g^2$. So $(F_1,\ F_2)$, $(g^1,\ g^2)$ are partitioned forms of $F,\ g$ respectively.

So, the active system at $\bar{x}$ is

$$\begin{aligned} Dx &= d, \\ F_1 x &= g^1, \end{aligned} \tag{4.9}$$

and as $\bar{x}$ is a BFS of (4.8), it is the unique solution for (4.9). Now suppose $\bar{x}$ can be expressed as $\alpha x^1 + (1-\alpha)x^2$, where $x^1,\ x^2 \in K$ and $0 < \alpha < 1$. From this we see that if either of x^1 or x^2 satisfies any of the inequality constraints in $F_1 x \geq g^1$ as a strict inequality, then so does $\bar{x}$, leading to a contradiction. So, both $x^1,\ x^2$ must satisfy each of these constraints as equations, and hence both of them must be feasible to (4.9). But $\bar{x}$ is the unique solution of (4.9), and this implies that $x^1 = x^2 = \bar{x}$; that is, $\bar{x}$ is an extreme point of K, that is, it satisfies the conditions for being an extreme point as stated in Definition 4.1.

Now suppose $\bar{x}$ is an extreme point of K, that is, satisfies the conditions in Definition 4.1. Suppose (4.9) is the active system at $\bar{x}$. Let the remaining constraints in (4.8) that $\bar{x}$ satisfies as strict inequalities form $F_2 x \geq g^2$, so we have $F_2\bar{x} > g^2$. Suppose $\bar{x}$ is not the unique solution for (4.9). As (4.9) is a system of linear equations, if its solution is not unique, the dimension of its set of solutions must be ≥ 1, so there must be a whole straight line through $\bar{x}$ all points on which are feasible to (4.9). Let the parametric representation of this straight line be $\{x : x = \bar{x} + \lambda a : \lambda$ takes all real values$\}$, where $a \neq 0$.

As $\bar{x}$ satisfies $F_2\bar{x} > g^2$, we can find an $\epsilon > 0$ such that $F_2(\bar{x} + \lambda a) \geq g^2$ for all $-\epsilon \leq \lambda \leq \epsilon$. $\bar{x} + \lambda a$ already satisfies (4.9) for all λ real. Therefore, we conclude that both $\bar{x} - \lambda a$ and $\bar{x} + \lambda a$ are feasible to (4.8), and hence contained in

K, and $\bar{x} = [(\bar{x} - \lambda a) + (\bar{x} + \lambda a)]/2$. Also, as $a \neq 0$, $\bar{x} - \lambda a \neq \bar{x} + \lambda a$. Thus we can express $\bar{x}$ as a convex combination of two distinct points of K, contradicting the assumption that $\bar{x}$ is an extreme point. Hence, $\bar{x}$ must be the unique solution of (4.9); that is, it is a BFS of (4.8). □

The standard form system is defined and explained with the numerical example, Example 4.6 above. In the following theorem we will specialize the definition of BFS to a system in standard form.

Theorem 4.3. Simpler characterization of BFS of a system in standard form: *Let $\bar{x} = (\bar{x}_j)$ be a feasible solution for the system of constraints*

$$\begin{aligned} Ax &= b, \\ x &\geq 0 \end{aligned} \tag{4.10}$$

in standard form, where A is a matrix of order $m \times n$. Then $\bar{x}$ is a BFS of (4.10) iff $\{A_{.j} : j \text{ such that } \bar{x}_j > 0\}$ is linearly independent.

Proof. Let r be the number of $\bar{x}_j$ which are >0. Rearrange the variables in (4.10) and in $\bar{x}$, so that we have $\bar{x}_j > 0$ for $j = 1$ to r and $\bar{x}_j = 0$ for $j = r+1$ to n. Let A_1 be the $m \times r$ matrix with column vectors $A_{.1}, \dots, A_{.r}$ and let A_2 be the $m \times (n-r)$ matrix with column vectors $A_{.r+1}, \dots, A_{.n}$. Then $(A_1 \vdots A_2)$ is a partitioned form of A.

As the constraints $x_j \geq 0$ for $j = r+1$ to n are all active at $\bar{x}$, the active system at $\bar{x}$ is the $(m+n-r) \times n$ system

$$\begin{pmatrix} A_1 & \vdots & A_2 \\ \dots & & \dots \\ 0 & \vdots & I \end{pmatrix} x = \begin{pmatrix} b \\ \dots \\ 0 \end{pmatrix}. \tag{4.11}$$

$\bar{x}$ is a BFS of (4.10) iff it is the unique solution of (4.11), which holds iff the set of column vectors in the coefficient matrix in (4.11) is linearly independent; that is, iff the set of column vectors of A_1 is linearly independent. And the set of column vectors of A_1 is the set of columns of A corresponding to positive $\bar{x}_j$. □

4.6.1 Nondegenerate, Degenerate BFSs for Systems in Standard Form

Consider the system (4.10) in standard form, where A is a matrix of order $m \times n$ and rank m. Then from Theorem 4.3, the number of positive variables in any BFS can be at most m.

If $\bar{x}$ is a BFS of (4.10) in which the number of positive variables is m, then the number of active constraints at $\bar{x}$ is $m + (n - m) = n$; therefore, the active system at $\bar{x}$ is a square system, and $\bar{x}$ is a nondegenerate BFS by definition.

If $\bar{x}$ is a BFS of (4.10) in which the number of positive variables is $r < m$, then the number of active constraints at $\bar{x}$ is $m+(n-r) > n$; therefore, $\bar{x}$ is a degenerate BFS by definition.

Hence some LP books mention the following as definitions.

Nondegenerate, Degenerate BFSs of (4.10) in Standard Form: When the coefficient matrix A is of order $m \times n$ and rank m, a feasible solution $\bar{x}$ of (4.10) is said to be:

- A BFS of (4.10) if $\{A_{.j} : j \text{ such that } \bar{x}_j > 0\}$ is linearly independent
- Then it is a nondegenerate BFS if the number of positive $\bar{x}_j = m$, degenerate if this number is $< m$.

4.6.2 *Basic Vectors and Bases for a System in Standard Form*

Consider the system in standard form

$$\begin{aligned} Ax &= b, \\ x &\geq 0, \end{aligned} \tag{4.12}$$

where A is a matrix of order $m \times n$. If (4.12) is feasible and the rank of A is $r < m$, then there are $m - r$ redundant equality constraints in (4.12), and they can be identified through pivot steps and eliminated one by one without affecting the set of feasible solutions of (4.12). So, without any loss of generality, we will assume that A is of full row rank, that is, its rank is m.

The concepts of *basic vector, basis* are linear algebra concepts and apply to systems of linear equations only. So, in (4.12), they apply to the equality constraints $Ax = b$.

A *basic vector* for (4.12) is a vector x_B of m variables in (4.12) satisfying the property that the set of column vectors $A_{.j}$ associated with them is linearly independent. Given a basic vector x_B for (4.12), variables in it are called 1st basic variable, 2nd basic variable, ..., mth basic variable in the order in which they appear in this vector, and the submatrix B of A consisting of column vectors of basic variables in this order, a square submatrix of order m, is known as the *basis* associated with x_B.

Let x_D be the vector of nonbasic variables (i.e., those not in x_B) and D the $m \times (n - m)$ matrix of column vectors $A_{.j}$ associated with them. Then after rearranging the variables, (4.12) can be written in detached coefficient tableau form as

Basic	Nonbasic	
x_B	x_D	RHS
B	D	b

all variables ≥ 0

The *basic solution* of (4.12) wrt the basic vector x_B is obtained by fixing each nonbasic variable at its lower bound 0, and then using the remaining equations to solve for the values of the basic variables to satisfy the equality constraints. It is

$$\begin{aligned} \text{Nonbasic vector} \quad x_D &= 0, \\ \text{Basic vector} \quad x_B &= B^{-1}b. \end{aligned} \tag{4.13}$$

Also, x_B, B are said to be

Feasible basic vector, feasible basis if the basic solution is feasible to (4.12) (i.e., if $B^{-1}b \geq 0$), then this basic solution is called a *basic feasible solution* or *BFS*

Infeasible basic vector, infeasible basis otherwise

Nondegenerate if all the entries in $B^{-1}b$ are nonzero

Degenerate if at least one entry in $B^{-1}b$ is 0

It can be verified that these definitions of nondegenerate, degenerate BFSs agree with the definitions mentioned earlier.

If $\bar{x} = (\bar{x}_j)$ is a nondegenerate BFS for (4.12), there are exactly m positive $\bar{x}_j$ in it, so $\bar{x}$ is the basic solution corresponding to the unique basic vector $x_B = (x_j : j$ such that $\bar{x}_j > 0)$. Thus each nondegenerate BFS of (4.12) corresponds to a unique basic vector defined by the vector of positive variables in it.

On the other hand, if $\bar{x} = (\bar{x}_j)$ is a degenerate BFS of (4.12), the vector $\Gamma = (x_j : j$ such that $\bar{x}_j > 0)$ does not have enough number of variables to constitute a basic vector for (4.12). If Γ has r variables in it, it is necessary to add $m-r$ variables associated with 0-values of $\bar{x}_j$ to Γ (while keeping linear independence of column vectors associated with them) to make it into a basic vector for (4.12). Usually these $(m - r)$ additional variables can be selected in many different ways. That is why a degenerate BFS is usually associated with many different basic vectors for (4.12).

Example 4.7. Consider the following system of constraints in standard form given as a detached coefficient tableau.

x_1	x_2	x_3	x_4	x_5	b
1	1	4	12	2	16
0	1	1	3	−4	4

$x_j \geq 0$ for all j

Here $m = 2$ and rank of the coefficient matrix is 2, and so every basic vector consists of two variables. The vector (x_3, x_4) is not a basic vector for this system, as their set of column vectors $\{(4,1)^T, (12,3)^T\}$ is linearly dependent.

The vector $x_B = (x_1, x_2)^T$ is a basic vector, and the basic solution associated with it $\bar{x}^1 = (12,4,0,0,0)^T$ is feasible, hence $\bar{x}^1$ is a BFS, and as both the basic variables are positive in it, it is a nondegenerate BFS. So, x_B is the unique basic vector corresponding to it.

The solution $\bar{x}^2 = (0,0,4,0,0)^T$ is a degenerate BFS of this system. Verify that (x_1, x_3), (x_2, x_3), (x_3, x_5) are all the different basic vectors of the system associated with it. ■

4.6.3 BFSs for Systems in Standard Form for Bounded Variables

The system of constraints in the standard form for bounded variable LPs is

$$\begin{aligned} Ax &= b, \\ \ell_j \le x_j \le u_j, &\quad j = 1,\dots,n, \end{aligned} \tag{4.14}$$

where A is an $m \times n$ matrix of rank m, and $\ell = (\ell_j) < u = (u_j)$. Some of the ℓ_j may be $-\infty$ or 0, some of the u_j may be $+\infty$.

Let $\bar{x} = (\bar{x}_j)$ be a feasible solution of (4.14). For some j, $\bar{x}_j$ may be ℓ_j (the lower bound on x_j, if it is finite), or u_j (the upper bound on x_j, if it is finite), or satisfy $\ell_j < \bar{x}_j < u_j$.

For convenience in notation let us assume that the variables are rearranged so that

$$\begin{aligned} \bar{x}_j &= \ell_j \quad \text{for } j = 1 \text{ to } r, \\ \ell_j < \bar{x}_j &< u_j \quad \text{for } j = r+1 \text{ to } r+s, \\ \bar{x}_j &= u_j \quad \text{for } j = r+s+1 \text{ to } n. \end{aligned}$$

Let A_1 be the $m \times r$ submatrix of A consisting of its column vectors $A_{.1}$ to $A_{.r}$; A_2 the $m \times s$ submatrix of A consisting of its column vectors $A_{.r+1}$ to $A_{.r+s}$; and A_3 the $m \times (n-r-s)$ submatrix of A consisting of its column vectors $A_{.r+s+1}$ to $A_{.n}$. As the bound constraints $x_j \ge \ell_j$ are active for $j = 1$ to r and the bound constraints $x_j \le u_j$ are active for $j = r+s+1$ to n, the active system at $\bar{x}$ in partitioned form is

$$\begin{pmatrix} A_1 & \vdots & A_2 & \vdots & A_3 \\ \cdots & & \cdots & & \cdots \\ I & \vdots & 0 & \vdots & 0 \\ \cdots & & \cdots & & \cdots \\ 0 & \vdots & 0 & \vdots & I \end{pmatrix} x = \begin{pmatrix} b \\ \cdots \\ \ell' \\ \cdots \\ u' \end{pmatrix},$$

where the I are unit matrices of appropriate orders, and $\ell' = (\ell_1, \dots, \ell_r)^T$, $u' = (u_{r+s+1}, \dots, u_n)^T$. Clearly, the set of column vectors of the coefficient matrix of this system is linearly independent iff the set of column vectors in the matrix A_2 is linearly independent. This leads to the following result.

Result 4.1. *A feasible solution* $\bar{x} = (\bar{x}_j)$ *for the bounded variable system (4.14) is a BFS iff* $\{A_{.j} : j \text{ such that } \ell_j < \bar{x}_j < u_j\}$ *is linearly independent.* □

Also, if the number of j such that $\ell_j < \bar{x}_j < u_j$ is less than m, the number of active bound constraints will exceed $n - m$, and the number of constraints in the active system at $\bar{x}$ will exceed n. This leads to the following result.

Result 4.2. *If $\bar{x}$ is a BFS of (4.14), it is a nondegenerate BFS iff the number of j such that $\ell_j < \bar{x}_j < u_j$ is equal to m. If this number is $< m$, $\bar{x}$ is a degenerate BFS of (4.14).* □

Example 4.8. Consider the following system with coefficient matrix A of order 3×7.

x_1	x_2	x_3	x_4	x_5	x_6	x_7	b
1	−1	1	0	1	2	4	68
−1	1	1	1	2	4	2	77
1	2	1	1	2	2	−2	23

$2 \leq x_j \leq 10$ for all j

Consider the feasible solution $\bar{x} = (2, 2, 3, 4, 5, 10, 10)^T$ to this system. $2 < \bar{x}_j < 10$ holds for $j = 3, 4, 5$. However, the corresponding set of column vectors of the coefficient matrix $\{A_{.3},\ A_{.4},\ A_{.5}\} = \{(1,1,1)^T,\ (0,1,1)^T,\ (1,2,2)^T\}$ is linearly dependent, so $\bar{x}$ is not a BFS.

Now consider the feasible solution $\tilde{x} = (2, 2, 6, 7, 2, 10, 10)^T$. $2 < \tilde{x}_j < 10$ holds for $j = 3, 4$, and the corresponding set of column vectors $\{A_{.3},\ A_{.4}\}$ is linearly independent. So $\tilde{x}$ is a BFS for this system. But as the number of $\tilde{x}_j$ strictly between the bounds is $2 < m$ is 3, $\tilde{x}$ is a degenerate BFS for this system. ■

4.7 Purification Routine for Deriving a BFSs from a Feasible Solution for Systems in Standard Form

The Importance of the Purification Routine: It will be shown later that if an LP with its constraints in standard form has an optimum solution, then it also has an optimum solution which is a BFS for it. The first successful method for solving LPs on a commercial scale is one of the variants of the simplex method, and it always obtains an optimum BFS when optimum solutions exist. Variants of the simplex method have remained as the only commercially viable algorithms for solving LPs for a very long time, and in this time, practitioners have learnt the subtle advantages of optimum BFSs for practical implementation over other types of optimum solutions when they exist.

Recently, newer methods called *interior point methods* have emerged as reasonable competitors of the simplex method, particularly for large-scale sparse LP models. When alternate optimum solutions exist, these interior point methods do not normally converge to an optimum BFS. At that time the purification routine, which can obtain a BFS from a feasible solution, and an optimum BFS from an optimum feasible solution, has become very prominent. So, good knowledge of this routine, and how it works, is very important for anyone studying LP today.

Also, the main technique used in the purification routine is a fundamental technique used in proving many important theoretical results on convex polyhedra. Therefore, mathematically, the purification routine is a very important and useful technique.

What is the reason for the name of this procedure? Purification routine is the name of the procedure for finding a BFS from a feasible solution for a system of linear constraints. In this section we discuss the version of it used for systems in standard form. The name comes from a chemical analogy. Usually, each substance in nature is a mixture of various individual compounds. In chemistry, "purification" is the name used for a process that separates a specific compound from such a mixture.

By the results stated in Sect. 4.13, it can be shown that if the set of feasible solutions of the system is bounded, then every feasible solution $\bar{x}$, is a convex combination of basic feasible solutions (or extreme point solutions) of the system. So, from the chemical analogy, we can view $\bar{x}$ as a "mixture" of these BFSs, and hence call the procedure for isolating one BFS from this mixture as "purification routine."

4.7.1 The Main Strategy of the Purification Routine

Let the LP under consideration be

$$\begin{aligned} \text{Minimize} \quad & cx \\ \text{subject to} \quad & Ax = b \\ & x \geq 0, \end{aligned} \tag{4.15}$$

where A is a matrix of order $m \times n$ and rank m. Suppose we are starting with a feasible solution $\bar{x} = (\bar{x}_j)$ for (4.15), which has r variables > 0 in it, let

$$J = \{j : \bar{x}_j > 0\}.$$

The method checks the set of column vectors $\{A_{.j} : j \in J\}$ for linear independence. If it is linearly independent, $\bar{x}$ itself is a BFS, the method terminates.

Otherwise, $\bar{x}$ is not a BFS. The number of inactive constraints in (4.15) at the feasible solution $\bar{x}$ is r (because $\bar{x}$ satisfies the constraint $x_j \geq 0$ for each $j \in J$ as a strict inequality), and so the system of active constraints in (4.15) at $\bar{x}$ is $m+(n-r)$; these active constraints define the active system at $\bar{x}$ to be

$$\begin{aligned} Ax &= b, \\ x_j &= 0. \text{ for } j \notin J \end{aligned} \tag{4.16}$$

As $\bar{x}$ is not a BFS, this system has alternate solutions, and so to move from this active system corresponding to $\bar{x}$ into another corresponding to a different feasible solution, say $\hat{x}$, which will have a unique solution, we need to make sure that the active system at $\hat{x}$ has additional equality constraints than those in (4.16); that is, besides the variables x_j for $j \in \{1,\dots,n\}\backslash J$, $\hat{x}$ must make some of the variables x_j for $j \in J$ equal to 0 also.

Another way of looking at this is the following: $\bar{x}$ is not a BFS because the set $\{A_{.j} : j \text{ such that } \bar{x}_j > 0\}$ is linearly dependent. So, to move from $\bar{x}$ into another feasible solution $\hat{x}$, which is closer to being a BFS than $\bar{x}$, we should make sure that the set $\{A_{.j} : j \text{ such that } \hat{x}_j > 0\}$ is a proper subset of $\{A_{.j} : j \text{ such that } \bar{x}_j > 0\}$; that is, while keeping all the variables x_j for $j \in \{1,\dots,n\}\backslash J$ at 0, $\hat{x}$ must make one or more variables x_j for $j \in J$ equal to 0 also.

That is why each step in the purification procedure moves from the current feasible solution into another at which the set $\{j : x_j > 0 \text{ in the solution}\}$ becomes a strictly smaller set.

When we check whether the set $\{A_{.j} : j \text{ such that } \bar{x}_j > 0\}$ is linearly independent; if it is not, that procedure will produce a linear dependence relation for this set of vectors; that is, a vector $(\alpha_j : j \in J) \neq 0$ such that $\sum_{j \in J} \alpha_j A_{.j} = 0$. Using this linear dependence relation, the purification routine produces a new feasible solution $\tilde{x}$ in which the number of variables with positive values is $\leq (r-1)$ and the whole process is repeated with $\tilde{x}$.

So, the routine has to terminate with a BFS after no more than r of these steps. Next we provide the details of the routine.

Let K denote the set of feasible solutions of (4.15). Geometrically, a step of the purification routine applied on a point $\bar{x}$ contained in a face F of dimension $d > 0$ of K leads to a point $\tilde{x}$ lying in a face of dimension $\leq d-1$ of K (and also of F).

4.7.2 General Step in the Purification Routine

Let $\bar{x} = (\bar{x}_j)$ be the current feasible solution, and $J = \{j : \bar{x}_j > 0\}$. Suppose $|J| = r$.

Check $\{A_{.j} : j \in J\}$ for linear independence. Any of the available algorithms for this can be used, but if this set is not linearly independent, the algorithm must produce a linear dependence relation for the set. See Murty (2004) of Chap. 1 for an algorithm, and this algorithm is illustrated in Example 4.10 given below.

If $\{A_{.j} : j \in J\}$ is linear independent, $\bar{x}$ is a BFS, terminate. Otherwise, let

$$\sum_{j \in J} \alpha_j A_{.j} = 0 \tag{4.17}$$

be the linear dependence relation obtained, where $(\alpha_j : j \in J) \neq 0$. As $\bar{x} = (\bar{x}_j)$ is feasible to (4.15) and $\bar{x}_j = 0$ for all $j \notin J$, we have

$$\sum_{j \in J} \bar{x}_j A_{.j} = b. \tag{4.18}$$

Multiplying (4.17) by a parameter λ and adding to (4.18) we get

$$\sum_{j \in J} (\bar{x}_j + \lambda \alpha_j) A_{.j} = b. \tag{4.19}$$

So, if we define the vector $x(\lambda) = (x_j(\lambda))$, where $x_j(\lambda) = \bar{x}_j + \lambda \alpha_j$ for $j \in J$, 0 otherwise, then $x(\lambda)$ satisfies the system of equations $Ax = b$. We know that $\bar{x}_j > 0$ for all $j \in J$. But in the linear dependence relation (4.17), α_j may have any sign. For $x_j(\lambda) = \bar{x}_j + \lambda \alpha_j$ to be ≥ 0

if $\alpha_j < 0$ we need $\lambda \leq -\bar{x}_j/\alpha_j$,
if $\alpha_j > 0$ we need $\lambda \geq -\bar{x}_j/\alpha_j$.
So, for $x(\lambda)$ to satisfy the nonnegativity constraints

λ must be $\leq -\bar{x}_j/\alpha_j$ for all j such that $\alpha_j < 0$, i.e., λ must be $\leq$ minimum$\{-\bar{x}_j/\alpha_j : j$ such that $\alpha_j < 0\}$.
λ must be $\geq -\bar{x}_j/\alpha_j$ for all j such that $\alpha_j > 0$, i.e., λ must be $\geq$ maximum$\{-\bar{x}_j/\alpha_j : j$ such that $\alpha_j > 0\}$.

We will now consider three separate cases.

Case 4.1. The vector $(\alpha_j : j \in J)$ has both positive and negative quantities.

In this case define

$$\theta_1 = \max\{-\bar{x}_j/\alpha_j : j \text{ such that } \alpha_j > 0\},$$
$$\theta_2 = \min\{-\bar{x}_j/\alpha_j : j \text{ such that } \alpha_j < 0\},$$

and verify that both θ_1, θ_2 are finite, $\theta_1 < 0$ and $\theta_2 > 0$, so we have $\theta_1 < \theta_2$, and for all $\theta_1 \leq \lambda \leq \theta_2$, $x(\lambda)$ will be feasible to (4.15). Also, verify that

$$\bar{x} = [(-\theta_1)/(\theta_2 - \theta_1)]x(\theta_2) + [(\theta_2)/(\theta_2 - \theta_1)]x(\theta_1).$$

So, if $\alpha = (-\theta_1)/(\theta_2 - \theta_1)$, then it can be verified that $0 < \alpha < 1$, and that the above expression becomes $\bar{x} = \alpha x(\theta_1) + (1 - \alpha)x(\theta_2)$, that is, it expresses the initial feasible solution $\bar{x}$ as a convex combination of two feasible solutions $x(\theta_1)$, $x(\theta_2)$. Also, for any $t = 1$ or 2, it can be verified that $x_j(\theta_t) = 0$ for the index j that ties for the maximum (or minimum) in the definition of θ_t above. Therefore, each of $x(\theta_1)$, $x(\theta_2)$ is a feasible solution of (4.15) in which the number of positive variables is $\leq (r - 1)$. So, we have the following result in this case.

Result 4.3. *From this one step in the purification procedure, in this case when the coefficient vector obtained for the linear dependence relation has components of both signs, we get an expression for the feasible solution $\bar{x}$ with r positive variables in it at the beginning of this step, as a convex combination of two distinct feasible solutions, in each of which the number of positive variables is $\leq (r - 1)$.* □

Case 4.2. The vector $(\alpha_j : j \in J) \geq 0$, but not 0.

In this case define θ_1 as under Case 1, verify that it is finite and <0. Here, $x(\lambda)$ will be feasible to (4.15) for all $\lambda \geq \theta_1$.

Define the vector $y = (y_j) \in R^n$ by $y_j = 0$ for $j \notin J$ and $y_j = \alpha_j$ for $j \in J$. So, $y \geq 0$, and as $\sum_{j \in J} \alpha_j A_{.j} = 0$, we have $Ay = 0$. So, y satisfies $Ay = 0,\ y \geq 0$; such vectors y are called *homogeneous solutions corresponding to the original system* (4.15).

The homogeneous system corresponding to (4.15) is obtained by changing the variable names in (4.15) from x to a different symbol like y (mainly to keep the distinction between the two systems), and then changing all the RHS constants to 0. So, this homogeneous system corresponding to (4.15) is $Ay = 0,\ y \geq 0$. Feasible solutions of the homogeneous system are called homogeneous solutions corresponding to the original system (4.15). See Sect. 4.13 for a full discussion.

In this case, verify that $\bar{x} = x(\theta_1) + (-\theta_1)y$, and that the number of positive variables in $x(\theta_1)$ is $\leq (r - 1)$.

Case 4.3. The vector $(\alpha_j : j \in J) \leq 0$, but not 0.

In this case define θ_2 as under Case 1, and verify that it is finite and >0. Here, $x(\lambda)$ will be feasible to (4.15) for all $\lambda \leq \theta_2$.

Define the vector $y = (y_j) \in R^n$ by $y_j = 0$ for $j \notin J$ and $y_j = -\alpha_j$ for $j \in J$. So, $y \geq 0$, and as $\sum_{j \in J} \alpha_j A_{.j} = 0$, we have $Ay = 0$. So, here this y is a homogeneous solutions corresponding to the original system (4.15).

In this case, verify that $\bar{x} = x(\theta_2) + (\theta_2)y$, and that the number of positive variables in $x(\theta_2)$ is $\leq (r - 1)$. From Cases 2 and 3, we have the following result.

Result 4.4. *From this one step in the purification procedure, in these cases when the coefficient vector obtained for the linear dependence relation has all components of the same sign, we get an expression for the feasible solution $\bar{x}$ with r positive variables in it at the beginning of this step, as the sum of a feasible solution in which the number of positive variables is $\leq (r - 1)$, and a positive multiple of a homogeneous solution corresponding to the original system.* □

As $(\alpha_j : j \in J) \neq 0$, at least one among θ_1, θ_2 given above is finite. If θ_1 $[\theta_2]$ is finite, verify that $x_j(\theta_1)$ $[x_j(\theta_2)]$ will be 0 for the j that attains the maximum [minimum] in the definition of θ_1 $[\theta_2]$. So, define $\bar{\lambda}$ to be one among θ_1 or θ_2, whichever is finite, then $x(\bar{\lambda})$ is a feasible solution of (4.15) in which the number of positive variables is $\leq (r - 1)$.

The purification routine repeats this general step with $x(\bar{\lambda})$ as the new feasible solution.

As the number of positive variables in the current feasible solution decreases by one or more in each step, the routine should terminate in at most r (r is the number of positive variables in the initial feasible solution) steps with a BFS.

Note 4.2. Results 4.3 and 4.4 are very useful to prove fundamental results known as *resolutions theorems for convex polyhedra*; that is why we derived their complete statements here. These resolution theorems are given as Exercises 4.4 and 4.5; they can be proved easily by mathematical induction using a purification step and Results 4.3 and 4.4.

Example 4.9. **Computing the bounds θ_1, θ_2 in purification:** Here we will illustrate the computation of the bounds for λ to keep $x(\lambda)$ nonnegative in purification. Three separate cases are discussed corresponding to the signs of the coefficients of λ in $x(\lambda)$.

Case 4.4. Suppose $x(\lambda) = (6, 27, 40, 16, 21, 56, 65, 80)^T + \lambda(-2, -3, -5, 0, 3, 7, 5, 10)^T$. Here the coefficients of λ in $x(\lambda)$ take both positive and negative values.

For $x_1(\lambda) = 6 - 2\lambda$ to be ≥ 0, we need $\lambda \leq 6/2$. In the same way, to keep all of $6 - 2\lambda$, $27 - 3\lambda$, $40 - 5\lambda \geq 0$, the parameter λ should be $\leq \theta_2 = \min\{6/2, 27/3, 40/5\} = 3$.

For $x_5(\lambda) = 21 + 3\lambda$ to be ≥ 0, we need $\lambda \geq -21/3$. In the same way, to keep all of $21 + 3\lambda$, $56 + 7\lambda$, $65 + 5\lambda$, $80 + 10\lambda \geq 0$, the parameter λ should be $\geq \theta_1 = \max\{-21/3, -56/7, -65/5, -80/10\} = -7$.

So, $x(\lambda) \geq 0$ for all $-7 \leq \lambda \leq 3$. Also notice that while $x_j(0) > 0$, $x_j(-7) = 0$ for $j = 5$, the index that attains the max in the definition of θ_1. Similarly, $x_j(3) = 0$ for $j = 1$, the index that attains the min in the definition of θ_2. Thus the number of positive components in each of $x(\theta_1)$, $x(\theta_2)$ is $\leq -1 +$ (the corresponding number in $x(0)$). Also, verify that in this case the initial feasible solution $x(0) = 0.7x(3) + 0.3x(-7)$ verifying Result 4.3.

Case 4.5. Suppose $x(\lambda) = (20, 36, 50, 34, 6, 8)^T + \lambda(-2, -6, -5, -2, 0, 0)^T$. As in Case 4.4, because all the coefficients of λ are ≤ 0, we see that $x(\lambda) \geq 0$ if $\lambda \leq \min\{20/2, 36/6, 50/5, 34/2\} = \theta_2 = 6$. Also verify that $x(6)$ has one less positive component than $x(0)$.

Also, in this case the homogeneous solution defined in Result 4.2 is $y = (2, 6, 5, 2, 0, 0)^T$ and the initial feasible solution $x(0) = x(6) + 6y$.

Case 4.6. Suppose $x(\lambda) = (11, 13, 30, 40, 6, 18)^T + \lambda(0, 0, 10, 20, 1, 9)^T$. As in Case 4.4, because all the coefficients of λ are ≥ 0, we see that $x(\lambda) \geq 0$ if $\lambda \geq \max\{-30/10, -40/20, -6/1, -18/9\} = \theta_1 = -2$. Also verify that $x(-2)$ has two less positive component than $x(0)$.

Also, in this case the homogeneous solution defined in Result 4.4 is $y = (0, 0, 10, 20, 1, 9)^T$ and the initial feasible solution $x(0) = x(-2) + 2y$. ■

Example 4.10. Here we give an illustration of the purification routine. Consider the feasible solution $\bar{x} = (1, 2, 10, 3, 0, 0)^T$ of the following system in standard form:

x_1	x_2	x_3	x_4	x_5	x_6	b
1	0	1	−3	−6	12	2
−1	2	1	−1	17	18	10
0	1	1	−2	8	−5	6
0	1	1	−2	−9	−6	6

$x_j \geq 0$ for all j

Let A denote the 4×6 coefficient matrix for this system. The set of indices of positive components in $\bar{x}$ is $J = \{1, 2, 3, 4\}$.

We need to check if the set of column vectors $\{A_{.1},\ A_{.2},\ A_{.3},\ A_{.4}\}$ is linearly independent. We will use the method discussed in Murty (2004) of Chap. 1 for this.

This method can be applied to check the linear independence of either a set of row vectors or a set of column vectors. However, as the method is based on row operations, it requires that we write each vector in the set as a row vector (even if the original vectors are column vectors) in a tableau. Then put a memory matrix (a full unit matrix) on the left side of this tableau, and label the columns of the memory matrix with the labels for the elements in the set in the same order in which they appear as rows in the tableau. Here is the initial tableau with the memory matrix on the left.

$A_{.1}$	$A_{.2}$	$A_{.3}$	$A_{.4}$					
				PC				
1	0	0	0	**[1]**	−1	0	0	PR
0	1	0	0	0	2	1	1	
0	0	1	0	1	1	1	1	
0	0	0	1	−3	−1	−2	−2	

Now perform a GJ pivot step in each row of the tableau one after the other. If all these pivot steps are carried out, the original set of vectors is linearly independent, terminate.

If the pivot step in a row cannot be carried out because there is no nonzero entry in that row, the vector in that row in the memory matrix at that time gives the coefficients in a linear dependence relation for the set. On our tableau, we will perform the pivot steps in top to bottom order, so we select row 1 as the pivot row (PR) and the first column in the tableau as the pivot column (PC). PR and PC indicate the pivot row and pivot column, respectively, for the pivot step performed in a tableau, and the pivot element will be boxed. Continuing, we get the following series of tableaus.

Memory matrix								
						PC		
1	0	0	0	1	−1	0	0	
0	1	0	0	0	2	**[1]**	1	PR
−1	0	1	0	0	2	1	1	
3	0	0	1	0	−4	−2	−2	
1	0	0	0	1	−1	0	0	
0	1	0	0	0	2	1	1	
−1	−1	1	0	0	0	0	0	
3	2	0	1	0	0	0	0	

The updated third row in the tableau has no nonzero element, so we conclude that the set $\{A_{.1},\ A_{.2},\ A_{.3},\ A_{.4}\}$ is linearly dependent and $(-1, -1, 1, 0)$ is the vector of coefficients in a linear dependence relation. So the linear dependence relation is

$-A_{.1} - A_{.2} + A_{.3} = 0$. And from the feasibility of $\bar{x}$ we have $A_{.1} + 2A_{.2} + 10A_{.3} + 3A_{.4} = b$. Using these two equations, we get

$$(1-\lambda)A_{.1} + (2-\lambda)A_{.2} + (10+\lambda)A_{.3} + 3A_{.4} = b,$$

where λ is a parameter. This shows that $x(\lambda) = (1-\lambda,\ 2-\lambda,\ 10+\lambda,\ 3,\ 0,\ 0)^T$ satisfies the equality constraints in the original system. Computing the bounds for the parameter λ to keep $x(\lambda) \geq 0$, we get $\theta_1 = \max\{-10/1\} = -10$, $\theta_2 = \min\{1/1,\ 2/1\} = 1$. So, $x(\lambda) \geq 0$, and hence feasible for the original system, for all $-10 \leq \lambda \leq 1$.

Fixing λ at one of the bounds, $\theta_2 = 1$ say, we get the new feasible solution $\tilde{x} = x(1) = (0, 1, 11, 3, 0, 0)^T$ for the original system with the new index set of positive variables to be $\{2, 3, 4\}$.

Verify that the set of indices of variables positive at $\tilde{x}$ is $\{2, 3, 4\}$, which is a proper subset of the corresponding set $\{1, 2, 3, 4\}$ for the feasible solution $\bar{x}$ we started with.

So, now we need to check the set $\{A_{.2},\ A_{.3},\ A_{.4}\}$ for linear independence, to check whether $\tilde{x}$ is a BFS. For this we use the same method as before, and get the following series of tableaus.

Memory matrix							
$A_{.2}$	$A_{.3}$	$A_{.4}$					
					PC		
1	0	0	0	2	[1]	1	PR
0	1	0	1	1	1	1	
0	0	1	−3	−1	−2	−2	
			PC				
1	0	0	0	2	1	1	
−1	1	0	[1]	−1	0	0	PR
2	0	1	−3	3	0	0	
1	0	0	0	2	1	1	
−1	1	0	1	−1	0	0	
−1	3	1	0	0	0	0	

As the updated row 3 has no nonzero entry, we know that $\{A_{.2},\ A_{.3},\ A_{.4}\}$ is linearly dependent, hence $\tilde{x}$ is not a BFS too. The linear dependence relation here is $-A_{.2} + 3A_{.3} + A_{.4} = 0$. Also, as $\tilde{x}$ is feasible, we have $A_{.2} + 11A_{.3} + 3A_{.4} = b$. Combining these two equations we get

$$(1-\lambda)A_{.2} + (11+3\lambda)A_{.3} + (3+\lambda)A_{.4} = b,$$

where λ is a parameter, that is, $x(\lambda) = (0,\ 1-\lambda,\ 11+3\lambda,\ 3+\lambda,\ 0,\ 0)^T$ satisfies the system of equality constraints in the original system. Computing the bounds, we see that $x(\lambda) \geq 0$ and hence feasible for the original system for all $-3 \leq \lambda \leq 1$.

Fixing λ at one of the bounds, -3 say, we get the new feasible solution $\hat{x} = x(-3) = (0,4,2,0,0,0)^T$ for the original system with the new index set of positive variables to be $\{2,3\}$. It can be verified that the set $\{A_{.2},\ A_{.3}\}$ is linearly independent. So, $\hat{x}$ is a BFS of the original system. As the number of positive components in it is $2 < m = 4 =$ number of equality constraints in the system, and as the coefficient matrix of this system can be verified to be of full row rank, this is a degenerate BFS. ■

Summary

We discussed the purification routine for obtaining a BFS from a feasible solution to a system in standard form. The method can easily be adopted to systems in other forms, using the definition of BFSs for a general system given in Sect. 4.5. We will give an example to illustrate this point.

4.7.3 Purification Routine for Systems in Symmetric Form

$$\begin{aligned} Ax &\geq b, \\ x &\geq 0. \end{aligned} \tag{4.20}$$

Adopting the purification routine to the system (4.20) in which all constraints are inequalities, and all the variables are nonnegative variables (known as the system in *symmetric form*, the reason for this name is explained in the duality Chap. 5) is particularly easy. Suppose A is of order $m \times n$. Introduce slack variables and convert all the inequality constraints into equations, obtaining the following system in standard form.

$$\begin{aligned} Ax - Is &= b, \\ x, s &\geq 0, \end{aligned} \tag{4.21}$$

where I is the unit matrix of order m and $s = (s_1, \dots, s_m)^T$ is the slack vector. If $\bar{x}$ is a feasible solution for (4.20), then $(\bar{x}, \bar{s} = A\bar{x} - b)$ is a feasible solution for (4.21) and vice versa. Also from the definition for the BFS of a general system given in Sect. 4.6, it can be verified that $\bar{x}$ is a BFS for (4.20) iff $(\bar{x},\ \bar{s} = A\bar{x} - b)$ is a BFS for (4.21) and vice versa. And $\bar{x}$ is a nondegenerate (degenerate) BFS for (4.20) iff $(\bar{x},\ \bar{s} = A\bar{x} - b)$ is a nondegenerate (degenerate) BFS for (4.21).

If $\bar{x}$ is a feasible solution to (4.20), which is not a BFS, apply the purification routine described above for systems in standard form to obtain a BFS for (4.21) beginning with the feasible solution $(\bar{x},\ \bar{s} = A\bar{x} - b)$ for it. If $(\hat{x},\ \hat{s} = A\hat{x} - b)$ is the BFS for (4.21) that has been obtained in this routine, then $\hat{x}$ is a BFS for (4.20).

Example 4.11. Consider the feasible solution $\bar{x} = (\bar{x}_1, \bar{x}_2)^T = (1/3, 4/3)^T$ for the system (4.22) in variables x_1, x_2, which is in symmetric and not standard form. We will check whether it is a BFS, and if not, obtain a BFS from it. After the original system, we give the system in standard form obtained by introducing the slack variables x_3, x_4, x_5.

$$\begin{aligned} x_1 + 2x_2 &\geq 2, \\ 2x_1 + x_2 &\geq 2, \\ -x_1 + x_2 &\leq 2, \\ x_1, x_2 &\geq 0. \end{aligned} \tag{4.22}$$

$$\begin{aligned} x_1 + 2x_2 - x_3 &= 2, \\ 2x_1 + x_2 - x_4 &= 2, \\ -x_1 + x_2 + x_5 &= 2, \\ x_1, \ldots, x_5 &\geq 0. \end{aligned} \tag{4.23}$$

To avoid confusion, we will denote solutions of (4.23) by $X = (x_1, x_2, x_3, x_4, x_5)^T$. As $x_3 = x_1 + 2x_2 - 2$, etc., the feasible solution $\bar{x} = (1/3, 4/3)^T$ for (4.22) corresponds to $\bar{X} = (1/3, 4/3, 1, 0, 1)^T$. The index set of positive components in $\bar{X}$ is $\{1, 2, 3, 5\}$, and the set of column vectors in (4.23) of corresponding variables is $\{(1, 2, -1)^T, (2, 1, 1)^T, (-1, 0, 0)^T, (0, 0, 1)^T\} = \{A_{.1}, A_{.2}, A_{.3}, A_{.5}\}$ say. It can be verified that a linear dependence relation for this set is

$$A_{.1} - 2A_{.2} - 3A_{.3} + 3A_{.5} = 0.$$

From the feasible solution $\bar{X}$ we get

$$(1/3)A_{.1} + (4/3)A_{.2} + A_{.3} + A_{.5} = b,$$

where b denotes the RHS vector in (4.23). Combining these equations we get

$$(\lambda + (1/3))A_{.1} + (-2\lambda + (4/3))A_{.2} + (-3\lambda + 1)A_{.3} + (3\lambda + 1)A_{.5} = b$$

So, the solution $X(\lambda) = ((\lambda + (1/3)), (-2\lambda + (4/3)), (-3\lambda + 1), 0, 3\lambda + 1)^T$ satisfies the equality constraints in (4.23). It is also nonnegative if $\theta_1 = (-1/3) \leq \lambda \leq (1/3) = \theta_2$.

By putting $\lambda = (-1/3)$, we get $X(-1/3) = X^1 = (0, 2, 2, 0, 0)^T$, which is a BFS of (4.23), and this corresponds to the BFS $x^1 = (0, 2)^T$ of (4.22).

By fixing $\lambda = (1/3)$, we get $X(1/3) = X^2 = (2/3, 2/3, 0, 0, 2)^T$, which is a BFS of (4.23), and this corresponds to the BFS $x^2 = (2/3, 2/3)^T$ of (4.22).

In Fig. 4.10, we show the feasible solution for (4.22) and the various points obtained during this routine. The initial feasible solution $\bar{x}$ is not a BFS (i.e., extreme

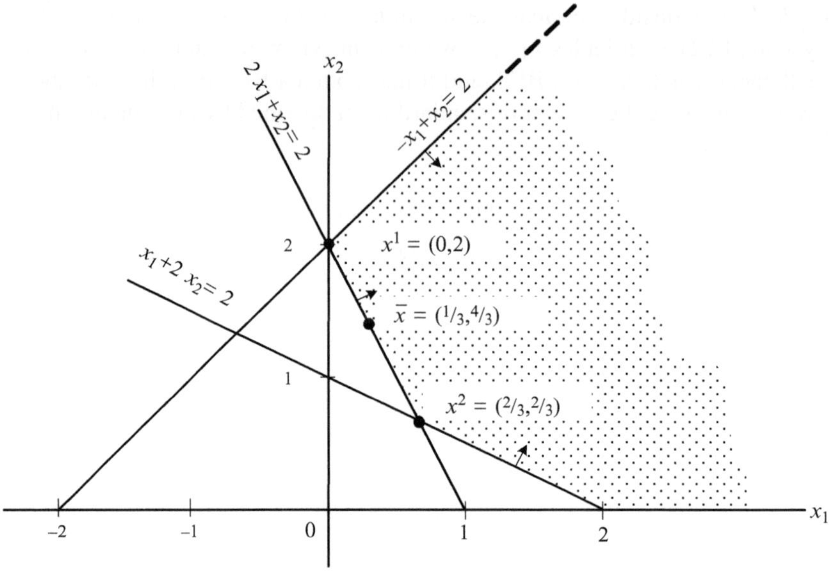

Fig. 4.10 Part of the feasible region is shaded

point solution). After one application of the purification routine, we move to one of the BFSs x^1 or x^2 depending on whether the parameter λ is fixed at its lower bound θ_1 or upper bound θ_2. ■

We will now derive an important property of the purification routine for LP, and use it to derive some fundamental results for LP in standard form.

Theorem 4.4. *Consider the LP in standard form*

$$\begin{aligned} \text{Minimize} \quad z(x) &= cx \\ Ax &= b \\ x &\geq 0, \end{aligned} \tag{4.24}$$

where A is a matrix of order $m \times n$. Let $\bar{x}$ be an optimum solution for this LP, which is not a BFS. Then all solutions obtained in the purification routine beginning with $\bar{x}$ to obtain a BFS for (4.24) will also be optimum solutions of (4.24).

Proof. Suppose $\bar{x}$ has r positive components. For notational convenience, assume that the variables in (4.24) have been rearranged so that $\bar{x}_1, \dots, \bar{x}_r$ are all >0 and $\bar{x}_{r+1}, \dots, \bar{x}_n$ are all $= 0$. So

$$\bar{x}_1 A_{.1} + \dots + \bar{x}_r A_{.r} = b$$

and $z(\bar{x}) = \bar{z} = c_1\bar{x}_1 + \cdots + c_r\bar{x}_r$ is the optimum objective value in (4.24). If $\{A_{.1}, \ldots, A_{.r}\}$ is linearly dependent, let a linear dependence relation for it be

$$\alpha_1 A_{.1} + \cdots + \alpha_r A_{.r} = 0,$$

where $(\alpha_1, \ldots, \alpha_r) \neq 0$. Define $\alpha_{r+1} = \cdots = \alpha_n = 0$ and let $\alpha = (\alpha_1, \ldots, \alpha_n)^T$. Then the new solution obtained by the purification routine in this step is

$$x(\lambda) = \bar{x} + \lambda\alpha = (\bar{x}_1 + \lambda\alpha_1, \ldots, \bar{x}_r + \lambda\alpha_r, 0, \ldots, 0)^T.$$

Since $\bar{x}_1, \ldots, \bar{x}_r$ are all >0, $x(\lambda)$ will remain nonnegative and hence feasible to (4.24) for all $-\epsilon \leq \lambda \leq +\epsilon$ when ϵ is a sufficiently small positive number.

The objective value at $x(\lambda)$, $z(x(\lambda)) = c(\bar{x} + \lambda\alpha) = c\bar{x} + \lambda c\alpha$.

If $c\alpha > 0$, then $z(x(-\epsilon)) = c\bar{x} - \epsilon(c\alpha) < c\bar{x}$, and as $x(-\epsilon)$ is feasible to (4.24), this contradicts the optimality of $\bar{x}$. Similarly, if $c\alpha < 0$, then $z(x(\epsilon)) = c\bar{x} - \epsilon(c\alpha) < c\bar{x}$, again there is a contradiction to the optimality of $\bar{x}$. So, under the hypothesis that $\bar{x}$ is optimal to (4.24), $c\alpha$ must be $= 0$.

Therefore, for all λ, $z(x(\lambda)) = c\bar{x}$, that is, if $x(\lambda)$ is feasible to (4.24), it is also optimal; that is, all new solutions obtained by the purification routine are also optimal. The same argument continues to hold for all steps of the purification routine. □

Theorem 4.5. *If an LP in standard form has a feasible solution, then it has a BFS.*

Proof. Starting from a feasible solution $\bar{x}$ for it, you can apply the purification routine to get a BFS for it. □

Theorem 4.6. *If an LP in standard form has a optimum feasible solution, then it has a BFS which is also optimal.*

Proof. Starting from a optimum feasible solution $\bar{x}$, by Theorem 4.4, a BFS which is optimal can be obtained by applying the purification routine. □

Note 4.3. The results stated in Theorems 4.5 and 4.6 are stated in terms of LPs in standard form, because in general these results may not be true. The results may fail to hold because an LP with constraints in general form may not have a BFS at all. For example, consider the following LP in two variables x_1, x_2.

$$\begin{aligned} \text{Minimize} \quad & 6x_1 \\ \text{subject to} \quad & x_1 \geq 0 \\ & x_2 \quad \text{unrestricted.} \end{aligned}$$

The set of feasible solutions of this LP is a half-space in R^2 that has no BFSs and hence no optimal BFSs.

But in practice, LPs in standard form and bounded variable LPs are the most important. The results in these theorems hold well.

Exercises

4.7.1. Suppose $\bar{x}$ is a feasible solution to the general system of linear constraints given below. Generalize the purification method discussed above to find a BFS of this system if one exists, or determine that the system has no BFSs, beginning with $\bar{x}$.

$$\begin{aligned} Dx &= d, \\ Fx &\geq g. \end{aligned}$$

4.7.2. Verify that $\bar{x} = (4, 6, 2, 8)^T$ is a feasible solution to the following system of constraints. Check whether it is a BFS. If not, apply purification to get a BFS from it.

x_1	x_2	x_3	x_4	RHS
1	−1	2	0	= 2
2	1	1	3	≥40
2	−1	3	1	≤16
1	1	0	2	≥26
1	2	0	0	≥6

$x_1, x_2 \geq 2;\ x_3, x_4 \geq 0$

4.7.3. Check whether $\bar{x} = (4, 4, 4, 4, 5, 0, 0)^T$ is a BFS for the following bounded variable system. If it is not, obtain a BFS from it using purification.

x_1	x_2	x_3	x_4	x_5	x_6	x_7	RHS
1	0	1	1	1	3	5	=17
2	0	2	2	1	6	−7	=29
0	2	−2	2	1	8	−2	=13
0	1	−1	1	1	−3	1	=9

$0 \leq x_j \leq 5$ for all j

4.7.4. For the following system, check whether the solution $\bar{x} = (5, 5, 5, 5, 0, 0, 0)^T$ is feasible. If so, obtain a BFS from it if it is not already.

x_1	x_2	x_3	x_4	x_5	x_6	x_7	RHS
1	2	1	3	1	−1	2	=35
2	0	−2	2	2	1	−1	=10
1	1	0	2	−1	1	2	=20
0	1	1	1	1	2	−1	=15

$x_j \geq 0$ for all j

4.7.5. For the LP given below, $\bar{x} = (1, 2, 3, 4, 5, 0, 0)^T$ is known to be an optimum feasible solution. Obtain an optimum BFS for it.

$$\begin{aligned}\text{Minimize } z(x) &= x_1 + x_2 + x_3 + 2x_4 + 2x_5 + 2x_6 + 4x_7\\ \text{subject to } x_1 + x_2 + 2x_4 + x_5 + x_7 &= 16\\ x_2 + x_3 + x_4 + 2x_5 + 5x_7 &= 19\\ x_1 + x_3 + x_4 + x_5 + 2x_6 + 2x_7 &= 13\\ x_j \geq 0 \quad &\text{for all } j.\end{aligned}$$

4.7.6. For each of the following systems in standard form, check whether the feasible solutions given are extreme point solutions or not. If not, starting from each of them, obtain an extreme point solution of the system. All $x_j \geq 0$ in each system.

x_1	x_2	x_3	x_4	x_5	x_6	x_7	x_8	x_9	b
1	1	−2	0	3	1	0	0	0	9
0	−1	0	−1	−1	0	1	0	0	−6
0	0	1	1	−1	0	0	1	0	1
1	−1	2	2	−3	0	0	0	1	1
2	−1	1	2	−2	1	1	1	1	5

Feasible sol. $(3, 2, 1, 2, 2, 0, 0, 0, 0)^T$.

x_1	x_2	x_3	x_4	x_5	x_6	b
1	−1	2	2	2	−1	≤ 7
−1	2	−1	0	2	−1	$=0$
1	2	2	1	2	−1	≤ 12
−1	2	2	2	−1	−1	≤ 11

Feasible sol. $(1, 2, 3, 0, 0, 0)^T$.
Feasible sol. $(1, 2, 3, 1, 0, 0)^T$.

x_1	x_2	x_3	x_4	x_5	x_6	b
0	−1	1	3	8	−7	2
1	0	1	0	9	11	14
1	−1	0	−1	12	2	−4
−1	2	−1	−2	3	8	2

Feasible sol. $\bar{x} = (8, 10, 6, 2, 0, 0)^T$.

4.7.7. (1) Let Γ denote the set of feasible solutions of the following system of constraints. Check whether $\bar{x} = (10, 15, 25)^T$ is an extreme point of Γ.

$$\begin{array}{rrrcr} x_1 & +2x_2 & +x_3 & = & 65,\\ x_1 & +3x_2 & -x_3 & \geq & -10,\\ -x_1 & +2x_2 & & \leq & 40,\\ & 2x_2 & +x_3 & \geq & 55,\\ x_1 & & & \leq & 10,\\ x_1 & +2x_2 & +2x_3 & \geq & 70,\\ x_1, & x_2, & x_3 & \geq & 0.\end{array}$$

(2) Consider the feasible solution $\bar{x} = (3,4,6,5,0,0,0)^T$ to the following LP. Compute the objective function value $\bar{z}$ at $\bar{x}$. Check whether $\bar{x}$ is an extreme point of the set of feasible solutions to this LP. Why?

If $\bar{x}$ is not an extreme point solution for the problem, find an extreme point solution from $\bar{x}$, making sure that in each step of your method the objective value decreases.

x_1	x_2	x_3	x_4	x_5	x_6	x_7	$-z$	b
1	1	−1	1	2	−3	1	0	6
−1	−2	5	−4	−3	4	−2	0	−1
1	2	−5	4	7	0	−3	0	1
−1	2	−11	8	0	8	3	0	−21
2	−2	−4	3	0	7	4	1	0

$x_j \geq 0$ for all j, min z

4.7.8. Check whether $\bar{x} = (9, 2, 8, 1, 5, 0, 5)^T$ is an extreme point solution of the following system.

x_1	x_2	x_3	x_4	x_5	x_6	x_7	$-z$	b
1	0	0	0	−1	1	−1	0	−1
0	1	0	0	−1	−1	1	0	2
0	0	1	0	−1	0	−1	0	−2
0	0	0	1	0	1	1	0	6
−2	2	1	2	−4	2	6	1	0

min z; $0 \leq x_j \leq 10$ for $j = 1$ to 4;
$0 \leq x_j \leq 5$ for $j = 5, 6, 7$

4.7.9. Is $\bar{x} = (7,2,5,3)^T$ an extreme point of the set of feasible solutions of the following system?

$$\begin{array}{rrrrrr}
x_1 & & -x_3 & +x_4 & \geq & 5, \\
 & x_2 & & -x_4 & \geq & -1, \\
x_1 & -x_2 & +x_3 & & \geq & 10, \\
x_1 & +2x_2 & -3x_3 & +x_4 & \geq & -1, \\
3x_1 & +2x_2 & -2x_3 & -x_4 & \geq & 5, \\
x_1 & +3x_2 & +4x_3 & +2x_4 & \leq & 50.
\end{array}$$

4.7.10. Let K denote the set of feasible solutions of the following system of constraints in standard form. Check whether $\bar{x} = (0,10,0,15,0,5)^T$ is an extreme point of K. If it is not, find the smallest dimension face of K containing $\bar{x}$, and express it in parametric form.

x_1	x_2	x_3	x_4	x_5	x_6	b
3	1	−1	1	3	−2	15
7	0	5	1	−1	−1	10
−2	0	2	1	8	−1	10

$x_j \geq 0$ for all j

4.7.11. Consider the following system of equations.

x_1	x_2	x_3	x_4	x_5	x_6	x_7	
1	0	−1	1	−1	−2	12	5
0	1	1	2	2	3	7	26
1	1	0	3	1	1	−2	31
0	1	1	2	2	3	8	26

Verify that $\bar{x} = (10,\ 15,\ 2,\ 1,\ 2,\ 1,\ 0)^T$ is a solution to this system of equations. Starting from $\bar{x}$, find a nonnegative solution to this system of equations in which the number of positive components is ≤ 4, showing all your work.

4.7.12. (a) Consider the following system of linear equations in nonnegative variables. Check whether $\bar{x} = (8, 10, 6, 2, 0, 0)^T$ is a BFS of this system. If $\bar{x}$ is not a BFS, obtain a BFS from it showing all your work carefully.

x_1	x_2	x_3	x_4	x_5	x_6	RHS
0	−1	1	3	8	−7	2
1	0	1	0	9	11	14
1	−1	0	−1	12	2	−4
−1	2	−1	−2	3	8	2

$x_j \geq 0$ for all j

(b) Let K denote the set of feasible solutions of the following system of linear constraints. Verify that $\bar{x} = (2, 1, 3, 4, 2)^T \in K$.

x_1	x_2	x_3	x_4	x_5	
1	0	0	0	1	$=4$
1	1	0	0	2	$=7$
0	1	1	0	2	≥ 8
0	0	1	1	2	≤ 11
0	0	0	1	1	≤ 8
1	1	1	1	1	≤ 30
2	−1	−1	−2	1	≥ -25
−1	−1	−1	1	1	≤ 20

and $x_j \geq 0$ for all j

Check whether $\bar{x}$ is an extreme point of K or not.

If $\bar{x}$ is not an extreme point of K, check whether it is contained on an edge of K, and if so obtain its parametric representation.

(c) Give the definition of a BFS (basic feasible solution) to the following general system of linear constraints

$$\begin{aligned} A_{i.}x &= b_i, \quad i = 1, \ldots, m, \\ &\geq b_i, \quad i = m+1, \ldots, m+p, \end{aligned}$$

where each $A_{i.}$ is a row vector in R^n, and any bound restrictions on the variables are already included among the inequality constraints. Prove that every BFS by this definition is an extreme point of the set of feasible solutions by the geometric definition of extreme points, and vice versa.

4.8 Edges, One-Dimensional Faces, Adjacency of Extreme Points, Extreme Directions

Every one-dimensional face or edge of a convex polyhedron either has

- **2** extreme points on it
- Or **1** extreme point
- Or no extreme points on it

In the first case, the edge is the line segment joining the two extreme points on it, which are themselves said to be an *adjacent pair* of extreme points. Such edges are called *bounded edges*.

In the second case, the edge is a half-line called *extreme half-line* or *unbounded edge* beginning at the extreme point on it. Its direction is said to be an *extreme direction* for the convex polyhedron.

In the third case, the edge is an entire straight line. In LP applications, we mostly meet the first two cases, that is why in this chapter we discuss only those two cases.

Definition 4.3. Geometric Definition of Adjacency: Two extreme points $x^1,\ x^2$ of a convex polyhedron $K \subset R^n$ are adjacent if every point $\bar{x}$ on the line segment joining them satisfies: if any pair of points $x^3, x^4 \in K$ satisfy $\bar{x} = \alpha x^3 + (1-\alpha)x^4$ for some $0 < \alpha < 1$, then x^3 and x^4 must both lie also on the line segment joining x^1 and x^2.

Definition 4.4. Algebraic Definition of Adjacency: Let K be the set of feasible solutions of

$$\begin{aligned} Dx &= d, \\ Fx &\geq g, \end{aligned} \tag{4.25}$$

where the inequality constraints include all the bound restrictions on individual variables, if any. Let $x^1,\ x^2$ be two BFSs of (4.25), and let $\bar{x}$ be an interior point of the line segment joining them (i.e., a point of the form $\alpha x^1 + (1-\alpha)x^2$ for some $0 < \alpha < 1$, e.g., $\bar{x} = (x^1 + x^2)/2$). Let (S) denote the active system at $\bar{x}$, that is, the system of all active constraints in (4.25) at $\bar{x}$ treated as a system of equations. $x^1,\ x^2$ are adjacent extreme points of K iff the set of all solutions of (S) is one-dimensional, that is, it is the straight line joining $x^1,\ x^2$. This happens iff the rank of the set of column vectors of the variables in (S) is one less than its cardinality.

Definition 4.5. Algebraic Definition of Adjacency for Systems in Standard Form: Let K be the set of feasible solutions of the system in standard form

$$\begin{aligned} Ax &= b, \\ x &\geq 0. \end{aligned} \tag{4.26}$$

Let x^1, x^2 be two BFSs of (4.26). Let $\bar{x}$ be some point in the interior of the line segment joining x^1, x^2 (i.e., a point of the form $\alpha x^1 + (1-\alpha)x^2$ for some $0 < \alpha < 1$, e.g., $\bar{x} = (x^1 + x^2)/2$). Then, x^1, x^2 are adjacent on K iff the set of column vectors $\{A_{.j} : j \text{ such that } \bar{x}_j > 0\}$ is one less than its cardinality.

All these definitions can be proved to be equivalent to one another using arguments very similar to those used in proving Theorems 4.2 and 4.3 in Sect. 4.6. These definitions lead us to the following procedure.

4.8.1 How to Check if a Given Feasible Solution is on an Edge

Let $\bar{x}$ be a feasible solution of the general system (4.25). Obtain the active system (S) of (4.25) at $\bar{x}$. If $p =$ the rank of the set of column vectors of variables in (S), is $= q =$ the number of column vectors in this set, then $\bar{x}$ is in fact an extreme point and will definitely be contained on every edge through it.

If $p = q - 1$, then $\bar{x}$ is on an edge, it is an interior point of that edge, and the extreme point(s) of this edge and its extreme direction, if any, can all be found by finding the parametric representation of the straight line which is the set of solutions of (S), and the subset of this straight line that satisfies all the inactive inequality constraints in (4.25) at $\bar{x}$ (this is very similar to finding the bounds for the parameter λ to keep the parametric solution feasible in the purification routine). If $p \leq q-2$, $\bar{x}$ is not contained on any edge of the set of feasible solutions of (4.25).

For the system in standard form (4.26), the procedure is similar. Let $\bar{x}$ be a feasible solution of (4.26) and let $p =$ the rank of $\{A_{.j} : j \text{ such that } \bar{x}_j > 0\}$ and $q =$ number of column vectors in this set. If $p = q$, $\bar{x}$ is a BFS of (4.26), and if $p = q - 1$, $\bar{x}$ is an interior point of an edge; that edge and all the relevant information about it can be found by applying one step of the purification routine beginning with $\bar{x}$. If $p \leq q - 2$, $\bar{x}$ is not contained on any edge of the set of feasible solutions of (4.26).

Example 4.12. For the following system, is the feasible solution $\bar{x} = (5, 10, 3, 0, 0)^T$ on an edge?

x_1	x_2	x_3	x_4	x_5	b
1	−1	−1	1	−2	−8
0	1	2	3	8	16
−1	1	1	8	−9	8

$x_j \geq 0$ for all j

The index set of positive components in $\bar{x}$ is $J = \{1, 2, 3\}$. Let A be the 3×5 coefficient matrix of this system, and denote its RHS vector by b. We need to find the rank of $\{A_{.1},\ A_{.2},\ A_{.3}\}$. We apply the method used earlier in Sect. 4.7. PC and PR indicate pivot column and pivot row, respectively, and the pivot elements are boxed.

As we have been able to carry out only two pivot steps, the rank of $\{A_{.1},\ A_{.2},\ A_{.3}\}$ is 2, which is one less than the number of column vectors in this set, so $\bar{x}$ is on an edge. To get the edge in parametric form, we apply one step of the purification routine. The linear dependence relation and the equation corresponding to the feasibility of $\bar{x}$ are

Memory matrix						
$A_{.1}$	$A_{.2}$	$A_{.3}$				
			PC			
1	0	0	$\boxed{1}$	0	−1	PR
0	1	0	−1	1	1	
0	0	1	−1	2	1	
				PC		
1	0	0	1	0	−1	
1	1	0	0	$\boxed{1}$	0	PR
1	0	1	0	2	0	
1	0	0	1	0	−1	
1	1	0	0	1	0	
−1	−2	1	0	0	0	

$$-A_{.1} - 2A_{.2} + A_{.3} = 0$$
$$5A_{.1} + 10A_{.2} + 3A_{.3} = b$$

So, $(5 - \lambda)A_{.1} + (10 - 2\lambda)A_{.2} + (3 + \lambda)A_{.3} = b$. Therefore, $x(\lambda) = (5 - \lambda,\ 10 - 2\lambda,\ 3 + \lambda,\ 0,\ 0)^T$ satisfies the equality constraints in the system, $x(\lambda) \geq 0$ and is feasible to the system for all $\theta_1 = -3 \leq \lambda \leq 5 = \theta_2$. So, one parametric representation of the edge containing $\bar{x}$ is $\{x(\lambda) = (5-\lambda,\ 10-2\lambda,\ 3+\lambda,\ 0,\ 0)^T : -3 \leq \lambda \leq 5\}$. In this representation $\bar{x}$ is obtained by fixing $\lambda = 0$.

The points $x(-3) = (8, 16, 0, 0, 0)^T$ and $x(5) = (0, 0, 8, 0, 0)^T$ are the two extreme points on this edge, so it is a bounded edge, which is the line segment joining $x(-3)$ and $x(5)$. So, another parametric representation of this same edge is $\{\alpha x(-3) + (1-\alpha)x(5) = (\alpha(8, 16, 0, 0, 0)^T + (1-\alpha)(0, 0, 8, 0, 0)^T : 0 \leq \alpha \leq 1\}$. In this representation, verify that you get $\bar{x}$ by setting $\alpha = 5/8$. ■

Example 4.13. Check whether the feasible solution $\bar{x} = (2, 4, 8,\ 12, 0, 0)^T$ of the following system in standard form is on an edge.

x_1	x_2	x_3	x_4	x_5	x_6	b
1	0	0	−1	1	17	−10
0	1	−1	0	2	18	−4
0	0	1	−1	3	19	−4

$x_j \geq 0$ for all j

The index set of positive components in $\bar{x}$ is $J = \{1, 2, 3, 4\}$. Let A be the 3×6 coefficient matrix of this system, and denote its RHS vector by b. We need to find the rank of $\{A_{.1},\ A_{.2},\ A_{.3},\ A_{.4}\}$. We apply the method used earlier in Sect. 4.7. PC and PR indicate pivot column and pivot row, respectively, and the pivot elements are boxed.

Memory matrix							
$A_{.1}$	$A_{.2}$	$A_{.3}$	$A_{.4}$				
				PC			
1	0	0	0	$\boxed{1}$	0	0	PR
0	1	0	0	0	1	0	
0	0	1	0	0	−1	1	
0	0	0	1	−1	0	−1	
					PC		
1	0	0	0	1	0	0	
0	1	0	0	0	$\boxed{1}$	0	PR
0	0	1	0	0	−1	1	
1	0	0	1	0	0	−1	
						PC	
1	0	0	0	1	0	0	
0	1	0	0	0	1	0	
0	1	1	0	0	0	$\boxed{1}$	PR
1	0	0	1	0	0	−1	
1	0	0	0	1	0	0	
0	1	0	0	0	1	0	
0	1	1	0	0	0	1	
1	1	1	1	0	0	0	

So the rank of $\{A_{.1},\ A_{.2},\ A_{.3},\ A_{.4}\}$ is 3, which is one less than the number of elements in it; so $\bar{x}$ is on an edge. Applying one step of purification, we have

$$A_{.1} + A_{.2} + A_{.3} + A_{.4} = 0,$$
$$2A_{.1} + 4A_{.2} + 8A_{.3} + 12A_{.4} = b.$$

So, $(2 + \lambda)A_{.1} + (4 + \lambda)A_{.2} + (8 + \lambda)A_{.3} + (12 + \lambda)A_{.4} = b$, where λ is a parameter. Hence $x(\lambda) = (2 + \lambda,\ 4 + \lambda,\ 8 + \lambda,\ 12 + \lambda,\ 0,\ 0)^T$ is a solution to the equality constraints in our system for all values of λ. $x(\lambda)$ is nonnegative, hence feasible to the system for all $\lambda \geq -2 = \theta_1$.

So, the edge containing $\bar{x}$ in parametric representation is $\{x(\lambda) = (2+\lambda,\ 4+\lambda,\ 8+\lambda,\ 12+\lambda,\ 0,\ 0)^T : \lambda \geq -2\}$.

It is an unbounded edge with its direction (coefficient vector of the parameter λ in the parametric representation) as $y = (1,1,1,1,0,0)^T$. This y is an extreme direction for this system. $x(-2) = (0,2,6,10,0,\ 0)^T$ is the extreme point on this unbounded edge. Another parametric representation of this unbounded edge is $\{x(-2) + \alpha y : \alpha \geq 0\}$ with the direction y given above. ■

Example 4.14. Check whether the feasible solution $\bar{x} = (2,3,4,\ 4,0,0)^T$ of the following system in standard form is on an edge.

x_1	x_2	x_3	x_4	x_5	x_6	x_7	b
1	−1	−1	−5	3	1	−1	−25
1	1	7	−1	−8	0	2	29
−1	2	5	7	2	−2	2	52
−1	0	−3	3	4	−4	2	2

$x_j \geq 0$ for all j

Memory matrix

$A_{.1}$	$A_{.2}$	$A_{.3}$	$A_{.4}$					
				PC				
1	0	0	0	$\boxed{1}$	1	−1	−1	PR
0	1	0	0	−1	1	2	0	
0	0	1	0	−1	7	5	−3	
0	0	0	1	−5	−1	7	3	
							PC	
1	0	0	0	1	1	−1	−1	
1	1	0	0	0	2	1	$\boxed{-1}$	PR
1	0	1	0	0	8	4	−4	
5	0	0	1	0	4	2	−2	
0	−1	0	0	1	−1	2	0	
−1	−1	0	0	0	−2	−1	1	
−3	−4	1	0	0	0	0	0	
3	−2	0	1	0	0	0	0	

The index set of positive components in $\bar{x}$ is $J = \{1,2,3,4\}$. Let A be the 3×7 coefficient matrix of this system, and denote its RHS vector by b. We need to find the rank of $\{A_{.1},\ A_{.2},\ A_{.3},\ A_{.4}\}$. We apply the method used earlier in Sect. 4.7. PC and PR indicate pivot column and pivot row, respectively, and the pivot elements are boxed.

So the rank of $\{A_{.1},\ A_{.2},\ A_{.3},\ A_{.4}\}$ is 2, which is two less than the number of elements in it; so $\bar{x}$ is not on any edge of the set of feasible solutions of this system; that is, it must be contained on a face of dimension 2 or higher of this set. ■

Example 4.15. $x^1 = (0,2)^T, x^2 = (2/3, 2/3)^T, x^3 = (2,0)^T$

$$
\begin{aligned}
x_1 + 2x_2 &\geq 2,\\
2x_1 + x_2 &\geq 2,\\
-x_1 + x_2 &\leq 2,\\
x_1, x_2 &\geq 0
\end{aligned}
$$

are extreme points of the set of feasible solutions of the system given above. Check whether the pairs of extreme points $\{x^1,\ x^2\}$ and $\{x^1,\ x^3\}$

x_1	x_2	x_3	x_4	x_5	b
1	2	−1	0	0	2
2	1	0	−1	0	2
−1	1	0	0	1	2

$x_j \geq 0$ for all j

are adjacent. We used this same system in Example 4.11, and the set of feasible solutions of this system can be seen in Fig. 4.10. From the figure we see that $x^1,\ x^2$ are adjacent but $x^1,\ x^3$ are not. We will now show how this conclusion can be reached algebraically. As in Example 4.11, we introduce slack variables $x_3,\ x_4,\ x_5$ and put this problem in standard form given above.

Memory matrix $A_{.1}$	$A_{.2}$	$A_{.3}$	$A_{.5}$				
						PC	
1	0	0	0	1	2	**[−1]**	PR
0	1	0	0	2	1	1	
0	0	1	0	−1	0	0	
0	0	0	1	0	0	1	
					PC		
−1	0	0	0	−1	−2	1	
1	1	0	0	3	**[3]**	0	PR
0	0	1	0	−1	0	0	
1	0	0	1	1	2	0	
				PC			
−1/3	2/3	0	0	1	0	1	
1/3	1/3	0	0	1	1	0	
0	0	1	0	**[−1]**	0	0	PR
1/3	−2/3	0	1	−1	0	0	
−1/3	2/3	1	0	0	0	1	
1/3	1/3	1	0	0	1	0	
0	0	−1	0	1	0	0	
1/3	−2/3	−1	1	0	0	0	

We denote solutions of this standard form by the symbol $X = (x_1, \ldots, x_5)^T$. Using $x_3 = x_1 + 2x_2 - 2$, $x_4 = 2x_1 + x_2 - 2$, $x_5 = 2 + x_1 - x_2$, we know that x^1, x^2, x^3 correspond to $X^1 = (0, 2, 2, 0, 0)^T$, $X^2 = (2/3, 2/3, 0, 0, 2)^T$, $X^3 = (2, 0, 0, 2, 4)^T$, respectively.

Also, from arguments mentioned in Example 4.11, we know that x^t is a BFS of the original system iff the corresponding X^t is a BFS of the standard form and that x^t, x^s are adjacent BFSs of the original system iff the corresponding X^t, X^s are adjacent BFSs of the standard form. So, as we are given that x^1, x^2, x^3 are BFSs of the original system, we know that X^1, X^2, X^3 are BFSs of the standard form.

To check whether X^1, X^2 are adjacent, we check whether their midpoint $\bar{X} = (X^1 + X^2)/2 = (1/3,\ 4/3,\ 1,\ 0,\ 1)^T$ is on an edge. The index set of positive components in $\bar{X}$ is $J = \{1, 2, 3, 5\}$. Denoting the coefficient matrix of the standard form by A, we need to find the rank of $\{A_{.1},\ A_{.2},\ A_{.3},\ A_{.5}\}$. PC and PR denote pivot column, pivot row, respectively, in the tableaus given above, and the pivot elements are boxed.

So, the rank of the set $\{A_{.1},\ A_{.2},\ A_{.3},\ A_{.5}\}$ is 3, one less than its cardinality, so $\bar{X}$ is on an edge, and hence X^1, X^2 are adjacent for the system in standard form; consequently, the corresponding x^1, x^2 are adjacent in the original system.

The midpoint of X^1, X^3 is $\tilde{X} = (X^1 + X^3)/2 = (1, 1, 1, 1, 2)^T$, in which all the variables x_1 to x_5 are >0, so we need to find the rank of $\{A_{.1}, \ldots, A_{.5}\}$. We have already seen that the rank of a subset of this set is 3, and as each vector in this set is from R^3, the rank of this set cannot exceed 3. So, its rank is 3, while its cardinality is 5. So, rank of this set is $< -1+$ its cardinality, so $\tilde{X}$ is not on an edge for the standard form, that is, X^1, X^3 are not adjacent. Consequently, the corresponding x^1, x^3 are not adjacent in the original system. ■

Exercises

4.8.1. Find all BFSs of the following bounded variable system in four variables: $-6 \le x_1 \le 2, -5 \le x_2 \le 3, 5 \le x_3 \le 10$, and $-20 \le x_4 \le -10$.

Given two BFSs, x^1, x^2 of the above system develop conditions to check whether they are adjacent BFSs. Apply your conditions on $x^1 = (-6, 3, 10, -20)^T$, $x^2 = (-6, 3, 5, -10)^T$.

4.8.2. Check whether $(4, 9, 0, 3, 0, 0)^T$ is an extreme point of the set of feasible solutions of the following system. Why? If not is it on an edge? If it is, is the edge bounded or unbounded?

$$\begin{aligned}
x_1 + x_2 - 3x_4 + 3x_5 + x_6 &= 4 \\
x_1 + 2x_2 - 5x_4 + 5x_5 + 3x_6 &= 7 \\
-x_2 + x_3 + 2x_4 - 5x_5 + x_6 &= -3 \\
x_j \ge 0 \quad &\text{for all } j
\end{aligned}$$

4.8.3. Let K denote the convex polyhedron, which is the set of feasible solutions of the system of linear inequalities $Ax \geq b$, where the coefficient matrix A is of order $m \times n$. Give the conditions under which a feasible solution for this system, $\bar{x}$, is not an extreme point of K, but lies on an edge of K. Remember that here the only constraints on x are those in $Ax \geq b$, other than this no other constraints exist on x in the definition of K.

Use your conditions to check whether $\bar{x} = (5, 10, 15, 20)^T$ is a nonextreme point feasible solution on an edge of the set of feasible solutions, K, of the following system of constraints. If $\bar{x}$ is on an edge of K, is the edge bounded or unbounded? Why (explain clearly). Find all the extreme points of K on this edge.

x_1	x_2	x_3	x_4	
1	1	1	1	≥ 50
0	1	1	1	≤ 45
0	0	1	1	≥ 35
3	−1	2	1	≥ 40
1	3	1	−1	≤ 40

4.8.4. Prove that Definitions 4.3 and 4.4 are equivalent for system (4.25).

4.8.5. Prove that Definitions 4.5 and 4.3 are equivalent for the system (4.26) in standard form.

4.8.6. Consider the following system of constraints with RHS vector $b = (b_1, b_2, b_3, b_4)^T$.

x_1	x_2	x_3	x_4	x_5	x_6	x_7	b
1	0	0	−1	1	1	1	b_1
1	1	0	0	2	−1	1	b_2
0	−1	1	1	0	−1	1	b_3
0	0	0	1	0	−1	1	b_4

$x_j \geq 0$ for all j

(1) Let $b = (3, 9, 0, 1)^T$. Verify that $x^1 = (1, 2, 1, 1, 3, 0, 0)^T$ is a feasible solution of the system. Check whether x^1 is a BFS, and if not whether it is on an edge. If it is on an edge, determine whether that edge is bounded or unbounded, find all the extreme points on it and its extreme direction if it has any, and a representation of it in terms of its extreme points and extreme direction.
(2) Let $b = (1, 0, 1, 0)^T$. Answer all the questions in (1) for the feasible solution $x^2 = (1, 1, 2, 2, 0, 2, 0)^T$ in this case.
(3) Consider the feasible solution $x^3 = (1, 1, 1, 1, 1, 1, 0)^T$ when $b = (2, 3, 0, 0)^T$ for this system. Can x^3 be on an edge of the set of feasible solutions of the system in this case? If not, starting from x^3, find a point on an edge of the set of feasible solutions in this case.

4.8.7. $\bar{x}$ is a point in a convex polyhedron K which is not an extreme point of K. Can $\bar{x}$ lie on two different edges of K?

4.8.8. Show that two edges cannot intersect at a point other than an extreme point.

4.8.9. (1) Let $K = \{x : Ax = b, \ x \geq 0\}$ be a convex polyhedron. Let $H = \{x : dx = d_0\}$ a hyperplane. Show that every extreme point of $K \cap H$ must be either an extreme point of K lying on H or the point of intersection of an edge of K with H (Murty 1971).

(2) Similarly show that every edge of $K \cap H$ is either an edge of K lying in H or the intersection of a two-dimensional face of K with H.

4.8.10. Consider the following system of constraints:

x_1	x_2	x_3	x_4	
1	−1	0	1	=8
0	1	−1	2	=21
−2	2	−2	6	≤60
−1	2	−2	5	≥51
1	1	1	1	≤25

$x_j \geq 0$ for all j.

Check whether $\bar{x} = (1, 3, 2, 10)^T$ is a BFS for this system.

Transform this into a system of linear equations in nonnegative variables. Find the feasible solution of the transformed system that corresponds to $\bar{x}$, and continue denoting it by the same symbol $\bar{x}$. Check whether $\bar{x}$ is a BFS of this system, and if not, whether $\bar{x}$ lies on an edge for the system. If so, is that edge bounded or unbounded? Why?

If that edge is bounded, find the two extreme points of the system on that edge. If it is unbounded, find the unique extreme point on that edge, and the direction of this edge, and an algebraic representation of that unbounded edge.

4.9 Adjacency in a Primal Simplex Pivot Step

The primal simplex method considers the LP in standard form

$$\begin{aligned} \text{Minimize } z &= cx \\ \text{subject to } Ax &= b \\ x &\geq 0, \end{aligned} \tag{4.27}$$

where A is a matrix of order $m \times n$ and rank m, and $b \neq 0$. Let K denote the set of feasible solutions of this LP.

Starting with an initial primal feasible basic vector for (4.27), the primal simplex algorithm performs a sequence of primal simplex steps on this problem, maintaining primal feasibility throughout, until a basic vector satisfying one of the termination conditions, optimality or unboundedness, is obtained.

For details of the primal simplex algorithm in the original and most elementary version (i.e., using canonical tableaus), see, for example, Chap. 4 in Murty (2005b) of Chap. 1. There, only one version, the primal simplex algorithm, is discussed; hence the algorithm is sometimes referred to as the simplex algorithm without the "primal" adjective. In this book we will also discuss some other versions of the algorithm (e.g., the "dual simplex algorithm"); hence here we will refer to that algorithm as the "primal simplex algorithm."

In each primal simplex pivot step, the primal simplex algorithm enters a single nonbasic variable into the present feasible basic vector. There are three different possible outcomes of this pivot step, which we discuss as three separate cases below. In each of these cases, our discussion will refer to the following canonical tableau.

To keep the discussion simple, we will assume that the primal feasible basic vector at the beginning of this pivot step is $x_B = (x_1, \dots, x_m)$, with $x_1, \dots, x_m$ as the basic variables in that order, and that the entering variable is x_{m+1}. So, the values of other nonbasic variables $x_{m+2}, \dots, x_n$ will all remain $= 0$ through this pivot step, so we ignore them. Also, it will be easier to discuss each outcome using canonical tableaus, which we will use. We provide illustrative numerical examples for some of the cases.

In the following discussion, PC (pivot column) indicates the updated column of the entering variable x_{m+1}, and the relative cost coefficient in it $\bar{c}_{m+1} < 0$ by the entering variable choice rule in the primal simplex algorithm (this comes from the result on the optimality criterion in the simplex method discussed in Chap. 4 in Murty (2005b) of Chap. 1, and the same result is proved again in Theorem 4.7 in Sect. 4.12 coming later in this chapter).

Canonical tableau wrt $x_B = (x_1, \dots, x_m)$

BV	Basic			PC	Other		Updated
	x_1	$\dots$	x_m	x_{m+1}	nonbasics	$-z$	RHS
x_1	1	$\dots$	0	$\bar{a}_{1,m+1}$	$\dots$	0	$\bar{b}_1$
$\vdots$	$\vdots$	$\ddots$	$\vdots$	$\vdots$		$\vdots$	$\vdots$
x_m	0	$\dots$	1	$\bar{a}_{m,m+1}$	$\dots$	0	$\bar{b}_m$
$-z$	0	$\dots$	0	$\bar{c}_{m+1}$	$\dots$	1	$-\bar{z}$

$x_j \geq 0$ for all j, min z

"BV" is the abbreviation for "basic variable in this row." So, the present BFS is $\bar{x} = (\bar{x}_1, \dots, \bar{x}_m, \bar{x}_{m+1}, \dots, \bar{x}_n)^T = (\bar{b}_1, \dots, \bar{b}_m, 0, \dots, 0)^T$ with objective value $= \bar{z}$.

So, what we do in this step is keep all other nonbasic variables $x_{m+2}, \dots, x_n$ fixed at 0, and give the entering variable x_{m+1} a value λ, which is a parameter.

Then re-evaluating the values of the present basic variables x_1 to x_m to satisfy the equality constraints in the system, we get the solution

$$\begin{aligned} x(\lambda) &= (\bar{b}_1 - \bar{a}_{1,m+1}\lambda, \ldots, \bar{b}_m - \bar{a}_{m,m+1}\lambda,\ \lambda,\ 0, \ldots,\ 0)^T \\ &= \bar{x} + \lambda(-\bar{a}_{1,m+1}, \ldots, -\bar{a}_{1,m+1},\ 1,\ 0, \ldots, 0)^T, \end{aligned}$$

with objective value $\bar{z} + \lambda\bar{c}_{m+1} = z(\lambda)$. Now we consider the various cases that can occur in this step.

Case 4.7. No positive entry in the pivot column: In this case, $(\bar{a}_{1,m+1}, \ldots, \bar{a}_{m,m+1}) \leq 0$. So, there are no ratios to compute in the minimum ratio substep in this primal simplex pivot step.

As all the entries in the PC are ≤ 0, $x(\lambda)$ remains ≥ 0 for all $\lambda \geq 0$; so $x(\lambda)$ is feasible to this problem for all $\lambda \geq 0$, and as $\lambda \to \infty$, $z(\lambda) \to -\infty$. That is why in this case we terminate the algorithm with the conclusion that the minimum value of z on the set of feasible solutions of this problem, K, is $= -\infty$. Hence when there is no positive entry in the pivot column, we say that *the unboundedness termination criterion has been satisfied.*

The half-line $\{x(\lambda) : \lambda \geq 0\}$ is an unbounded edge or extreme half-line of K beginning with the extreme point $\bar{x}$, along which $z \to -\infty$. The direction of this unbounded edge is $y = (-\bar{a}_{1,m+1}, \ldots, -\bar{a}_{1,m+1},\ 1,\ 0, \ldots, 0)^T$. Verify that this direction y satisfies $Ay = 0,\ y \geq 0$, and $cy = \bar{c}_{m+1} < 0$.

The system $Ay = 0,\ y \geq 0$, obtained from the system of constraints in the original problem by changing all the RHS constants in it to 0 is called the *homogeneous system corresponding to the original system.* To avoid confusion between the two different systems, we are denoting the vector of variables in the homogeneous system by the symbol y. The direction of every unbounded edge of K satisfies this homogeneous system, that is why all those directions are called *homogeneous solutions* corresponding to the original system.

Example 4.16. Consider the following canonical tableau for an LP wrt the basic vector $(x_1,\ x_2)$, in which the nonbasic variable x_3 has been selected as the entering variable in the primal simplex algorithm. In this example, $m = 2$ and the entering variable is $x_{m+1} = x_3$. We only show the columns of $x_1,\ x_2,\ x_3$ (the basic and entering variables) in this canonical tableau.

BV	x_1	x_2	x_3	Other nonbasics	$-z$	Updated RHS
			PC			
x_1	1	0	−2	...	0	0
x_2	0	1	−1	...	0	3
$-z$	0	0	−3	...	1	−10

$x_j \geq 0$ for all j, min z

The present BFS is $\bar{x} = (0, 3, 0, \dots, 0)^T$ with objective value $\bar{z} = 10$. The PC has no positive entry, so the unboundedness termination criterion has been satisfied in this step. In this case, the unbounded edge generated is

$$\{x(\lambda) = (0 + 2\lambda,\ 3 + \lambda,\ \lambda,\ 0, \dots, 0)^T :\ \lambda \geq 0\}.$$

Its direction $\bar{y} = (2, 1, 1, 0, \dots, 0)^T$ satisfies $c\bar{y} = \bar{c}_3 = -3 < 0$. The objective value $z(\lambda) = 10 - 3\lambda \to -\infty$, as λ increases along this unbounded edge.

The homogeneous system corresponding to the original system is given in the following tableau.

y_1	y_2	y_3	Other nonbasics	RHS
1	0	-2	$\dots$	0
0	1	-1	$\dots$	0

$y_j \geq 0$ for all j

It can be verified that the direction $\bar{y} = (2, 1, 1, 0, \dots, 0)^T$ of the unbounded edge above is feasible to this homogeneous system. Also, the objective value at $\bar{y} = c\bar{y} = (0, 0, -3, \dots)\bar{y} = -3 = \bar{c}_3$, which is negative. ■

Case 4.8. The PC has at least one positive entry, and it has a positive entry in at least one row i in which $\bar{b}_i = 0$

For $t = 1$ to m, the value of the basic variable x_t in $x(\lambda)$ is $\bar{b}_t + \lambda \bar{a}_{t,m+1}$, and to keep it non-negative, we need

$$\lambda \leq (\bar{b}_t / \bar{a}_{t,m+1}) \text{ for all } t \text{ such that } \bar{a}_{t,m+1} > 0.$$

The quantity $(\bar{b}_t / \bar{a}_{t,m+1})$ is called the *ratio* in this primal simplex pivot step in row t; it is computed only if the entry $\bar{a}_{t,m+1}$ in the pivot column (PC) in row t is positive. So, to keep $x(\lambda) \geq 0$, the maximum value we can give to the parameter λ when the PC has at least one positive entry is $\theta = \text{minimum}\{(\bar{b}_t / \bar{a}_{t,m+1}) :$ over t such that $\bar{a}_{t,m+1} > 0\}$. This quantity θ is called the *minimum ratio* in this pivot step in the primal simplex algorithm, because it is the minimum of the ratios given above. It is ≥ 0 because all $\bar{b}_t \geq 0$. So, $x(\lambda)$ is feasible to the problem for all $0 \leq \lambda \leq \theta$.

In Case 4.7 discussed above, we considered the situation where PC has no positive entry, there we found that $x(\lambda)$ remains feasible to the problem for all $0 \leq \lambda \leq +\infty$. So, to keep conformity, we define the minimum ratio θ to be $= +\infty$ when the PC has no positive entry, that is, when there is no ratio to compute.

In this case, we have the row index i such that $\bar{b}_i = 0$ and $\bar{a}_{i,m+1} > 0$, so using the above definition of the minimum ratio, $\theta = 0$ in this case. In this case, $x(\lambda)$ is feasible to the problem, only for $\lambda = 0$, and we see that $x(0) = \bar{x}$, the present BFS.

So, in this step we do not obtain a new feasible solution to the problem, but remain at the same BFS as at the beginning of this step, and there will be no change at all in the objective value. That is why a pivot step like this where the minimum ratio is 0 is called a *degenerate pivot step*. In a degenerate pivot step, the basic vector changes by one variable (the entering variable replaces the basic variable x_t, where t is any of the indices that ties for the minimum ratio in its definition). The basic variable replaced by the entering variable is called the *dropping basic variable* in this pivot step. But there is no change in the BFS or the objective value, and the new basic vector is another basic vector that corresponds to the present BFS.

With the new basic vector, the primal simplex algorithm now goes to the next primal simplex pivot step.

Example 4.17. Consider the following canonical tableau for an LP wrt the basic vector (x_1, x_2), the top one among the two given below, in which the nonbasic variable x_3 has been selected as the entering variable in the primal simplex algorithm. In this example, $m = 2$ and the entering variable is $x_{m+1} = x_3$. We show only the columns of x_1, x_2, x_3 (the basic and entering variables) in this canonical tableau. PC and PR indicate pivot column and pivot row, respectively, and the pivot element is boxed

BV	x_1	x_2	x_3	Other nonbasics	$-z$	Updated RHS	Ratio
			PC				
x_1	1	0	−2	...	0	1	
x_2	0	1	$\boxed{1}$	...	0	0	0 PR
$-z$	0	0	−3	...	1	−10	$\theta = 0$
x_1	1	2	0	...	0	1	
x_3	0	1	1	...	0	0	
$-z$	0	3	0	...	1	−10	

$x_j \geq 0$ for all j, min z

Here the PC has a positive entry in row 2 with updated RHS constant = 0, so the ratio in this row is 0; hence in this step the minimum ratio θ is 0, and this is a degenerate step. The minimum ratio is attained in row 2, so x_2, the present basic variable in row 2, is the dropping basic variable in this pivot step, and it will be replaced by the entering variable x_3. The canonical tableau after this pivot step is the second one shown. In this pivot step, there is no change in the objective value or the primal BFS, but we get a new basic vector (x_1, x_3) representing it after this pivot step. ■

Case 4.9. The PC has at least one positive entry, and it does not have a positive entry in any row i in which $\bar{b}_i = 0$

Let the index set of variables that are >0 in the present BFS $\bar{x}$ wrt the present basic vector $x_B = (x_1, \ldots, x_m)$ be $J = \{i : \bar{b}_i > 0\}$. $\bar{J} = \{i : \bar{b}_i = 0\}$, is the index set of all basic variables only, which are 0 in $\bar{x}$. Define $\bar{J}_1, \bar{J}_2$

$$\bar{J}_1 = \{i : \bar{b}_i = 0 \text{ and } \bar{a}_{i,m+1} = 0\},$$
$$\bar{J}_2 = \{i : \bar{b}_i = 0 \text{ and } \bar{a}_{i,m+1} < 0\}.$$

So, the PC has nonzero entries only among rows $i \in J \cup \bar{J}_2$. From linear algebra we know that the PC, $\bar{A}_{.m+1}$, the updated column of the entering variable x_{m+1}, is the coefficient vector in the representation of the original column $A_{.m+1}$ of x_{m+1} as a linear combination of the present basic columns $A_{.1}$ to $A_{.m}$; that is,

$$A_{.m+1} = \text{original column vector of } x_{m+1} = \sum_{i \in J \cup \bar{J}_2} \bar{a}_{i,m+1}(A_{.i}).$$

Since $\{A_{.i} : i \in J \cup \bar{J}_2\}$ is linearly independent, the above equation provides a unique linear dependence relation for the set of original columns of x_i for $i \in J \cup \bar{J}_2 \cup \{m+1\}$, so the rank of this set is one less than its cardinality.

From the condition satisfied in this case, the updated RHS constant $\bar{b}_i$ is >0 for all i such that the PC has a positive entry in row i, so all the ratios will be strictly positive, so the minimum ratio in this step $\theta > 0$, and it occurs in a row $i \in J$. Hence the dropping basic variable will be an x_i for $i \in J$. In the primal simplex algorithm, such a pivot step is called a *nodegenerate pivot step*.

As $\bar{a}_{i,m+1} < 0$ for all $i \in \bar{J}_2$ and $=0$ for all $i \in \bar{J}_1$, after this pivot step the updated RHS constant will remain 0 for all $i \in \bar{J}_1$, and >0 for all $i \in \bar{J}_2$.

So in the BFS $\tilde{x}$ obtained after this pivot step, all basic variables x_i for $i \in \bar{J}_1$ will be 0, and >0 for all $i \in \bar{J}_2$. Also, in $\tilde{x}$ the entering variable x_{m+1} will have a value equal to the minimum ratio, which is >0. In the BFS $\bar{x}$ before this pivot step, all $\bar{x}_i$ for $i \in J$ are >0.

So, at $(\bar{x} + \tilde{x})/2$, the index set of positive x_i is $J \cup \bar{J}_2 \cup \{m+1\}$. As we have already seen above that the rank of $\{A_{.i} : i \in J \cup \bar{J}_2 \cup \{m+1\}\}$ is one less than its cardinality, we conclude that $\bar{x}$ and $\tilde{x}$ are adjacent extreme points of K, and the parametric representation of the edge joining them is $\{x(\lambda) = (\bar{b}_1 - \bar{a}_{1,m+1}\lambda, \dots, \bar{b}_m - \bar{a}_{m,m+1}\lambda,\ \lambda,\ 0, \dots, 0)^T : 0 \le \lambda \le \theta\}$.

So, in every nondegenerate step, the primal simplex algorithm moves along a bounded edge of the set of feasible solutions of the problem from one end to the other.

Example 4.18. Consider the following canonical tableau for an LP wrt the basic vector (x_1 to x_6), in which the nonbasic variable x_7 has been selected as the entering variable in the primal simplex algorithm. In this example, $m = 6$ and the entering variable is $x_{m+1} = x_7$. We show only the columns of x_1 to x_7 (the basic and entering variables) in this canonical tableau. The PR (pivot row) is the row containing the boxed pivot element.

The present BFS $\bar{x} = (0, 0, 0, 6, 2, 4, 0, \cdots, 0)^T$, with objective value 10. The set $J = \{4, 5, 6\}$ is the present index of rows with positive $\bar{b}_i$. The set $\bar{J}_1 = \{1\}$. The set $\bar{J}_2 = \{2, 3\} =$ index of rows with $\bar{b}_i = 0$ and $\bar{a}_{i,m+1} < 0$.

The canonical tableau after this pivot step is shown at the bottom. The new BFS $\tilde{x}$ is $(0, 2, 4, 2, 0, 4, 2, 0, \cdots, 0)^T$ with objective value $= 6$. You can verify that the index of positive components in $(\bar{x} + \tilde{x})/2$ is $J \cup \bar{J}_2 \cup \{m+1\}$.

BV	x_1	x_2	x_3	x_4	x_5	x_6	x_7	Others	$-z$	$\bar{b}$	Ratio
							PC				
x_1	1	0	0	0	0	0	0	...	0	0	
x_2	0	1	0	0	0	0	−1	...	0	0	
x_3	0	0	1	0	0	0	−2	...	0	0	
x_4	0	0	0	1	0	0	2	...	0	6	3
x_5	0	0	0	0	1	0	$\boxed{1}$	...	0	2	2 PR
x_6	0	0	0	0	0	1	0	...	0	4	
$-z$	0	0	0	0	0	0	−2	...	1	−10	$\theta = 2$
x_1	1	0	0	0	0	0	0	...	0	0	
x_2	0	1	0	0	1	0	0	...	0	2	
x_3	0	0	1	0	2	0	0	...	0	4	
x_4	0	0	0	1	−2	0	0	...	0	2	
x_7	0	0	0	0	1	0	$\boxed{1}$	...	0	2	
x_6	0	0	0	0	0	1	0	...	0	4	
$-z$	0	0	0	0	2	0	0	...	1	−6	

$x_j \geq 0$ for all j, min z ■

4.10 How to Obtain All Adjacent Extreme Points of a Given Extreme Point?

Consider the system in standard form

$$\begin{aligned} Ax &= b, \\ x &\geq 0, \end{aligned} \tag{4.28}$$

where A is a matrix of order $m \times n$ and rank m, and $b \neq 0$. Let K denote the set of feasible solutions of this system. Let x_B be a feasible basic vector for this system associated with the BFS $\bar{x}$. Here we discuss how to generate all the adjacent extreme points of $\bar{x}$ in K, and in general all the edges (bounded or unbounded) of $\bar{x}$ on K, just to complete our discussion of extreme points and edges of convex polyhedra.

In applications, one does not often encounter the problem of enumerating the extreme points and edges incident at a given extreme point of a given convex polyhedron represented by linear constraints. Hence the material in this section is mainly of interest to people who are studying the mathematical aspects of convex polyhedra.

There are two important cases to consider.

Case 4.10. The basic vector x_B and the BFS $\bar{x}$ are nondegenerate.

In this case all the m basic variables in x_B are strictly >0 in $\bar{x}$ (i.e., in the canonical tableau wrt the basic vector x_B, all updated RHS constants are >0) and x_B is the unique basic vector corresponding to $\bar{x}$. This is the nice and easy case for this problem. All the edges of K incident at $\bar{x}$ can be generated efficiently because there are exactly $n - m$ of them. To generate them, first create a list for edges of K incident at $\bar{x}$, and a list for adjacent extreme points of $\bar{x}$ (initially both the lists are $\emptyset$) and do the following for each nonbasic variable x_j (i.e., a variable not in the basic vector x_B).

Perform the primal simplex pivot step of entering that nonbasic variable x_j into the basic vector x_B. By the nondegeneracy of x_B, this must generate an edge of K incident at $\bar{x}$ as under Case 1 or Case 3 of the previous Sect. 4.9. Put this edge in the list of edges. If the edge is bounded, the other extreme point on it is an adjacent extreme point of $\bar{x}$ on K, put it in the list of adjacent extreme points.

When this work is completed for all nonbasic variables x_j not in x_B, the list of edges contains all the bounded and unbounded edges of K incident at $\bar{x}$, and the list of adjacent extreme points contains all the adjacent vertices of $\bar{x}$ on K. Terminate.

Case 4.11. The Basic Vector x_B and the BFS $\bar{x}$ Are Degenerate.

In this case some of the m basic variables in x_B are $= 0$ in $\bar{x}$ (i.e., in the canonical tableau wrt the basic vector x_B, some updated RHS constants are $=0$). The degree of degeneracy of $\bar{x}$ refers to how large [$m -$ (number of positive variables in $\bar{x}$)] is. Depending on how large this degree of degeneracy of $\bar{x}$ is, the number of adjacent extreme points of $\bar{x}$ and the number of edges incident at $\bar{x}$ may be huge (i.e., it may grow exponentially with $n - m$).

Mathematically, an "efficient algorithm" for this problem in this case refers to one in which the effort to generate the next adjacent extreme point of $\bar{x}$ after r of them have already been obtained requires a computational effort that is bounded above by a polynomial in r. No such algorithm is known at the moment. This is one *open problem* in the mathematical study of convex polyhedra.

We will discuss an enumerative method for this problem, it may not be efficient because it may generate the same adjacent extreme point many (could be exponential in $n - m$) times, depending on the degree of degeneracy of $\bar{x}$.

Here is the direct enumeration procedure: Let $J = \{j : \bar{x}_j > 0\}$, $|J| = r$. In this case there are many basic vectors of (4.28) corresponding to the BFS $\bar{x}$. All variables x_j for $j \in J$ are basic variables in every basic vector of (4.28) corresponding to $\bar{x}$.

Find out all subsets of $m - r$ variables among $\{x_j : j \notin J\}$, which when combined with the variables in $\{x_j : j \in J\}$ forms a basic vector for (4.28). This process needs the checking of $((n-r)!)/((m-r)!(n-m)!)$ sets of m column vectors each for linear independence (this is the enumerative part of this approach). It generates all the basic vectors of (4.28) corresponding to the BFS $\bar{x}$. Let $\mathcal{B}$ denote the set all these basic vectors. For each basic vector $x_B \in \mathcal{B}$ carry out the following procedure.

Procedure: For each nonbasic variable x_j (i.e., one not in the present basic vector x_B) perform the primal simplex pivot step of entering that nonbasic variable x_j into the basic vector x_B. If it leads to an unbounded edge (as in Case 4.7 of Sect. 4.9), store it in the list of unbounded edges incident at $\bar{x}$, if it is not there already.

If it leads to a nondegenerate pivot step, store the edge in the list of edges incident at $\bar{x}$ if it is not there already, and store the other extreme point on this edge in a list of adjacent extreme points of $\bar{x}$ if it is not there already.

If it leads to a degenerate pivot step, continue.

When this procedure is completed for all the basic vectors in $\mathcal{B}$, the lists of all the bounded edges, unbounded edges, adjacent extreme points contain all the bounded, unbounded edges, respectively, of K incident at $\bar{x}$, and all the adjacent extreme points of $\bar{x}$ on K. Terminate.

Example 4.19. In this example, we find the set of all basic vectors corresponding to a degenerate BFS for a system in standard form. Consider the following canonical tableau for a system in standard form.

BV	x_1	x_2	x_3	x_4	x_5	x_6	x_7	RHS
x_1	1	0	0	−1	−2	0	2	0
x_2	0	1	0	1	−1	2	1	4
x_3	0	0	1	1	2	3	3	5

$x_j \geq 0$ for all j.

The present basic vector (x_1, x_2, x_3) and the BFS $\bar{x}$ corresponding to it are degenerate. The positive basic variables x_2, x_3 will be basic variables in every basic vector corresponding to $\bar{x}$. The set of all basic vectors corresponding to $\bar{x}$ is $\{(x_1, x_2, x_3), (x_4, x_2, x_3), (x_5, x_2, x_3), (x_7, x_2, x_3)\}$. (x_6, x_2, x_3) is not a basic vector because the rank of the set of column vectors corresponding to these variables is only 2. ■

One encounters this problem of enumerating all the adjacent extreme points of a given extreme point only rarely in a few specialized applications, and when it does, the systems on which it has to be solved are usually small (i.e., m and $n - m$ are small). On such small systems, this algorithm is quite satisfactory to solve the problem.

The vertices and edges of a convex polyhedron put together define the *one-dimensional skeleton* or the *graph* of the polyhedron.

Another related problem is the problem of enumerating all the extreme points of a convex polytope specified by a system of linear constraints. As discussed earlier, mathematically, an "efficient algorithm" for this problem is defined to be one in which the effort to generate the next extreme point of the polytope after r of its extreme points have already been obtained requires a computational effort that is bounded above by a polynomial in r. No such algorithm is known at the moment if the polytope has degenerate extreme points. This is another *open problem* in the mathematical study of convex polytopes. References Khachiyan et al. (2006); Murty (1971, 2009); Murty and Chung (1995); Provan (1994) are all research papers dealing with this problem. (PORTA) is a software package for enumerating the extreme points of a convex polyhedron specified by a system of linear constraints.

Exercises

4.10.1. Let K be the set of feasible solutions of the system given below. Verify that $\bar{x} = (1, 3, 5, 0, 0, 0, 0, 0)^T$ is an extreme point of K. Generate all the edges of K incident at $\bar{x}$ and all the adjacent extreme points of $\bar{x}$ on K.

x_1	x_2	x_3	x_4	x_5	x_6	x_7	x_8	b
1	1	1	2	0	−3	1	−2	9
0	1	1	1	0	−2	0	−2	8
0	0	1	0	−1	−1	−1	−1	5

$x_j \geq 0$ for all j

4.10.2. Generate all the adjacent extreme points of $\bar{x} = (3, 4, 0, 0, 0, 0)^T$ on the set of feasible solutions of the following system.

x_1	x_2	x_3	x_4	x_5	x_6	b
1	0	0	1	1	−1	3
0	1	0	1	−1	1	4
0	0	1	−1	0	1	0

$x_j \geq 0$ for all j

4.11 Faces of Dimension ≥2 of a Convex Polyhedron

Because zero- and one-dimensional faces (extreme points and edges) are fundamental tools that the simplex method uses to solve LPs, we have discussed them and their important properties in some detail in the previous sections. Faces of higher dimensions do not play a major role in the study of LP, that is why we will discuss them only briefly in this section.

As the simplex method for LP operates on LPs in standard form, we discussed results on zero- and one-dimensional faces for convex polyhedra in terms of systems in standard form. For our discussion of faces of higher dimensions, however, we will state them in terms of a general system of linear constraints. Let K be the set of feasible solutions of a general system of linear constraints (4.29), where the inequality constraints include any sign restrictions and bounds on individual variables. Suppose D is of order $p \times n$ and F is of order $q \times n$.

$$\begin{aligned} Dx &= d, \\ Fx &\geq g. \end{aligned} \tag{4.29}$$

In general a face of K is the set of feasible solutions of a system obtained from (4.29) by taking a subset of inequality constraints in it and converting them to equality constraints, while leaving the other inequality constraints as they are. So, if

$(F_1,\ F_2)$ is a partition of F into submatrices $F_1,\ F_2$ of orders $q_1 \times n,\ q_2 \times n$, respectively, where $q_1 + q_2 = q$ and (g^1, g^2) is the corresponding partition of g into subvectors, then the set of feasible solutions of the system

$$\begin{aligned} Dx &= d \\ F_1 x &= g^1 \\ F_2 x &\geq g^2 \end{aligned} \tag{4.30}$$

is a face of K, and conversely for every face of K, there exists a partition of $F,\ g$ in this way so that it can be represented by a system of the type (4.30).

Let $\bar{x}$ be a feasible solution for (4.29). If $\bar{x}$ is contained in a face $\bar{F}$ of K of dimension r, then $\bar{x}$ is of course contained in every face F of K of dimension higher than r that contains $\bar{F}$ as a subset. So, the important question that this brings up is finding the smallest dimension face of K that contains $\bar{x}$, that smallest dimension face of K containing the given point $\bar{x}$ is of course unique. To find it, we find the subset of inequality constraints in (4.29), which are active at $\bar{x}$. Putting these together, suppose they form the subsystem $F_1 x \geq g^1$. Let the remaining inequality constraints in (4.29) which are inactive at $\bar{x}$ form the subsystem $F_2 x \geq g^2$. So, we have

$$\begin{aligned} F_1 \bar{x} &= g^1, \\ F_2 \bar{x} &> g^2. \end{aligned}$$

Recall that in Sect. 4.6, we defined the system of equality constraints $Dx = d$, $F_1 x = g^1$ as the *active system* of (4.29) at $\bar{x}$. Let $\bar{r}$ be the dimension of the set of solutions of this active system. Actually

$$\bar{r} = n - \left[\text{rank of the matrix } \begin{pmatrix} D \\ F_1 \end{pmatrix}\right],$$

it is the dimension of the smallest dimension face of K containing the point $\bar{x}$. That face itself is the set of feasible solutions of the system.

$$\begin{aligned} Dx &= d, \\ F_1 x &= g^1, \\ F_2 x &\geq g^2. \end{aligned}$$

4.11.1 Facets of a Convex Polyhedron

Faces of K whose dimension is $(-1 + (\text{dimension of } K))$ have a special name; they are called *facets* of K. Each facet F of K is the set of feasible solutions of a system obtained by changing one inequality constraint in (4.29) into an equation, and leaving all other constraints unchanged. So, the number of facets of K is $\leq$ the number of inequality constraints in the representation (4.29) of K.

Example 4.20. If Γ is the set of feasible solutions of the system in standard form

$$\begin{aligned} Ax &= b, \\ x &\geq 0, \end{aligned}$$

where A is a matrix of order $m \times n$, then every facet F of Γ corresponds to one of the nonnegativity restrictions; that is, every facet of K is the set of feasible solutions of a system

$$\begin{aligned} Ax &= b, \\ x_t &= 0, \\ x_j &\geq 0, \quad \text{for } j = 1, \ldots, t-1, t+1, \ldots, n \end{aligned}$$

for some $t \in \{1, \ldots, n\}$. So, this convex polyhedron Γ has at most n facets.

4.12 Optimality Criterion in the Primal Simplex Algorithm

Consider the LP in standard form

$$\begin{aligned} \text{Minimize} \quad z(x) &= cx \\ \text{subject to} \quad Ax &= b \\ x &\geq 0, \end{aligned} \tag{4.31}$$

where A is a matrix of order $m \times n$ and rank m. $c \in R^n$ is the row vector of *original cost coefficients* of the variables $x = (x_1, \ldots, x_n)^T$.

Let $\bar{x}$ be a BFS for this LP associated with the basic vector x_B, with objective value $z(\bar{x}) = \bar{z}$. For notational convenience, we assume that $x_B = (x_1, \ldots, x_m)$.

Canonical tableau wrt $x_B = (x_1, \ldots, x_m)$

BV	PC								Updated
	x_1	$\ldots$	x_m	x_{m+1}		x_n	$-z$		RHS
x_1	1	$\ldots$	0	$\bar{a}_{1,m+1}$	$\ldots$	$\bar{a}_{1n}$	0		$\bar{b}_1$
$\vdots$	$\vdots$	$\ddots$	$\vdots$	$\vdots$		$\vdots$	$\vdots$		$\vdots$
x_m	0	$\ldots$	1	$\bar{a}_{m,m+1}$	$\ldots$	$\bar{a}_{mn}$	0		$\bar{b}_m$
$-z$	0	$\ldots$	0	$\bar{c}_{m+1}$	$\ldots$	$\bar{c}_n$	1		$\bar{z}$

$x_j \geq 0$ for all j, min z

"BV" is the abbreviation for "basic variable in this row." So, the present BFS is $\bar{x} = (\bar{x}_1, \dots, \bar{x}_m, \bar{x}_{m+1}, \dots, \bar{x}_n)^T = (\bar{b}_1, \dots, \bar{b}_m, 0, \dots, 0)^T$ with objective value $z(\bar{x}) = \bar{z}$.

Entries in the objective row (bottom row in which $-z$ is the basic variable) in the canonical tableau are known as the *updated* (or *relative* or *reduced*) *cost coefficients* of the variables wrt the basic vector x_B.

Theorem 4.7. Optimality Criterion in the Primal Simplex Algorithm: *In the canonical tableau wrt the feasible basic vector x_B, if all the relative cost coefficients $\bar{c}_j \geq 0$ for all j, then the present BFS $\bar{x}$ is an optimum solution for this LP.*

Proof. As the canonical tableau is obtained from the original tableau after performing some GJ pivot steps, the equation corresponding to every row in it is satisfied by every feasible solution of the original LP. Let K denote the set of feasible solutions of the original LP. So, for all $x \in K$ we have

$$\bar{c}_{m+1}x_{m+1} + \cdots + \bar{c}_n x_n - z(x) = -\bar{z}$$
$$\text{i.e.,}\quad z(x) = \bar{z} + (\bar{c}_{m+1}x_{m+1} + \cdots + \bar{c}_n x_n)$$

As $\bar{c}_{m+1}, \dots, \bar{c}_n$ are all ≥ 0 by hypothesis, and all $x_j \geq 0$ for all $x \in K$, $(\bar{c}_{m+1}x_{m+1} + \cdots + \bar{c}_n x_n) \geq 0$ for all $x \in K$, that is, $z(x) \geq \bar{z} = z(\bar{x})$ for all $x \in K$. So, $\bar{x}$ minimizes $z(x)$ over $x \in K$. □

If the optimality criterion is not satisfied, some nonbasic variables x_j are associated with $\bar{c}_j < 0$, each of them is eligible to be selected as the entering variable into this canonical tableau by the entering variable choice rule of the primal simplex algorithm. If the resulting pivot step is nondegenerate, the objective value decreases below $\bar{z}$ in this pivot step, and the present basic vector x_B can never reappear as the algorithm continues.

If this pivot step is a degenerate pivot step, there is no change in the objective value or the BFS in this step. Then a sequence of degenerate pivot steps may follow, and in this case the algorithm may return to the present basic vector x_B, completing a *cycle*, and the algorithm could keep moving around in the same cycle indefinitely. This is the problem of *cycling* in the primal simplex algorithm, for examples of it, see Dantzig (1963); Dantzig and Thapa (1997) of Chap. 1; Murty (1983) of Chap. 2

There are several techniques developed to resolve the problem of cycling under degeneracy in the simplex algorithm for LP. One prominent technique is based on *perturbation of the RHS vector*; it replaces the RHS constants vector b in the original LP by $b(\epsilon) = b + (\epsilon, \epsilon^2, \dots, \epsilon^m)^T$, where ϵ is a small positive parameter that is not given a specific numerical value, but left as a small positive parameter.

It can be shown that the perturbed problem does not have any degenerate BFSs when the parameter ϵ is positive but sufficiently small; so in solving it by the primal simplex algorithm, all the pivot steps will be nondegenerate, and hence the objective value undergoes a strict decrease in every step and no basic vector can reappear; that is cycling cannot occur. Also, the application of the algorithm on the perturbed problem can be carried out without using the actual value of the parameter ϵ, but using only updated values of the original data elements. The only change in the algorithm involves replacing the *minimum ratio computation* for determining the dropping basic variable in each pivot step by a little more complicated *lexico-minimum ratio computation*. This lexico minimum ratio test identifies the dropping basic variable uniquely in every pivot step. As the total number of possible basic vectors is finite, these facts imply that the algorithm must terminate after a finite number of pivot steps with a feasible basic vector for the perturbed problem, in which either the optimality or the unboundedness criterion is satisfied. As this final basic vector is feasible to the perturbed problem for all ϵ positive and sufficiently small, it can be verified that it is also feasible when $\epsilon = 0$, that is feasible to the original problem. So, when you plug in $\epsilon = 0$ at termination, you will have a feasible basic vector for the original problem, which satisfies the same termination condition, optimality, or unboundedness. This leads to the following theorem.

Theorem 4.8. *Starting with a feasible basic vector for an LP (4.31), the primal simplex algorithm (with techniques for resolving cycling as necessary) terminates with either an optimum basic feasible solution or with an extreme half-line of the set of feasible solutions along which the objective function diverges to $-\infty$, after a finite number of pivot steps.* □

The lexico-minimum ratio version of the primal simplex algorithm is theoretically simple and very elegant. Unfortunately, to implement it, one needs the explicit inverse of the basis in every step. In practice, the basis inverse tends to be very dense; computing and updating it is computationally intensive, and time consuming, and prone to serious round-off error accumulation; that is why the basis inverse is not used in practice. So, in practice this and other techniques developed to resolve cycling are not used. That is why we will not discuss these techniques in this book. Readers interested in theory who want to learn these techniques can refer to Dantzig (1963); Dantzig and Thapa (1997); Schrijver (1986) all from Chap. 1; Murty (1983) of Chap. 2.

The problem of cycling is encountered in practical computation also; software systems for LP use simple heuristic techniques that work very well to prevent cycling; we will discuss them in later chapters.

If a feasible basic vector for (4.31) is not available to initiate the primal simplex algorithm, we convert the problem of finding an initial feasible basic vector itself into a Phase I LP problem. This involves making b into a nonnegative vector (for each i such that the original $b_i < 0$, multiply both sides of the ith constraint by -1), then form the Phase I problem by introducing artificial variables. Using the same symbols for the data elements, one Phase I model is

$$\begin{aligned} \text{Minimize} \quad w &= \sum_{i=1}^{m} t_i \\ \text{subject to} \quad Ax + It &= b \\ x, t &\geq 0, \end{aligned}$$

where I denotes the unit matrix of order m, and $t = (t_1, \dots, t_m)^T$ is the vector of artificial variables introduced.

As now $b \geq 0$, t is a feasible basic vector for this Phase I problem. As $t \geq 0$, the minimum value for w in this problem is ≥ 0. Beginning with t as the basic vector, this Phase I problem can be solved by the lexico-minimum ratio version of the Primal simplex algorithm. After a finite number of pivot steps it will terminate. At termination:

- If the minimum value of w is >0, we conclude that the original LP is infeasible
- If it is $=0$, the Phase I basic vector at termination leads to a feasible basic vector for the original LP, with which its solution by the same algorithm can be initiated; this part of the process is usually called Phase II.

The combined Phase I and II is known as the *primal simplex method* for solving an LP; it uses the primal simplex algorithm twice, first to solve the Phase I problem to find a feasible basic vector for the original problem, and then to solve the original problem in Phase II.

Theorem 4.9. *When the primal simplex method is applied to solve an LP in which the objective function is to be minimized, after a finite number of pivot steps, it terminates with one of three outcomes: (1) conclusion that the LP is infeasible, (2) with an extreme half-line in the set of feasible solutions, along which the objective function diverges to $-\infty$, (3) with an optimum basic feasible solution of the problem.* □

4.13 Boundedness of Convex Polyhedra

$$\begin{aligned} Dx &= d, \\ Fx &\geq g. \end{aligned} \tag{4.32}$$

Let K denote the set of feasible solutions of the general system of linear constraints (4.32) in $x = (x_j : j = 1 \text{ to } n)$, where the inequalities include all the bound constraints on individual variables, if any. An important mathematical question is to determine whether K is bounded or not; that is, develop conditions under which we can conclude that it is bounded or not. This question involves another system of constraints related to (4.32), known as the *homogeneous system* corresponding to (4.32).

In the homogeneous system corresponding to (4.32), there will be a variable corresponding to each variable x_j in (4.32); in order to distinguish the two systems we will denote that corresponding variable in the homogeneous system by y_j for $j = 1$ to n. So, the vector of variables in the homogeneous system is $y = (y_1, \dots, y_n)^T$.

The simple rules for obtaining the homogeneous system are: Change the variables x in (4.32) into y, and change *every* RHS constant in every constraint (including any bounds on individual variables) into 0. The result is the homogeneous system, which is

$$\begin{aligned} Dy &= 0, \\ Fy &\geq 0. \end{aligned} \tag{4.33}$$

Remember that the homogeneous system is always feasible because 0 is a feasible solution for it. So, the important thing to determine is whether it has a nonzero feasible solution.

Example 4.21. Consider the following system of constraints:

$$\begin{aligned} x_1 - x_2 + x_3 + x_4 - x_5 - x_6 &= 100 \\ 2x_1 + 2x_2 - 3x_3 - 4x_4 + x_5 - 7x_6 &\geq 20 \\ -3x_1 - 3x_2 + 4x_3 + x_4 - 3x_5 + x_6 &\leq -40 \\ -20 \leq -x_1 + 4x_2 - 7x_3 - 2x_4 + x_5 + x_6 &\leq 50 \\ x_1 \text{ unrestricted, } x_2 \geq 0, \ x_3 &\leq 0 \\ 0 \leq x_4 \leq 10, \ x_5 \geq -6, \ x_6 &\leq 15 \end{aligned}$$

For deriving the homogeneous system, one should remember the following points:

- $x_2 \geq 10$ in original system becomes $y_2 \geq 0$ in homogeneous system.
- $-10 \leq x_3 \leq 12$ in original system becomes $0 \leq y_3 \leq 0$, or in effect $y_3 = 0$ in homogeneous system.
- $6 \leq 2x_1 + 10x_2 \leq 15$ in original system becomes $0 \leq 2y_1 + 10y_2 \leq 0$, in effect $2y_1 + 10y_2 = 0$ in homogeneous system.

Remembering these, the homogeneous system corresponding to the above system is

$$\begin{aligned} y_1 - y_2 + y_3 + y_4 - y_5 - y_6 &= 0 \\ 2y_1 + 2y_2 - 3y_3 - 4y_4 + y_5 - 7y_6 &\geq 0 \\ -3y_1 - 3y_2 + 4y_3 + y_4 - 3y_5 + y_6 &\leq 0 \\ -y_1 + 4y_2 - 7y_3 - 2y_4 + y_5 + y_6 &= 0 \\ y_1 \text{ unrestricted, } y_2 \geq 0, \ y_3 &\leq 0 \\ y_4 = 0, \ y_5 \geq 0, \ y_6 &\leq 0 \end{aligned}$$

We now state several important classical results on convex polyhedra.

Result 4.5. *If $\bar{x}$ is a feasible solution of the original system (4.32) and $\bar{y}$ is a nonzero feasible solution of the homogeneous system (4.33) corresponding to it, it can be verified directly that $\bar{x} + \lambda\bar{y}$ is also feasible to (4.32) for all $\lambda \geq 0$. Thus every nonzero feasible solution of the homogeneous system (4.33) is the direction for a feasible half-line for (4.32) at every feasible solution of (4.32).*

Result 4.6. Fundamental Theorem about Unboundedness of Convex Polyhedra: *When K is the set of feasible solutions of (4.32) is nonempty, it is bounded iff 0 is the only feasible solution of the homogeneous system (4.33) associated with it.*

This can be proved algorithmically using Theorem 4.9.

Result 4.7. Carathéodory's Theorem: *If $\bar{x} \in R^m$ is in the convex hull of $\{a^1, \dots, a^n\}$, then it can be expressed as a convex combination of at most $m+1$ points among $\{a^1, \dots, a^n\}$.*

This result follows from Theorem 4.5.

Now we state several results relating to a system in standard form. We consider the system in standard form

$$\begin{aligned} Ax &= b, \\ x &\geq 0, \end{aligned} \tag{4.34}$$

where A is a matrix of order $m \times n$. The homogeneous system corresponding to it is

$$\begin{aligned} Ay &= 0, \\ y &\geq 0. \end{aligned} \tag{4.35}$$

Result 4.8. *Every feasible solution of (4.34) can be expressed as (a convex combination of BFSs of (4.34)) + (a homogeneous solution).*

Result 4.9. *Every nonzero homogeneous solution (i.e., nonzero feasible solution of (4.35)) can be expressed as a convex combination of extreme points of the normalized homogeneous system*

$$\begin{aligned} Ay &= 0, \\ \sum_j y_j &= 1, \\ y &\geq 0. \end{aligned} \tag{4.36}$$

BFSs of (4.36) are known as *extreme homogeneous solutions* corresponding to (4.34). Result 4.8 can be proved by induction on the number of positive variables in the feasible solution, using the argument in the purification routine. Result 4.9 follows by applying Result 4.8 to (4.36). Following Results 4.13.6 and 4.13.7 follow from Result 4.8. Result 4.12 follows from Theorem 4.8.

Result 4.10. *Every feasible solution of (4.34) can be expressed as (a convex combination of BFSs of (4.34)) + a non-negative multiple of (a convex combination of extreme homogeneous solutions).*

Result 4.11. *If the set of feasible solutions of (4.34) is bounded (i.e., if 0 is the only feasible solution of the corresponding homogeneous system (4.35)) then every feasible solution of (4.34) can be expressed as a convex combination of its BFSs.*

A bounded convex polyhedron is called a **convex polytope**. An equivalent statement of the above is: Every point in a convex polytope can be expressed as a convex combination of its extreme points.

Result 4.12. *If the LP in standard form*

$$\begin{aligned} \textit{Minimize} \quad z(x) &= cx \\ Ax &= b \\ x &\geq 0 \end{aligned}$$

has a feasible solution, then the objective function z is unbounded below in it iff there exists an extreme homogeneous solution y satisfying $cy < 0$.

4.14 Exercises

4.1. Consider the LP

$$\begin{aligned} \text{Minimize} \quad & cx \\ \text{subject to} \quad & Ax = b \\ & x \geq 0 \end{aligned}$$

where A is a matrix of order $m \times n$ and rank m. In this LP the number of variables is n, and the number of constraints is m, and we know that a nondegenerate BFS for this problem will have exactly $(n - m)$ variables with value 0.

Now consider the special case of the above LP, the $p \times q$ balanced transportation problem with p sources, q sinks, availability vector $(a_1, \dots, a_p)$ at the sources, requirement vector $(b_1, \dots, b_q)$ at the sinks, with data satisfying $\sum_i a_i = \sum_j b_j$, which has $n = pq$ variables x_{ij}, $m = p + q$ equality constraints in the model.

$$\begin{aligned} \text{Minimize} \quad & \sum_i \sum_j c_{ij} x_{ij} \\ \text{subject to} \quad & \sum_j x_{ij} = a_i \quad \text{for all } i = 1 \text{ to } p \\ & \sum_i x_{ij} = b_j \quad \text{for all } j = 1 \text{ to } q \\ & x_{ij} \geq 0 \quad \text{for all } i,\ j. \end{aligned}$$

We know that every nondegenerate BFS for this problem will have exactly $n - m + 1$ variables at 0, rather than $n - m$ as seen above for the general LP. What is the reason for this discrepancy?

4.2. (Amy Cohn) Consider the LP

$$\begin{aligned} \text{Minimize} \quad & cx \\ \text{subject to} \quad & Ax = b \\ & x \geq 0, \end{aligned}$$

where A is a matrix of order $m \times n$ and rank m. Consider a feasible basic vector x_B for this LP associated with the BFS $\bar{x}$.

(a) In the canonical tableau wrt this basic vector x_B suppose that two or more nonbasic variables have negative reduced cost. Does this mean that the current BFS $\bar{x}$ is not optimal to the LP? Why or why not?
(b) Suppose that we have selected the nonbasic variable with the most negative reduced cost to enter the basic vector x_B and moved to a new BFS with an improved objective value. Will this lead to the largest improvement in the objective value in one pivot step from the basic vector x_B? Why or why not?
(c) Use your answer to part (b) to develop an entering variable choice rule in the primal simplex algorithm that will always yield the largest possible improvement in objective value in every pivot step.
(d) Will the entering variable choice rule developed in (c) lead to finding the optimal solution in the fewest possible pivot steps? Why or why not?

4.3. (Yash Aneja) Consider the LP

$$\begin{aligned} \text{Minimize} \quad & cx \\ \text{subject to} \quad & Ax = b \\ & x \geq 0, \end{aligned}$$

where A is a matrix of order $m \times n$ and rank m. $\bar{x}$ is a feasible solution to this LP, which is not a BFS, with objective value $\bar{z} = c\bar{x}$. It is required to find a BFS to this LP, if one exists, which has the same objective value $\bar{z}$. Discuss an approach for doing this. Mention whether this approach is guaranteed to produce a BFS with objective value $\bar{z}$, or conclude that no such BFS exists.

4.4. Consider the set of feasible solutions K of the system: $Ax = b, \ \ x \geq 0$, where $A, \ b$ are given. We are given that K is bounded.

Let $\bar{x}$ be a feasible solution to this problem; and suppose that the number of positive variables in $\bar{x}$ is r. Using mathematical induction on r, and a purification step and Result 4.3, show that every point in K can be expressed as a convex combination of extreme points of K.

4.5. Let K be the set of feasible solutions of the system: $Ax = b, \ \ x \geq 0$, where A is of order $m \times n$. We know that K is unbounded. Using the approach outlined in

Exercise 4.4 and Result 4.4, show that every feasible solution of this system can be expressed in the form:

(a convex combination of extreme points of K) + (a nonnegative combination of extreme homogeneous solutions corresponding to this system).

4.6. Let K be the set of feasible solutions of the system: $Ax = b,\ x \geq 0$, where A is of order $m \times n$. The solution $\bar{x}$ is a feasible solution of this system satisfying the property that the number of positive variables in it is the smallest among all the feasible solutions of the system. Then prove that $\bar{x}$ must be an extreme point of K.

4.7. Let $\theta(x)$ be a convex function defined over the bounded polytope $K = \{x : Ax = b,\ x \geq 0\}$, where A is a given matrix. Prove that the global maximum of $\theta(x)$ over K is attained at an extreme point of K.

4.8. Consider the LP: minimize cx over $x \in K$, where K is a convex polytope specified as the convex hull of a given set of points $\{x^1, \dots, x^r\}$. Show how to solve this problem efficiently.

4.9. Consider the LP: Minimize some objective function $z = cx$ subject to the following constraints in standard form. Assuming that the problem has an optimum solution, prove that there must be an optimum solution of the problem in which at least one of the two variables x_6 or x_7 is 0.

x_1	x_2	x_3	x_4	x_5	x_6	x_7	
1	0	0	0	3	1	−1	13
0	1	0	0	7	−2	2	29
0	0	1	0	−2	0	0	45
0	0	0	1	8	1	−1	83

$x_j \geq 0$ for all j

4.10. Let K denote the set of feasible solutions of the following system of constraints in standard form. Check whether $\bar{x} = (0, 10, 0, 15, 0, 5)^T$ is an extreme point of K. If it is not, find the smallest dimension face of K containing $\bar{x}$, and express it in parametric form.

x_1	x_2	x_3	x_4	x_5	x_6	b
3	1	−1	1	3	−2	15
7	0	5	1	−1	−1	10
−2	0	2	1	8	−1	10

$x_j \geq 0$ for all j

4.11. Let **K** denote the set of feasible solutions of the following system of linear constraints

$$\begin{aligned} \mathrm{Ax} &= \mathrm{b}, \\ \mathrm{Dx} &\geq \mathrm{d}, \end{aligned} \tag{5.37}$$

where A, D are given $m \times n$ and $p \times n$ matrices, respectively.

Write the geometric definition for an $\bar{x} \in \mathbf{K}$ to be an extreme point of $\mathbf{K}$.

Give a complete proof to show that $\bar{x} \in \mathbf{K}$ is an extreme point as defined geometrically iff it is the unique solution of the system of linear equations consisting of the active constraints in (5.37) at $\bar{x}$ treated as equations.

Write down the conditions that must be satisfied for $\bar{x}$ to be an extreme point of K, in terms of the rank of some set of vectors.

State conditions on K, which will guarantee the existence of a point in K at which all the inequality constraints in the system will hold as strict inequalities.

4.12. Let K be the set of feasible solutions of the following system of constraints.

x_1	x_2	x_3	x_4	x_5	x_6	x_7	
1	0	1	1	3	1	−1	=8
0	1	−1	0	0	3	2	=−1
1	−1	2	2	4	2	−2	=11
−1	2	−3	−2	−4	4	3	=−12
2	1	1	3	7	−2	4	=17

$x_j \geq 0$ for all j

Check whether $\bar{x} = (2, 1, 2, 1, 1, 0, 0)^T$ is an extreme point of K. If not, obtain an extreme point of K from $\bar{x}$ showing all your work.

If $\bar{x}$ is not an extreme point of K, is it contained on an edge of K? Explain.

4.13. Let $K_0 = \{x : Ax \geq b\}$. The first constraint in the system defining K_0 is $A_{1.}x \geq b_1$. Discuss what happens to the set of feasible solutions of the system as b_1 decreases from its present value. Show that there may be a value $\bar{b}_1$ for b_1, which is $\leq$ its present value, such that when b_1 decreases below $\bar{b}_1$, there is no further change in the set of feasible solutions of the system. Find an LP model for computing this $\bar{b}_1$.

Similarly discuss what happens to the set of feasible solutions of the system as b_1 increases from its present value. Show that there may be a value $\hat{b}_1$ for b_1, which is $\geq$ its present value, such that when b_1 increases above $\hat{b}_1$, the set of feasible solutions of the system becomes empty. Find an LP model for computing this $\hat{b}_1$.

4.14. Let $\Gamma = \{x : Ax = b\}$, where A is a matrix of order $m \times n$ and rank m. Does Γ have an extreme point? Why?

4.15. Consider the following LP in which A is of order $m \times n$ and rank r.

$$\begin{aligned} \text{minimize } & cx \\ Ax &= b \ (II) \\ x &\geq 0 \end{aligned}$$

Assume that (II) also has at least one feasible solution. Then does it have at least one extreme point solution?

Let $\bar{x}$ be an extreme point solution of (II). What can you say about the number of its adjacent extreme point solutions in the two cases when $\bar{x}$ is *nondegenerate, degenerate*?

4.16. Consider the parametric RHS−LP

$$\begin{aligned} \min \quad & z = cx \\ \text{subject to} \quad & Ax = b + \lambda b^* \\ & x \geq 0, \end{aligned}$$

where λ is a real-valued parameter. The problem is known to have an optimum solution when $\lambda = \lambda_1$ and again when $\lambda = \lambda_2 > \lambda_1$. Prove that the problem must have an optimum solution for all $\lambda_1 \leq \lambda \leq \lambda_2$, and that the optimum objective value function $\bar{z}(\lambda)$ as a function of λ is a convex function in this interval.

References

Grünbaum B (1967) Covex polytopes. Wiley, NY

Khachiyan L, Boros E, Borys K, Elbassioni K, Gurvich V (2006) Generating all vertices of a polyhedron is Hard. RUTCOR, Rutgers University, NJ

Murty KG (1971) Adjacency on convex polyhedra. SIAM Rev 13(3):377–386

Murty KG (2009) A problem in enumerating extreme points, and an efficient algorithm for one class of polytopes. Optim Lett 3(2):211–237

Murty KG, Chung SJ (1995) Segments in enumerating facets. Math Program 70:27–45

PORTA (Polyhedron Representation Transformation Algorithm), a software package for enumerating the extreme points of a convex polyhedron specified by linear constraints, can be downloaded from the website: http://www.iwr.uni-heidelberg.de/groups/comopt/software/PORTA/

Provan JS (1994) Efficient enumeration of the vertices of polyhedra associated with network LPs. Math Program 63:47–64

Ziegler GM (1994) Lectures on polytopes. Springer, New York

Chapter 5
Duality Theory and Optimality Conditions for LPs

5.1 The Dual Problem

Associated with every linear programming problem, there is another linear program called its *dual*, involving a different set of variables, but sharing the same data. When referring to the dual problem of an LP, the original LP is called *the primal* or *the primal problem*. Together, the two problems are referred to as a *primal, dual pair* of linear programs. The names *primal, dual* for the two problems are coined by Tobias Dantzig, father of George Dantzig, around 1955 in conversations with his son.

A duality type result for systems of linear equations only (no inequalities) is the theorem of alternatives for systems of linear equations (Theorem 1.1 in Sect. 1.2); it has been known for a long time (by the eighteenth century or even earlier), but similar results for systems of linear constraints including linear inequalities were unknown until recently. These important duality-type results for systems of linear constraints including inequalities known as *either/or theorems* or *theorems of alternatives* started appearing in published literature beginning in mid-nineteenth century.

These theorems show that a given system of linear constraints has a feasible solution iff another system in a different set of variables but constructed using the same data as the original system has no feasible solution. The first such necessary and sufficient condition for feasibility of a system involving linear inequalities seems to be that of Gordon in an 1873 paper:

> **Gordon's Theorem:** *The system* $Ax < 0$ *has a solution iff the alternate system* $yA = 0$, $y \geq 0$ *has no nonzero solution.*

At the end of the nineteenth and the first half of twentieth centuries, many such results were published by Farkas, Minkowski, Stienke, Motzkin, and several others. The most famous among them, that appeared in 1894, is:

> **Farkas' Lemma:** *The system* $Ax = b$, $x \geq 0$ *has a feasible solution iff the alternate system* $yA \leq 0$, $yb > 0$ *has no feasible solution.*

For a proof of Farkas' lemma based on the duality theorem for LPs, see Exercise 5.3 at the end of this chapter; for a direct proof of Farkas' lemma, see Exercises 5.36 and 5.37.

K.G. Murty, *Optimization for Decision Making: Linear and Quadratic Models*, International Series in Operations Research & Management Science 137,
DOI 10.1007/978-1-4419-1291-6_5,

The fundamental notion of *duality*, and the term itself, were introduced by von Neumann in conversations with Dantzig in 1947, and stated in a working paper written the same year. Gale, Kuhn, and Tucker independently formulated the duality theorem of linear programming and proved it rigorously using Farkas' lemma in 1951; and they received the John von Neumann Theory prize of INFORMS in 1980 for this work.

Duality plays a big role in LP theory, in this chapter we will discuss some of the most important theoretical aspects of duality with algorithmic and computational applications. Duality also plays a big role in nonlinear programming, see Bazaraa et al. (2006).

In Chap. 5 of Murty (2005b) of Chap. 1, we derived the dual of an LP in a certain form, purely from economic arguments, and explained how the dual variables can be interpreted as the marginal values associated with the RHS constants in the LP model, using the example of the Fertilizer problem that was discussed earlier in Sect. 1.3.1 and Examples 2.5 and 4.3. We reproduce that material here for the sake of completeness.

5.2 Deriving the Dual by Rational Economic Arguments

In this problem formulated in Sects. 1.3.1 and 1.7.1 of Chap. 1, the fertilizer manufacturer has a daily supply of 1,500 tons of RM 1, 1,200 tons of RM 2, and 500 tons of RM 3 from the company's quarries at a cost of \$50, \$75, \$60 per ton, respectively, for RM1, RM2, RM3. Presently these supplies can be used to manufacture Hi-ph or Lo-ph fertilizers to make profit. Relevant data from Sects. 1.3.1 and 1.7.1 is tabulated.

Item	Tons required to make one ton of		Maximum amount available daily (tons)
	Hi-ph	Lo-ph	
RM 1	2	1	1,500
RM 2	1	1	1,200
RM 3	1	0	500
Net profit (\$)/ton made	15	10	

The LP model for this problem is:

$$
\begin{array}{lrcrl}
\text{Maximize } z(x) = & 15x_1 & + & 10x_2 & \quad \text{Item} \\
\text{S. to} & 2x_1 & + & x_2 \leq 1500 & \quad \text{RM 1} \\
 & x_1 & + & x_2 \leq 1200 & \quad \text{RM 2} \\
 & x_1 & & \leq 500 & \quad \text{RM 3} \\
 & x_1 \geq 0, & & x_2 \geq 0, &
\end{array}
\tag{5.1}
$$

where the decision variables are

$$x_1 = \text{the tons of Hi-ph made per day},$$
$$x_2 = \text{the tons of Lo-ph made per day}.$$

There is a detergent company in the area that needs supplies of RM 1, 2, and 3. The detergent manufacturer wants to persuade the fertilizer manufacturer to give up fertilizer making, and instead sell the supplies of RM 1, 2, 3 to the detergent company. Being very profit conscious, the fertilizer manufacturer will not agree to this deal unless the prices offered by the detergent manufacturer for each of these raw materials fetch at least as much income as each of the options in the fertilizer making business.

In this problem, the fertilizer maker is measuring money in *net profit dollar* units (i.e., after subtracting the cost of raw materials and manufacturing costs from sales revenue dollars). Let the offer made by the detergent manufacturer be

$$\pi_i = \text{price/ton for RM}i, i = 1, 2, 3$$

in these same money units (i.e., as the cost of RM1, RM2, RM3 per ton is \$50, \$75, \$60, respectively, in real-life \$, the detergent manufacturer offers to pay $\$50 + \pi_1$, $\$75 + \pi_2$, $\$60 + \pi_3$ per ton of RM1, RM2, RM3, respectively). With this understanding, we will continue our discussion in the fertilizer maker's net profit dollar units for money, and dollar will refer to these units. Clearly, these prices π_1, π_2, π_3 have to be ≥ 0 for the deal to be acceptable to the fertilizer manufacturer.

Now consider the Hi-ph fertilizer making process. Manufacturing one ton of this fertilizer yields a net profit of \$15, and uses up 2 tons RM 1, 1 ton RM 2, and 1 ton RM 3. The same basket of raw materials fetches a price of $2\pi_1 + \pi_2 + \pi_3$ from the detergent manufacturer. So, the fertilizer manufacturer will not find the price vector $\pi = (\pi_1, \pi_2, \pi_3)$ acceptable unless $2\pi_1 + \pi_2 + \pi_3 \geq 15$.

Similar economic analysis with the Lo-ph fertilizer process leads to the constraint $\pi_1 + \pi_2 \geq 10$.

With the price vector π, the cost to the detergent company of acquiring the daily raw material supply is $1500\pi_1 + 1200\pi_2 + 500\pi_3$, and the detergent manufacturer would clearly like to see this minimized. Thus to make it acceptable to the fertilizer manufacturer, the price vector $\pi = (\pi_1, \pi_2, \pi_3)$ that the detergent manufacturer offers for the supplies of RM 1, 2, 3, should be an optimum solution for

$$\begin{array}{lllll}
\text{Minimize } v(\pi) = & 1500\pi_1+ & 1200\pi_2+ & 500\pi_3 & \\
\text{s. to} & 2\pi_1+ & \pi_2+ & \pi_3 & \geq 15 \\
 & \pi_1+ & \pi_2 & & \geq 10 \\
 & \pi_1, & \pi_2, & \pi_3 & \geq 0 \\
 & & \text{all } \pi_i \geq 0. & &
\end{array} \quad (5.2)$$

The LP (5.2) is the dual of (5.1), and vice versa. This pair of problems is a primal–dual pair of LPs. When considering the primal (5.1), the variables in its dual (5.2) are called the *dual variables*, and the slacks in (5.2) corresponding to the inequality constraints in it are called the *dual slack variables*.

As the first constraint in (5.2) comes from the economic analysis of the Hi-ph manufacturing process, this dual constraint is said to correspond to the Hi-ph primal variable x_1. Likewise, the second dual constraint in (5.2) corresponds to the primal variable x_2. In the same way, the dual variable π_1, the detergent manufacturer's price for the item RM 1, is associated with the RM 1 (first) primal constraint in (5.1). Similarly the dual variables π_2 *and* π_3 are associated with the second (RM 2) and third (RM 3) primal constraints in (5.1), respectively. Thus there is a dual variable associated with each primal constraint, and a dual constraint corresponding to each primal variable. Also, verify the following facts.

1. The coefficient matrix in the detergent manufacturer's problem (5.2) is just the transpose of the coefficient matrix in the fertilizer manufacturer's problem (5.1), and vice versa.
2. The RHS constants in (5.2) are the objective coefficients in (5.1), and vice versa.
3. Each variable in (5.1) leads to a constraint in (5.2), and vice versa.
4. LP (5.1) is a maximization problem in which the constraints are $\leq$ type; and (5.2) is a minimization problem in which the constraints are $\geq$ type.

5.2.1 Dual Variables are Marginal Values

The *marginal value of RM* i in the fertilizer manufacturer's problem is the rate of change in the maximum net profit per unit change in the availability of RM i from its present value; thus it is the net worth of one additional unit of RM i over the present supply, for $i = 1, 2, 3$, to the fertilizer manufacturer.

Hence, if the detergent manufacturer offered to buy RM i at a price $\geq$ its marginal value, for $i = 1, 2, 3$, the fertilizer manufacturer would find the deal acceptable. Being cost conscious, the detergent manufacturer wants to make the price offered for any raw material to be the smallest value that will be acceptable to the fertilizer manufacturer. Hence, in an optimum solution of (5.2), the π_i will be the marginal value of RM i, for $i = 1, 2, 3$, in (5.1). Thus the dual variables are the marginal values of the items associated with the constraints in the primal problem. These marginal values depend on the data, and may change if the data does.

5.2.2 The Dual of the General Problem in This Form

Now consider the general LP in the same form, it is

$$\begin{aligned} \text{Maximize } z(x) &= cx \\ \text{subject to } Ax &\leq b \\ x &\geq 0, \end{aligned} \tag{5.3}$$

where A is an $m \times n$ matrix. LPs in this form are said to be in *symmetric form*. From similar arguments it can be seen that the marginal values of this LP are the dual variables in the dual of this problem given below. Let π_i denote the dual variable associated with the ith constraint in this LP, $i = 1$ to m. If we write the vector of dual variables as a column vector, the statement of the dual problem will involve A^T as the coefficient matrix. But usually the vector of dual variables is written as the row vector $\pi = (\pi_1, \dots, \pi_m)$. Using it, the dual of the above LP is

$$\begin{aligned} \text{Minimize} \quad v(\pi) &= \pi b \\ \text{subject to} \quad \pi A &\geq c \\ \pi &\geq 0. \end{aligned} \tag{5.4}$$

We will discuss some of the relationships between the primal and dual problems after we discuss the dual of the LP in general form in the next section.

5.3 Rules for Writing the Dual of a General LP

In this section we describe the rules for writing the dual of an LP in general form. This dual is also based on economic arguments similar to those given above for LPs in symmetric form.

Remember that in describing the simplex method for solving an LP in (Murty (2005b) of Chap. 1), we required the transformation of the LP into standard form first. To write the dual of a general LP, there is really no need to transform it in any way first. We will describe how to write the dual of any LP as it is given, without carrying out any transformation on it. This, however, requires remembering the *right* and *wrong types of inequalities* for an LP, as defined below:

For a minimization problem: Any inequality constraint or bound constraint of the $\geq$ type is the right type; all other inequality constraints and bounds of the $\leq$ type are of the wrong type.

For a maximization problem: Any inequality constraint or bound constraint of the $\leq$ type is the right type; all other inequality constraints and bounds of the $\geq$ type are of the wrong type.

The question is: how does one remember these definitions easily? I always found the following helpful. I think of optimizing a single variable subject to any single bound constraint on it, for example: *optimize* x_1 subject to $x_1 \geq 6$.

If *optimize* in this statement is replaced by *maximize*, this problem has no optimum solution at all, suggesting that for a maximization problem "$\geq$" is of the wrong type (hence the right type is "$\leq$" for maximization). But if *optimize* is replaced by *minimize*, then this problem has an optimum solution $x_1 = 6$, suggesting that for a minimization problem "$\geq$" is of the right type (hence "$\leq$" is wrong type for minimization).

Here, the LP whose dual we are deriving is called the primal LP. Let a *primal constraint* refer to any condition in the primal involving two or more variables (including equations), or any nonzero lower or upper bound condition on a single variable. Hence the only primal conditions not counted as constraints for this purpose are the ≥ 0 or ≤ 0 restrictions on individual variables, which we call as *sign restrictions*. Remember that any primal variable for which both a lower and an upper bound constraints are included in the list of constraints is treated as an unrestricted variable, because the bounds already spell out the range within which this variable can vary.

For convenience in referring to them, we will assume that all the primal constraints are put in the form of a detached coefficient constraint tableau, each row corresponding to a constraint, each column corresponding to a variable; and b the RHS constants vector. Suppose x_j, associated with the column vector $\mathcal{A}_j$ in the tableau, and objective coefficient c_j, $j = 1$ to n, are the primal variables. Here are the rules for writing the dual of this LP.

1. *Defining dual variables:* Associate a separate dual variable to each primal constraint (i.e., row of the constraint tableau). Let π denote the row vector of dual variables in this order.
2. *Dual objective function:* The dual objective function is $\pi b +$ (any constant term in the primal objective function). If primal is a minimization problem, dual is a maximization problem and vice versa.
3. *Sign restrictions on dual variables:* Dual variables associated with (1) primal equality constraints are unrestricted in sign, (2) right (wrong) type of primal inequality constraints are nonnegative (non-positive) variables in the dual problem.
4. *Dual constraints:* There is a dual constraint corresponding to each primal variable. That, associated with x_j, is (1) $\pi \mathcal{A}_j = c_j$ if x_j is an unrestricted variable in primal, (2) $\pi \mathcal{A}_j \geq c_j$ or $\pi \mathcal{A}_j \leq c_j$ whichever is the right (wrong) type for the dual problem if x_j has the nonnegativity (nonpositivity) restriction in the primal.

Example 5.1. Consider the following LP: minimize $z' = -3x_1 - 4x_2 + 5x_3 - 6x_5 + 7x_6 - 35$ subject to constraints given below. The constant term -35 in the objective function z' does not depend on the feasible solution, so finding an optimum solution minimizing z' is the same as that minimizing $z = z' - (-35)$. In the sequel we consider minimizing z. The problem with constraints arranged in a detached coefficient tableau are shown next.

$$
\begin{aligned}
x_1 + x_2 - x_3 - 2x_4 + 3x_6 &= -17 \\
-x_1 - x_2 + 2x_3 - 4x_4 + 6x_5 - 3x_6 &\geq -18 \\
2x_1 - 3x_2 + 3x_3 - x_4 + 4x_5 &\leq 40 \\
x_2 \geq 0,\ x_3 \leq 0,\ 2 \leq x_4 \leq 15,\ 0 \leq x_5 \leq 6,\quad x_6 &\geq 3, \\
x_1 \text{ unrestricted}
\end{aligned}
$$

x_1	x_2	x_3	x_4	x_5	x_6		RHS	Associated dual var.
1	1	−1	−2	0	3	=	−17	π_1
−1	−1	2	−4	6	−3	$\geq$	−18	π_2
2	−3	3	−1	4	0	$\leq$	40	π_3
0	0	0	1	0	0	$\geq$	2	π_4
0	0	0	1	0	0	$\leq$	15	π_5
0	0	0	0	1	0	$\leq$	6	π_6
0	0	0	0	0	1	$\geq$	3	π_7
−3	−4	5	0	−6	7	= z minimize		
x_1, x_4, x_6 unrestricted; $x_2, x_5 \geq 0$, $x_3 \leq 0$								

The dual problem in detached coefficient form is

π_1	π_2	π_3	π_4	π_5	π_6	π_7		RHS	Assoc. var.
1	−1	2	0	0	0	0	=	−3	x_1
1	−1	−3	0	0	0	0	$\leq$	−4	x_2
−1	2	3	0	0	0	0	$\geq$	5	x_3
−2	−4	−1	1	1	0	0	=	0	x_4
0	6	4	0	0	1	0	$\leq$	−6	x_5
3	−3	0	0	0	0	1	=	7	x_6
−17	−18	40	2	15	6	3	= v maximize		
π_1 unrestricted, $\pi_2, \pi_4, \pi_7 \geq 0$, and $\pi_3, \pi_5, \pi_6 \leq 0$.									

Verify the following

1. The coefficient matrix of the dual constraints is the transpose of the coefficient matrix of the primal constraints.
   The primal RHS constants vector is the dual objective coefficient vector and vice versa.
2. The dual of the dual problem is the primal problem. That is why the relationship between a primal LP and its dual is said to be a *symmetric relationship*. ■

5.3.1 Complementary Pairs in a Primal, Dual Pair of LPs

Each inequality constraint in the primal LP corresponds to its own slack variable; in fact, after rearranging the terms, if the inequality is put in the form

$$\text{(some expression in the variables)} \geq 0,$$

the left-hand side of this inequality is the expression for the primal slack variable corresponding to this inequality. This primal inequality constraint corresponds to a sign-restricted dual variable in the dual problem. This primal slack variable and the corresponding sign-restricted dual variable are said to correspond to each other; together they form a pair known as a *complementary pair of variables* in this primal, dual pair of LPs. In the same way, each dual slack variable and the corresponding sign-restricted primal variable correspond to each other, and together they form a complementary pair of variables.

Thus each complementary pair consists of a primal (or dual) slack variable and its corresponding sign-restricted dual (or primal) variable, and in writing a complementary pair, both quantities in it are always written in a nonnegative form (i.e., if a quantity is nonpositive, its negative would be entered in its place in the complementary pair). Also, in writing each complementary pair, usually the primal quantity is written first, then the dual quantity is written next.

Example 5.2. Here we will list the various complementary pairs in the primal, dual pair of LPs from Example 5.1. They are: $(-x_1-x_2+2x_3-4x_4+6x_5-3x_6+18,\ \pi_2)$, $(40-2x_1+3x_2-3x_3+x_4-4x_5,\ -\pi_3)$, $(x_4-2,\ \pi_4)$, $(15-x_4,\ -\pi_5)$, $(6-x_5,\ -\pi_6)$, $(x_6-3,\ \pi_7)$, $(x_2,\ -4-\pi_1+\pi_2+3\pi_3)$, $(-x_3,\ -\pi_1+2\pi_2+3\pi_3-5)$, $(x_5,\ -6-6\pi_2-4\pi_3-\pi_6)$. ■

5.3.2 What Is the Importance of Complementary Pairs?

Complementary pairs have a very important role in the optimality conditions for an LP. A condition known as the *complementary slackness condition* or *property* states that a pair of primal, dual feasible solutions is optimal to the respective problems iff at least one quantity in every complementary pair for these solutions is 0 (this is discussed in Theorem 5.5 in the next section). The name *complementary pair* actually refers to this property, because if we know that one of the quantities in the pair is positive, the other must be 0 for optimality to hold.

Also, given an arbitrary pair of primal, dual feasible solutions, the *duality gap* in this pair, defined as a measure of how far these solutions are from being optimal to the respective problems, is shown to be equal to the sum of the products of various complementary pairs in this pair (see Theorem 5.6 in the next section).

5.3.3 Complementary Pairs for LPs in Standard Form

Consider the primal LP

$$\begin{aligned} \text{Minimize } z &= cx \\ \text{subject to } Ax &= b \\ x &\geq 0, \end{aligned} \tag{5.5}$$

where A is a matrix of order $m \times n$. Associating the dual variable π_i to the ith constraint in (5.5), we have the row vector of dual variables $\pi = (\pi_1, \dots, \pi_m)$. Then the dual problem is

$$\begin{aligned} \text{Maximize} \quad v &= \pi b \\ \text{subject to} \quad \pi A_{.j} &\le c_j, \quad j = 1 \text{ to } n \\ \pi \quad & \text{unrestricted}, \end{aligned} \tag{5.6}$$

or in matrix notation the dual problem is

$$\begin{aligned} \text{Maximize} \quad v &= \pi b \\ \text{subject to} \quad \pi A &\le c \\ \pi \quad & \text{unrestricted}. \end{aligned}$$

The various complementary pairs in this primal, dual pair of LPs are $(x_j, \bar{c}_j = c_j - \pi A_{.j})$, $j = 1$ to n. The dual slack $c_j - \pi A_{.j}$ corresponding to x_j is usually denoted by $\bar{c}_j$; we will show later that it is exactly the relative (or reduced) cost coefficient of the primal variable x_j computed in the simplex algorithm.

Example 5.3. Consider the LP in standard form in the following detached coefficient tableau

x_1	x_2	x_3	x_4	x_5	x_6	RHS	Associated dual var.
1	2	3	−2	1	16	17	π_1
0	1	−4	1	1	1	2	π_2
0	0	1	−2	1	0	1	π_3
3	11	−15	10	4	57	$= z$ minimize	

$x_j \ge 0$ for all j

Here is the dual problem in detached coefficient tableau form.

π_1	π_2	π_3	RHS		Associated primal var.
1	0	0	$\le$	3	x_1
2	1	0	$\le$	11	x_2
3	−4	1	$\le$	−15	x_3
−2	1	−2	$\le$	10	x_4
1	1	1	$\le$	4	x_5
16	1	0	$\le$	57	x_6
17	2	1	$= v$ maximize		

So, the complementary pairs in this primal, dual pair are $(x_1, \bar{c}_1 = 3 - \pi_1)$, $(x_2, \bar{c}_2 = 11 - 2\pi_1 - \pi_2)$, $(x_3, \bar{c}_3 = -15 - 3\pi_1 + 4\pi_2 - \pi_3)$, $(x_4, \bar{c}_4 = 10 + 2\pi_1 - \pi_2 + 2\pi_3)$, $(x_5, \bar{c}_5 = 4 - \pi_1 - \pi_2 - \pi_3)$, $(x_6, \bar{c}_6 = 57 - 16\pi_1 - \pi_2)$. ■

5.3.4 Complementary Pairs for LPs in Symmetric Form

Consider the primal LP

$$\begin{aligned} \text{Minimize } z &= cx \\ \text{subject to } Ax &\geq b \\ x &\geq 0, \end{aligned} \tag{5.7}$$

where A is a matrix of order $m \times n$. Associating the dual variable π_i to the ith constraint in (5.7), we have the row vector of dual variables $\pi = (\pi_1, \cdots, \pi_m)$. Then the dual problem is

$$\begin{aligned} \text{Maximize } v &= \pi b \\ \text{subject to } \pi A_{.j} &\leq c_j, \quad j = 1 \text{ to } n \\ \pi &\geq 0, \end{aligned} \tag{5.8}$$

or in matrix notation the dual problem is

$$\begin{aligned} \text{Maximize } v &= \pi b \\ \text{subject to } \pi A &\leq c \\ \pi &\geq 0. \end{aligned}$$

When all the primal constraints are inequalities of the right type and all the primal variables are nonnegative variables, the dual is also in the same form; that is why LPs in this form are said to be in *symmetric form*. The constraints in the primal are $A_{i.}x \geq b_i$, $i = 1$ to m; hence the primal slack variables are $A_{i.}x - b_i$ themselves for $i = 1$ to m. Likewise the dual slack variables are $c_j - \pi A_{.j}$, $j = 1$ to n. So, we see that the various complementary pairs in this primal, dual pair of LPs are $(A_{i.}x - b_i, \pi_i)$, $(x_j, c_j - \pi A_{.j})$, $i = 1$ to m, $j = 1$ to n.

Example 5.4. Consider the LP in symmetric form in the following detached coefficient tableau

x_1	x_2	x_3	RHS	Associated dual var.
1	1	−1	≥ 10	π_1
0	3	7	≥ 13	π_2
6	7	0	$= z$ minimize	

$x_j \geq 0$ for all j

Here is the dual problem in detached coefficient tableau form.

π_1	π_2	RHS	Associated primal var.
1	0	≤ 6	x_1
1	3	≤ 7	x_2
-1	7	≤ 0	x_3
10	13	$= v$ maximize	

$$\pi_1, \pi_2 \geq 0.$$

So, the various complementary pairs in this pair of primal, dual LPs are $(x_1 + x_2 - x_3 - 10, \pi_1)$, $(3x_2 + 7x_3 - 13, \pi_2)$, $(x_1, 6 - \pi_1)$, $(x_2, 7 - \pi_1 - 3\pi_2)$, $(x_3, \pi_1 - 7\pi_2)$. ■

5.3.5 *Complementary Pairs for LPs in Bounded Variable Standard Form*

Consider the primal LP

$$\begin{aligned} \text{Minimize} \quad & z = cx \\ \text{subject to} \quad & Ax = b \\ & \ell \leq x \leq k, \end{aligned}$$

where A is a matrix of order $m \times n$ and $\ell = (\ell_j)$, $k = (k_j)$ are finite vectors satisfying $\ell < k$. The problem in matrix notation in detached coefficient form is

x	RHS	Associated dual var.
A	$= b$	$\pi = (\pi_1, \cdots, \pi_m)$
I	$\geq \ell$	$\mu = (\mu_1, \cdots, \mu_n)$
I	$\leq k$	$\nu = (\nu_1, \cdots, \nu_n)$
c	$= z$ minimize.	

where I is the unit matrix of order n. Associating with the primal constraints, the dual variables as shown in the above tableau; the dual problem in matrix notation is

$$\begin{aligned} \text{Maximize} \quad & v = \pi b + \mu \ell + \nu k \\ \text{s. to} \quad & \pi A + \mu + \nu = c \\ & \pi \text{ unrestricted}, \quad \mu \geq 0, \quad \nu \leq 0. \end{aligned}$$

Denoting ν as $-\delta = (-\delta_1, \cdots, -\delta_n)$, the dual problem is

$$\begin{aligned} \text{Maximize } v &= \pi b + \mu\ell - \delta k \\ \text{s. to } \pi A + \mu - \delta &= c \\ \pi \text{ unrestricted, } &\mu, \delta \geq 0. \end{aligned}$$

There are n dual constraints, if you show them individually, the dual problem becomes

$$\begin{aligned} \text{Maximize } v &= \pi b + \mu\ell - \delta k \\ \text{s. to } \pi A_{.j} + \mu_j - \delta_j &= c_j, \quad j = 1, \cdots, n \\ \pi \text{ unrestricted, } &\mu, \delta \geq 0. \end{aligned}$$

The various complementary pairs in this primal, dual pair of LPs are $(x_j - \ell_j, \mu_j)$, $j = 1, \cdots, n$; $(k_j - x_j, \delta_j)$, $j = 1, \cdots, n$.

Exercises

5.3.1. Write the duals of the following problems, list all the complementary pairs in each of them; write the dual of the dual and verify that it is the original problem.

(a) Maximize $-17x_1 + 13x_3 - 25x_4 + 3x_5$ subject to

$$\begin{aligned} -3x_2 + 7x_3 - 20x_4 + 6x_5 - x_6 + x_7 &= -29 \\ 5x_1 - 9x_3 + 15x_5 - 66x_6 &\geq -13 \\ 2x_1 + 5x_2 + 9x_3 + 22x_4 + 34x_5 + 7x_6 + 2x_7 &\leq 133 \\ x_1 \geq 0,\ x_2 \leq 0,\ -2 \leq x_3 \leq 12,\ 0 \leq x_4 \leq 10, &\quad x_5 \leq 37 \\ x_6 \geq -8,\ x_7 \text{ unrestricted.} \end{aligned}$$

(b) Minimize $3x_1 + 6x_2 - 7x_3 + 9x_4$ subject to $-7 \leq x_1 \leq 7, 8 \leq x_2 \leq 38, 0 \leq x_3 \leq 6, -12 \leq x_4 \leq 0$.

(c) Maximize $2x_1 - 3x_2 + 4x_3 - 5x_4 - 6x_5 - 4x_6 + 10x_7 + 2x_8$ subject to (in the following tableau blank entries are 0)

x_1	x_2	x_3	x_4	x_5	x_6	x_7	x_8	RHS
1	1	1	1	−2	3		−3	≥ 28
	1	1	1	2	1	−4	−1	≤ 20
		1	1	1	−5	3	4	$= 15$
			1	−3	2	−1	1	$= 8$

all $x_j \geq 0$.

(d) Minimize $3x_1 + 5x_2 - 7x_3 + 4x_5 - 5x_6$ subject to $6x_1 + 2x_2 + 9x_3 + 3x_4 + 11x_5 + 7x_6 = 150$ and $5 \leq x_j \leq 25$ for all j.

(e) Minimize $z(x) = \sum_{j=1}^{n} c_j x_j$ subject to $\ell_j \leq x_j \leq k_j$ for all j and $\sum_{j=1}^{n} a_{ij} x_j \geq b_i$ for $i = 1$ to m.

5.4 Duality Theory and Optimality Conditions for LP

In this section we derive results relating the primal and dual LPs, and optimality conditions for them.

Theorem 5.1. *Duals of equivalent LPs are equivalent.*

Proof. We know that LPs can be transformed from one form (standard form, symmetric form, etc.) to another by simple transformations like introducing slack variables to make an inequality into an equation, replacing an equation by an equivalent pair of opposite inequalities, expressing an unrestricted variable as a difference of two nonnegative variables, multiplying both sides of an inequality by -1 and changing its direction, row operations on equality constraints, etc. These transformations maintain the problems equivalent. When this is done, this theorem says that the dual of the transformed problem is equivalent to the dual of the original.

We will first show that the theorem holds when an LP in general form is transformed into symmetric form. Consider the LP in general form:

$$\begin{aligned} \text{Minimize} \quad z(x) &= cx \\ \text{s. to} \quad Dx &= d \\ Fx &\geq g, \end{aligned} \tag{5.9}$$

where D is a matrix of order $m \times n$ and F is a matrix of order $p \times n$. Its dual is

$$\begin{aligned} \text{Minimize} \quad v(\pi, \mu) &= \pi d + \mu g \\ \text{s. to} \quad \pi D + \mu F &= c \\ \mu &\geq 0, \end{aligned} \tag{5.10}$$

where $\pi = (\pi_1, \cdots, \pi_m)$, $\mu = (\mu_1, \ldots, \mu_p)$ are row vectors of dual variables.

To transform (5.9) into symmetric form, we need to replace the system of equations $Dx = d$ by the opposing pair of inequalities $Dx \geq d$, $Dx \leq d$; and express the vector of unrestricted variables x as difference of two vectors of nonnegative variables, say, $x = x^+ - x^-$ with x^+, x^- both ≥ 0. This leads to the equivalent transformed problem given in the following detached coefficient tableau using matrix notation.

x^+	x^-	RHS	Associated dual vector
D	$-D$	$\geq d$	δ
$-D$	D	$\geq -d$	γ
F	$-F$	$\geq g$	ν
c	$-c$	$= z$ minimize	

$x^+, x^- \geq 0$

Associating the dual vectors (all row vectors) as indicated in the tableau, we see that the dual of this transformed problem is

$$\begin{aligned} \text{Maximize} \quad & (\delta - \gamma)d + \nu g \\ \text{s. to} \quad & (\delta - \gamma)D + \nu F \leq c \\ & -[(\delta - \gamma)D + \nu F] \leq -c \\ & \delta,\ \gamma,\ \nu \geq 0. \end{aligned} \tag{5.11}$$

The second constraint here is the same as $[(\delta - \gamma)D + \nu F] \geq c,$ which together with the first is equivalent to $(\delta - \gamma)D + \nu F = c$. Also, when $\delta,\ \gamma \geq 0,\ \delta - \gamma = \pi$ is a vector of unrestricted variables. Then we see that (5.11) is equivalent to the dual (5.10) of the original LP (5.9).

Using similar simple arguments, we can verify that the theorem holds in general. □

Using Theorem 5.1, we see that the subsequent duality results that we will discuss need to be proved for any one form of LP, then by this theorem they hold for LP in any form.

Theorem 5.2. *In a primal, dual pair of LPs, both problems may be infeasible.*

Proof. By example. Consider the primal LP

$$\begin{aligned} \text{Minimize} \quad z = \ & 2x_1 - 4x_2 \\ \text{s. to} \quad & x_1 - x_2 = 1 \\ & -x_1 + x_2 = 2 \\ & x_1,\ x_2 \geq 0. \end{aligned}$$

Its dual is

$$\begin{aligned} \text{Maximize} \quad v = \ & \pi_1 + 2\pi_2 \\ \text{s. to} \quad & \pi_1 - \pi_2 \leq 2 \\ & -\pi_1 + \pi_2 \leq -4 \\ & \pi_1,\ \pi_2 \geq 0. \end{aligned}$$

When we add the two primal equality constraints we get the *fundamental inconsistent equation* $0 = 3$; so the primal is infeasible. When we add the two dual "$\leq$" inequalities, we get the *fundamental inconsistent inequality* $0 \leq -2$; so the dual is infeasible too. □

5.4.1 The Importance of Good Lower Bounding Strategies in Solving Optimization Problems

A general problem considered in optimization theory is of the form

$$\text{Minimize } z(x) \qquad \text{Subject to some specified constraints on } x. \tag{5.12}$$

Optimization theory is concerned with developing techniques useful to handle problems like (5.12). As maximizing a function $z'(x)$ is the same as minimizing $-z'(x)$, there is no need to discuss maximization problems separately, that is why much of optimization literature is written in terms of minimization.

Let z^* denote the unknown minimum value of $z(x)$ in (5.12). To size up the unknown quantity z^*, it will be helpful to get both an upper bound and a lower bound for it if these bounds can be computed by techniques that are much simpler than computing z^* directly.

Also, a lower bound like $-\infty$ (or the negative of a very large positive number) is really of no value; the quality (i.e., usefulness) of a bound improves as the difference between it and the unknown z^* gets smaller.

If $\bar{x}$ is any feasible solution for (5.12), clearly $z^* \leq z(\bar{x})$, and hence $z(\bar{x})$ is an upper bound for z^*. The process of computing a feasible solution for (5.12) is part and parcel of solving (5.12) itself, it happens automatically in any successful algorithm for solving (5.12), and the quality of the upper bound improves automatically as better and better solutions are obtained in the algorithm. That is why computing an upper bound for z^* is not considered a great achievement.

However, computing a lower bound for z^* does not appear to be a necessary part of an algorithm for solving (5.12). But a good lower bound for z^* when combined with upper bounds for z^* obtained during the algorithm can provide very valuable information about the unknown value of z^*.

That is why an efficient technique for computing a nontrivial good quality lower bound for z^* is very valuable. In optimization literature, techniques like this are usually called *lower bounding strategies*.

Lower bounding strategies in minimization problems correspond to upper bounding strategies in maximization problems, but as much of optimization literature is in terms of minimization, both these are usually clubbed under lower bounding strategies.

An effective lower bounding strategy for LPs comes from duality theory; the result known as the *weak duality theorem* is the first significant result proved in LP theory. It was proved by Gale, Kuhn, and Tucker in the early 1950s, and they received the Von Neumann Theory Prize of INFORMS (American Society for OR and Management Science) in 1980 for their contribution. We will discuss this next.

Theorem 5.3. The Weak Dualty Theorem: *Consider a primal, dual pair of LPs in which the primal is the minimization problem with objective function $z(x)$ and the dual is the maximization problem with objective function $v(\pi)$. If $\bar{x}$, $\bar{\pi}$ are, respectively, primal, dual feasible solutions, then $z(\bar{x}) \geq v(\bar{\pi})$.*

So, for any dual feasible solution $\bar{\pi}$, $v(\bar{\pi})$ is a lower bound for the minimum objective value in the primal. Likewise, for any primal feasible solution $\bar{x}$, $z(\bar{x})$ is an upper bound for the maximum objective value in the dual.

Proof. We will consider the primal, and dual in symmetric form.

Primal (P)	Dual (D)
Minimize $z(x) = cx$	Maximize $v(\pi) = \pi b$
s. to $Ax \geq b$	s. to $\pi A \leq c$
$x \geq 0$	$\pi \geq 0$

where A is a matrix of order $m \times n$, and $x = (x_1, \dots, x_n)^T$, $\pi = (\pi_1, \dots, \pi_m)$ are vectors of primal, dual variables, respectively.

As $\bar{x}$ is primal feasible, we have $A\bar{x} \geq b$. Also, as $\bar{\pi}$ is dual feasible, $\bar{\pi} \geq 0$, that is why we can multiply both sides of $A\bar{x} \geq b$ on the left by $\bar{\pi}$, leading to

$$\bar{\pi} A\bar{x} \geq \bar{\pi} b. \tag{5.13}$$

As $\bar{\pi}$ is dual feasible, we have $\bar{\pi} A \leq c$. Also, as $\bar{x}$ is primal feasible, $\bar{x} \geq 0$, that is why we can multiply both sides of $\bar{\pi} A \leq c$ on the right by $\bar{x}$, leading to

$$c\bar{x} \geq \bar{\pi} A\bar{x}. \tag{5.14}$$

Combining (5.13) and (5.14), we have $c\bar{x} \geq \bar{\pi} b$, that is, $z(\bar{x}) \geq v(\bar{\pi})$. So, the result in this theorem is true for this primal, dual pair of LPs (P), (D).

Any LP can be transformed into this symmetric form, and by Theorem 5.1 the result in this theorem holds for that LP and its dual. Also, this theorem can be proved for LP in any form directly using arguments similar to those used above. □

Corollaries of the Weak Duality Theorem: All the following results are direct consequences of the weak duality theorem, Theorem 5.3.

1. *Sufficient Optimality Criterion for LP:* Let $\bar{x}$ be a feasible solution for an LP. If you can find a feasible solution $\bar{\pi}$ for its dual, which also satisfies primal objective value at $\bar{x}$ = dual objective value at $\bar{\pi}$, then $\bar{x}$ is an optimum solution for the primal, and $\bar{\pi}$ is an optimum solution of the dual.

Example 5.5. Consider the fertilizer manufacturer's problem, and its dual discussed in Sect. 5.2, reproduced below.

Primal (5.1)	Dual (5.2)
Max. $z(x) = 15x_1 + 10x_2$	Min. $v(\pi) = 1500\pi_1 + 1200\pi_2 + 500\pi_3$
s. to $2x_1 + x_2 \leq 1500$	s. to $2\pi_1 + \pi_2 + \pi_3 \geq 15$
$x_1 + x_2 \leq 1200$	$\pi_1 + \pi_2 \geq 10$
$x_1 \leq 500$	$\pi_1, \pi_2, \pi_3 \geq 0$
$x_1, x_2 \geq 0$	

Let $\bar{x} = (300, 900)^T$. We verify that $\bar{x}$ is primal feasible and that $z(\bar{x}) = \$13{,}500$. Given the dual solution $\bar{\pi} = (5, 5, 0)$, we find that it is dual feasible and that $v(\bar{\pi}) = \$13{,}500$, the same as $z(\bar{x})$. By the weak duality theorem, this information is enough to conclude that $\bar{x}$, $\bar{\pi}$ are optimum solutions of the primal and the dual. ■

2. *Dual Infeasibility and Primal Unboundedness:* If the primal is feasible, and the objective value to be minimized in it, $z(x) \to -\infty$ on its feasible solution set, then the dual must be infeasible.
3. *Primal Infeasibility and Dual Unboundedness:* If the dual is feasible, and the objective value to be maximized in it, $v(\pi) \to \infty$ on its feasible solution set, then the primal must be infeasible.
4. *Bounds on Objective Values:* In a primal, dual pair of LPs, (the minimum objective value in the minimization problem in the pair) is $\geq$ (the maximum objective value in the maximization problem in the pair).

5.4.2 Definition of the Dual Solution Corresponding to Each Primal Basic Vector for an LP in Standard Form

Consider the primal LP in standard form

$$\begin{aligned} \text{Minimize} \quad z &= cx \\ \text{subject to} \quad Ax &= b \\ x &\geq 0, \end{aligned} \tag{5.15}$$

Basic			Nonbasic						
x_1	$\dots$	x_m	x_{m+1}	$\dots$	x_j	$\dots$	x_n	$-z$	RHS
a_{11}	$\dots$	a_{1m}	$a_{1,m+1}$	$\dots$	a_{1j}	$\dots$	a_{1n}	0	b_1
$\vdots$		$\vdots$	$\vdots$		$\vdots$		$\vdots$	$\vdots$	$\vdots$
a_{m1}	$\dots$	a_{mm}	$a_{m,m+1}$	$\dots$	a_{mj}	$\dots$	a_{mn}	0	b_m
c_1	$\dots$	c_m	c_{m+1}	$\dots$	c_j	$\dots$	c_n	1	0

$x_j \geq 0$ for all j

where $A = (a_{ij})$ is a matrix of order $m \times n$, and rank m. A vector of m primal variables$x_B = (x_1, \cdots, x_m)$ say is a *basic vector* for this LP if the set of column vectors in A of primal basic variables in x_B is linearly independent; that is, if the square matrix B consisting of those column vectors is nonsingular, in this case B is said to be the *basis* corresponding to the basic vector x_B. We can rearrange the variables in the detached coefficient tableau representation of this problem into basic, nonbasic parts as shown in the tableau given above.

$$B = \begin{pmatrix} a_{11} & \dots & a_{1m} \\ \vdots & & \vdots \\ a_{m1} & \dots & a_{mm} \end{pmatrix}.$$

The vector $c_B = (c_1, \ldots, c_m)$, the subvector of the original cost vector c corresponding to the basic variables in x_B is known as the *basic cost vector* corresponding to the basic vector x_B. Denoting the vector of nonbasic variables (i.e., those primal variables not in x_B) by the symbol x_D, the submatrix of A corresponding to them by D, and the vector of their cost coefficients by c_D, the tableau in partitioned form in matrix notation is

x_B	x_D	$-z$	RHS
B	D	0	b
c_B	c_D	1	0

The *primal basic solution* corresponding to the basic vector x_B is defined by fixing all the nonbasic variables at their lower bound 0, and then using the equality constraints to solve for the values of the basic variables. It is

$$\begin{aligned} \bar{x}_D &= 0, \\ \bar{x}_B &= B^{-1}b, \\ \text{objective value } z = c_B\bar{x}_B + c_D\bar{x}_D &= (c_B B^{-1})b. \end{aligned}$$

This primal basic solution is only guaranteed to satisfy the equality constraints in (5.15), it may or may not satisfy the nonnegativity constraints. This solution is called a *basic feasible solution (BFS)* if it is also nonnegative, that is, if $B^{-1}b \geq 0$. In this case, x_B and B are said to be the *primal feasible basic vector and basis*, respectively.

If $B^{-1}b \not\geq 0$ (i.e., it has at least one negative component), then x_B, B are said to be *primal infeasible basic vector, basis*, respectively.

Even though the basic vector x_B consists of primal variables only and has nothing to do directly with the dual problem, it is convenient to also associate a dual basic solution corresponding to it. This is done by the following:

> the *dual basic solution* corresponding to the primal basic vector x_B is defined to be the unique solution of the system of dual constraints corresponding to the basic variables in x_B treated as equations, ignoring all the other dual constraints.

The dual of (5.15) is

$$\begin{aligned} &\text{Maximize } v(\pi) = \pi b \\ &\text{subject to } \pi A_{.j} \leq c_j, \quad j = 1, \ldots, n, \end{aligned}$$

where the jth constraint here corresponds to x_j. So, the system of dual constraints corresponding to basic variables in $x_B = (x_1, \cdots, x_m)$ treated as equations is

$$\pi A_{.j} = c_j, \quad j = 1, \ldots, m$$

or $\pi B = c_B$. This has the solution $\bar{\pi} = c_B B^{-1}$, which is defined as the dual basic solution associated with the primal basic vector x_B.

With this definition, the dual slack variable in the dual constraint corresponding to the primal variable x_j is $c_j - \bar{\pi} A_{.j} = c_j - (c_B B^{-1}) A_{.j} = \bar{c}_j$. We will show shortly that this dual slack variable is exactly the relative (or reduced) cost coefficient of x_j in the canonical tableau wrt the basic vector x_B.

So, the dual basic solution $\bar{\pi} = c_B B^{-1}$ is feasible to the dual problem iff all the dual slacks $\bar{c}_j = c_j - (c_B B^{-1}) A_{.j} \geq 0$ for all $j = 1$ to n. In this case we say that the primal basic vector x_B and the corresponding basis B are *dual feasible* if all these $\bar{c}_j \geq 0$ for all j, *dual infeasible* if at least one $\bar{c}_j < 0$.

To get the canonical tableau wrt the basic vector x_B, we take $-z$ as the permanent basic variable in the objective row. With this additional basic variable, the basis becomes

$$\mathcal{B} = \begin{pmatrix} B & 0 \\ c_B & 1 \end{pmatrix}.$$

Its inverse, known as the *inverse tableau* corresponding to the basic vector x_B or $(x_B, -z)$, contains the negative of the dual basic solution in its last row.

$$\mathcal{B}^{-1} = \begin{pmatrix} B^{-1} & 0 \\ -c_B B^{-1} & 1 \end{pmatrix} = \begin{pmatrix} B^{-1} & 0 \\ -\bar{\pi} & 1 \end{pmatrix}. \tag{5.16}$$

The canonical tableau wrt this basic vector can be obtained by multiplying the original tableau on the left by the inverse tableau $\mathcal{B}^{-1}$. So it is

Canonical tableau wrt basic vector $(x_B, -z)$

Basic			Nonbasic							
x_1	...	x_m	x_{m+1}	...	x_j	...	x_n	$-z$	RHS
1	...	0	$\bar{a}_{1,m+1}$	...	$\bar{a}_{1j}$	...	$\bar{a}_{1n}$	0	$\bar{b}_1$
$\vdots$		$\vdots$	$\vdots$		$\vdots$		$\vdots$	$\vdots$	$\vdots$
0	...	1	$\bar{a}_{m,m+1}$	...	$\bar{a}_{mj}$	...	$\bar{a}_{mn}$	0	$\bar{b}_m$
0	...	0	$\bar{c}_{m+1}$	...	$\bar{c}_j$	...	$\bar{c}_n$	1	0

where updated RHS column $(\bar{b}, -\bar{z})$ = (inverse tableau) (original RHS column) is

$$\begin{pmatrix} \bar{b} \\ -\bar{z} \end{pmatrix} = \mathcal{B}^{-1} \begin{pmatrix} b \\ 0 \end{pmatrix}, \tag{5.17}$$

or $\bar{b} = B^{-1} b$ and $\bar{z} = c_B B^{-1} b = \bar{\pi} b$. Also, for $j = 1$ to n, updated column of $x_j = (\bar{A}_{.j}, \bar{c}_j) =$ (inverse tableau) (original column of x_j), so

$$\begin{pmatrix} \bar{A}_{.j} \\ \bar{c}_j \end{pmatrix} = \mathcal{B}^{-1} \begin{pmatrix} A_{.j} \\ c_j \end{pmatrix}, \tag{5.18}$$

$$\bar{A}_{.j} = B^{-1} A_{.j}, \ \bar{c}_j = c_j - \bar{\pi} A_{.j} = c_j - (c_B B^{-1}) A_{.j}. \tag{5.19}$$

So, we see that the reduced (relative) cost coefficient of the variable x_j wrt the basic vector x_B is the dual slack in the dual constraint corresponding to x_j, and the formula for it is

$$\bar{c}_j = \text{(last row of inverse tableau) (original column of } x_j). \tag{5.20}$$

5.4.3 *Properties Satisfied by the Primal and Dual Basic Solutions Corresponding to a Primal Basic Vector*

The definition of the dual basic solution corresponding to the primal basic vector x_B satisfies the following properties:

1. In $\bar{x}$, the primal basic solution corresponding to x_B, basic variables in x_B are the only ones that can take nonzero values. In the dual basic solution $\bar{\pi}$ corresponding to x_B, the dual slack variables $\bar{c}_j$ in dual constraints corresponding to basic variables in x_B are all 0, because $\bar{\pi}$ is defined so as to satisfy $\bar{\pi}B = c_B$. So, in the primal, dual solution pair $(\bar{x}, \bar{\pi})$, every complementary pair $(\bar{x}_j, \bar{c}_j)$, $j = 1$ to n, contains at least one 0 quantity. This property called the *complementary slackness property* is an important optimality condition for the LP (5.15), as explained later.
2. The primal objective value $z(\bar{x})$ and the dual objective value $v(\bar{\pi})$ are both equal (both equal to $c_B B^{-1} b$).
3. For each $j = 1$ to n, the value of the dual slack $\bar{c}_j$ in the dual constraint corresponding to x_j, in the dual solution $\bar{\pi}$, is exactly the reduced (relative) cost coefficient of x_j wrt the basic vector x_B.
   The reduced cost coefficients $\bar{c}_j$ for all basic variables should be 0 by definition of $\bar{\pi}$. The definition of the dual basic solution corresponding to the primal basic vector x_B (making the slacks in the dual constraints corresponding to basic variables $= 0$) comes from this property.
4. In the primal simplex algorithm, the criterion for the primal feasible basic vector x_B to be optimal is that all reduced cost coefficients $\bar{c}_j$ wrt it must be ≥ 0. We see that this is exactly the condition for the dual basic solution $\bar{\pi}$ wrt x_B, to be dual feasible.
   Hence this definition of the dual basic solution wrt x_B makes it possible to interpret the optimality criterion of the primal simplex algorithm as the dual feasibility criterion.

Example 5.6. Consider the following LP, which is a model for a diet problem with six foods. x_1 to x_6 represent the quantities of foods one to six included in the daily diet; x_7, x_8 are the slack variables corresponding to the two constraints.

Original tableau

x_1	x_2	x_3	x_4	x_5	x_6	x_7	x_8	$-z$	RHS
1	0	2	2	1	2	−1	0	0	9
0	1	3	1	3	2	0	−1	0	19
35	30	60	50	27	22	0	0	1	0

$x_j \geq 0$ for all j, minimize z

The number of equality constraints, m, in this problem is 2, and they clearly form a linearly independent constraint set. So, the number of basic variables in every basic vector for this problem will be 2.

1. First consider the vector (x_1, x_7). This is not even a basic vector as the set of column vectors of x_1, x_7 form a linearly dependent set.

2. Now consider the vector of variables (x_1, x_8). As the set of column vectors of x_1, x_8 is linearly independent, this is a basic vector for the problem. Call it $x_{B_1} = (x_1, x_8)$. The associated basic cost vector is $c_{B_1} = (35, 0)$; nonbasic vector $x_{D_1} = (x_2, x_3, x_4, x_5, x_6, x_7)$. The

$$\text{Basis} = B_1 = \begin{pmatrix} 1 & 0 \\ 0 & -1 \end{pmatrix}.$$

To get the primal basic solution $\bar{x}^1$ corresponding to x_{B_1}, we set $\bar{x}_{D_1} = 0$, and solve for $\bar{x}_{B_1}$ from the remaining system of equations, which is

x_1	x_8	RHS
1	0	9
0	−1	19

From this we see that $\bar{x}^1 = (9, 0, 0, 0, 0, 0, 0, -19)^T$, which is primal infeasible because it makes x_8 negative. So, x_{B_1}, B_1 are primal infeasible.

The dual problem is

π_1	π_2	RHS	Associated primal var.
1	0	≤ 35	x_1
0	1	≤ 30	x_2
2	3	≤ 60	x_3
2	1	≤ 50	x_4
1	3	≤ 27	x_5
2	2	≤ 22	x_6
−1	0	≤ 0	x_7
0	−1	≤ 0	x_8
9	19	$= v(\pi)$	Maximize

To get the dual basic solution associated with x_{B_1}, we need to solve the system consisting of dual constraints corresponding to x_1, x_8 as equations, which gives $\bar{\pi}^1 = (35, 0)$.

Substituting $\bar{\pi}^1 = (35, 0)$ in the dual constraints, we compute the corresponding vector of dual slacks as $\bar{c}^1 = (0, 30, -10, -20, -8, -48, 35, 0)$, which is not ≥ 0. Hence $\bar{\pi}^1$ is dual infeasible, and so are x_{B_1}, B_1.

The extended basis, its inverse, corresponding to the basic vector $(x_1, x_8, -z)$ are

$$\mathcal{B}_1 = \begin{pmatrix} 1 & 0 & 0 \\ 0 & -1 & 0 \\ 35 & 0 & 1 \end{pmatrix}, \quad \mathcal{B}_1^{-1} = \begin{pmatrix} 1 & 0 & 0 \\ 0 & -1 & 0 \\ -35 & 0 & 1 \end{pmatrix},$$

and we verify that the last row of the inverse tableau is $(-\bar{\pi}^1, 1)$. Also, the canonical tableau of the problem wrt x_{B_1} is

Canonical tableau wrt basic vector x_{B_1}

BV	x_1	x_2	x_3	x_4	x_5	x_6	x_7	x_8	$-z$	RHS
x_1	1	0	2	2	1	2	−1	0	0	9
x_8	0	−1	−3	−1	−3	−2	0	1	0	−19
$-z$	0	30	−10	−20	−8	−48	35	0	1	−315

Here "BV" stands for "the basic variable in this row." Verify that the relative cost coefficients in the canonical tableau are equal to the dual slacks computed earlier. Also, verify that the updated columns in the canonical tableau are the same as those obtained from the formulae given above.

3. Now consider the primal basic vector $x_{B_2} = (x_1, x_2)$ for this problem. We have

$$\text{Basis} = B_2 = \begin{pmatrix} 1 & 0 \\ 0 & 1 \end{pmatrix}, \quad c_{B_2} = (35, 30).$$

So, we compute $\bar{x}^2 = (9, 19, 0, 0, 0, 0, 0, 0)^T$ is the primal basic solution wrt x_{B_2}; $\bar{\pi}^2 = (35, 30)$ is the dual basic solution wrt x_{B_2}; $\bar{c}^2 = (0, 0, -100, -50, -98, -108, 35, 30)$ is the dual slack vector at $\bar{\pi}^2$.

As $\bar{x}^2$ satisfies the nonnegativity restrictions on the primal variables, it is a BFS for the problem, and x_{B_2}, B_2 are primal feasible. But as $\bar{c}^2$ has negative entries, x_{B_2}, B_2 are dual infeasible.

4. Now consider the primal basic vector $x_{B_3} = (x_5, x_6)$ for this problem. The basis, basic cost vector, extended basis, and its inverse are

$$B_3 = \begin{pmatrix} 1 & 2 \\ 3 & 2 \end{pmatrix}, \quad c_{B_3} = (27, 22), \quad \mathcal{B}_3 = \begin{pmatrix} 1 & 2 & 0 \\ 3 & 2 & 0 \\ 27 & 22 & 1 \end{pmatrix}$$

$$\mathcal{B}_3^{-1} = \begin{pmatrix} -1/2 & 1/2 & 0 \\ 3/4 & -1/4 & 0 \\ -3 & -8 & 1 \end{pmatrix}.$$

The primal basic solution wrt x_{B_3} is $\bar{x}^3 = (0,0,0,0,5,2,0,0)^T$. From the last row of the inverse tableau and (5.16) we conclude that the dual basic solution wrt x_{B_3} is $\bar{\pi}^3 = (3,8)$. The vector of dual slacks at $\bar{\pi}^3$ is $\bar{c}^3 = (32, 22, 30, 36, 0, 0, 3, 8)$.

As $\bar{x}^3, \bar{c}^3$ are both ≥ 0, we conclude that this basic vector x_{B_3} is both primal and dual feasible for this problem. ■

5.4.4 The Duality Theorem of LP

We will now discuss the *duality theorem*, the most fundamental result in the theory of linear programming, linking an LP and its dual. In some books this theorem is also referred to as the *strong duality theorem* of LP, as its statement strengthens the statement in the weak duality theorem (Theorem 5.3). Remember that the statement of the weak duality theorem and of Corollary 4 under it only guarantees a "$\geq$" relationship in the two optimum objective values; this theorem guarantees that this "$\geq$" always holds as "$=$." Also, Corollaries 2, 3 of the weak duality theorem only guarantee if the objective value is unbounded in one of the problems in a primal, dual pair, the other problem must be infeasible; this theorem shows the converse of these statements also hold.

Theorem 5.4. The Duality Theorem of LP: *In a primal, dual pair of LPs, (1) if one has an optimum solution, the other does also, and the two optimum objective values are equal, (2) if one of the problems is feasible and has the objective unbounded ("below" if the problem is a minimization problem, "above" if it is a maximization problem), then the other problem is infeasible and vice versa.*

Proof. Consider the problem in the pair that has an optimum solution, as the primal problem, in the form of a minimization problem in standard form

$$\begin{aligned} \text{Minimize } z(x) &= cx \\ \text{subject to } Ax &= b \\ x &\geq 0, \end{aligned} \tag{5.21}$$

where A is a matrix of order $m \times n$ and rank m. So, this is the primal. The other problem in the pair is the dual of (5.21). From the hypothesis, the primal (5.21) has an optimum solution, so by Theorem 4.9, it has a BFS which is optimum for it. So, when the simplex method is applied to solve (5.21), it will terminate with an optimum basis for it that satisfies the optimality criterion of the primal simplex algorithm.

So, by the discussion in property 4 stated above, the dual basic solution corresponding to optimum basic vector is going to be dual feasible, and has the same dual objective value as the optimum primal objective value. Hence that dual feasible basic solution is an optimum solution of the dual problem, and the two optimum objective values are equal, establishing (1).

Statement (2) in the theorem follows from the weak duality theorem and statement (1). □

Other Proofs of the Duality Theorem. The proof given above for the duality theorem is based on the simplex method for LP. For an alternate proof of the duality theorem using the Farkas' lemma, see Exercise 5.34. Farkas' lemma itself can be proved directly as shown in Exercises 5.36 and 5.37.

5.4.5 Optimality Conditions for LP

Theorem 5.5. The Complementary Slackness Theorem: *In a primal, dual pair of LPs, let $(\bar{x}, \bar{\pi})$ be a primal, dual feasible solution pair. They are optimal to the respective problems if at least one quantity in every complementary pair is 0 in $(\bar{x}, \bar{\pi})$. These conditions are known as complementary slackness conditions for optimality.*

Proof. Consider the primal and dual in symmetric form given below

Primal (P)	Dual (D)
Minimize $z(x) = cx$	Maximize $v(\pi) = \pi b$
subject to $Ax \geq b$	subject to $\pi A \leq c$
$x \geq 0$	$\pi \geq 0$

where A is an $m \times n$ matrix. The various complementary pairs in this primal, dual pair of problems are

$$(x_j,\ c_j - \pi A_{.j}),\ j = 1, \dots, n; \quad (A_{i.}x - b_i,\ \pi_i),\ i = 1, \dots, m. \tag{5.22}$$

As $\bar{x}$, $\bar{\pi}$ are primal and dual feasible, all the quantities $\bar{x}_j$, $c_j - \bar{\pi} A_{.j}$, $A_{i.}\bar{x} - b_i$, $\bar{\pi}_i$ are all ≥ 0 for all i, j.

Therefore, for each j, $\bar{x}_j(c_j - \bar{\pi} A_{.j}) \geq 0$ and is 0 iff at least one quantity in the complementary pair $(\bar{x}_j,\ c_j - \bar{\pi} A_{.j})$ is 0. So, $c\bar{x} - \bar{\pi} A\bar{x} = \sum_{j=1}^{n} \bar{x}_j(c_j - \bar{\pi} A_{.j}) = 0$ iff one quantity in every complementary pair $(\bar{x}_j,\ c_j - \bar{\pi} A_{.j})$ is 0 for each $j = 1$ to n.

Similarly we see that $\bar{\pi} A\bar{x} - \bar{\pi} b = 0$ iff at least one quantity in every complementary pair $(A_{i.}\bar{x} - b_i,\ \bar{\pi}_i)$ is 0 for each $i = 1$ to m.

As shown in the proof of the weak Duality Theorem 5.3, $c\bar{x} - \bar{\pi} A\bar{x}$ and $\bar{\pi} A\bar{x} - \bar{\pi} b$ are both ≥ 0; therefore, their sum $c\bar{x} - \bar{\pi} b = 0$ iff each of $c\bar{x} - \bar{\pi} A\bar{x}$ and $\bar{\pi} A\bar{x} - \bar{\pi} b$ is 0.

As $\bar{x}$, $\bar{\pi}$ are optimal to the respective problems, by the Duality Theorem 5.4, $c\bar{x} - \bar{\pi} b = 0$. So by the above, both $c\bar{x} - \bar{\pi} A\bar{x}$ and $\bar{\pi} A\bar{x} - \bar{\pi} b$ must be 0. By the above, this implies that one quantity in every complementary pair is 0 at $(\bar{x}, \bar{\pi})$. □

Proof of this theorem for LPs in other forms follows from Theorem 5.1, or can be derived directly using arguments similar to those used above for LP in the symmetric form.

Now we will summarize the complementary slackness (*CS*) conditions for optimality for some of the other forms in which LPs appear commonly. In all these problems, A is a matrix of order $m \times n$.

LP in standard form: Here are the primal and dual

Primal (P)	Dual (D)
Minimize $z(x) = cx$	Maximize $v(\pi) = \pi b$
subject to $Ax = b$	subject to $\pi A \leq c$.
$x \geq 0$	

So, the dual constraints are $\bar{c}_j = c_j - \pi A_{.j} \geq 0$ for $j = 1$ to n. The CS conditions are: for each $j = 1$ to n at least one quantity in each complementary pair $(x_j, \bar{c}_j = c_j - \pi A_{.j})$ should be 0, or equivalently $x_j \bar{c}_j = 0$ for all j.

LP in bounded variable standard form: Here are the primal and dual. Here ℓ, k are finite lower and upper bound vectors for the variables and $\ell < k$.

Primal (P)	Dual (D)
Minimize $z(x) = cx$	Maximize $v(\pi, \mu, \delta) = \pi b + \ell\mu - k\delta$
subject to $Ax = b$	subject to $\pi A + \mu - \delta = c$
$\ell \leq x \leq k$	$\mu, \delta \geq 0$.

The CS conditions are: at least one quantity must be 0 in each of the complementary pairs $\{(x_j - \ell_j, \mu_j), j = 1$ to n; and $(k_j - x_j, \delta_j), j = 1$ to $n\}$, or equivalently for all $j = 1$ to n

$$(x_j - \ell_j)\mu_j = 0; \ (k_j - x_j)\delta_j = 0$$

LP in general form: Here the primal and dual are given below. All bounds on individual variables are included as constraints in the statement of the primal problem.

Primal (P)	Dual (D)
Minimize $z(x) = cx$	Maximize $v(\pi, \mu) = \pi b + \mu g$
s. to $Dx = d$	s. to $\pi D + \mu F = c$
$Fx \geq g$	$\mu \geq 0$.

Here suppose F is a matrix of order $p \times q$. Then the CS conditions are: at least one quantity must be 0 in each complementary pair $(F_{i.}x - g_i, \mu_i), i = 1$ to p, or equivalently $(F_{i.}x - g_i)\mu_i = 0$ for all i.

5.4.6 Necessary and Sufficient Optimality Conditions for LP

Consider a primal, dual pair of LPs. From the weak duality theorem (Theorem 5.3), duality theorem (Theorem 5.4), and the complementary slackness theorem (Theorem 5.5), we know that a necessary and sufficient condition for a given pair of primal, dual feasible solutions to be optimum to the respective problems is either (1) the primal and dual objective values at this pair of solutions are equal, or (2) the pair of primal and dual feasible solutions satisfy the complementary slackness conditions for optimality (i.e., in this pair of solutions, at least one quantity in every complementary pair is 0).

Also, given a feasible solution to an LP, a necessary and sufficient condition for it to be optimal to the LP is the existence of a feasible solution to its dual that satisfies the complementary slackness conditions with it.

A basic vector for an LP in standard form is an *optimum basic vector* iff it is primal and dual feasible (because the primal and dual basic solutions associated with a primal basic vector always satisfy the complementary slackness conditions by the way the dual basic solution associated with a primal basic vector is defined).

5.4.7 Duality Gap, a Measure of Distance from Optimality

Now we state an important theorem that provides a measure of how far away a given pair of primal, dual feasible solutions are from being optimal to the respective problems, and relates this measure to the violation of the CS property in this pair.

In a primal, dual pair of LPs, suppose $z(x)$ is the primal objective function to be minimized, and $v(\pi)$ the dual objective function to be maximized. Let $(\bar{x}, \bar{\pi})$ be a primal, dual feasible solution pair. From the weak duality theorem (Theorem 5.3), $z(\bar{x}) \geq v(\bar{\pi})$. If $d = z(\bar{x}) - v(\bar{\pi}) = 0$, then we conclude that $(\bar{x}, \bar{\pi})$ are optimal to the respective problems.

This $d = z(\bar{x}) - v(\bar{\pi})$ is known as the *duality gap* in the primal, dual pair of feasible solutions $(\bar{x}, \bar{\pi})$.

If the duality gap $d > 0$, from the weak duality theorem (Theorem 5.3) we know that

- $z(\bar{x}) \geq$ the minimum objective value in the primal problem by at most d;
- $v(\bar{\pi}) \leq$ the maximum objective value in the dual problem by at most d.

So, the duality gap d provides a measure of how far $\bar{x}$, $\bar{\pi}$ are from being optimal as a pair to their respective problems.

Theorem 5.6. *Let d be the duality gap in a pair $(\bar{x}, \bar{\pi})$ of primal and dual feasible solutions to a primal, dual pair of LPs. Then d = the total complementary slackness (CS) violation in the pair $(\bar{x}, \bar{\pi})$, which is defined as the sum of the products of the various complementary pairs in $(\bar{x}, \bar{\pi})$.*

Proof. Consider the primal and dual in symmetric form given below

Primal (P)	Dual (D)
Minimize $z(x) = cx$	Maximize $v(\pi) = \pi b$
subject to $Ax \geq b$	subject to $\pi A \leq c$
$x \geq 0$	$\pi \geq 0$

where A is an $m \times n$ matrix. Then the duality gap in the given solutions $(\bar{x}, \bar{\pi})$ is $d = c\bar{x} - \bar{\pi}b$.

The various complementary pairs in this primal, dual pair of problems are

$$(x_j,\ c_j - \pi A_{.j}),\ j = 1, \dots, n;\ \ (A_{i.}x - b_i,\ \pi_i), i = 1, \dots, m. \tag{5.23}$$

The CS property requires that the product of each complementary pair be 0; so the violation of the CS property in the pair $(\bar{x}, \bar{\pi})$ is $\sum_{j=1}^{n} x_j(c_j - \pi A_{.j}) + \sum_{i=1}^{m}(A_{i.}x - b_i)\pi_i = (c\bar{x} - \bar{\pi}A\bar{x}) + (\bar{\pi}A\bar{x} - \bar{\pi}b) = c\bar{x} - \bar{\pi}b = d$, the duality sgap.

Proof of this theorem for LPs given in other forms is based on similar arguments. □

5.4.8 *Using CS Conditions to Check the Optimality of a Given Feasible Solution to an LP*

CS conditions can often be used to conclusively establish whether a given feasible solution for an LP is optimal or not. We illustrate with an example.

Example 5.7. Consider the LP given in detached coefficient form for clarity. Check whether $\bar{x} = (6, 0, -1, 0, 2)^T$ is an optimum solution for it.

x_1	x_2	x_3	x_4	x_5		RHS
1	1	−1	2	−1	$\geq$	5
−2	0	2	−1	3	$\geq$	−8
1	0	0	0	0	$\geq$	6
0	−3	0	3	0	$\geq$	−5
0	0	5	−1	7	$\geq$	7
−3	1	3	0	5	$= z$ minimize	

The first thing to do is to check whether $\bar{x}$ is feasible to the problem, and classify the primal inequality constraints as active, inactive, at $\bar{x}$. If $\bar{x}$ is infeasible, the process stops with that conclusion.

Here $\bar{x}$ is feasible, constraints 1, 2, 3 are active inequality constraints, and constraints 4, 5 are inactive inequality constraints at $\bar{x}$. Associating the dual variables π_1 to π_5 to primal constraints from top to bottom, the dual of this problem is

π_1	π_2	π_3	π_4	π_5	RHS
1	−2	1	0	0	= −3
1	0	0	−3	0	= 1
−1	2	0	0	5	= 3
2	−1	0	3	−1	= 0
−1	3	0	0	7	= 5
	All π_i			≥ 0	
5	−8	6	−5	7	$= v$ maximize

The CS conditions for this pair of problems are

$$\begin{aligned}(x_1 + x_2 - x_3 + 2x_4 - x_5 - 5)\pi_1 &= 0,\\ (-2x_1 + 2x_3 - x_4 + 3x_5 + 8)\pi_2 &= 0,\\ (x_1 - 6)\pi_3 &= 0,\\ (-3x_2 + 3x_4 + 5)\pi_4 &= 0,\\ (5x_3 - x_4 + 7x_5 - 7)\pi_5 &= 0.\end{aligned}$$

If $\bar{x}$ is an optimum solution of the primal, we can find a dual feasible solution that satisfies the above CS conditions together with $\bar{x}$. If such a dual feasible solution $\bar{\pi}$ exists, we conclude that $\bar{\pi}_4 = \bar{\pi}_5 = 0$ from the fourth and fifth CS conditions. Substituting 0 values for π_4, π_5 in the dual constraints, the remaining system is

$$\begin{aligned}\pi_1 - 2\pi_2 + \pi_3 &= -3,\\ \pi_1 &= 1,\\ -\pi_1 + 2\pi_2 &= 3,\\ 2\pi_1 - \pi_2 &= 0,\\ -\pi_1 + 3\pi_2 &= 5,\\ \pi_1, \pi_2, \pi_3 &\geq 0.\end{aligned}$$

If the linear equations among these have a unique solution, we can compute it and check whether it satisfies the remaining inequality constraints among the above constraints. Here, the linear equations in this system do have the unique solution $\bar{\pi} = (1, 2, 0, 0, 0)$ which satisfies the nonnegativity restrictions.

As $\bar{x}$ is primal feasible, and $\bar{\pi}$ is dual feasible, and together they satisfy the CS conditions, we conclude that $\bar{x}$ is indeed an optimum solution for the primal. ■

Example 5.8. Check whether $\bar{x} = (12, 7, 2, 1, 0)^T$ is an optimum solution for the following LP in standard form.

x_1	x_2	x_3	x_4	x_5	$-z$	RHS
1	−2	2	1	3	0	3
0	1	−1	−2	4	0	3
0	0	2	5	−1	0	9
−6	9	−5	10	−25	1	0

$x_j \geq 0$ for all j, minimize z

We verify that $\bar{x}$ is feasible to the problem. To check optimality, we write the dual problem and the complementary slackness conditions. Let π_1, π_2, π_3 be the dual variables associated with the equality constraints from top to bottom.

$$\begin{aligned} \text{Maximize } & 3\pi_1 + 3\pi_2 + 9\pi_3 \\ \text{s. to } & \pi_1 \leq -6 \\ & -2\pi_1 + \pi_2 \leq 9 \\ & 2\pi_1 - \pi_2 + 2\pi_3 \leq -5 \\ & \pi_1 - 2\pi_2 + 5\pi_3 \leq 10 \\ & 3\pi_1 + 4\pi_2 - \pi_3 \leq -25. \end{aligned}$$

The CS conditions are

$$\begin{aligned} (-6 - \pi_1)x_1 &= 0, \\ (9 + 2\pi_1 - \pi_2)x_2 &= 0, \\ (-5 - 2\pi_1 + \pi_2 - 2\pi_3)x_3 &= 0, \\ (10 - \pi_1 + 2\pi_2 - 5\pi_3)x_4 &= 0, \\ (-25 - 3\pi_1 - 4\pi_2 + \pi_3)x_5 &= 0. \end{aligned}$$

As $\bar{x}_1, \bar{x}_2, \bar{x}_3, \bar{x}_4$ are all > 0, $\bar{x}$ is an optimum solution to the primal iff there exists a dual feasible solution that satisfies the first four dual constraints as equations. This gives us the following system of linear equations.

$$\begin{aligned} \pi_1 &= -6, \\ -2\pi_1 + \pi_2 &= 9, \\ 2\pi_1 - \pi_2 + 2\pi_3 &= -5, \\ \pi_1 - 2\pi_2 + 5\pi_3 &= 10. \end{aligned}$$

This system of four equations in three unknowns has the unique solution $\bar{\pi} = (-6, -3, 2)$. $\bar{\pi}$ also satisfies the remaining fifth dual constraint. So, $(\bar{x}, \bar{\pi})$ together are primal, dual feasible and satisfy the CS conditions, so they are optimal to their respective problems. ■

Example 5.9. Check whether $\bar{x} = (10, 5, 0, 0, 0)^T$ is an optimum solution for the following LP in standard form.

x_1	x_2	x_3	x_4	x_5	$-z$	RHS
2	1	−1	2	1	0	25
2	0	2	−1	3	0	20
8	6	−10	20	−2	1	0

$x_j \geq 0$ for all j, minimize Z

We verify that $\bar{x}$ is feasible to the problem. To check optimality, we write the dual problem and the complementary slackness conditions. Let π_1, π_2 be the dual variables associated with the equality constraints from top to bottom.

$$\begin{aligned} &\text{Maximize } 25\pi_1 + 20\pi_2 \\ &\text{S. to } \bar{c}_1 = 8 - 2\pi_1 - 2\pi_2 \geq 0, \\ &\bar{c}_2 = 6 - \pi_1 \geq 0, \\ &\bar{c}_3 = -10 + \pi_1 - 2\pi_2 \geq 0, \\ &\bar{c}_4 = 20 - 2\pi_1 + \pi_2 \geq 0, \\ &\bar{c}_5 = -2 - \pi_1 - 3\pi_2 \geq 0. \end{aligned}$$

The CS conditions are $x_j \bar{c}_j = 0$ for all $j = 1$ to 5. As $\bar{x}_1, \bar{x}_2 > 0$, if $\bar{x}$ is optimum to the primal, we should be able find a solution to

$$\begin{aligned} 8 - 2\pi_1 - 2\pi_2 &= 0, \\ 6 - \pi_1 &= 0, \end{aligned}$$

which makes $\bar{c}_j \geq 0$ for $j = 3, 4, 5$. This system has the unique solution $\bar{\pi} = (6, -2)$, but this solution makes $\bar{c}_5 = -2 - \pi_1 - 3\pi_2 = -2 < 0$, so is not dual feasible. So there is no dual feasible solution that satisfies the CS conditions together with $\bar{x}$, hence we conclude that $\bar{x}$ is feasible but not optimal to the primal LP. ■

Example 5.10. Check whether $\bar{x} = (10, 15, 0, 0, 0)^T$ is an optimum solution for the following LP in standard form.

x_1	x_2	x_3	x_4	x_5	$-z$	RHS
1	0	1	−2	1	0	10
0	1	−1	2	2	0	15
1	1	1	1	−1	0	25
1	4	−2	4	20	1	0

$x_j \geq 0$ for all j, minimize z

We verify that $\bar{x}$ is feasible to the problem. To check optimality, we write the dual problem and the complementary slackness conditions. Let π_1, π_2, π_3 be the dual variables associated with the equality constraints from top to bottom.

As only $\bar{x}_1, \bar{x}_2$ are > 0, you can verify that the CS conditions yield only two equations given below that a dual solution must satisfy as equations in order to satisfy the CS conditions together with $\bar{x}$

$$\pi_1 + \pi_3 = 1,$$
$$\pi_2 + \pi_3 = 4.$$

These equations do not have a unique solution. So, in this case the CS conditions have not yielded enough number of equations in the dual variables, to identify a dual solution for satisfying the CS conditions together with $\bar{x}$, if this occurs it is not easy to check the optimality of $\bar{x}$ directly using the CS conditions only. In this case, to check if $\bar{x}$ is optimum, one can solve the primal LP using some LP algorithm, and after obtaining the optimum objective value in it, compare its value to the objective value at $\bar{x}$. ■

Exercises

5.4.1. Consider the following LP in standard form. Find the primal and dual basic solutions associated with the basic vectors $x_{B_1} = (x_1, x_2, x_3)$, $x_{B_2} = (x_2, x_3, x_4)$ for this problem. Mention whether each of these basic vectors is primal feasible or not, primal nondegenerate or not, and dual feasible or not. Compute the inverse tableau for this problem with respect to x_{B_1}.

x_1	x_2	x_3	x_4	x_5	
1	0	1	0	−1	20
0	1	0	1	1	10
0	0	1	−3/2	0	5
1	1	1	2	0	$= z$, minimize

$x_j \geq 0$ for all j

5.4.2. Consider the following LP

$$\begin{array}{llllllllll}
\text{Minimize} \\
z(x) = & 12x_1 & -x_2 & +3x_3 & +5x_4 & & -8x_6 & & \\
\text{s. to} & x_1 & -2x_2 & +x_3 & -x_4 & +x_5 & -x_6 & +x_7 & = -4 \\
& x_1 & +x_2 & & +2x_4 & -x_5 & -x_6 & -2x_7 & = 9 \\
& -2x_1 & +x_2 & & & -x_5 & +x_6 & +x_7 & = 5
\end{array}$$

$$x_j \geq 0, \; j = 1 \text{ to } 7.$$

Write the dual of this problem and all the complementary pairs in these primal, dual problems.

Is $\bar{\pi} = (3, 4, -1)$ dual feasible in this problem? Compute all the relative cost coefficients with respect to $\bar{\pi}$.

Mention all the primal variables that have to be equal to 0 in every optimum solution of the primal LP if it is known that $\bar{\pi}$ is an optimum dual solution.

In the system of *equality constraints* in the primal problem, substitute 0 for each primal variable determined above, and obtain the general solution for the remaining system. What kind of geometric object is the set of all solutions to the remaining system?

Identify nonnegative solutions of the remaining system. Are these optimum solutions of the primal LP? Why?

5.4.3. Check whether the basic vector (x_4, x_2, x_6) is primal feasible for the following LP. Is it dual feasible? Is it optimal?

x_1	x_2	x_3	x_4	x_5	x_6	b
1	1	1	1	1	1	9
1	1	0	1	1	0	5
1	0	0	1	0	0	2
3	−2	1	−5	4	−2	$= z$, minimize

$x_j \geq 0$ for all j.

5.4.4. Write the dual of the following LP, and the complementary slackness optimality conditions for this primal, dual pair. Using them check whether $\pi = (-2, -3, 1)$ is an optimum solution of the dual problem. If it is, derive an optimum solution of the original LP using it and the optimality conditions.

x_1	x_2	x_3	x_4	x_5	x_6	x_7	b
1	2	−2	1	0	3	1	−5
−2	1	−1	2	1	1	2	1
1	0	2	2	2	2	−2	8
10	0	9	−2	−1	−1	−10	$= z$, minimize

$x_j \geq 0$ for all j.

5.4.5. Ice cream is a mixture of four ingredients, and its flavor is measured in units of mmm/gallon. A certain brand of ice cream is required to have flavor $\geq$1 mmm/gallon. Data on the cost and mmm content of the ingredients is given below. Formulate the problem of determining a minimum cost composition for acceptable ice cream.

Write the dual of this problem and solve this dual geometrically. Find the optimum composition from the dual optimum using the complementary slackness optimality conditions.

Ingredient	1	2	3	4
Cost ($/gallon)	2	0.7	1	1.5
mmm/gallon	3	0	0.5	1

5.4.6. For the following LP check whether $\hat{x} = (5, 2, 3, 1, 0, 0, 0)^T$ is an optimum solution using the complementary slackness optimality conditions.

x_1	x_2	x_3	x_4	x_5	x_6	x_7	b
1	−1	1	0	1	0	2	6
0	2	1	−1	−1	2	0	6
0	0	2	1	0	1	2	7
0	0	0	2	1	−1	1	2
−2	4	6	3	−1	10	5	$= z(x)$, minimize

$x_j \geq 0$ for all j

5.4.7. Consider the balanced transportation problem with the following data (the availability and requirement data is shown in bold letters to distinguish it from cost data), and dual solution $(\tilde{u}, \tilde{v})$. Verify that $(\tilde{u}, \tilde{v})$ is dual feasible. Assuming that $(\tilde{u}, \tilde{v})$ is an optimum dual solution, find an optimum solution of the primal transportation problem.

Source i	Unit shipping cost to mkt. j				Availability	$\tilde{u}_i$
	1	2	3	4		
1	11	10	8	5	**200**	5
2	6	8	5	2	**100**	2
3	11	5	6	5	**150**	3
Requirement	**50**	**50**	**200**	**150**		
$\tilde{v}_j$	4	2	3	0		

5.4.8. A company has divided their marketing area into six zones based on the characteristics of the shoppers, their economic status, etc. They want to appoint a director for each zone to run the marketing effort there. They have already selected six candidates to fill the positions. The total annual sales in a zone would depend on which candidate is appointed as director there. Based on the candidates skills, demeanor, and background, it is estimated that \$ c_{ij} million in annual sales will be generated in zone j if candidate i is appointed as director there, and this (c_{ij}) data is given below for $i,\ j = 1$ to 6. The problem is to decide which zone each candidate should be assigned to, to maximize the total annual sales in all the zones (each zone gets one candidate and each candidate goes to one zone).

Formulate this problem and write its dual using u_i, v_j as the dual variables corresponding to the constraints associated with the ith candidate, jth zone, respectively.

	c_{ij} = annual sales volume in \$million if candidate i is assigned to zone j					
Zone j =	1	2	3	4	5	6
Candidate i = 1	1	2	6	10	17	29
2	3	4	8	11	20	30
3	5	7	9	12	22	33
4	13	14	15	16	23	34
5	18	19	21	24	25	35
6	26	27	28	31	32	36

Many people would consider this a very easy problem to solve, and would use a method which can aptly be called the *greedy method.*

The method proceeds this way. As we want to maximize total sales, look for the cell in the table fetching the largest annual sales; this is the cell corresponding to the assignment of candidate 6 to zone 6, yielding an annual sales rate of $36 million. So, assign candidate 6 to zone 6. This selection achieves the maximum possible sales at this stage with this one allocation, hence it is known as a *greedy choice*. As candidate 6 is already assigned to zone 6, strike off the row of candidate 6 and the column of zone 6 from the table and handle the remaining problem the same way, one allocation at a time, using the greedy choice at each stage. Verify that the solution obtained by this method is $\{(C_1, Z_1), (C_2, Z_2), (C_3, Z_3), (C_4, Z_4), (C_5, Z_5), (C_6, Z_6)\}$, where for $i = 1$ to 6, C_i denotes candidate i, Z_i denotes zone i, and (C_i, Z_i) indicates that candidate i is assigned to zone i. This fetches a total annual sales rate of $36 + 25 + 16 + 9 + 4 + 1 = \$91$ million.

Using $\tilde{u} = (0, 2, 4, 9, 16, 23)$, $\tilde{v} = (1, 2, 5, 7, 9, 13)$ as the dual solution, show that the greedy solution obtained above actually minimizes the total sales volume, instead of maximizing it.

5.5 How Various Algorithms Solve LPs

As we have seen in the previous section, to conclude that a given feasible solution to an LP is in fact an optimum solution, there are three conditions to be met. These are as follows:

Primal feasibility
Dual feasibility
Complementary slackness property.

The *primal simplex algorithm* starts with an initial BFS and searches only among BFSs for an optimum solution; so it maintains primal feasibility throughout. Also, as it only searches among basic solutions, the primal and dual solutions in every step of this algorithm satisfy the complementary slackness property. It passes through an edge of the set of feasible solutions to go from one BFS to a better one in its search. So, it operates only on the one-dimensional boundary of the set of feasible solutions, and is therefore classified as a *one-dimensional boundary method*. The search process in this algorithm has the aim of attaining dual feasibility, in fact the *optimality termination condition* in this algorithm is the dual feasibility condition and the *unboundedness termination condition* establishes dual infeasibility.

The *dual simplex algorithm* that we will discuss in the next chapter starts with a dual feasible basic vector and maintains dual feasibility throughout. Also, as it passes through basic vectors only, the primal and dual solutions in every step are basic solutions, and hence satisfy the complementary slackness property always. The termination condition in this algorithm is either primal feasibility or establishing primal infeasibility. When one of these conditions is reached, this algorithm terminates.

The *IPMs* (interior point methods) that we discuss in later chapters operate only in the interior (or relative interior) of the set of feasible solutions. The most popular among them for computer implementations are the *primal–dual IPMs*, which are initiated with a primal–dual pair of interior feasible solutions. They trace a path of primal-dual interior feasible pairs, trying to reduce the complementary slackness violation (i.e., the duality gap in the pair) to 0.

5.6 How to Check if an Optimum Solution is Unique

5.6.1 Primal and Dual Degeneracy of a Basic Vector for an LP in Standard Form

Consider the LP in standard form

$$\begin{aligned} \text{Minimize } z(x) &= cx \\ \text{subject to } Ax &= b \\ x &\geq 0, \end{aligned}$$

where A is a matrix of order $m \times n$ and rank m. Let x_B be a basic vector for this problem. From the definition discussed in Chap. 4, we can classify this basic vector, and the associated basis B and the corresponding primal and dual basic solutions as

- *Primal nondegenerate* if all basic variables have nonzero values in its primal basic solution, that is, if every entry in the basic values vector $B^{-1}b$ is nonzero
- *Primal degenerate* if at least one basic variable a has zero value in its primal basic solution, that is, if at least one entry in the basic values vector $B^{-1}b$ is zero
- *Dual nondegenerate* if none of the nonbasic dual slacks $\bar{c}_j$ have 0-value at its dual basic solution, that is, if $\bar{c}_j = c_j - c_B B^{-1} A_{.j}$ is nonzero for every nonbasic variables x_j, where c_B is the basic cost vector corresponding to x_B
- *Dual degenerate* if at least one of the nonbasic dual slacks $\bar{c}_j$ has 0-value at its dual basic solution, that is, if $\bar{c}_j = c_j - c_B B^{-1} A_{.j}$ is zero for at least one nonbasic variables x_j, where c_B is the basic cost vector corresponding to x_B.

Example 5.11. From the discussion of Sect. 5.4, we know that the values of the basic variables in the basic solution corresponding to a basic vector for an LP in standard form are the updated RHS constants in the canonical tableau wrt this basic vector; and that the values of dual slacks at the dual basic solution are the relative cost coefficients. In the following, we provide canonical tableaus wrt the basic vector

(x_1, x_2) for four different LPs. So, in all these tableaus, the nonbasic variables are x_3, x_4, x_5. In each tableau "BV" is an abbreviation for " basic variable in row."

Tableau 1

BV	x_1	x_2	x_3	x_4	x_5	$-z$	RHS
x_1	1	0	1	−1	1	0	3
x_2	0	1	−1	1	1	0	4
$-z$	0	0	−2	3	2	1	−10

$x_j \geq 0$ for all j, minimize z.

Tableau 2

BV	x_1	x_2	x_3	x_4	x_5	$-z$	RHS
x_1	1	0	1	−1	1	0	5
x_2	0	1	−1	1	1	0	0
$-z$	0	0	2	3	2	1	−10

$x_j \geq 0$ for all j, minimize z.

Tableau 3

BV	x_1	x_2	x_3	x_4	x_5	$-z$	RHS
x_1	1	0	1	−1	1	0	3
x_2	0	1	−1	1	1	0	5
$-z$	0	0	2	3	0	1	−10

$x_j \geq 0$ for all j, minimize z.

Tableau 4

BV	x_1	x_2	x_3	x_4	x_5	$-z$	RHS
x_1	1	0	1	−1	1	0	5
x_2	0	1	−1	1	1	0	0
$-z$	0	0	2	3	0	1	−10

$x_j \geq 0$ for all j, minimize z.

For the problem in Tableau 1, the basic vector (x_1, x_2) is primal feasible and primal nondegenerate as both $\bar{b}_1, \bar{b}_2$ are > 0. It is dual infeasible because $\bar{c}_3 < 0$, but dual nondegenerate as all nonbasic $\bar{c}_j$ are nonzero.

For the problem in Tableau 2, the basic vector (x_1, x_2) is primal feasible and primal degenerate as both $\bar{b}_1, \bar{b}_2$ are ≥ 0 but one of them is 0. It is dual feasible and dual nondegenerate as all nonbasic $\bar{c}_j$ are > 0.

For the problem in Tableau 3, the basic vector (x_1, x_2) is primal feasible and primal nondegenerate as both $\bar{b}_1, \bar{b}_2$ are > 0. It is dual feasible and dual degenerate as all nonbasic $\bar{c}_j$ are ≥ 0, but one of them is 0.

For the problem in Tableau 4, the basic vector (x_1, x_2) is primal feasible and primal degenerate as both $\bar{b}_1, \bar{b}_2$ are ≥ 0, but one of them is 0. It is dual feasible and dual degenerate as all nonbasic $\bar{c}_j$ are ≥ 0, but one of them is 0. ■

5.6.2 Sufficient Conditions for Checking the Uniqueness of Primal and Dual Optimum Solutions

Theorem 5.7. *Consider the LP in standard form and let x_B be an optimum basic vector for it. Let x_D be the vector of nonbasic variables (i.e., those not in x_B). Let the basic, nonbasic partition of the canonical tableau wrt x_B be ("BV" is abbreviation for "basic vector")*

Canonical tableau

BV	x_B	x_D	$-z$	Updated RHS
x_B	I	$\bar{D}$	0	$\bar{b}$
$-z$	0	$\bar{c}_D$	1	$-\bar{z}$

$x_j \geq 0$ *for all* j, *min* z

If $\bar{c}_D > 0$ (i.e., all nonbasic relative cost coefficients are positive or x_B is dual nondegenerate), then $\bar{x} = (\bar{x}_B, \bar{x}_D) = (\bar{b}, 0)$ is the unique primal optimum solution for this LP.

If $\bar{b} > 0$ (all updated RHS constants > 0, or x_B is primal nondegenerate), then the dual optimum solution is unique for this problem.

Proof. If $\bar{c}_D > 0$, by the CS property, we know that x_D must be 0 in every optimum solution of the primal. When x_D is fixed at 0, $x_B = \bar{b}$ is the only possible solution for the equality constraints in this LP, because the basis B corresponding to x_B is a nonsingular square matrix. So, $\bar{x}$ is the only optimum solution for this LP in this case.

If $\bar{b} > 0$, by the CS property the dual constraints corresponding to the basic variables in x_B must hold as equations (because the corresponding dual slacks must be 0 for optimality), in any optimum solution of the dual. The coefficient matrix of the dual variables in these equations is B^T, as it is square and nonsingular; the dual basic solution corresponding to x_B is the only dual optimum solution in this case. □

Theorem 5.7 states only that $c_D > 0$ is a sufficient condition for the BFS corresponding to the basic vector x_B to be the unique optimum solution for the LP considered there. This means that even if some of the nonbasic relative cost coefficients in $\bar{c}_D$ are 0, that BFS may still be the unique optimum solution for that LP.

The following describes a simple procedure to conclusively check the uniqueness of the optimum solution for an LP when some of the relative cost coefficients in $\bar{c}_D$ are zero (but this procedure requires solving another LP).

5.6.3 *Procedure to Check if the BFS Corresponding to an Optimum Basic Vector x_B is the Unique Optimum Solution*

Let the original LP be

$$\begin{aligned} \text{Minimize} \quad & z = cx \\ \text{subject to} \quad & Ax = b \\ & x \geq 0, \end{aligned} \tag{5.24}$$

where A is a matrix of order $m \times n$ and rank m, with an optimum basic vector x_B, and the canonical tableau wrt which is denoted using the notation in Theorem 5.7. For simplicity, let $x_B = (x_1, \ldots, x_m)$, $x_D =$ the vector of nonbasic variables $= (x_{m+1}, \ldots, x_n)$, and $\bar{x}$ the associated BFS.

As x_B is an optimum basic vector, the nonbasic relative cost coefficients $\bar{c}_{m+1}, \ldots, \bar{c}_n$ are all ≥ 0. If they are all > 0, by Theorem 5.7, $\bar{x}$ is the unique optimum solution for this LP, terminate.

If some of $\bar{c}_{m+1}, \ldots, \bar{c}_n$ are 0, for convenience in notation, let us assume that $\bar{c}_{m+1}, \ldots, \bar{c}_{m+r}$ are $= 0$ and $\bar{c}_{m+r+1}, \ldots, \bar{c}_n$ are all > 0.

Then by the CS theorem, Theorem 5.5, we know that x_{m+r+1} to x_n must all equal 0 in every optimum solution for this LP, and that any primal feasible solution with x_{m+r+1} to $x_n = 0$ is optimal to this LP. So the *optimum face* of this LP is the set of feasible solutions of

$$\begin{aligned} Ax &= b \\ x &\geq 0 \\ x_{m+r+1}, \ldots, x_n &= 0. \end{aligned} \tag{5.25}$$

So, the problem of checking if $\bar{x}$ is the unique optimum solution for this LP boils down to checking whether any of the variables $x_{m+1}, \ldots, x_{m+r}$ can take a positive value on this face.

To check this, we can try to enter each of the nonbasic variables $x_{m+r+1}, \ldots, x_n$ with 0 relative cost coefficients into the basic vector x_B. If any one of them enters with a nondegenerate pivot step, it will lead to an alternate optimum BFS. If one of them has a nonpositive updated column in the canonical tableau wrt x_B, then entering that nonbasic variable at a value λ, where λ is a nonnegative parameter, leads to an unbounded edge every point on which is an optimum solution for this LP.

But if each of these nonbasic variables $x_{m+1}, \ldots, x_{m+r}$ enter the basic vector x_B with a degenerate pivot step only, then we can solve the following LP

$$\text{Maximize} \quad \sum_{j=m+1}^{m+r} x_j \quad \text{s. to constraints in (5.25).} \tag{5.26}$$

If the optimum objective value in (5.26) is 0, clearly $\bar{x}$ is the unique optimum solution for (5.24). On the other hand, if the optimum objective value in (5.26) is >0, an optimum solution for (5.26) is an alternate optimum solution for (5.24).

We will now provide several illustrative examples. For simplicity, in each example, we state the LP under discussion using an optimum canonical tableau for it.

Example 5.12. Consider the following LP in standard form, for which the optimum canonical tableau wrt the basic vector (x_1, x_2, x_3) is given. "BV" refers to " basic variable in row." θ is the minimum ratio in the pivot step.

BV	x_1	x_2	x_3	x_4	x_5	x_6	x_7	$-z$	$\bar{b}$	Ratio
x_1	1	0	0	−1	1	1	2	0	0	
x_2	0	1	0	$\boxed{1}$	−1	2	1	0	2	2
x_3	0	0	1	2	2	4	3	0	6	3
$-z$	0	0	0	0	0	10	20	1	−100	$\theta = 2$

$x_j \geq 0$ for all j, min z

The current BFS $\bar{x} = (0, 2, 6, 0, 0, 0, 0)^T$ is an optimum solution with optimum objective value 100. Nonbasic variables x_4, x_5 have 0 relative cost coefficients, but x_6, x_7 have positive relative cost coefficients. So, for this LP any feasible solution in which $x_6 = x_7 = 0$ is an optimum solution.

We find that x_4 enters the basic vector (x_1, x_2, x_3) with a nondegenerate pivot step, yielding an alternate optimum BFS $\hat{x} = (2, 0, 2, 2, 0, 0, 0)^T$ wrt the new basic vector (x_1, x_4, x_3). Its canonical tableau is

BV	x_1	x_2	x_3	x_4	x_5	x_6	x_7	$-z$	$\bar{b}$	Ratio
x_1	1	1	0	0	0	3	3	0	2	
x_4	0	1	0	1	−1	2	1	0	2	
x_3	0	−2	1	0	4	0	1	0	2	
$-z$	0	0	0	0	0	10	20	1	−100	

Other alternate optimum BFSs can be obtained by performing pivot steps in columns of nonbasic variables x_1 to x_5 in one of these basic vectors. Also, any convex combination of optimum BFSs is also optimum. ■

Example 5.13. Consider the following LP in standard form for which the optimum canonical tableau wrt the basic vector (x_1, x_2, x_3) is given. "BV" refers to " basic variable in row."

BV	x_1	x_2	x_3	x_4	x_5	x_6	x_7	$-z$	$\bar{b}$	Ratio
x_1	1	0	0	1	−1	1	2	0	0	
x_2	0	1	0	1	−1	2	1	0	2	2
x_3	0	0	1	2	0	4	3	0	6	3
$-z$	0	0	0	0	0	10	20	1	−100	$\theta = 2$

$x_j \geq 0$ for all j, min z

The current BFS $\bar{x} = (0, 2, 6, 0, 0, 0, 0)^T$ is an optimum solution with optimum objective value 100. Nonbasic variables $x_4,\ x_5$ have 0 relative cost coefficients, but $x_6,\ x_7$ have positive relative cost coefficients. So, for this LP any feasible solution in which $x_6 = x_7 = 0$ is an optimum solution.

The variable x_4 enters the basic vector $(x_1,\ x_2,\ x_3)$ with a degenerate pivot step, hence this pivot step does not yield an alternate BFS, it just gives another basic vector corresponding to the same BFS $\bar{x}$.

But the updated column vector of x_5 is ≤ 0. Fixing the nonbasic variables $x_4,\ x_6,\ x_7$ at 0 and making $x_5 = \lambda$, a nonnegative parameter, yields the following solution to the system of equations in this problem:

$$x(\lambda) = (\lambda, 2 + \lambda, 6, 0, \lambda, 0, 0)^T,$$

which remains nonnegative and hence feasible to this LP for all $\lambda \geq 0$. And the objective value does not change as $\bar{c}_5 = 0$. Hence $\{x(\lambda) : \lambda \geq 0\}$ is an unbounded edge of the set of feasible solutions of this LP every point on which is optimal to the LP. ■

Example 5.14. Now consider the following LP in standard form, for which the optimum canonical tableau wrt the basic vector $(x_1,\ x_2,\ x_3)$ is given. "BV" refers to "basic variable in row."

BV	x_1	x_2	x_3	x_4	x_5	x_6	x_7	$-z$	$\bar{b}$	Ratio
x_1	1	0	0	1	1	1	2	0	0	
x_2	0	1	0	1	−1	2	1	0	2	
x_3	0	0	1	2	2	4	3	0	6	
$-z$	0	0	0	0	0	10	20	1	−100	

$x_j \geq 0$ for all j, min z

The current BFS $\bar{x} = (0, 2, 6, 0, 0, 0, 0)^T$ is an optimum solution with optimum objective value 100. Nonbasic variables $x_4,\ x_5$ have 0 relative cost coefficients, but $x_6,\ x_7$ have positive relative cost coefficients. So, for this LP any feasible solution in which $x_6 = x_7 = 0$ is an optimum solution.

We find that both $x_4,\ x_5$ enter the basic vector $(x_1,\ x_2,\ x_3)$ with degenerate pivot steps, so neither of these pivot steps lead to an alternate optimum BFS.

In this case, the existence of an alternate optimum solution for this LP can be checked by solving the following LP:

BV	x_1	x_2	x_3	x_4	x_5	$-u$	$\bar{b}$	Ratio
x_1	1	0	0	1	1	0	0	
x_2	0	1	0	1	−1	0	2	
x_3	0	0	1	2	2	0	6	
$-u$	0	0	0	1	1	1	0	

$x_j \geq 0$ for all j, max u

If the maximum value of u in this problem is 0, then $\bar{x}$ is the unique optimum solution for this LP. If the maximum value of u in this LP is > 0, then an optimum solution of this LP (or any feasible solution of this LP with $u > 0$) together with $x_6,\ x_7 = 0$ gives an alternate optimum solution for the original LP. ■

5.6.4 The Optimum Face for an LP

Definition 5.1. The *optimum face* for any LP is the set of its optimum solutions.

Consider the LP in standard form

$$\begin{aligned} \text{Minimize} \quad z(x) &= cx \\ \text{subject to} \quad Ax &= b \\ x &\geq 0. \end{aligned}$$

Let K denote the set of feasible solutions of this problem. Given any optimum solution x^* for this LP, its optimum face is the set of feasible solutions of the following system of constraints:

$$\begin{aligned} Ax &= b \\ cx &= cx^* \\ x &\geq 0. \end{aligned}$$

As cx^* is the minimum value of cx over K, we have $cx \geq cx^*$ for all $x \in K$. So, the hyperplane $\{x : cx = cx^*\}$ contains the convex polyhedron K on one side, and it has a nonempty intersection with K (the optimum solution x^* is in K and also on this hyperplane). A hyperplane like this is called a *supporting hyperplane* for the convex polyhedron K, and a *face* of K by definition is the intersection of K with a supporting hyperplane. That is why the set of feasible solutions of the above system in a face of K, and hence it is called the optimum face for the LP under discussion.

If we are given more detailed information about the optimum solution, then we can give simpler characterizations of the optimum face using the complementary slackness optimality conditions. For example, if we are given an optimum BFS for this LP associated with an optimum basic vector, x_B say, let $\bar{c}^* = (\bar{c}^*_j)$ be the vector of relative cost coefficients in the canonical tableau wrt the basic vector x_B. Then let $J^* = \{j : \bar{c}^*_j > 0\}$. Then, from the CS optimality conditions we know that the optimum face of this LP is the set of feasible solutions of the system of constraints.

$$\begin{aligned} Ax &= b \\ x_j &= 0 \quad \text{for all } j \in J^* \\ x &\geq 0. \end{aligned}$$

Now, suppose we are not given an optimum BFS for this LP, but are given some optimum dual solution $\pi^* = (\pi^*_i)$ associated with it. Then let $\bar{c}^* = c - \pi^* A = (\bar{c}^*_j)$, the associated vector of dual slacks, and $J^* = \{j : \bar{c}_j^* > 0\}$. Then the optimum face of this LP is again the set of feasible solutions of the system of constraints given above.

Example 5.15. Now consider the LP in standard form from Example 5.14, in which the objective function being minimized is $z = 100 + 10x_6 + 20x_7$. Given that $x^* = (0,2,6,0,0,0,0)^T$ is an optimum solution to this LP, the optimum objective value is $z^* = 100 + 10x_6^* + 20x_7^* = 100$. So, the optimum face of this LP is its set of all feasible solutions satisfying $100 + 10x_6 + 20x_7 = 100$ or $10x_6 + 20x_7 = 0$, that is, the set of feasible solutions of

x_1	x_2	x_3	x_4	x_5	x_6	x_7	$\bar{b}$
1	0	0	1	1	1	2	0
0	1	0	1	−1	2	1	2
0	0	1	2	2	4	3	6
0	0	0	0	0	10	20	0

$x_j \geq 0$ for all j

If we are given more detailed information such as (x_1, x_2, x_3) is an optimum basic vector or that $\pi^* = (0,0,0)$ is an optimum dual solution, then the corresponding vector of relative cost coefficients or dual slacks is $\bar{c}^* = (0,0,0,0,0,10,20)$, and from the CS optimality conditions, a simpler representation of the optimum face is the following system of constraints

x_1	x_2	x_3	x_4	x_5	$\bar{b}$
1	0	0	1	1	0
0	1	0	1	−1	2
0	0	1	2	2	6

$x_j \geq 0$ for all $j = 1$ to 5,

$x_6 = x_7 = 0.$

Representations through a system of constraints for optimum faces for LPs in forms other than the standard form can be written down in a similar way using the complementary slackness optimality conditions, given a primal or dual optimum solution. ∎

5.7 Mathematical Equivalence of LP to the Problem of Finding a Feasible Solution of a System of Linear Constraints Involving Inequalities

An LP involves optimizing a linear objective function subject to linear constraints involving inequalities. On the surface, the problem of finding an optimum feasible solution appears to be mathematically harder than finding an arbitrary feasible solution.

But we will show that these two problems are mathematically equivalent, in the sense any algorithm to find a feasible solution to a system of linear constraints involving inequalities can be used directly to solve LPs and vice versa.

Consider an LP in general form in which there may be equality constraints on the variables, inequality constraints, and bounds on individual variables. By combining the bounds on individual variables with the inequality constraints, the problem is in the form

$$\begin{aligned} \text{Minimize} \quad & f\xi \\ \text{subject to} \quad & F\xi = h \\ & G\xi \geq g. \end{aligned} \tag{5.27}$$

Let π, μ be dual vectors corresponding to the constraints in the two lines in (5.27). Its dual is

$$\begin{aligned} \text{Maximize} \quad & \pi h + \mu g \\ \text{subject to} \quad & \pi F + \mu G = f \\ & \mu \geq 0. \end{aligned} \tag{5.28}$$

If ξ, (π, μ) are primal, dual feasible solutions, by the weak duality theorem $f\xi - \pi h - \mu g \geq 0$. So, if we take the system containing both the primal and dual constraints and the additional constraint $f\xi - \pi h - \mu g \leq 0$, then any feasible solution to this system must satisfy this additional constraint as an equation, and by the duality theorem it will be a primal, dual pair of optimum solutions. This system is

$$\begin{aligned} F\xi &= h, \\ \pi F + \mu G &= f, \\ G\xi &\geq g, \\ \mu \geq 0; \quad -f\xi + \pi h + \mu g &\geq 0. \end{aligned} \tag{5.29}$$

So, solving the LP (5.27) involving optimization is equivalent to finding a feasible solution to the system of linear constraints (5.29).

Conversely, given a system of linear constraints involving inequalities, the problem of finding a feasible solution to it can be formulated as an LP through the phase I formulation used in the primal simplex method. In fact, such phase I formulations offer the most practical techniques for solving linear constraints involving inequalities.

5.8 Marginal Values and the Dual Optimum Solution

Consider an LP in standard form

$$\begin{aligned} \text{Minimize} \quad & z = cx \\ \text{subject to} \quad & Ax = b \\ & x \geq 0, \end{aligned} \tag{5.30}$$

where A is a matrix of order $m \times n$. *Marginal values (MVs)* are defined as rates of change of the optimum objective value in this LP per unit change in the RHS constants from their current values. So, to describe them mathematically, define the optimum objective value function in this LP as a function of the RHS constants vector b, as b varies while A, c remain fixed at present values, as $f(b)$.

Then mathematically, the MV wrt b_i in this LP is exactly equal to the partial derivative $\partial f(b)/\overline{\partial b_i}$ when it exists. The definition of this partial derivative as a limit is

$$\stackrel{\lim}{\epsilon \rightarrow 0} (\delta(\epsilon)/\epsilon),$$

where $\delta(\epsilon) = f(b_1, \ldots, b_{i-1}, b_i + \epsilon, b_{i+1}, \ldots, b_m) - f(b)$.

A simple proof of the following theorem can be seen in Murty (1983) of Chap. 2. The following theorem follows from the results in Sect. 5.6 (Theorem 5.7).

Theorem 5.8. Sufficient Condition for the Existence of Marginal Values: *If the LP (5.30) has a primal nondegenerate optimum BFS, then MVs wrt b_i exist for all i, and the unique optimum dual solution is the vector of MVs for (5.30).* □

If (5.30) has an optimum BFS, but no nondegenerate optimum BFS, then the dual may have alternate optimum solutions; in this case MVs do not exist for it. But in this case, (5.30) will have *positive, negative marginal values*. These positive (negative) MVs are *directional derivatives*, they are limits as defined above, but when ϵ tends to 0 through positive (negative) values only. Detailed proofs of these results can be given based on the results discussed earlier in this chapter, and on the properties satisfied by the canonical tableaus obtained in the primal simplex algorithm for solving LPs. All these results are discussed in great detail in Chap. 8 of Murty (1983) of Chap. 2. The treatment there also contains the formulae for the values of these positive and negative marginal values and procedures for computing them, which involves solving a separate linear programming problem for each of these values.

See Exercise 6.73 at the end of Chap. 6 for an example of an LP in which the marginal value wrt an RHS constant b_1 does not exist, but positive and negative values exist and their values are different.

However, these directional derivatives have never come into practical use for several reasons: one is the computational burden of computing them, the other is the difficulty of deciding whether the computed 0-value in the BFS, of a basic variable, is actually 0, or a small positive value rounded off to 0 in computation with bounded precision.

Hence all the results on the positive and negative marginal values remain interesting theoretically, but with no practical application so far. So, we will not discuss them here; those interested in them for theoretical research should refer to that chapter in Murty (1983) of Chap. 2.

5.9 Summary of Optimality Conditions for Continuous Variable Nonlinear Programs and Their Relation to Those for LP

In discussing LPs so far, we did not talk about any "types of linear programs," because all LPs are of just one type. The optimality conditions for LP discussed earlier in this chapter apply equally to all LPs, and they completely characterize the optimum solutions of any LP. In fact, the termination conditions in algorithms for LP are based on these optimality conditions.

But the situation in quadratic and general continuous variable nonlinear optimization problems is very different. These problems belong to two types. At the moment we have complete characterizations of optimum solutions for only one type of these continuous optimization problems, those in which the objective function to be minimized is a convex function and the set of feasible solutions is a convex set (these are called *convex programming problems*). For the other type called the *non-convex programming problems*, we only have partial characterizations of optimality, which we will explain.

In mathematics, conditions that a solution of an optimization problem must satisfy in order to be an optimum solution are called *necessary optimality conditions*. This means that a feasible solution not satisfying these necessary optimality conditions is definitely not an optimum solution, but one satisfying them may not be optimal.

Conditions which guarantee that every solution satisfying them is an optimum solution are called *sufficient optimality conditions*. So, sufficient optimality conditions are in general stronger conditions than necessary optimality conditions.

5.9.1 Global Minimum (Maximum), Local Minimum (Maximum), and Stationary Points

A major difference between linear and general continuous variable optimization models appears in the classification of various types of "optimum solutions." In LP, feasible solutions are classified into only two classes, those which are optimal and those which are not. In LP optimum, solutions are all of one type, they all correspond to the same optimum objective value.

In contrast, in QP, and continuous variable nonlinear programs, "optimum solutions" are classified into three types, and in some of these, they may correspond to different objective values.

Feasible solutions that correspond to the true minimum (or maximum, as required in the problem) value of the objective function are called *global minimum (or maximum)*, respectively. The original problem requires finding these global optima. Clearly all global minima correspond to the same minimum objective value in a minimization problem. In the same way, in a maximization problem, all global maxima correspond to the true maximum objective value in the problem.

A feasible solution of the problem that corresponds to the minimum (maximum) objective value in a small neighborhood around it is called a *local minimum (maximum)*, respectively, and such local optima may not be global optima in general continuous optimization problems. In contrast, in LP, every local optimum is also guaranteed to be a global optimum.

In fact, some types of quadratic programs and general continuous variable nonlinear programs may have local optima, which are not global optima. And different local minima may correspond to different objective values. Also, in these QPs, and general continuous variable optimization problems, we do not have any useful optimality conditions to characterize the global optima. The only optimality conditions known are characterizations of local optima.

We will now summarize these optimality conditions for a nonlinear program in which all functions (objective and constraint functions) are assumed to be continuously differentiable (twice continuously differentiable if the Hessian matrix appears in the conditions). As the problem of maximizing a function $z(x)$ is the same as that of minimizing $-z(x)$, we will state the conditions for a minimization problem.

For any real-valued function $f(x)$ of $x = (x_1, \cdots, x_n)^T$, $H(f(x))$ denotes its *Hessian matrix* (the $n \times n$ matrix of second partial derivatives) at the point x, and $\nabla f(x)$ denotes the row vector of its partial derivatives.

A square matrix D of order n is said to be a

- *PSD (positive semidefinite)* matrix if $y^T D y \geq 0$ for all $y \in R^n$,
- *NSD (negative semidefinite)* matrix if $y^T D y \leq 0$ for all $y \in R^n$,
- *PD (positive definite)* matrix if $y^T D y > 0$ for all $0 \neq y \in R^n$,
- *ND (negative definite)* matrix if $y^T D y < 0$ for all $0 \neq y \in R^n$.

See Sects. 9.1 and 9.2 for efficient methods to check whether any given square matrix belongs to one of these classes.

Optimality Conditions for an Unconstrained Minimization Problem: Consider the problem: Minimize $\theta(x)$ over $x \in R^n$.

- *First order necessary optimality conditions for a given point $\bar{x}$ to be a local minimum:* $\nabla\theta(\bar{x}) = 0$.
- *Second order necessary optimality conditions for a given point $\bar{x}$ to be a local minimum:* $\nabla\theta(\bar{x}) = 0$ and $H(\theta(\bar{x}))$ is PSD.
- *Sufficient optimality conditions for a given point $\bar{x}$ to be a local minimum:* $\nabla\theta(\bar{x}) = 0$ and $H(\theta(\bar{x}))$ is PD.

We will now discuss optimality conditions for minimization problems involving constraints. In the literature, the first-order necessary optimality conditions given below are often called *KKT (Karush, Kuhn, Tucker) necessary optimality conditions*, and any point satisfying them is called a *stationary point* or *KKT point* for the problem.

Optimality Conditions for an Equality Constrained Minimization Problem: Consider the problem: Minimize $\theta(x)$, subject to $h_i(x) = 0$ for $i = 1$ to m.

Denote $h(x) = (h_i(x), \dots, h_m(x))^T$. For $i = 1$ to m, associate a *Lagrange multiplier* π_i to the ith constraint in the problem and construct the *Lagrangian* for the problem $L(x, \pi) = \theta(x) - \sum_{i=1}^{m} \pi_i h_i(x) = \theta(x) - \pi h(x)$, where $\pi = (\pi_1, \cdots, \pi_m)$ is the row vector of Lagrange multipliers for the problem.

The following optimality conditions for a given feasible solution $\bar{x}$ to be a local minimum are proved under the assumption that the constraint functions $h_i(x)$ satisfy a regularity condition at the point $\bar{x}$, which is $\{\nabla h_i(\bar{x}) : i = 1 \text{ to } m\}$, a linearly independent set of vectors. Under these conditions, the point $\bar{x}$ is called a *regular point*. Conditions like this that depend on the constraint functions are called *constraint qualifications* in the literature. The optimality conditions given below hold when either the constraints are linear (in this special case, no additional constraint qualifications are needed) or when some constraint qualifications hold at the given feasible solution $\bar{x}$. See references Murty (1988) of Chap. 2; Bazaraa et al. (2006). The first order optimality conditions given below for this problem are often referred to as the *Lagrange multiplier optimality conditions*.

- *First order necessary optimality conditions for a given point $\bar{x}$ to be a local minimum:* Feasibility, that is, $h(\bar{x}) = 0$, and there exists a Lagrange multiplier vector $\bar{\pi}$, which together with $\bar{x}$ satisfies

$$\nabla_x L(\bar{x}, \bar{\pi}) = \nabla\theta(\bar{x}) - \bar{\pi}\nabla h(\bar{x}) = \nabla\theta(\bar{x}) - \sum_{i=1}^{m} \bar{\pi}_i \nabla h_i(\bar{x}) = 0.$$

  So, this condition states that the objective gradient at $\bar{x}$ is a linear combination of constraint gradients at $\bar{x}$.

- *Second order necessary optimality conditions for a given point $\bar{x}$ to be a local minimum:* Feasibility, that is, $h(\bar{x}) = 0$, and there exists a Lagrange multiplier vector $\bar{\pi}$, which together with $\bar{x}$ satisfies

$$\nabla_x L(\bar{x}, \bar{\pi}) = \nabla\theta(\bar{x}) - \bar{\pi}\nabla h(\bar{x}) = 0 \quad \text{and}$$
$$y^T H_x(L(\bar{x}, \bar{\pi}))y \geq 0$$
$$\text{for all } \; y \in \{y : \nabla h_i(\bar{x})y = 0, \; i = 1, \dots, m\}.$$

- *Sufficient optimality conditions for a given point $\bar{x}$ to be a local minimum:* Feasibility, that is, $h(\bar{x}) = 0$, and there exists a Lagrange multiplier vector $\bar{\pi}$, which together with $\bar{x}$ satisfies:

$$\nabla_x L(\bar{x}, \bar{\pi}) = \nabla\theta(\bar{x}) - \bar{\pi}\nabla h(\bar{x}) = 0 \quad \text{and}$$
$$y^T H_x(L(\bar{x}, \bar{\pi}))y > 0$$
$$\text{for all } \; 0 \neq y \in \{y : \nabla h_i(\bar{x})y = 0, \; i = 1, \dots, m\}.$$

Optimality Conditions for a General Constrained Minimization Problem: Consider the problem: Minimize $\theta(x)$, subject to $h_i(x) = 0$ for $i = 1$ to m, and $g_p(x) \geq 0$ for $p = 1$ to t.

Denote $h(x) = (h_1(x), \cdots, h_m(x))^T$, $g(x) = (g_1(x), \cdots, g_t(x))^T$. For $i = 1$ to m, associate a Lagrange multiplier π_i to the ith equality constraint, and for $p = 1$ to t associate a Lagrange multiplier μ_p to the pth inequality constraint in the problem, and construct the *Lagrangian* for the problem $L(x, \pi, \mu) = \theta(x) - \sum_{i=1}^{m} \pi_i h_i(x) - \sum_{p=1}^{t} \mu_p g_p(x) = \theta(x) - \pi h(x) - \mu g(x)$, where $\pi = (\pi_1, \cdots, \pi_m)$, $\mu = (\mu_1, \cdots, \mu_t)$ are the row vectors of Lagrange multipliers for the problem.

- *First order necessary optimality conditions for a given point $\bar{x}$ to be a local minimum:* Feasibility, that is, $h(\bar{x}) = 0$, $g(\bar{x}) \geq 0$, and there exist Lagrange multiplier vectors $\bar{\pi}$, $\bar{\mu}$, which together with $\bar{x}$ satisfy

$$\nabla_x L(\bar{x}, \bar{\pi}, \bar{\mu}) = \nabla\theta(\bar{x}) - \bar{\pi}\nabla h(\bar{x}) - \bar{\mu}\nabla g(\bar{x})$$

$$= \nabla\theta(\bar{x}) - \sum_{i=1}^{m} \bar{\pi}_i \nabla h_i(\bar{x}) - \sum_{p=1}^{t} \bar{\mu}_p \nabla g_p(\bar{x}) = 0.$$

$$\bar{\mu} \geq 0$$

$$\bar{\mu}_p g_p(\bar{x}) = 0 \quad \text{for all } p$$

The last set of conditions here are called the *complementary slackness optimality conditions*. They are equivalent to $\bar{\mu} g(\bar{x}) = 0$.

- *Second order necessary optimality conditions for a given point $\bar{x}$ to be a local minimum:* Feasibility, that is, $h(\bar{x}) = 0$, $g(\bar{x}) \geq 0$, and there exist Lagrange multiplier vectors $\bar{\pi}, \bar{\mu}$, which together with $\bar{x}$ satisfy

$$\nabla_x L(\bar{x}, \bar{\pi}, \bar{\mu}) = \nabla\theta(\bar{x}) - \bar{\pi}\nabla h(\bar{x}) - \bar{\mu}\nabla g(\bar{x})$$

$$= \nabla\theta(\bar{x}) - \sum_{i=1}^{m} \bar{\pi}_i \nabla h_i(\bar{x}) - \sum_{p=1}^{t} \bar{\mu}_p \nabla g_p(\bar{x}) = 0.$$

$$\bar{\mu} \geq 0$$

$$\bar{\mu}_p g_p(\bar{x}) = 0 \quad \text{for all } p$$

$$y^T H_x(L(\bar{x}, \bar{\pi}, \bar{\mu}))y \geq 0 \quad \text{for all } y \in$$

$$\{y : \nabla h(\bar{x})y = 0,\ \nabla g_p(\bar{x})y = 0 \quad \text{for all } p \in P(\bar{x})\}$$

where $P(\bar{x}) = \{p : g_p(\bar{x}) = 0\}$.

- *Sufficient optimality conditions for a given point $\bar{x}$ to be a local minimum:* Feasibility, that is, $h(\bar{x}) = 0$, $g(\bar{x}) \geq 0$, and there exist Lagrange multiplier vectors $\bar{\pi}, \bar{\mu}$, which together with $\bar{x}$ satisfy

$$\nabla_x L(\bar{x}, \bar{\pi}, \bar{\mu}) = \nabla(\theta(\bar{x})) - \bar{\pi}\nabla(h(\bar{x})) - \bar{\mu}\nabla(g(\bar{x}))$$

$$= \nabla(\theta(\bar{x})) - \sum_{i=1}^{m} \bar{\pi}_i \nabla(h_i(\bar{x})) - \sum_{p=1}^{t} \bar{\mu}_p \nabla(g_p(\bar{x})) = 0.$$

$$\bar{\mu} \geq 0$$

$$\bar{\mu}_p g_p(\bar{x}) = 0 \quad \text{for all } p$$

$$y^T H_x(L(\bar{x}, \bar{\pi}, \bar{\mu}))y > 0 \quad \text{for all } 0 \neq y \in T_1,$$

where $T_1 = \{y : \nabla h(\bar{x})y = 0 \text{ and } \nabla g_p(\bar{x})y = 0 \text{ for all } p \in P_1, \text{ and } \nabla g_p(\bar{x})y \geq 0 \text{ for all } p \in P_2\}$, $P_1 = P(\bar{x}) \cap \{p : \bar{\mu}_p > 0\}$, $P_2 = P(\bar{x}) \cap \{p : \bar{\mu}_p = 0\}$.

Necessary and Sufficient Optimality Conditions for a Given Point $\bar{x}$ to be a Global Minimum: If the problem is a convex programming problem, that is, if $\theta(x)$ is a convex function, $h(x)$ are all affine functions and $g(x)$ are all concave functions, then the first order optimality conditions (KKT optimality conditions) are both necessary and sufficient for $\bar{x}$ to be a global minimum to the problem. So for a convex programming problem, every stationary point (i.e., KKT point) is a global minimum.

The concepts of various types of optimum solutions in a nonconvex programming problem are illustrated in Fig. 5.1 for the problem of optimizing a real-valued function $\theta(x)$ of one variable $x \in R^1$ subject to $a \leq x \leq b$, where $[a, b]$ is a

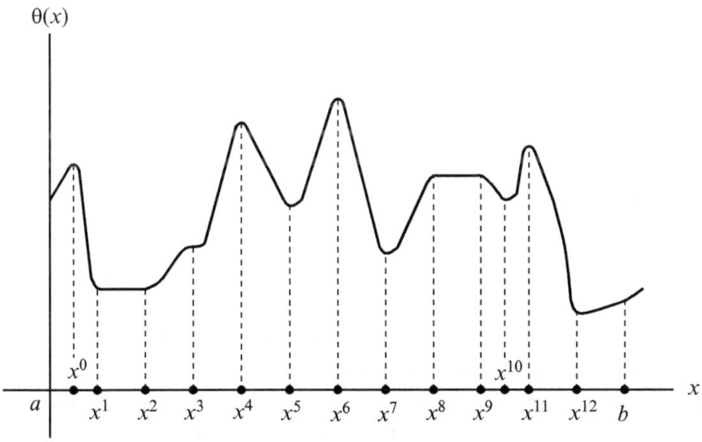

Fig. 5.1 Various types of minima and maxima of a general objective function

given closed interval in R^1. Figure 5.1 contains a plot of $\theta(x)$ on this interval, with the variable plotted on the horizontal axis, and the values of $\theta(x)$ plotted on the vertical axis.

The points $a,\ x^5,\ x^7,\ x^{10},\ x^{12}$ are local minima, $x^0,\ x^4,\ x^6,\ x^{11},\ b$ are local maxima, x^{12} is the global minimum, and x^6 is the global maximum for $\theta(x)$ in this interval. In the interval $x^1 \le x \le x^2$, the function $\theta(x)$ is a constant, and every point in this interval is a local minimum for $\theta(x)$. In the interval $x^8 \le x \le x^9$ the function $\theta(x)$ is again a constant, and every point in this interval is a local maximum for $\theta(x)$.

At the point x^3, the derivative of $\theta(x)$ is 0, and so x^3 is a stationary point for this problem, even though it is neither a local minimum or maximum.

5.9.2 Relationship to Optimality Conditions for LP Discussed Earlier

Consider the LP in standard form

$$\begin{aligned} &\text{Minimize } cx \\ &\text{subject to } Ax = b \\ &\qquad\qquad\quad x \ge 0, \end{aligned} \tag{5.31}$$

where A is a matrix of order $m \times n$. To derive the optimality conditions given above to this LP, we associate the Lagrange multipliers $\pi = (\pi_1, \dots, \pi_m)$, $\mu = (\mu_1, \dots, \mu_n)$ to the two sets of constraints in (5.31) in that order. The Lagrangian is $L(x,\ \pi,\ \mu) = cx - \pi(Ax - b) - \mu x$.

Therefore, the first order necessary optimality conditions or the KKT optimality conditions for a point $\bar{x}$ to be optimal to this LP are that there exist Lagrange multiplier vectors $\bar{\pi},\ \bar{\mu}$, which together with $\bar{x}$ satisfy

$$A\bar{x} = b, \quad \bar{x} \ge 0$$

$$\frac{\partial L(\bar{x}.\bar{\pi},\ \bar{\mu})}{\partial x} = c - \bar{\pi}A - \bar{\mu} = 0, \quad \bar{\mu} \ge 0, \quad \bar{\pi} \text{ unrestricted.}$$

$$\bar{\mu}\bar{x} = 0.$$

The conditions in the first line are the primal feasibility conditions for $\bar{x}$. Conditions in the second line are the dual feasibility conditions for the Lagrange multiplier vector $\bar{\pi}$. And $\bar{\mu} = c - \bar{\pi}A$ is the vector of dual slacks in the dual feasible solution $\bar{\pi}$, this is the vector which we denoted by $\bar{c}$ (the relative cost vector when the dual solution is $\bar{\pi}$) in earlier sections. The conditions in the third line are the complementary slackness optimality conditions discussed in earlier sections. Thus the duality theorem and the complementary slackness theorem together provide a proof that the first order necessary optimality conditions are both necessary and sufficient optimality conditions for LPs.

5.10 Exercises

5.1. A woman is trying to get as much vitamin K as possible in her diet to improve her health. She started eating a special breed of avocado (BA) and a tropical fruit from Brazil (BF), which are excellent sources of vitamin K. Both BA and BF also contain potassium and leutein. There is a maximum daily limit (MDL) for potassium and a minimum daily requirement (MDR) for leutein from BA and BF in her diet. Also, BA contains a type of fat and BF contains sugars; for these reasons, she has to limit the quantities of these foods in her diet. Data on the composition of these foods and all the limits is given below. Also, in this problem ignore any Vit K, potassium, and leutein she may get in foods other than BA and BF that she eats.

Nutrient	Composition (units/unit)		Limit
	BA	BF	
Vit K	65	35	Maximize
Potassium	6	8	≤ 48 (MDL)
Leutein	1	1	≥ 2 (MDR)
Maximum in diet	6	4	

Formulate the problem of determining how much BA, BF to include in her daily diet as an LP. Solve this LP by the geometric method and determine the marginal values of the MDL on potassium, and the MDR on leutein, and explain their interpretation.

5.2. Consider the following LP. Denoting the dual variables associated with constraints 1–5 from *top* to *bottom* as π_1 to π_5, write the dual problem.

Given the optimum dual solution, $\bar{\pi} = (-4, 3, 4, 5, 0)$, find an optimum solution to the primal using the CS conditions. Show that the set of optimum solutions of the primal is a line segment in the x-space. Find its parametric representation.

x_1	x_2	x_3	x_4	x_5	RHS
1	−1	0	1	0	$= 6$
0	−1	0	0	−2	≥ -4
0	0	1	0	−1	≥ 5
0	0	0	2	−2	≥ 14
0	−2	6	0	−4	≤ 25
−4	1	4	6	−20	$= z$ minimize

all $x_j \geq 0$

5.3. Consider the following LP

$$\begin{aligned} \text{Minimize } & 0x \\ \text{subject to } & Ax = b \\ & x \geq 0, \end{aligned} \tag{5.32}$$

where A is an $m \times n$ matrix. As all the cost coefficients of the variables are 0, verify that every feasible solution of (5.32) as an optimum solution for it.

Applying the duality results to this LP, show that if the system of constraints in (5.32) has no feasible solution x, then there exists a linear combination of equality constraints in (5.32) in which all the variables x_j have nonpositive coefficients, but the RHS constant is strictly positive. This result is known popularly as *Farkas' lemma*, it is a theorem of alternatives for the system of constraints in (5.32).

5.4. *The Balanced LP Problem (BLP).* BLP arises in situations that call for equitable distribution or sharing of resources, commonly encountered in economic and military contexts. The set of feasible solutions for BLP is a convex polyhedron $K = \{x : Ax = b,\ x \geq 0\}$, where $x = (x_1, \dots, x_n)^T \in R^n$, and $A = (a_{ij})$, $b = (b_i)$ are given matrices of orders $m \times n$, $m \times 1$, respectively.

In this problem, we are given an objective function $cx = \sum_{j=1}^{n} c_j x_j$ and a subset of column indices $J \subset \{1, \dots, n\}$, where $c_j > 0$ for each $j \in J$. The aim of BLP is to find a *balanced solution*, which is defined as a feasible solution $x \in K$ that minimizes the difference $(\max\{c_j x_j : j \in J\} - \min\{c_j x_j : j \in J\})$.

(1) Formulate the BLP as an LP problem. (2) If there exists a feasible solution $x \in K$ satisfying $c_j x_j$ has the same value, say t, for all $j \in J$, it is called a *totally balanced solution*. Show that any totally balanced solution is an optimum solution for the BLP. If $J = \{1, \dots, n\}$, show that a totally balanced solution exists iff $b_i / (\sum_{j=1}^{n} a_{ij}/c_j)$ has the same value for all i, which is easy to check without solving any LP. (3) Let y be a scalar parameter, and consider the following parametric RHS LP in variables $x = (x_j)$ and $z \in R^1$: min z subject to $x \in K$ and $(y/c_j) \leq x_j \leq (z/c_j)$ for all $j \in J$. Let $x(y)$, $z(y)$ denote an optimum solution, optimum objective value of this parametric LP as a function of the parameter y. Show that $z(y)$ is a PL convex function in y, and that there exists a value of the parameter y such that $x(y)$ solves the BLP (Ahuja 1997).

5.5. Consider the following LP

x_1	x_2	x_3	x_4	x_5	x_6	RHS	
1	1	1	1	1	−1	$\geq$	50
0	1	1	1	−1	2	$\geq$	45
0	0	1	1	2	2	$\geq$	35
0	0	0	1	1	1	$\geq$	20
1	3	6	10	7	14	=	$z(x)$, minimize

$$x_j \geq 0 \quad \text{for all } j.$$

The solution $\bar{x} = (5, 10, 15, 20, 0, 0)$ is a feasible solution for this LP with objective value $z(\bar{x}) = 325$.

(1) Check whether 325 is the minimum objective value in this LP, using the optimality conditions in the simplex algorithm showing all your work very clearly.
(2) Check whether 325 is the minimum objective value in this LP, using the CS optimality conditions, showing all your work.

Comment on why (1) and (2) are the same.

5.6. Gordon's theorem of alternatives states that the system $Ax < 0$ for a given matrix A of order $m \times n$ has a solution iff the alternate system $yA = 0,\ y \geq 0$ has no nonzero solution y. Prove it using the duality theorem of LP (hint: apply the strategy discussed in Exercise 5.3).

5.7. A is an $m \times n$ matrix, A^T is the transpose of A, c is a row vector in R^n, and b is a column vector in R^m. Consider the following LP in which there are $n + m$ constraints in $n+m$ nonnegative variables written in the form of two column vectors, $x = (x_1, \ldots, x_n)^T$ and $y = (y_1, \ldots, y_m)^T$:

$$\begin{array}{lrrcr} \text{Minimize} & f = cx & -b^T y & & \\ \text{subject to} & Ax & & \geq & b \\ & & -A^T y & \geq & -c^T \\ & x, & y & \geq & 0. \end{array}$$

Prove that this LP in $n + m$ variables is either infeasible or is feasible, and the optimum value for the objective function f will be 0 in this case.

5.8. Let $A,\ b$ be matrices of orders $m \times n,\ m \times 1$, respectively. Let $x \in R^n,\ \pi \in R^m$ be a column vector, row vector of variables and consider the two systems:

$$Ax = b, \quad x \geq 0, \tag{5.33}$$

$$\pi A \leq 0, \quad \pi b = 1. \tag{5.34}$$

Farkas' lemma states that either (5.33) has a solution x or (5.34) has a solution π, but not both.

Using Farkas' lemma, prove the corresponding result when the first system is $Ax = b$ without the non-negativity requirement on the variables.

5.9. Consider the fertilizer problem discussed in Sect. 5.2. There are three raw materials $RM_1,\ RM_2,\ RM_3$ available in maximum quantities of 1,500, 1,200, 500 tons/day, respectively. The input/output requirements for the two products HI-ph, Lo-ph are summarized below.

Raw	Input (tons/ton) to make		Max. (tons)
material	Hi-ph	Lo-ph	available/day
RM_1	2	1	1,500
RM_2	1	1	1,200
RM_3	1	0	500

Now the prices have gone up quite a bit. The prices/ton of $RM_1,\ RM_2,\ RM_3$ are, respectively, \$4,500, \$4,000, and \$3,500. Manufacturing costs of Hi-ph, Lo-ph other than the costs of these three raw materials are \$600, 400/ton, respectively. The selling prices of Hi-ph, Lo-ph are \$17,600, \$9,025/ton, respectively.

Write the model for maximizing the companies net profit/day. Letting $x_1,\ x_2$ denote the tons of Hi-ph, Lo-ph produced/day; the optimum solution of the model is $(\bar{x}_1,\ \bar{x}_2) = (500, 500)^T$.

Find the constraints in the model that are active at $\bar{x}$.

Assuming that exactly the same constraints remain active at the optimum solution when small perturbations occur in the availability of any of the raw materials, find the marginal values of the three raw materials in this problem, using the small perturbation technique discussed in this chapter.

The company's chemist has developed a new fertilizer and named it *lushgarden*. To make one ton of lushgarden needs one ton each of RM_1, RM_2, RM_3 and incurs total other manufacturing cost of \$500. Determine the minimum sale price/ton of lushgarden at which it will be competitive with Hi-ph, Lo-ph, explaining clearly how you obtained it.

5.10. Let A, b be given matrices of orders $m \times n$, $m \times 1$, respectively. Consider the LP

$$\begin{aligned} \text{Minimize } z(x) &= cx \\ \text{subject to } \quad Ax &= b \\ x &\geq 0. \end{aligned} \tag{5.35}$$

(1) Is there any relationship between "this LP having more than one optimum solution" and "degeneracy." If so explain what this relationship is very clearly.

(2) Suppose $\bar{x}$ is a given optimum solution of this LP. If $\bar{x}$ is not a basic feasible solution (BFS) of this LP, show that this LP must have alternate optimum solutions.

Also, in this case, describe briefly a procedure that can be used to obtain an optimum BFS for this LP.

(3) Suppose we are given an optimum basic vector x_B for this LP. Let $\mathbf{P}$ = Set of i such that the entry in row i in the updated RHS column with respect to x_B is 0, $\mathbf{Q}$ = set of j such that x_j is a nonbasic variable with zero relative cost coefficient in the canonical tableau with respect to x_B, $\mathbf{S}$ = set of j such that x_j is a nonbasic variable with nonzero relative cost coefficient in the canonical tableau with respect to x_B.

Prove that the set of optimum solutions of this LP is unique iff the optimum objective value in the following LP is zero:

$$\begin{aligned} \text{Maximize } & \sum_{j \in Q} x_j \\ \text{subject to } \quad & Ax = b \\ & x_j = 0 \quad \text{for all } j \in S \\ & x \geq 0. \end{aligned}$$

5.11. Let $b \in R^n$ be a given vector. In defining its positive and negative parts, we have expressed b as $x - y$, where x, $y \geq 0$. Consider the LP: minimize $cx + dy$ subject to $x - y = b$, x, $y \geq 0$.

Write the dual of this LP and express the condition when this dual is feasible or infeasible, and relate this to the primal having an optimum solution or having an unbounded objective function on its set of feasible solutions. Also, verify the complementary slackness property when both problems have optimum solutions.

What conditions on c, d will guarantee that $x_j y_j = 0$ at an optimum solution of the original primal LP?

5.12. Consider the LP: minimize $z = cx$ subject to $b \leq Ax \leq d$, where A is a matrix of order $m \times n$ and $b < d$ are both finite vectors, and any bound restrictions on individual variables are included in the constraints.

Assume that this LP has a feasible solution. Show that the objective function z is unbounded below on the set of feasible solutions of this LP iff c is not in the linear hull of the row vectors of A.

5.13. Let $A = (a_{ij})$ be a 2×4 matrix, and consider the LP: minimize $z(x) = cx$, subject to $(-1, -1)^T \leq Ax \leq (1, 1)^T$, with the variables x unrestricted in sign.

Using symbols $\mu_1, \mu_2; \delta_1, \delta_2$ for the dual variables, write the dual of this LP.

Show that the system of dual constraints can be simplified and expressed in terms of two unrestricted variables only. Use this to show that the dual problem has a feasible solution iff the cost row c can be expressed as a linear combination of the rows of A. Using this show that the original LP has an optimum solution iff the cost row c is in the linear hull of the rows of A.

5.14. Consider the following LP in standard form.

x_1	x_2	x_3	x_4	x_5	x_6	$-z$	RHS
1	−3	1	1	1	2	0	16
1	5	1	2	0	−3	0	1
0	2	1	−4	0	5	0	14
7	12	5	26	3	−16	1	0

$x_j \geq 0$ for all j, minimize z

Write its dual and the complementary slackness optimality conditions that a pair of primal and dual feasible solutions have to satisfy in order to be optimal solutions to the respective problems.

Use these conditions to check whether $\bar{x} = (3, 0, 4, 0, 5, 2)^T$ is an optimum solution to this LP.

5.15. Consider the following LP in which A, D, c are $m \times n$, $p \times n$, $1 \times n$ matrices, respectively. Write its dual using symbols $\pi = (\pi_i)$, $\mu = (\mu_t)$ to denote the dual variables.

$$
\begin{aligned}
\text{Minimize } z &= cx \\
\text{subject to } Ax &= b \\
Dx &\geq d \\
x &\geq 0.
\end{aligned}
$$

Given $\bar{x}$ and $(\bar{\pi}, \bar{\mu})$, a pair of primal and dual feasible solutions, prove directly that $c\bar{x}$ is $\geq$ the dual objective value at $(\bar{\pi}, \bar{\mu})$.

5.16. A is a given matrix of order $m \times n$. Consider the following problems:

Problem 1. Find a vector $b \in R^m$ satisfying the property that the system $Ax \leq b$ has at least one feasible solution $x \in R^n$.

Problem 2. Find a vector $b \in R^m$ satisfying the property that the system $Ax \leq b$ has no feasible solution $x \in R^n$.

Describe the simplest and most efficient algorithms for each of Problems 1 and 2.

Which of Problems 1 or 2 can be solved using direct linear algebra operations (i.e., matrix arithmetic, solving linear equations but no inequalities, finding rank, etc.) only? Why?

Are the two problems of the same order of difficulty? If not, explain the reasons for the difference.

5.17. Consider the system of homogeneous linear constraints: $Ax = 0, x \geq 0$, where A is a given $m \times n$ matrix.

Using the duality theorem show that it has a nonzero feasible solution iff the system in variables $\pi = (\pi_1, \dots, \pi_m)$: $\pi A > 0$ has no feasible solution.

5.18. Consider the LP: minimize $z = cx$ subject to $Ax = b, \ell \leq x \leq k$, where $A = (a_{ij})$ is a given $m \times n$ matrix, $c = (c_j) \in R^n$, $b = (b_i) \in R^m$, and $\ell = (\ell_j) < k = (k_j)$ are given finite vectors in R^n.

Write the dual of this LP. Show why the dual problem always has a feasible solution, irrespective of what values the primal data elements take.

5.19. Optimum objective value as a function of the cost vector in an LP. Consider the following LP where A is a matrix of order $m \times n$: minimize $z(x) = cx$ subject to $Ax = b, \ x \geq 0$.

In this LP A, b remain fixed, but the cost vector c varies, and we denote by $g(c)$ the optimum objective value as a function of the cost vector c. Show that $g(c)$ is a concave function.

5.20. Consider the LP: minimize $z = cx$, subject to $Ax \geq b$.

Assuming that the LP has at least one feasible solution, prove that it has an optimum solution if and only if the cost vector c is in the Pos cone (i.e., nonnegative hull) of the set of row vectors of A.

5.21. Let A be a matrix of order $m \times n$. Consider the LP: minimize $z(x) = cx$ subject to $Ax = b, x \geq 0$. Suppose we know that this LP has an optimum solution. From this, what can we conclude about the dual of this LP?

Now consider the modified primal with an additional constraint: minimize $z(x) = cx$ subject to $Ax = b, A_{m+1.}x = b_{m+1}, x \geq 0$. Suppose we know that this modified primal is infeasible. Then prove that the dual of this modified primal is feasible, but its objective function is unbounded above in it.

5.22. Consider the following primal linear program, where A is an $m \times n$ matrix.

$$\begin{aligned} \text{Minimize} \quad z(x) &= cx \\ \text{subject to} \quad Ax &\geq b \\ x &\geq 0. \end{aligned}$$

It is known that this problem has at least one feasible solution. Prove that the set of feasible solutions of either the primal or the dual or both must be nonempty and unbounded.

5.23. Consider the following single constraint bounded variable LP:

$$\begin{aligned} \min \quad z(x) &= \sum_{j=1}^{n} c_j x_j \\ \text{s. to} \quad \sum_{j=1}^{n} a_j x_j &= b \\ \ell_j \leq x_j &\leq k_j \quad \text{for all } j. \end{aligned}$$

(1) Write the dual of this problem.
(2) If $\bar{x} = (\bar{x}_j)$ is a feasible solution for the original LP, prove that it is an optimum solution iff there exists a real number $\bar{\pi}$ satisfying, for all $j = 1$ to n,

$$\begin{aligned} \bar{\pi} a_j < c_j \quad &\text{implies} \quad \bar{x}_j = \ell_j \\ \bar{\pi} a_j > c_j \quad &\text{implies} \quad \bar{x}_j = k_j \\ \ell_j < \bar{x}_j < k_j \quad &\text{implies} \quad \bar{\pi} a_j = c_j. \end{aligned}$$

5.24. Consider the LP: minimize $z(x) = d^T x$ subject to $Dx \geq d$, $x \geq 0$, where d is a column vector in R^m and D is a square matrix of order m satisfying $D = D^T$.

If $x^* \geq 0$ satisfies $Dx^* = d$, prove that x^* is an optimum solution of this LP.

5.25. Consider the LP in standard form where $A_{m \times n}$ has rank m: minimize $z = cx$, subject to $Ax = b$, $x \geq 0$.

The solution $\tilde{x} = (\tilde{x}_j)$ is a given optimum solution of the above LP and $\tilde{J} = \{j : \tilde{x}_j > 0\}$. Then prove that every feasible solution $\hat{x} = (\hat{x}_j)$ for this LP that satisfies $\{j : \hat{x}_j > 0\} \subset \tilde{J}$ must also be an optimum solution of this LP.

5.26. Consider the LP: minimize $z = cx$, subject to $\sum_{j=1}^{n} a_j x_j \leq b$ and $0 \leq x_j \leq u$ for all j, where $u > 0$ is a given common upper bound for all the variables and all a_j, c_j, b are > 0.

Write the dual of this LP. Prove that there exists a $j_1 \in \{1, \ldots, n\}$ such that there is an optimum solution $(\bar{x}) = (\bar{x}_j)$ for this LP, where

$$\begin{aligned} &\text{for all } j \quad \text{for which } (c_j/a_j) > (c_{j_1}/a_{j_1}),\ \bar{x}_j = 0, \\ &\text{for all } j \quad \text{for which } (c_j/a_j) < (c_{j_1}/a_{j_1}),\ \bar{x}_j = u. \end{aligned}$$

5.27. Consider the system of linear equations $Ax = b,\ x \geq 0,$ where $A,\ b$ are given below. Using Farakas' lemma discussed at the beginning of this chapter, show that the system $A^T y \leq 0,\ b^T y > 0$ must have a solution y (R. Saigal).

$$A = \begin{pmatrix} 1 & 2 \\ -1 & 2 \\ -1 & 0 \end{pmatrix},\quad b = \begin{pmatrix} 1 \\ 0 \\ 5 \end{pmatrix}.$$

5.28. Let D be a diagonal matrix with positive diagonal entries, and e the column vector of all 1's of appropriate size. Consider the nonlinear program: maximize $x^T x$ subject to $x^T x = e^T Dx$.

Write the KKT optimality conditions for this problem, and then find all the KKT points for it. Determine which of these problems are local maxima, and among these find the global maximum for this problem (R. Saigal).

5.29. Let $Q,\ A,\ B$ be $m \times n,\ p \times m,\ q \times n$ matrices and $c,\ d$ are row vectors in $R^m,\ R^n$ respectively. Consider the following quadratic program in variables $x = (x_1, \ldots, x_m)^T,\ y = (y_1, \ldots, y_n)^T$, known as a *bilinear programming problem* (the reason for this name is the following: if x or y is fixed, the remaining problem is an LP in the remaining variables):

Minimize $z(x, y) = cx + x^T Qy + dy$, subject to $Ax = a,\ By = b,\ x,\ y \geq 0$.

Write the KKT necessary optimality conditions for this problem.

Fix $x = \bar{x} \geq 0$ and write the KKT conditions for the remaining LP in y. Suppose $\bar{y}$ is an optimum solution of the remaining LP in y.

Fix $y = \bar{y}$ and write the KKT conditions for the remaining LP in x. Suppose $\bar{x}$ happens to be an optimum solution of the remaining LP in x.

Then show that $(\bar{x},\ \bar{y})$ is a KKT point for the original bilinear program (R. Saigal).

5.30. Consider the problem: minimize $5x_1^2 + 6x_2^2 - \lambda \log(x_1) - \lambda \log(x_2)$ subject to $x_1 + x_2 = 1,\ x_1,\ x_2 \geq 0,$ where λ is a positive constant.

Ignoring the "$x_1,\ x_2 \geq 0$" constraints for the moment, write the KKT necessary optimality conditions for the remaining problem. Find a solution to this KKT system as a function of the parameter λ, and check that this solution satisfies the sufficiency condition for being a local minimum to this problem. Verify that this solution also satisfies the ignored constraints $x_1,\ x_2 \geq 0$. Investigate what happens to this solution as λ tends to 0.

5.31. Consider the problem: minimize $(1/2)||Dx||^2$ subject to $x + A^T y = c$, where D is a square diagonal matrix with positive diagonal entries and A, c are an $m \times n$ matrix, and column vector in R^n, respectively.

Write the KKT necessary optimality conditions for this problem. Under what conditions will this KKT system have a solution? If these conditions are satisfied, find all solutions of this KKT system (R. Saigal).

5.32. Let $f(x)$ be a real-valued function defined on R^n, and $h(x) = (h_1(x), \dots, h_m(x))$, all the functions being continuously differentiable. Consider the problem: minimize $f(x)$ subject to $h(x) = 0$, and $\ell \leq x \leq u$, where ℓ, u are given bound vectors.

Write the necessary optimality conditions for this problem. Show that these conditions are equivalent to finding a feasible solution $x = (x_j)$, satisfying, for $j = 1$ to n

$$\begin{aligned}\frac{\partial f(x)}{\partial x_j} - \sum_{i=1}^{m} \lambda_i \frac{\partial h_i(x)}{\partial x_j} &= 0 \quad \text{if } \ell_j < x_j < u_j \\ &\geq 0 \quad \text{if } \ell_j = x_j \\ &\leq 0 \quad \text{if } x_j = u_j.\end{aligned}$$

where $\lambda = (\lambda_1, \dots, \lambda_m)^T$ is the vector of Lagrange multiplier vectors associated with the equality constraints in the problem (R. Saigal).

5.33. Consider the LP (5.36) where A is an $m \times n$ matrix.

$$\min z(x) = cx \quad \text{s.to} \quad Ax = b, \quad x \geq 0 \tag{5.36}$$

The vector $x_B = (x_1, \dots, x_m)$ is a dual feasible but primal infeasible basic vector for (5.36). $\bar{x} = (\bar{x}_1, \dots, \bar{x}_n)^T$ is the basic solution of (5.36) corresponding to the basic vector x_B and $\bar{z} = c\bar{x}$. We have

$$\begin{aligned}\bar{x}_j &< 0 \quad \text{for } j = 1 \text{ to } r, \\ \bar{x}_j &\geq 0 \quad \text{for } j = r + 1 \text{ to } n.\end{aligned}$$

(1) Show that $\bar{z}$ is the optimum objective value in the LP (5.37):

$$\min z(x) = cx \text{ s.to } Ax = b, \text{ and } x_j \geq 0 \text{ for } j = r + 1 \text{ to } n \tag{5.37}$$

(2) Show that $\bar{z}$ is a lower bound for the minimum objective value in (5.36).

5.34. Direct Proof of the Duality Theorem Using Farkas' Lemma. Consider the Primal LP, where A is an $m \times n$ matrix.

$$\min z(x) = cx \quad \text{s. to} \quad Ax \geq b, \quad x \geq 0.$$

(1) Suppose we are given an optimum solution $\bar{x}$ for this LP. Let P be the index set of active constraints at $\bar{x}$, that is, $P = \{i : A_{i.}\bar{x} = b_i\}$.
Then show that the system of constraints:

$$A_{i.}y \geq 0 \text{ for all } i \in P, \quad y \geq 0, \quad cy < 0$$

cannot have a solution y.

Denote the vector of dual variables by the row vector π. Using the above result, and Farkas' Lemma show that there must exist a dual feasible solution $\bar{\pi}$ with dual objective value $v(\bar{\pi}) = \bar{\pi}b = c\bar{x}$, and then using the result in the weak duality theorem show that $\bar{\pi}$ is an optimum dual solution.

(2) If the primal is infeasible, show that Farkas' Lemma implies that the system $\mu A \leq 0$, $\mu \geq 0$, $\mu b > 0$ has a feasible solution μ. Using this show that in this case, if the dual has a feasible solution, then the dual objective function $v(\pi)$ is unbounded above on the dual feasible set.

(3) Show that (1) and (2) together provide a direct proof of the duality theorem for LPs in symmetric form, using Farkas' lemma, without having to appeal to the simplex algorithm (the proof given in Sect. 5.4 is based on the simplex algorithm).

5.35. Consider a primal LP and the following possibilities in it (each is separate from the others): (1) The ith constraint in it is of the type $\geq$. (2) The ith constraint in it is of the type $\leq$. (3) The ith constraint in it is of the type $=$. (4) The jth variable in it is sign restricted to be ≥ 0. (5) The jth variable in it is sign restricted to be ≤ 0. (6) The jth variable in it is unrestricted in sign. (7) The jth variable in it is restricted by a lower bound (LB) ℓ_j, and an upper bound (UB) u_j which are included among the primal constraints. (8) The jth variable in it only has an LB ℓ_j specified, which is included among the primal constraints. (9) The jth variable in it only has an UB u_j specified, which is included among the primal constraints.

Mention the effect of each of these possibilities in the primal LP, on the dual of this problem, assuming that the primal is a minimization (maximization) problem separately.

5.36. Tucker's lemma. Let A be a matrix of order $m \times n$. Tucker's lemma states that there exists a column vector $\bar{x} \in R^n$ and row vector $\bar{\pi} \in R^m$ satisfying $A\bar{x} \geq 0$; $\bar{\pi}A = 0$, $\bar{\pi} \geq 0$; and $A_{1.}\bar{x} + \bar{\pi}_1 > 0$.

To prove Tucker's lemma by induction, set up the following:

Induction hypothesis. Tucker's lemma holds for all matrices of order $(m-1)\times n$.

Let $\mathcal{A}$ be the matrix of order $(m-1)\times n$ obtained by removing the mth row $A_{m.}$ from A. Then under the induction hypothesis, we know that there exists a column vector $x' \in R^n$ and row vector $u' \in R^{m-1}$ satisfying

$$\mathcal{A}x' \geq 0;\ \ u'\mathcal{A} = 0,\ \ u' \geq 0;\ \ u'_1 + A_{1.}x' > 0.$$

If $A_{m.}x' \geq 0$, show that $\bar{x} = x'$, $\bar{\pi} = (u', 0)$ satisfies the conditions required under Tucker's Lemma.

If $A_{m.}x' < 0$, define $\lambda_i = (A_{i.}x')/(A_{m.}x')$, and $B_{i.} = A_{i.} + \lambda_i A_{m.}$ for $i = 1$ to $m-1$, and let B be the $(m-1)\times n$ matrix whose rows are $B_{i.}$ for $i = 1$ to $m-1$.

Now by the induction hypothesis, there exists an x'', u'' satisfying

$$Bx'' \geq 0;\ \ u''B = 0,\ \ u'' \geq 0;\ \ u''_1 + B_{1.}x'' > 0.$$

Defining $\bar{x} = x'' - x'(A_{m.}x'')/(A_{m.}x')$, $\bar{\pi} = (u'', \sum_{i=1}^{m} \lambda_i u_i'')$, show that this $\bar{x}$, $\bar{\pi}$ satisfy the requirements in Tucker's lemma. Using these complete the proof of Tucker's lemma by induction.

5.37. Direct Proof of Farkas' Lemma. Let A be a given $m \times n$ matrix and b a column vector in R^m. Farkas' lemma states that either system (5.38) given below has a solution x, or system (5.39) has a solution π, but not both.

$$Ax = b, \quad x \geq 0 \tag{5.38}$$

$$\pi A \leq 0, \quad \pi b > 0 \tag{5.39}$$

(1) Show that it is impossible for both systems (5.38) and (5.39) to be feasible, by showing that if both of them have feasible solutions, it leads to the **fundamental inconsistent inequality** $0 > 1$.

(2) Assume that (5.38) has no feasible solution π. Let $\bar{A} = \begin{pmatrix} b^T \\ -A^T \end{pmatrix}$, a matrix of order $(n+1) \times m$.

Since (5.39) is infeasible, in every solution of $\bar{A}y \geq 0$, we must have $\bar{A}_{1.}y = 0$. Using this and Tucker's lemma of Exercise 5.36, obtain a direct proof of Farkas' lemma.

5.38. Let A be a given matrix of order $m \times n$. Using Tucker's lemma (Exercise 5.36), show that there exists a column vector $\bar{x} \in R^n$ and row vector $\bar{\pi} \in R^m$ satisfying $A\bar{x} \geq 0$; $\bar{\pi}A = 0$, $\bar{\pi} \geq 0$ and $A\bar{x} + \bar{\pi}^T > 0$.

5.39. Gordon's Theorem of Alternatives. Let A be a given matrix of order $m \times n$. Use the result in Exercise 5.38 to show that exactly one of the following systems (5.40) and (5.41) has a feasible solution, and the other is infeasible.

$$Ax > 0, \tag{5.40}$$

$$\pi A = 0, \ \pi \geq 0, \ \pi \neq 0 \tag{5.41}$$

5.40. We are given a primal linear program with coefficient matrix A, in which the constraints are numbered 1 to m with $\{1, \ldots, m\}$ partitioned as $M_1 \cup M_2 \cup M_3$, and the variables have subscripts 1 to n with $\{1, \ldots, n\}$ partitioned as $N_1 \cup N_2 \cup N_3$. Here is the primal:

> Minimize $z = cx$, subject to: $A_{i.}x - b_i \geq 0$ for $i \in M_1$, ≤ 0 for $i \in M_2$, and $= 0$ for $i \in M_3$; and $x_j \geq 0$ for $j \in N_1$, ≤ 0 for $j \in N_2$, and unrestricted for $j \in N_3$.

Write the dual problem using π_i as the dual variable corresponding to the ith constraint in the primal for $i = 1$ to m.

Given $\bar{x}$, $\bar{\pi}$ which are primal, dual feasible, respectively, using the weak duality theorem show that they are optimum solutions to the respective problems iff $(A_{i.}\bar{x} - b_i)\bar{\pi} = 0$ for all $i \in M_1 \cup M_2$ and $\bar{x}_j(\bar{\pi}A_{.j} - c_j) = 0$ for all $j \in N_1 \cup N_2$.

5.41. The matrix A of order $m \times n$ and the column vector b in R^m are given. Consider the LP: Minimize $z = cx$, subject to $Ax = b$. Show directly, that for this LP, either every feasible solution is an optimum solution; or there exists no optimum solution.

Write the dual of this LP, and show how you can get the same conclusion from the duality theorem.

5.42. K is a convex polytope of full dimension in R^n with extreme points $\{d^1, ..., d^N\}$. Let $\bar{d}$ be an interior point of K, for example $\bar{d} = (\sum_{r=1}^N d^r)/N$. Consider the following LP with variables α_r, $r = 1$ to N: Minimize $z = \sum_{r=1}^N \alpha_r ||d^r||^2$ subject to $\sum_{r=1}^N \alpha_r d^r = \bar{d}$, $\sum_{r=1}^N \alpha_r = 1$, and $\alpha_r \geq 0$ for all r.

(i) Denoting the variables in the dual of this LP by $\beta_1, \ldots, \beta_n, \delta$ in that order, write the dual of this LP.
(ii) Assume that $\alpha_B = (\alpha_1, ..., \alpha_{n+1})$ is an optimum basic vector for the primal. Also assume that $(\bar{\beta} = (\bar{\beta}_1, \ldots, \bar{\beta}_n), \bar{\delta})$ is the corresponding optimum dual solution. Show that $(1/2)\bar{\beta}$ is the center of an n-dimensional sphere that contains all the points $d^1, \ldots, d^{n+1}$ on its boundary.

References

Ahuja RK (1997) The balanced linear programming problem. Eur J Oper Res 101:29–38
Bazaraa MS, Sherali HD, Shetty CM (2006) Nonlinear programming, theory and algorithms, 3rd edn. Wiley, NY

Chapter 6
Revised Simplex Variants of the Primal and Dual Simplex Methods and Sensitivity Analysis

Consider an LP in standard form in n nonnegative variables subject to m equality constraints that the variables are required to satisfy, in which we assume that the rank of the coefficient matrix is m, without any loss of generality. So, $n \geq m$. In LP models to be solved in practice, typically n will be much larger than m.

The original form of the primal simplex method developed in 1947, discussed in LP textbooks (e.g., see Chap. 4 of Murty (2005b) of Chap. 1), is a wonderful educational tool to explain the principles behind the method to beginners. Starting with a primal feasible basic vector, this method goes through a series of steps, in each step one GJ pivot step is carried out to replace one basic variable in the basic vector by a specially selected nonbasic variable. All the work in a step of this method is also discussed in our Sects. 4.9 and 4.12.

But as a computational method for solving LP models in practice, this original form of the primal simplex method is highly inefficient as it updates every one of the $n + 1$ column vectors in the canonical tableau in every pivot step to solve the LP under consideration. In this chapter, we will discuss the much more efficient revised simplex variants of the primal simplex method that are based on the same fundamental theory, but in these variants we only need to update a smaller number, $m+2$, column vectors of the basis inverse in every pivot step. We will discuss several of the early (developed in 1950s and 1960s) variants of this method to describe the ideas used to make the simplex method a practically valuable tool to solve LP models in decision making.

Instead of updating the entire canonical tableau in every pivot step, these revised simplex variants update the much smaller basis inverse in every pivot step, while leaving the original system unchanged throughout the algorithm. The revised GJ method for solving linear equations is discussed in Sect. 1.2, and the revised simplex variants use the same formulas discussed there to execute all the operations of the simplex algorithm using the basis inverse and the original tableau.

K.G. Murty, *Optimization for Decision Making: Linear and Quadratic Models*, International Series in Operations Research & Management Science 137, DOI 10.1007/978-1-4419-1291-6_6,

6.1 Primal Revised Simplex Algorithm Using the Explicit Basis Inverse

To solve a general LP using these variants, we begin with transforming the problem into standard form using the simple transformations discussed in Sect. 4.1 of Murty(2005b) of Chap. 1. After these transformations if necessary, let the problem in standard form be

$$\begin{aligned} \text{Minimize} \quad z(x) &= cx \\ \text{subject to} \quad Ax &= b \\ x &\geq 0, \end{aligned} \tag{6.1}$$

Original Tableau

x_1	...	x_j	...	x_n	$-z$	RHS
a_{11}	...	a_{1j}	...	a_{1n}	0	b_1
$\vdots$		$\vdots$		$\vdots$	$\vdots$	$\vdots$
a_{m1}	...	a_{mj}	...	a_{mn}	0	b_m
c_1	...	c_j	...	c_n	1	0

$x_j \geq 0$ for all j, minimize z

where the coefficient matrix A is a matrix of order $m \times n$ and rank m.

This algorithm (the revised primal simplex algorithm) needs as input an initial primal feasible basic vector. Let x_B be the initial primal feasible basic vector for (6.1), B is the $m \times m$ associated basis consisting of the column vectors of A corresponding to the basic variables in x_B, and c_B is the row vector of basic cost coefficients. The extended basis corresponding to $(x_B, -z)$ is

$$\mathcal{B} = \begin{pmatrix} B & 0 \\ c_B & 1 \end{pmatrix}.$$

The inverse tableau corresponding to $(x_B, -z)$ is

$$\mathcal{B}^{-1} = \begin{pmatrix} B & 0 \\ c_B & 1 \end{pmatrix}^{-1} = \begin{pmatrix} B^{-1} & 0 \\ -\pi & 1 \end{pmatrix}, \tag{6.2}$$

where $\pi = c_B B^{-1}$ is the dual basic solution corresponding to the basic vector $(x_B, -z)$. The vector of values of the basic variables in the primal basic solution corresponding to this basic vector is

$$\text{updated RHS vector} = \mathcal{B}^{-1} \begin{pmatrix} b \\ 0 \end{pmatrix} = \begin{pmatrix} \bar{b} \\ -\bar{z} \end{pmatrix},$$

where $\bar{b} \geq 0$ by primal feasibility of x_B. All this information is displayed in the form of a tableau, which we also call the *inverse tableau.* "BV" stands for "basic variables."

Inverse tableau wrt x_B

BV	Inverse Tableau		Basic values
x_B	B^{-1}	0	$\bar{b}$
$-z$	$-\pi$	1	$-\bar{z}$

Using the formulas derived in Sects. 1.2 and 5.4 ((5.16) to (5.20)), here we list the various steps in the iteration of the simplex algorithm when $(x_B, -z)$ is the basic vector.

The fact that $(x_B, -z)$ is a basic vector encountered in the primal simplex algorithm automatically implies that it is primal feasible. The algorithm is initiated with a primal feasible basic vector and it maintains primal feasibility throughout (through the selection rule using the primal simplex minimum ratio test for the dropping basic variable that will be replaced by the entering variable in the basic vector in each pivot step). So, checking primal feasibility of the basic vector is not usually mentioned in every step of the algorithm.

But, in practice, periodically after every few steps, the present primal solution is checked for primal feasibility to make sure that the inevitable accumulation of round-off errors in computation have not moved the present solution, $\bar{x}$, too far away from primal feasibility. A typically used measure of primal infeasibility of $\bar{x}$ is $||A\bar{x}-b||$; under exact computation it should be 0 if $\bar{x}$ is primal feasible. However, in digital computation with finite precision, we incur round-off errors, and so in practical computation a small positive tolerance is selected and $\bar{x}$ is considered primal feasible as long as $||A\bar{x}-b|| \leq$ the selected tolerance. So, in solving (6.1) using the revised primal simplex algorithm, if $\bar{x}$ is the present primal solution and $||A\bar{x}-b||$ is higher than the specified tolerance for it, it is an indication that the round-off error accumulation in the present basis inverse is unacceptable. In this case, the actual basis matrix corresponding to the present basic vector is retrieved from the original tableau and inverted again (this operation is called *reinversion of the basis*). With the new basis inverse, continue the pivot step with the present basic vector.

6.1.1 Steps in an Iteration of the Primal Simplex Algorithm When $(x_B, -z)$ is the Primal Feasible Basic Vector

1. *Compute relative cost coefficients of nonbasic variables:* From (5.18) in Sect. 5.4 we know that the relative cost coefficient of the variable x_j is

$$\bar{c}_j = (-\pi, 1)\begin{pmatrix} A_{.j} \\ c_j \end{pmatrix},$$

that is, the dot product of the last row of the inverse tableau with the column vector of x_j in the original tableau.

2. *Check optimality criterion:* The optimality criterion is $\bar{c}_j \geq 0$ for all the variables x_j in the nonbasic vector x_D. If it is satisfied, the present BFS $(x_D, x_B) = (0, \bar{b})$, objective value $z = \bar{z}$, is an optimum solution for the problem. π from the last row of the present basis inverse is an optimum dual solution, so we have optimum solutions for both the primal and the dual problems. Terminate.
If the optimality criterion is not satisfied, go to Step 3.
3. *Select the entering variable:* Let $E = \{j : \bar{c}_j < 0\}$. Each x_j for $j \in E$ is eligible to enter the present basic vector x_B in this iteration. In this step we select one of them, say x_s, as the *entering variable* in this iteration.
There are several entering variable selection rules. *Dantzig's rule* selects one that has the most negative $\bar{c}_j$. Another rule is to select the first one that is eligible, as the $\bar{c}_j$'s are being computed in Step 1 (if this rule is used, after the entering variable is selected, the remaining $\bar{c}_j$'s are not even computed).
Computer programmers have developed a variety of heuristic strategies and rules for this selection that have been successful in speeding up the simplex algorithm in large-scale computational tests on randomly generated LP models (see Fourer (2005); Harris (1973) and Greenberg (1978)), and incorporated the best among these strategies in commercial LP Solver systems.
4. *Compute the updated column of the entering variable:* From (5.18) in Sect. 5.4 we know that the updated column of the entering variable x_s is

$$\begin{pmatrix} \bar{A}_{.s} \\ \bar{c}_s \end{pmatrix} = \mathcal{B}^{-1} \begin{pmatrix} A_{.s} \\ c_s \end{pmatrix},$$

that is, the inverse tableau times the column vector of x_s in the original tableau. This column is known as the *entering column* or the *pivot column* in this iteration.
5. *Check the unboundedness criterion:* If $\bar{A}_{.s}$, the updated column of the entering variable x_s, is ≤ 0, the objective function $z(x)$ is unbounded below on the set of feasible solutions of this problem, terminate. Define $x(\lambda)$, a feasible solution depending on a parameter λ, by

$$(x_B(\lambda),\ x_s(\lambda),\ \text{all other nonbasic variables}) = (\bar{b} - \lambda \bar{A}_{.s},\ \lambda,\ 0),$$

with objective value $z(\lambda) = \bar{z} + \bar{c}_s\lambda$. As $\bar{A}_{.s} \leq 0$, $x(\lambda)$ remains feasible to the problem for all $\lambda \geq 0$, and as $\lambda \to \infty$, $z(\lambda) \to -\infty$. So $\{x(\lambda) : \lambda \geq 0\}$ is an extreme half-line of the set of feasible solutions of this problem, along which $z \to -\infty$.
If this unboundedness criterion is not satisfied, go to Step 6.
6. *Minimum ratio test to determine the dropping basic variable, and pivot step to update the inverse tableau:* Enter the pivot column $\begin{pmatrix} \bar{A}_{.s} \\ \bar{c}_s \end{pmatrix} = (\bar{a}_{1s}, \bar{a}_{2s}, \ldots, \bar{a}_{ms}, \bar{c}_s)^T$ by the side of the inverse tableau and compute the ratios $\bar{b}_i/\bar{a}_{is}$ in all rows i where $\bar{a}_{is} > 0$. Identify the minimum, θ, of these ratios, and select a row

r where the minimum ratio is attained as the *pivot row*. If there is a tie for the minimum ratio, the pivot row is selected among those tied by some selection rule. In commercial LP solver systems, the best selection rule determined heuristically through computational experimentation is usually incorporated.

The present basic variable in the pivot row is the *dropping basic variable* in this iteration; it will be replaced by the entering variable to get the next basic vector. Perform the pivot step on the inverse tableau with the pivot column (PC), pivot row (PR) selected, this updates the inverse tableau into the inverse tableau wrt the new basic vector.

With the new basic vector and the inverse tableau corresponding to it, go to the next iteration.

Example 6.1.

Original tableau

x_1	x_2	x_3	x_4	x_5	x_6	x_7	$-z$	b
1	0	0	0	−1	1	1	0	2
0	1	0	0	1	−1	1	0	1
0	0	1	0	2	20	1	0	5
0	0	0	1	0	−1	1	0	0
0	0	1	1	−1	29	−8	1	0

$x_j \geq 0$ for all j, minimize z

First inverse tableau

Basic var.	Inverse tableau					Basic values	PC x_5	Ratios	
x_1	1	0	0	0	0	2	−1		
x_2	0	1	0	0	0	1	1 (boxed)	1/1	PR
x_3	0	0	1	0	0	5	2	5/2	
x_4	0	0	0	1	0	0	0		
$-z$	0	0	−1	−1	1	−5	−3	Min. $= \theta = 1$	

PC pivot column, *PR* pivot row

We will now provide a numerical example of the primal simplex algorithm, with detailed explanation of the various steps in it. Consider the LP in the original tableau given above.

Notice that the RHS vector in the original tableau is nonnegative and $x_B = (x_1, x_2, x_3, x_4)$ is a primal feasible basic vector. The augmented basis corresponding to $(x_B, -z)$ is the square matrix consisting of the column vectors of $x_1, x_2, x_3, x_4, -z$ in the original tableau given above. Its inverse can be obtained by the formulae in (6.2). This yields the first inverse tableau given above.

The primal BFS corresponding to x_B is $\bar{x} = (\bar{x}_1 \text{ to } \bar{x}_7)^T = (2, 1, 5, 0; 0, 0, 0)^T$, with objective value $\bar{z} = 5$. The dual basic solution corresponding to x_B is $\bar{\pi} = (0, 0, 1, 1)$ from the last row of the inverse tableau.

The relative cost coefficients of the nonbasic variables, $x_5,\ x_6,\ x_7$, $(\bar{c}_5,\ \bar{c}_6,\ \bar{c}_7)=$ (−3, 10, −10), are obtained by taking the dot product of the last row of the inverse tableau with the original columns of these variables.

The optimality criterion of the primal simplex algorithm is not satisfied because $\bar{c}_5,\ \bar{c}_7$ are negative and $x_5,\ x_7$ are eligible to enter the basic vector x_B. Among these, suppose x_5 is selected as the entering variable.

The updated column of x_5 is its original column multiplied on the left by the inverse tableau, and this is $(-1, 1, 2, 0, -3)^T$. As there are positive entries in this column, the unboundedness criterion is not satisfied.

So, the updated column of x_5 is the pivot column, and we have to carry out the minimum ratio test to identify the pivot row.

We will now explain the rationale behind the minimum ratio test in this pivot step for this example. Its purpose is to make sure that the next basic vector obtained after the pivot step will also be primal feasible, that is, to maintain primal feasibility throughout the algorithm.

x_1	x_2	x_3	x_4	$-z$	x_5	
1	0	0	0	0	−1	2
0	1	0	0	0	1	1
0	0	1	0	0	2	5
0	0	0	1	0	0	0
0	0	1	1	1	−1	0

$x_6 = x_7 = 0$

At this stage the simplex algorithm fixes all the nonbasic variables other than the entering variable at their present value of 0, gives the entering variable a tentative value, say λ which is a nonnegative parameter, and obtains the solution of the remaining system as a function of λ. Suppose this solution is denoted by $x(\lambda)$ and its objective value by $z(\lambda)$. In this example, the remaining system after fixing the nonbasic variables $x_6,\ x_7$ at 0 and rearranging variables is given above.

Multiplying this system on the left by the present inverse tableau converts it into

x_1	x_2	x_3	x_4	$-z$	x_5	
1	0	0	0	0	−1	2
0	1	0	0	0	1	1
0	0	1	0	0	2	5
0	0	0	1	0	0	0
0	0	0	0	1	−3	−5

$x_6 = x_7 = 0$

In this tableau, the column vector of x_5 is its updated column computed above. From this system we read out the solution $x(\lambda)$ discussed above to be

$$\begin{aligned}(x_1(\lambda), \ldots, x_7(\lambda))^T &= (2+\lambda,\ 1-\lambda,\ 5-2\lambda,\ 0,\ \lambda,\ 0,\ 0)^T,\\ z(\lambda) &= 5-3\lambda. \end{aligned} \tag{6.3}$$

The coefficient of λ in $z(\lambda)$, -3, is $\bar{c}_5$, the relative cost coefficient of the entering variable x_5. Thus, $\bar{c}_5$ is the rate of change in the objective value per unit change in the value of the nonbasic variable x_5 from its current value of 0 in the present BFS. This is the reason for selecting the entering variable among those with negative relative cost coefficients, because it helps to get a better solution with smaller value for z.

In the solution in (6.3), $z(\lambda)$ decreases as λ increases from 0, and so to get maximum possible reduction in this iteration, we should give λ the maximum possible value. As λ increases, the basic values, $x_1(\lambda)$, $x_4(\lambda)$ (which are $2+\lambda$, and 0) remain ≥ 0 always. These are the values of basic variables in rows in which the pivot column has entries ≤ 0.

However, the basic values $x_2(\lambda)$, $x_3(\lambda)$ (which are $1-\lambda$, $5-2\lambda$) will become <0 if λ becomes too large. For $x_2(\lambda) = 1-\lambda$ to be ≥ 0, we need $\lambda \leq 1/1 = 1$. This is the ratio in row 2 in which x_2 is the basic variable. This row has a positive entry in the pivot column. For $x_3(\lambda) = 5-2\lambda$ to be ≥ 0, we need $\lambda \leq 5/2$, and this is the ratio in row 3 in which x_3 is the basic variable, this row also has a positive entry of 2 in the pivot column. Thus each ratio computed in the ratio test is an upper bound for the value of λ to keep $x(\lambda)$ in (6.3) to be ≥ 0, that is, feasible. Hence the maximum possible value that we can give to λ in this iteration is the smallest of these upper bounds for it, that is, the minimum ratio, $\theta = \min\{1/1, 5/2\} = 1$ here.

Hence the minimum ratio test determines the maximum possible value for the entering variable that keeps the new solution $x(\lambda)$ primal feasible.

When $\lambda =$ the minimum ratio θ, the basic variable x_2 in whose row the minimum ratio is attained, will reach a value of 0. Any increase in λ beyond the minimum ratio θ will make $x_2(\lambda) < 0$, that is why x_2 is called a *blocking basic variable* in this step. When we fix $\lambda = \theta$, this blocking basic variable becomes 0, and can be dropped from the present basic vector to become a nonbasic variable at that time, and its place in the basic vector given to the entering variable.

As every basic vector for an LP always consists of the same number of basic variables, when the entering variable is brought into the present basic vector, one of the present basic variables must be dropped from it. Candidates eligible to be dropping basic variables are those whose value becomes 0 in $x(\theta)$.

The inverse tableau corresponding to the new basic vector is obtained by performing a pivot step with the updated column of the entering variable as the pivot column and the row of the dropping basic variable as the pivot row.

In the solution $x(\lambda)$ in (6.3), if we put $\lambda = 0$, we get the present BFS $\bar{x}$. If we put $\lambda = \theta$, the minimum ratio, we get the BFS associated with the next basic vector $x_{\hat{B}}$, with its objective value of $5 - 3\theta = 2$, as $\theta = 1$ here. The line segment joining these two solutions is an **edge** of the set of feasible solutions of the problem. So, in this iteration, we can say that the simplex algorithm travels from one end of this edge, $x(0)$, to the other end, $x(\theta)$, attaining a strict decrease in the objective value in this process. Geometrically, this move is like moving from the point $\bar{x}$ to the point $\hat{x}$ in that figure, along the edge joining them, see Fig. 6.1.

Dantzig's entering variable choice rule selects the entering variable to be one among the nonbasic variables at this stage that has the most negative relative cost

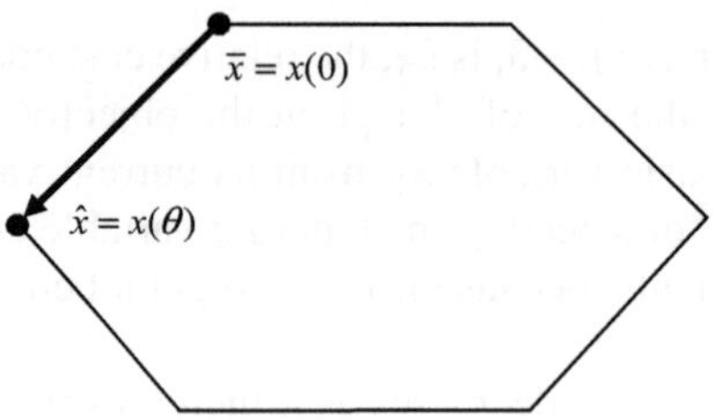

Fig. 6.1 The move along an edge, taken by the primal simplex algorithm in a nondegenerate pivot step from the BFS $\bar{x}$ to an adjacent BFS $\hat{x}$ illustrated on a two-dimensional polytope. The objective value undergoes a strict improvement in this move

coefficient. So, under this rule, the entering variable corresponds to the best rate of change in the objective value per unit change in the value of the entering variable from its present value of 0.

Continuing with the example, the updated column of x_5, the pivot column, is entered on the first inverse tableau, and the ratios are computed on the tableau itself. Row 2, in which the minimum ratio is achieved, is the pivot row, and the pivot element is boxed. As the minimum ratio is $\theta = 1 > 0$, this pivot step is a nondegenerate pivot step. Performing the pivot step yields the second inverse tableau given below.

Second inverse tableau

Basic var.	Inverse tableau					Basic values	PC x_7	Ratios	
x_1	1	1	0	0	0	3	2	3/2	
x_5	0	1	0	0	0	1	1	1/1	
x_3	0	−2	1	0	0	3	−1		
x_4	0	0	0	1	0	0	$\boxed{1}$	0/1	PR
$-z$	0	3	−1	−1	1	−2	−7	Min. $= \theta = 0$	

PC pivot column, *PR* pivot row

The new BFS is $\hat{x} = (3,0,3,0,1,0,0)^T$ with an objective value of $\hat{z} = 2$. The relative cost coefficients of the nonbasic variables here are $(\bar{c}_2, \bar{c}_6, \bar{c}_7) = (3,7,-7)$. So, x_7 is the entering variable in this iteration. Its updated column is $(2,1,-1,1,-7)^T$. It is the pivot column, and the minimum ratio is zero, and the pivot row is row 4. This pivot step is a degenerate pivot step. Performing this pivot step leads to the next inverse tableau.

Third inverse tableau

Basic var.	Inverse tableau					Basic values
x_1	1	1	0	−2	0	3
x_5	0	1	0	−1	0	1
x_3	0	−2	1	1	0	3
x_7	0	0	0	1	0	0
$-z$	0	3	−1	6	1	−2

Verify that the BFS and objective value are $\hat{x}$, $\hat{z}$, no change since the pivot step was a degenerate pivot step. The relative cost coefficients of the nonbasic variables wrt the new basic vector are $(\bar{c}_2, \bar{c}_4, \bar{c}_6) = (3, 7, 0)$.

As they are all ≥ 0, the optimality criterion is satisfied, and the present BFS $\hat{x} = (3, 0, 3, 0, 1, 0, 0)^T$ is an optimum solution with an optimum objective value of $\hat{z} = 2$. The dual solution $\hat{\pi} = (0, -3, 1, -6)$ from the third inverse tableau is an optimum dual solution.

The primal basic vector here is primal degenerate, as the primal basic variable x_7 is 0 in the associated BFS. So, as discussed in Chap. 5, we cannot interpret the optimum dual solution ($\hat{\pi}$ here), as the marginal value vector in this problem. ■

Example 6.2. This example illustrates the unboundedness termination. Consider the following LP.

Original tableau

x_1	x_2	x_3	x_4	x_5	x_6	$-z$	b
0	0	1	1	−1	−5	0	7
1	0	0	−1	−1	−3	0	9
0	1	0	−1	−1	0	0	1
−1	−1	−1	10	6	4	1	0

$x_j \geq 0$ for all j, minimize z

We will initiate the primal simplex algorithm on this problem with the primal feasible basic vector $x_B = (x_3, x_1, x_2)$. The inverse tableau corresponding to it is given below.

Inverse tableau

Basic var.	Inverse tableau				Basic values
x_3	1	0	0	0	7
x_1	0	1	0	0	9
x_2	0	0	1	0	1
$-z$	1	1	1	1	17

The primal BFS associated with x_B is $\bar{x} = (9, 1, 7, 0, 0, 0)^T$ with an objective value of $\bar{z} = -17$.

The relative cost coefficients of the nonbasic variables are $(\bar{c}_4, \bar{c}_5, \bar{c}_6) = (9, 3, -4)$. As $\bar{c}_6$ is the only negative relative cost coefficient, we select x_6 as the entering variable.

The updated column of x_6 is $(-5, -3, 0, -4)^T$. There is no positive entry in this column. The unboundedness criterion is satisfied, and hence the objective function $z(x)$ is unbounded below in this problem, and there is no finite optimum solution.

Actually, the solution $x(\lambda)$ obtained by fixing the nonbasic variables x_4, x_5 at 0, giving the value λ to the entering variable x_6, and then finding the values of the basic variables as functions of λ, is the solution of the following system corresponding to $x_6 = \lambda$.

x_3	x_1	x_2	$-z$	x_6	
1	0	0	0	−5	7
0	1	0	0	−3	9
0	0	1	0	0	1
0	0	0	1	−4	17

It is

$$x(\lambda) = (9 + 3\lambda,\ 1,\ 7 + 5\lambda,\ 0,\ 0,\ \lambda)^T \text{ with an objective value of}$$
$$z(\lambda) = -17 - 4\lambda.$$

The solution $x(0)$ is the present BFS $\bar{x}$. $\{x(\lambda) : \lambda \geq 0\}$ is a half-line beginning with the present BFS, every point on which is feasible to the primal problem, so it is a primal feasible half-line. As $\lambda \to \infty$, $z(\lambda) \to -\infty$ along this half-line.

For example, if you want a feasible solution with an objective value of -4017, $z(\lambda) = -17 - 4\lambda$ is attained when $\lambda = 1{,}000$, this corresponds to the point $x(1000) = (1009, 1, 5007, 0, 0, 1000)^T$ on this half-line.

In the same manner, for any objective value $\leq -17 = z(\bar{x})$, we can find a $\lambda > 0$ such that the point $x(\lambda)$ on this half-line is feasible to this problem and attains that objective value. So, we terminate with the unboundedness conclusion. ■

6.1.2 *Practical Consequences of Satisfying the Unboundedness Criterion*

Suppose the unboundedness criterion is satisfied while solving an LP model for a practical problem. Negative cost is profit, so this implies that we have found a way for making an unlimited profit! This infinite profit will be achieved if one implements the feasible solution at the infinite end of the half-line identified by the algorithm along which the cost function $z(x)$ diverges to $-\infty$.

Clearly, some of the variables will have the value $+\infty$ in that solution. For a simple example of this, consider a scheme popularly known in business circles as a *daisy link.* Here we are able to buy crude oil at a cost say of $115/barrel from a Middle East supplier, and sell it to a local company at $125/barrel. To make an infinite profit from this deal, we have to buy and sell an infinite number of barrels at these rates.

In practice, we are always limited by finite resources, and hence it is impossible to implement a solution in which some variables are $+\infty$. Thus, while infinite profit is a mathematical possibility, if the unboundedness criterion is satisfied while solving an LP model for a practical problem, it is probably an indication that some constraints on the decision variables have been forgotten in constructing the model, or omitted from it intentionally, or there may be serious errors in the cost coefficients used in the model. So, one should review the model very carefully, and look for errors and omissions.

6.1.3 Features of the Simplex Algorithm

The primal simplex algorithm discussed in this section requires a primal feasible basic vector initially, and maintains primal feasibility of the basic vector throughout. Also, in every iteration, the primal and dual basic solutions satisfy all the complementary slackness conditions. So, of the three conditions required for optimality mentioned in Chap. 5 (primal and dual feasibility, and complementary slackness) it maintains two (primal feasibility and complementary slackness) and strives to attain the third one, dual feasibility.

6.2 Revised Primal Simplex Method (Phase I, II) with Explicit Basis Inverse

This is the version to solve an LP in standard form if an initial feasible basic vector is not known. The input needed for this method is the LP in standard form.

First make all the RHS constants in all the rows in the original tableau nonnegative (if any of them is <0, multiply all the elements in that row by -1). Let the original tableau be

Original Tableau

x_1	$\dots$	x_j	$\dots$	x_n	$-z$	RHS
a_{11}	$\dots$	a_{1j}	$\dots$	a_{1n}	0	b_1
$\vdots$		$\vdots$		$\vdots$	$\vdots$	$\vdots$
a_{m1}	$\dots$	a_{mj}	$\dots$	a_{mn}	0	b_m
c_1	$\dots$	c_j	$\dots$	c_n	1	0

$x_j \geq 0$ for all j, minimize z

A basic vector for this problem is said to be a *unit basic vector* if its basis is the unit matrix. As the RHS constants vector in the original tableau is ≥ 0, a unit basic vector will be primal feasible.

Search for a unit basic vector in the original tableau. If a full unit basic vector, x_B, is found, starting with x_B as the initial feasible basic vector, apply the revised simplex algorithm discussed in Sect. 6.1 to solve the problem. This is called Phase II of the primal simplex method.

6.2.1 Setting Up the Phase I Problem

Suppose a full unit basic vector has not been obtained in the original tableau. So, at this stage we do not even have a feasible solution for the problem. The simplex method divides the task of solving the problem into two phases. In Phase I it focuses on finding a BFS for the problem, ignoring the original objective function. If a BFS is obtained in Phase I, starting with it the original objective function is minimized in Phase II.

Phase I uses additional variables called **artificial variables** because they are not part of the original problem. When the original problem is in standard form, Phase I is set up so that it is also an LP in standard form, but with an initial feasible basic vector containing these artificial variables, so starting with that the Phase I problem can be solved by the primal simplex algorithm discussed in Sect. 6.1. We will now discuss how to set up the Phase I problem.

Suppose in the process of searching for the unit basic vector in the original problem, some variables corresponding to ith column vector of the unit matrix for $i \in \{i_1, \dots, i_r\}$ have been selected into the basic vector, where $0 \leq r < m$. For each $i \in \{1, \dots, m\} \backslash \{i_1, \dots, i_r\}$, introduce a nonnegative artificial variable associated with the ith column vector of the unit matrix into the original tableau, and introduce these artificial variables into the basic vector. Let x^1_B be the resulting full unit basic vector. If $r = 0$, x^1_B consists of artificial variables only. If $r > 0$, x^1_B consists of some original problem variables and some artificial variables.

Phase I original tableau

Original			Artificial					
x_1	...	x_n	x_{n+1}	...	x_{n+m-r}	$-z$	$-w$	RHS
a_{11}	...	a_{1n}				0	0	b_1
⋮		⋮		Missing unit		⋮	⋮	⋮
a_{m1}	...	a_{mn}		vectors		0	0	b_m
c_1	...	c_n	0	...	0	1	0	0
0	...	0	1	...	1	0	1	0

All variables ≥ 0, minimize w

Make the cost coefficient of each artificial variable in the z-row equal to 0; this z will now be called the Phase II objective function. Introduce a new Phase I objective function $w =$ (sum of all the artificial variables), and enter this objective function row at the bottom (i.e., as the $m + 2$-th row) of the original tableau. The resulting tableau is called the Phase I original tableau.

In the Phase I problem, we minimize the Phase I objective function w starting with the unit feasible basic vector x^1_B, with the artificial variables introduced. During Phase I, the Phase II objective row, the $(m + 1)$th, is updated in every iteration, but the entries in it are not used in the algorithm.

As all artificial variables are nonnegative variables, the minimum value that their sum w can have is 0. If a solution in which $w = 0$ is obtained, all the artificial variables must have 0-values in it, and by dropping their 0-values, the solution becomes a feasible solution for the original problem.

As the value of w has to be reduced to 0 to reach a feasible solution of the original problem, at any stage of Phase I, the current value of w measures how far away we are from being feasible to the original problem. That is why $w =$ the Phase I objective function value = (sum of the values of all the artificial variables in the present solution) is known as an *infeasibility measure* for the present solution. If the minimum value of w in the Phase I problem is strictly >0, and this implies that the original problem is infeasible.

Phase I Inverse Tableau wrt x_B

BV	Inverse tableau			Basic values
x_B	B^{-1}	0	0	$\bar{b}$
$-z$	$-\pi$	1	0	$-\bar{z}$
$-w$	$-\sigma$	0	1	$-\bar{w}$

As the minimum value of w cannot be <0, Phase I will always have a finite optimum solution, that is, the unboundedness criterion can never be satisfied during Phase I.

Phase I inverse tableau wrt a basic vector x_B for the Phase I problem will be of the form given above. "BV" means "basic variable."

Here σ is the dual basic solution of the Phase I problem wrt the present basic vector x_B.

We denote the coefficients of x_j in the original Phase I objective row by d_j ($d_j = 0$ if x_j is an original problem variable; $d_j = 1$ if x_j is an artificial variable), and its Phase I relative cost coefficient by $\bar{d}_j$. $\bar{d}_j$ = dot product of the last row of the Phase I inverse tableau with the column vector of x_j in the Phase I original tableau, that is,

$$\bar{d}_j = (-\sigma, 0, 1) \begin{pmatrix} A_{.j} \\ c_j \\ d_j \end{pmatrix}.$$

If an artificial variable drops out of the basic vector and becomes a nonbasic variable during Phase I, its column vector in the original tableau is then completely deleted from further consideration because it is not needed any more. So, at any stage of Phase I, the only artificial variables left in the original tableau are those that are in the basic vector. So, during Phase I, the entering variable in every iteration will always be an original problem variable. So, we have

Phase I termination condition: The Phase I relative cost coefficient $\bar{d}_j \geq 0$ for all original problem variables x_j.

Variables eligible to enter the basic vector during Phase I are those original variables for which the Phase I relative cost coefficient $\bar{d}_j < 0$. When the Phase I problem is solved by the revised primal simplex algorithm discussed in the previous section beginning with the initial feasible basic vector x^1_B, there are three ways in which it can terminate.

1. *With the conclusion that the original LP is infeasible:* This happens if the minimum value of w in the Phase I problem is >0.

   In this case there is a technique called *infeasibility analysis*, which determines how the model for the original problem can be made feasible by making the smallest necessary changes in the data in the original tableau. This analysis is discussed in Sect. 6.4.
2. *Moving directly to Phase II:* This happens if the minimum value of w is 0, and the terminal basic vector of Phase I contains only original problem variables and no artificial variables.

   In this case, the final Phase I relative cost coefficient $\bar{d}_j$ for all original problem variables will be 0. To get the inverse tableau for the original problem wrt the present basic vector, all you need to do is to drop the last (Phase I objective) row from the Phase I terminal inverse tableau, and the last column from the augmented basis inverse in it. With this inverse tableau, begin Phase II to solve the original problem.
3. *Detecting some original problem variables x_j that have a constant value of 0 at every feasible solution of the original problem:* This case can occur when the minimum value $\bar{w}$ of w is 0, but there are still some artificial variables in the terminal Phase I basic vector with values of 0 in the corresponding BFS.

   In this case, some of the Phase I relative cost coefficients $\bar{d}_j$ for nonbasic original problem variables may be >0. Let $J = \{j : \bar{d}_j > 0\}$. As all remaining artificial variables are always basic variables, all the variables x_j for $j \in J$ must be original problem variables that are nonbasic.

   The equation represented by the updated Phase I objective row at this stage will be

$$\sum_{j \in J} \bar{d}_j x_j - w = \bar{w} = 0$$

   or $w = \sum_{j \in J} \bar{d}_j x_j$. This suggests that even though the present value of w is 0, if any of the variables x_j for $j \in J$ were to be made >0, then the value of w (the infeasibility measure) will become >0, that is, the solution will become infeasible to the problem. So, all these x_j for $j \in J$ are $= 0$ at every feasible solution of the original problem.

   We therefore fix all these x_j for $j \in J$ at 0 and delete them from the original tableau. After this, as all the remaining $\bar{d}_j$ are 0, the value of w will remain $= 0$ in any subsequent pivot steps carried out during Phase II.

   Then we drop the last (i.e., Phase I objective) row from the final Phase I inverse tableau (and also the Phase I original tableau), and the last column from the augmented basis inverse in it, and begin Phase II with the remaining inverse tableau to solve the original problem. The artificial basic variables may remain in the basic vector during Phase II iterations, but their values in the primal solution will never change from 0.

Example 6.3. Consider the following LP in standard form.

x_1	x_2	x_3	x_4	x_5	$-z$	b
2	3	1	−1	0	0	10
1	2	−1	0	1	0	5
1	1	2	0	0	0	4
1	2	3	0	0	1	0

$x_j \geq 0$ for all j, minimize z

The RHS constants vector is already >0. We only have the second unit vector in the column of x_5 in the tableau. So, we select x_5 as the basic variable in row 2, and introduce artificial variables t_1, t_3 as basic variables in rows 1, 3 for the initial basic vector. The Phase I original tableau is

Phase I original tableau

x_1	x_2	x_3	x_4	x_5	t_1	t_3	$-z$	$-w$	b
2	3	1	−1	0	1	0	0	0	10
1	2	−1	0	1	0	0	0	0	5
1	1	2	0	0	0	1	0	0	4
1	2	3	0	0	0	0	1	0	0
0	0	0	0	0	1	1	0	1	0

$x_j \geq 0$ for all j; artificials $t_1, t_3 \geq 0$; minimize w

Basic var.	Inverse tableau					Basic values	PC	Ratios
							x_1	
t_1	1	0	0	0	0	10	2	10/2
x_5	0	1	0	0	0	5	1	5/1
t_3	0	0	1	0	0	4	**[1]**	4/1
$-z$	0	0	0	1	0	0	1	
$-w$	−1	0	−1	0	1	−14	−3	$\bar{d} = (-3, -4, -3, 1, 0)$
							x_2	
t_1	1	0	−2	0	0	2	1	2/1
x_5	0	1	−1	0	0	1	**[1]**	1/1
x_1	0	0	1	0	0	4	1	4/1
$-z$	0	0	−1	1	0	−4	1	
$-w$	−1	0	2	0	1	−2	−1	$\bar{d} = (0, -1, 3, 1, 0)$
t_1	1	−1	−1	0	0	1		
x_2	0	1	−1	0	0	1		
x_1	0	−1	2	0	0	3		
$-z$	0	−1	0	1	0	−5		
$-w$	−1	1	1	0	1	−1	$\bar{d} = (0, 0, 0, 1, 1)$	

PC pivot column

The vector (t_1, x_5, t_3) is the initial feasible basic vector for the Phase I problem. The inverse tableau corresponding to it is the first in the series of tableaus given above. The value of w in the initial solution is 14; this is the infeasibility measure of the initial solution to the original problem. Subsequent inverse tableaus are listed one below the other, with the pivot elements in a box and all the information like $\bar{d}_j$s, the Phase I relative cost coefficients of x_1 to x_5, in a last row under each tableau.

Phase I terminates when the basic vector (t_1, x_2, x_1) is reached. As the value of w in the solution corresponding to this basic vector is $1 > 0$, the original problem has no feasible solution. ■

Example 6.4. Consider the following LP in standard form

Original Problem

x_1	x_2	x_3	x_4	x_5	x_6	$-z$	b
1	−1	0	0	2	0	0	0
−2	1	0	0	−2	0	0	0
1	0	1	0	1	−1	0	3
0	2	1	1	2	1	0	4
−40	−10	0	0	−7	−14	1	0

$x_j \geq 0$ for all j, minimize z

Again the original RHS constants vector is already >0. Here we only have the fourth unit vector in the column of x_4. So, we select x_4 as the basic variable in row 4, and introduce artificial variables t_1, t_2, t_3 as a basic variables in rows 1, 2, 3 for the initial basic vector.

Phase I original tableau

x_1	x_2	x_3	x_4	x_5	x_6	t_1	t_2	t_3	$-z$	$-w$	b
1	−1	0	0	2	0	1	0	0	0	0	0
−2	1	0	0	−2	0	0	1	0	0	0	0
1	0	1	0	1	−1	0	0	1	0	0	3
0	2	1	1	2	1	0	0	0	0	0	4
−40	−10	0	0	−7	−14	0	0	0	1	0	0
0	0	0	0	0	0	1	1	1	0	1	0

$x_j \geq 0$ for all j; artificials $t_1, t_2, t_3 \geq 0$; minimize w

Basic var.	First inverse tableau						Basic values	PC x_3	Ratios
t_1	1	0	0	0	0	0	0	0	
t_2	0	1	0	0	0	0	0	0	
t_3	0	0	1	0	0	0	3	[1]	3/1
x_4	0	0	0	1	0	0	4	1	4/1
$-z$	0	0	0	0	1	0	0	0	
$-w$	−1	−1	−1	0	0	1	−3	−1	

PC = Pivot column, $\bar{d} = (0, 0, -1, 0, -1, 1)$

The Phase I original tableau is given above.

The vector (t_1, t_2, t_3, x_4) is the initial feasible basic vector for the Phase I problem. The first inverse tableau is given above. The second inverse tableau obtained during Phase I is given below. The last row under each inverse tableau gives the $\bar{d} = (\bar{d}_j)$ vector wrt the present basic vector.

Basic var.	Second inverse tableau						Basic values
t_1	1	0	0	0	0	0	0
t_2	0	1	0	0	0	0	0
x_3	0	0	1	0	0	0	3
x_4	0	0	−1	1	0	0	1
$-z$	0	0	0	0	1	0	0
$-w$	−1	−1	0	0	0	1	0

$\bar{d} = (1, 0, 0, 0, 0, 0)$

Phase I terminates with $w = 0$. Artificial variables t_1, t_2 are still in the basic vector, but with 0-values in the solution. This final solution leads to the feasible solution $(x_1 \text{ to } x_6)^T = (0, 0, 3, 1, 0, 0)^T$ for the original problem. Now we need to go to Phase II.

We look for original problem variables with positive Phase I relative cost coefficients at Phase I termination. Only x_1 satisfies this property. So, $x_1 = 0$ in every feasible solution of the original problem. We fix $x_1 = 0$ and delete it from the problem. In Phase II we only consider variables x_2 to x_6 as candidates to enter the basic vector. From the Phase I terminal tableau we get the inverse tableau given above to initiate Phase II. The second inverse tableau is given below. In each inverse tableau, the vector $(\bar{c}_j : j = 2 \text{ to } 6)$ wrt the present basic vector is given in a bottom row.

Basic var.	First inverse tableau					Basic values	PC x_6	Ratios
t_1	1	0	0	0	0	0	0	
t_2	0	1	0	0	0	0	0	
x_3	0	0	1	0	0	3	−1	
x_4	0	0	−1	1	0	1	**2** (boxed)	1/2
$-z$	0	0	0	0	1	0	−14	

$(\bar{c}_2 \text{ to } \bar{c}_6) = (-10, 0, 0, -7, -14)$

PC = pivot column

Basic var.	Second inverse tableau					Basic values
t_1	1	1	0	0	0	0
t_2	0	1	0	0	0	0
x_3	0	0	1/2	1/2	0	7/2
x_6	0	0	$-1/2$	1/2	0	1/2
$-z$	0	0	-7	7	1	7

$(\bar{c}_2 \text{ to } \bar{c}_6) = (4, 0, 7, 0, 0)$

As $(\bar{c}_2 \text{ to } \bar{c}_6) \geq 0$, now Phase II terminates by satisfying the optimality criterion. The present BFS $\bar{x} = (\bar{x}_1 \text{ to } \bar{x}_6) = (0, 0, 7/2, 0, 0, 1/2)^T$ is an optimum solution to the problem with an optimum objective value of $\bar{z} = -7$.

Notice that even though the artificial variables t_1, t_2 were in the basic vector, their values in the solution remained 0 during Phase II. ■

6.3 How to Find a Feasible Solution to a System of Linear Constraints

If the system consists of linear equations only, then Gaussian elimination can be applied to find a feasible solution.

If the system involves linear inequalities and/or bounds on the variables, we put the system in standard form and apply Phase I of the primal simplex method to find a feasible solution.

Example 6.5. Find a feasible solution for the system of linear constraints given below.

$$
\begin{aligned}
x_1 + x_3 - x_4 &= 3 \\
x_1 + x_2 + 2x_3 &= 10 \\
x_1 + x_2 + x_3 - 2x_4 &\geq 14 \\
x_j &\geq 0, \quad \text{for all } j.
\end{aligned}
$$

Original system, standard form

x_1	x_2	x_3	x_4	x_5	b
1	0	1	-1	0	3
1	1	2	0	0	10
1	1	1	-2	-1	14

$x_j \geq 0$ for all j

Phase I original tableau

BV	x_1	x_2	x_3	x_4	x_5	t_1	t_2	t_3	$-w$	b
t_1	1	0	1	−1	0	1	0	0	0	3
t_2	1	1	2	0	0	0	1	0	0	10
t_3	1	1	1	−2	−1	0	0	1	0	14
	0	0	0	0	0	1	1	1	1	0

We introduce the slack variable x_5 corresponding to the last inequality, and get the system in standard form with all the RHS coefficients nonnegative. Then we introduce the artificial variables to get the Phase I problem with a unit feasible basis to apply the primal simplex method. Notice that there is no $-z$-row in this tableau, because there is no objective function to be optimized; our aim is to find a feasible solution of the original system of linear constraints. We now apply the Phase I of the primal simplex method. The various inverse tableaus obtained are given below one after the other. BV = present basic vector, PC = pivot column, with entering variable given, and $\bar{d} = (\bar{d}_j : j = 1 \text{ to } 5)$ is the present Phase I relative cost coefficient vector for the original variables.

Basic var.	Inverse tableau				Basic values	PC	Ratios	
						x_1		
t_1	1	0	0	0	3	[1]	3	
t_2	0	1	0	0	10	2	5	$\bar{d} = (-3, -2,$
t_3	0	0	1	0	14	1		$-4, 3, 1)$
$-w$	−1	−1	−1	1	−27	4	$\theta = 3$	
						x_2		
x_3	1	0	0	0	3	0		
t_2	−2	1	0	0	4	[1]	4	$\bar{d} = (1, -2,$
t_3	1	0	1	0	11	1	11	$0, -1, 1)$
$-w$	−5	−1	−1	1	−15	−2	$\theta = 4$	
						x_1		
x_3	1	0	0	0	3	[1]	3	
x_2	−2	1	0	0	4	−1		$\bar{d} = (-1, 0,$
t_3	3	−1	1	0	7	1	7	$0, 3, 1)$
$-w$	−9	1	−1	1	−7	−1	$\theta = 3$	
x_1	1	0	0	0	3			
x_2	−1	1	0	0	7			$\bar{d} = 0, 0,$
t_3	2	−1	1	0	4			$1, 2, 1)$
$-w$	−8	1	−1	1	−4			

The Phase I optimality criterion is now satisfied because all the Phase I relative cost coefficients are ≥ 0. The minimum value of $w = 4 > 0$. This implies that the original system has no feasible solution.

Notice how the value of w, the Phase I objective function, decreases from 27 to 15, to 7, and finally to its minimum value of 4 in the sequence of Phase I solutions obtained in the method. If the final value of the Phase I objective function w turned out to be 0, we would have concluded that the final solution obtained is a feasible solution of the original system. ■

6.4 Infeasibility Analysis

When an appropriate and correct mathematical model constructed for a real-world problem turns out to be infeasible, the mathematical and computational work on that model has been completed and the conclusion reached that the problem is infeasible with the data provided.

However, the real-world problem does not go away; it has to be modified and a solution found for it. Engineers are left with the task of figuring out what changes to make in the problem situation and data to make it feasible. Infeasibility analysis deals with these changes and best modifications to adopt.

Suppose the original model for the problem is the following LP in standard form:

$$\begin{aligned} \text{Minimize} \quad & z = cx \\ \text{subject to} \quad & Ax = b \\ & x \geq 0, \end{aligned} \tag{6.4}$$

which is infeasible. Let $A = (a_{ij})$ be an $m \times n$ matrix in this model.

The first thing to do is to check that this model is a reasonable representation of the real-world problem being solved. If there are some constraints in the model that should not be there, or if the values of some data elements have errors in them, etc., correcting these errors may fix the problem. Fixing such errors is not really an optimization, or a mathematical problem, but a task for the engineers who have intimate knowledge of the context of the real-world problem being solved.

Suppose it turns out that the model (6.4) for the problem is fine and that the data in it is reasonably accurate. In this case, perhaps the only option available to make it feasible is to change the values of some data elements.

In most applications for which an LP model of the form (6.4) is suitable, the coefficient matrix A is an input–output (or technology) matrix, that is, elements in it measure how many units of one material are required as input to make one unit of another, etc. In these applications, making changes in the values of entries in A may require developing and adopting new technology, etc., which is usually hard.

On the other hand, the RHS constants $b = (b_i)$ typically represent production levels, sales volume commitments, material supply levels, etc. These quantities are relatively easier to change through negotiations with suppliers, customers, etc. That is why practitioners usually prefer to make changes in the RHS constants to make the model feasible, leaving the other data unchanged. This is what we will discuss in this section.

Making changes in the values of the RHS constants b_i usually involves some expenses, typically proportional to the amount of change, and may be at different rates for different i. In this case, the problem of determining an optimal way of changing b to make (6.4) feasible can itself be posed as another LP (see Exercise 6.24).

Here we will show that the final Phase I solution directly yields one way of modifying the $b = (b_i)$ to make (6.4) feasible, which minimizes the sum of absolute values of all the changes needed. Let the basic vector at Phase I termination be $x_B = (x_{j_1}, \dots, x_{j_m})$ associated with the basis B and the following inverse tableau

Final Phase I Inverse Tableau

BV	Inverse tableau			Basic values	Desired values
x_{j_1}		0	0	$\bar{b}_1$	$\bar{b}'_1$
$\vdots$	B^{-1}	$\vdots$	$\vdots$	$\vdots$	$\vdots$
x_{j_m}		0	0	$\bar{b}_m$	$\bar{b}'_m$
$-z$	$-\pi$	1	0	$-\bar{z}$	$-\bar{z}$
$-w$	$-\sigma$	0	1	$-\bar{w}$	0

Here $\bar{w} > 0$, the minimum value of the infeasibility measure, it is the sum of $\bar{b}_i$ over $i = 1$ to m such that x_{j_i} is an artificial variable.

To change the minimum value of w from $\bar{w}$ to 0, we need to modify the updated RHS constants vector to $\bar{b}' = (\bar{b}'_1, \dots, \bar{b}'_m)$, where for $i = 1$ to m

$$\bar{b}'_i = \begin{cases} \bar{b}_i & \text{if } x_{j_i} \text{ is an original problem variable,} \\ 0 & \text{if } x_{j_i} \text{ is an artificial variable.} \end{cases}$$

The corresponding change in the original RHS constants vector modifies the original $b = (b_i)$ to $b' = (b'_i)$, where for $i = 1$ to m

- $b'_i = b_i$ if x_{j_i}, the basic variable in the final Phase I inverse tableau in the ith row, is an original problem variable,
- $b'_i = b_i - \bar{b}_i$ if x_{j_i}, the basic variable in the final Phase I inverse tableau in the ith row, is an artificial variable with value $\bar{b}_i$.

Clearly $\sum |b_i - b'_i| = \bar{w}$, so modifying b to b' is a way of making (6.4) feasible by changing only the RHS constants to minimize the sum of absolute values of all the changes. Clearly this is an optimal way of changing b to make (6.4) feasible if (1) the cost rate per unit change in any b_i is the same for all i; and (2) for each i, the per unit cost of increasing any b_i, or decreasing it, is the same.

To get an optimum solution of the modified problem, it is not necessary to do Phase I on it again. In the final Phase I inverse tableau for the original problem, change the updated RHS vector (i.e., the basic values vector) from $(\bar{b}, -\bar{z}, -\bar{w})$ to $(\bar{b}', -\bar{z}, 0)$ (this is marked as the "desired values" vector on the tableau). With this as the final Phase I inverse tableau, go to Phase II to solve the modified original problem.

As an example, the original b-vector for the LP discussed in Example 6.2.1 is $(10, 5, 4)^T$. In the final Phase I solution obtained for this example, only the artificial variable t_1, basic variable in the first row, has positive value of 1. So, if the RHS constants vector in this problem is modified to $(9, 5, 4)^T$, the problem becomes feasible, as discussed earlier. The final Phase I inverse tableau for this modified problem is obtained from that of the original problem by changing the final value of the basic variable t_1 to 0. It is

Basic var.	Inverse tableau					Basic values
t_1	1	−1	−1	0	0	0
x_2	0	1	−1	0	0	1
x_1	0	−1	2	0	0	3
$-z$	0	−1	0	1	0	−5
$-w$	−1	1	1	0	1	0

and the primal simplex method now moves to Phase II to solve the modified original problem, beginning with this feasible basic vector.

6.5 Practical Usefulness of the Revised Simplex Method Using Explicit Basis Inverse

The revised simplex method using explicit basis inverse is a very useful method to solve small LPs by hand computation. It is also useful to solve LPs on the computer if the number of constraints in the problem (omitting bound constraints on individual variables) is not too large, that is, in the range of a few hundreds. Beyond that, this variant of the simplex method begins to suffer from the following problems:

Fill-in: Many LP models encountered in practice are *very sparse*, that is, the fraction of nonzero entries in the coefficient matrix in them is a very small fraction. Computer programmers try to take advantage of this sparcity to save the amount of space in the memory needed to store the data (it is only necessary to store the nonzero entries along with their locations in the matrix). Also arithmetic operations become much faster when a lot of the elements you have to deal with are 0 (e.g., to find the sum of a set of elements, it is only necessary to add the nonzero elements among them, thus finding the sum can be carried out very fast when a lot of elements in the set are 0). In fact, computer programs for solving LPs are able to solve large-scale models with only a small amount of computer time only because these models are very sparse. Programmers have built techniques into their programs to take maximum advantage of this sparcity.

Actually, if an LP model is *dense* (this is the opposite of sparse, i.e., a large fraction of entries in the data matrix in the model are nonzero), then even though it may be of moderate size, present day LP computer programs will take a lot of time to obtain its solution.

It has been observed that even though a basis is very sparse, its inverse tends to become *very dense* (i.e., the fraction of nonzero entries in it is close to 1). This phenomenon is called the *fill-in property* of the inverse, and it makes the explicit basis inverses in even very sparse LP models very dense. So storing the basis inverse occupies a lot of memory, and any computation using it becomes very time consuming and inefficient.

Effects of round-off errors: All our computers are digital computers that operate with fixed finite precision, so each arithmetic operation carried out with them involves rounding-off. So, in operations with a dense inverse round-off errors accumulate very rapidly. The result is that after only a few pivot steps in the revised simplex method using the explicit inverse, the product of a basis and its "inverse" in storage may be very different from the unit matrix. This leads to loss of precision, unreliable output, primal basic solutions not even close to being primal feasible, wrong decisions, etc.

For these reasons, the revised simplex method with explicit inverse is not used to solve LP models in which the number of constraints is larger than a few hundred.

For solving larger models versions of the revised simplex method using *factorizations of the basis inverse* instead of the explicit inverse have been developed. One of these early versions will be discussed in Sect. 6.7. These versions alleviate some of the problems with using the explicit inverse to some extent.

6.6 Cycling in the Simplex Method

Cycling is a phenomenon that can occur when the simplex method encounters a degenerate BFS along its path to the optimum. It is possible for the method to cycle through several basic vectors corresponding to the same degenerate BFS endlessly and never move away from this BFS. If this happens, it will never terminate and never reach an optimum solution for the problem. Numerical examples illustrating this possibility have been constructed, and these are discussed in most textbooks on LP (see Dantzig (1963), Dantzig and Thappa (1997), Schrijver (1986) of Chap. 1, Murty (1983) of Chap. 2).

One of the first theoretical techniques to resolve the problem of cycling and theoretically guarantee finite termination of the simplex method is the *lexicographic technique*. It involves replacing the minimum ratio test to select the dropping basic variable by a *lexicographic minimum ratio test*, and needs the explicit basis inverse in every pivot step. So, the revised simplex method using the explicit basis inverse is the ideal variant of the simplex method to incorporate the lexicographic technique into the computer program.

Details of the lexicographic simplex method, how and why it works, and proofs of its finite convergence are discussed in many LP textbooks see for example the references cited above.

However, because of the difficulties discussed in the previous section, this version of the simplex method is not used except in specialized programs meant only for LPs of small size. So, while the lexicographic technique is an elegant theoretical tool to eliminate the possibility of cycling, it is not used in practical computation.

Several other elegant theoretical techniques to resolve the possibility of cycling in the simplex method discussed in many LP textbooks have similar problems that make them unattractive for practical use.

Programmers take care of the possibility of cycling by heuristic techniques that have proved very effective in practice. One of the most commonly used techniques is to actually perturb the bounds on the variables by small quantities to guarantee strict improvement in the objective value in every iteration while making corrections for these perturbations at the end using sensitivity analysis techniques. This technique is explained in detail in Sect. 6.11.

6.7 Revised Simplex Method Using the Product Form of the Inverse

This is one of the earliest variants of the revised simplex method based on maintaining a factorization of the basis inverse instead of the explicit inverse. The matrix factors used in this implementation are called *pivot matrices*.

6.7.1 Pivot Matrices

Let $D = (d_{ij})$ be a matrix of order $p \times q$. Consider performing a GJ (Gauss-Jordan) pivot step on D with its sth column $D_{.s}$ as the pivot column (PC), rth row as the pivot row, and hence the element $d_{rs} \neq 0$ as the pivot element. Let $\bar{D}$ denote the matrix obtained from D after this pivot step.

Then there is a square matrix P of order $p \times p$ called the *pivot matrix* corresponding to this pivot step, satisfying the property that $\bar{D} = PD$; that is, carrying out this pivot step on a matrix is equivalent to left-multiplying it by the pivot matrix corresponding to this pivot step.

The pivot matrix P will be the same as the unit matrix I of order p except for one column known as its *eta column*, whose position, the rth in P, is the same as the position of the pivot row in D. Here is the formula for writing this η (eta) column from the pivot column (the boxed element is the pivot element). We also show the whole pivot matrix on the right, the column in it different from the unit column is its rth column, its η-column.

$$
\begin{array}{c|c||c}
\text{PC} & \eta\text{-col.} & \text{Pivot matrix } P \\
\hline
d_{1s} & -d_{1s}/d_{rs} & \\
\vdots & \vdots & \\
d_{r-1,s} & -d_{r-1,s}/d_{rs} & \\
\boxed{d_{rs}} & 1/d_{rs} & \\
d_{r+1,s} & -d_{r+1,s}/d_{rs} & \\
\vdots & \vdots & \\
d_{ms} & -d_{ms}/d_{rs} &
\end{array}
\quad
\begin{pmatrix}
1 \dots 0 & -d_{1s}/d_{rs} & 0 \dots 0 \\
\vdots \ddots \vdots & \vdots & \vdots \quad \vdots \\
0 \dots 1 & -d_{r-1,s}/d_{rs} & 0 \dots 0 \\
0 \dots 0 & 1/d_{rs} & 0 \dots 0 \\
0 \dots 0 & -d_{r+1,s}/d_{rs} & 1 \dots 0 \\
\vdots \quad \vdots & \vdots & \vdots \ddots \vdots \\
0 \dots 0 & -d_{ms}/d_{rs} & 0 \dots 1
\end{pmatrix}.
$$

To store any pivot matrix P, it is only necessary to store its eta vector and its position in P, because from this information P can be generated easily.

Example 6.6. Let D be the matrix $(A\vdots b)$ in the system of linear equations in the top tableau .

x_1	x_2	x_3	x_4	RHS b
7	0	−1	1	10
4	−6	**[2]**	2	4
5	−2	1	0	15
9	−3	0	2	12
2	−3	1	1	2
3	1	0	−1	13

Consider performing a GJ pivot step in the column of x_3 (the PC), with row 2 as the PR (pivot row). The pivot element is boxed. The pivot matrix corresponding to this pivot operation is

$$P = \begin{pmatrix} 1 & 1/2 & 0 \\ 0 & 1/2 & 0 \\ 0 & -1/2 & 1 \end{pmatrix}.$$

The tableau obtained after the pivot step is performed is given at the bottom of the original tableau. It can be verified that it is $PD = P(A\vdots b)$. ■

6.7.2 A General Iteration in the Revised Simplex Method Using the Product Form of the Inverse

Consider the revised simplex method started with a unit basic vector which may contain some artificial variables. In this case, the initial inverse tableau, denoted by P_0, differs from the unit matrix in the last row only if we begin with Phase II, or in the last two rows only if we begin with Phase I. So, P_0 is also like a pivot matrix, and it can be generated by storing only those one or two rows.

In this version, the inverse tableau is not computed explicitly, but the various pivot matrices corresponding to all the pivot steps carried out so far are stored in the order in which they occurred, the newest pivot matrix always joining the string on the left side of all those generated above. That is why at any stage, the sequence of pivot matrices in this order of occurrence is called the *current string of pivot matrices*. All the quantities needed to carry out the simplex method are generated using this string. Every time it is needed, each pivot matrix in the string is generated from its stored eta vector and its position.

If the method is started with a known feasible basic vector which is not a unit basic vector, the corresponding basis is inverted, and the various pivot matrices corresponding to the pivot steps in this process form the current string.

In the various steps in this iteration of the revised simplex method, the inverse tableau is used only twice: first to get its last row to compute the relative cost coefficients of the nonbasic variables, then the inverse tableau is used to compute the pivot column which is the updated column of the entering variable selected. We will show how each of these operations can be carried out using the pivot matrices in the string one by one in their proper order without ever computing the whole inverse tableau explicitly.

Let the string of pivot matrices at this stage be $P_r,\ P_{r-1}, \cdots, P_0$, in its order, each new pivot matrix generated in the algorithm joining the current string as its leftmost entry. Then the formula for the inverse tableau is $P_r P_{r-1} \dots P_1 P_0$. So, the last row of the inverse tableau is = (last row of the unit matrix of the same order as the inverse tableau)(inverse tableau) =

$$(0, \dots, 0, 1) P_r P_{r-1} \dots P_1 P_0 \tag{6.5}$$

This can be computed efficiently from left to right. First multiply the row vector $(0, \dots, 0, 1)$ by P_r on the right. Then multiply the resulting row vector by P_{r-1} on the right, and continue this way until multiplication by the entire string is completed.

Because of the special structure of each of the pivot matrices, multiplication of a vector by a pivot matrix takes about the same effort as computing the dot product of two vectors. The left-to-right scheme described above to compute the expression in (6.5) is commonly known as *left-to-right string multiplication.*

Now consider the operation of updating the entering column. Let x_s denote the entering variable, and let $\mathcal{A}_{.s}$ denote its column in the original tableau. Let $\bar{\mathcal{A}}_{.s}$ denote the updated column of x_s. We have

$$\bar{\mathcal{A}}_{.s} = P_r P_{r-1} \cdots P_1 P_0 \mathcal{A}_{.s}. \tag{6.6}$$

This can be computed efficiently by right to left string computation, which involves multiplying the column vector $\mathcal{A}_{.s}$ by P_0 on the left, then multiplying the resulting column by P_1 on the left, and continuing this way until the multiplication by all the pivot matrices in the string is complete.

6.7.3 Transition from Phase I to Phase II

Consider solving the LP in standard form

$$\begin{aligned} \text{Minimize} \quad z &= cx \\ \text{subject to} \quad Ax &= b \\ x &\geq 0, \end{aligned} \tag{6.7}$$

where A is a matrix of order $m \times n$ and rank m. In solving this LP by the revised simplex method using the product form of the inverse, during Phase I, pivot matrices will be of order $m + 2$, while in Phase II they will be of order $m + 1$.

The last row and column (i.e., the $(m + 2)$th row and column) in each of the pivot matrices during Phase I corresponds to the Phase I objective function, which is deleted when we move to Phase II. This $(m + 2)$th row and column are deleted at the end of Phase I, from each pivot matrix in the string, while moving to Phase II (in effect, this essentially amounts to deleting the $(m+2)$th element from the stored eta column of each pivot matrix stored at that stage). Then the present string becomes the string corresponding to the present basic vector for the Phase II problem, and you can begin Phase II with it.

Unlike the simplex method using canonical tableaus or the revised simplex method with the explicit form of the inverse, this revised simplex method using the *product form of the inverse (PFI)* is quite cumbersome for hand computation to solve even small size LPs. That is why we do not present a numerical example for this implementation here (such numerical examples can be seen in (Murty (1983) of Chap. 2). However, its computer implementations offer superior performance to solve LP models involving up to several hundreds of constraints.

Advantages of the product form of the inverse implementation: An eta column is said to be sparse if the fraction of nonzero entries in it is small; as this fraction approaches 1, the eta column is said to be dense. As long as eta columns of pivot matrices generated in the various steps remain sparse, the revised simplex method using the PFI (product form of the inverse) continues its superior performance. At the beginning of the method, the eta columns in pivot matrices tend to be sparse, but gradually they tend to fill-in more and more and become dense as r, the number of the step being performed increases. Let m be the number of constraints in the model being solved, and r the number of the current step in the method. As r/m exceeds 2, the eta columns in pivot matrices become dense, and the memory space needed to store the pivot matrices begins to approach that needed to store the explicit inverse itself and the advantages of the PFI disappear. At this stage, the method goes to an operation called *reinversion* discussed next.

6.7.4 Reinversions in the Revised Simplex Method Using PFI

Let x_B be the present basic vector in the current iteration, with B, $\mathcal{B}$, c_B the associated basis, augmented basis, basic cost vector, respectively, and let $\bar{x}$, $\bar{\pi}$ be the associated primal and dual basic solutions. $||Ax - b||$ and $||c_B - \bar{\pi}B||$ can be used as measures to monitor the effects of roundoff errors accumulated so far. If both these measures are smaller than specified tolerances, we continue with the iterations in the method.

However, if one or both of $||Ax - b||$, $||c_B - \bar{\pi}B||$ exceed their specified tolerances, it is an indication that the accumulation of roundoff errors is already a problem. In this case we go to this *reinversion*.

Reinversion involves getting rid of the present string of pivot matrices. Reinvert the present augmented basis $\mathcal{B}$, that is, carry out the pivot operations needed to compute $\mathcal{B}^{-1}$. But instead of generating the explicit inverse, generate the string of pivot matrices corresponding to the various pivot steps carried out to compute $\mathcal{B}^{-1}$. This new string consists of $m + 1$ or $m + 2$ pivot matrices (depending on whether we are currently in Phase II or Phase I). With this new string, resume the iterations of the revised simplex method for solving the LP model.

Also, when the number of pivot matrices in the current string becomes much larger than m (i.e., their ratio begins to exceed 2 and continuing to grow), then the storage space needed to store the eta vectors in all the pivot matrices in the present string may exceed the space needed to store the explicit inverse of the basis. At this time, carrying out a reinversion will replace the present string with a new string of $m + 1$ (or $m + 2$) pivot matrices.

6.8 Revised Simplex Method Using Other Factorizations of the Basis Inverse

The PFI implementation is the first implementation developed for the simplex method using matrix factorizations. It helps to reduce the effects of problems like fill-in, roundoff error accumulation, excessive computer time for each iteration, etc. to some extent, for values of m up to several hundreds. But for solving LP models in which m is much higher, even this PFI implementation is affected by the same problems.

Implementations of the simplex method using other types of matrix factorizations such as Cholesky factorization, LU-decomposition, etc. have been developed subsequently. Many of the high-performance commercial systems for solving linear programs available today are based on these newer factorizations. This area has become a specialized area of computer programming, and is outside the scope of this book. For details of these factorizations and implementations of the simplex method using them, see Fourer (2005), Bartels (1971), Bartels and Golub (1969) and Todd (1983).

6.9 Finding the Optimum Face of an LP (Alternate Optimum Solutions)

In practical decision making, operating conditions usually change from the time a mathematical model for the problem is constructed to the time an optimum solution is computed by solving the model. Because of these changes or for some other reasons that may have come up since the model construction, one may then find that a computed optimum solution is not quite suitable. In this situation, if the original model has alternate optimum solutions, it is always better to check whether one of them may be suitable to implement before going through the expense of solving the revised model.

As discussed in Chap. 5, in linear programming there are efficient techniques to check whether an optimum solution computed by the algorithm is the unique optimum, and to generate other optimum solutions when there are alternate optima, using the complementary slackness optimality conditions.

For an LP in standard form with x_B as an optimum basic vector, if the relative cost coefficients wrt x_B of all the nonbasic variables are >0, then the BFS associated with x_B is the *unique optimum solution* of this LP.

On the other hand, if some nonbasic relative cost coefficients wrt x_B are 0, by bringing the corresponding nonbasic variables into this basic vector as in the primal simplex algorithm, we can get alternate optimum BFSs for this LP. Also, any convex combination of optimum solutions is also optimal. And the optimum face for this LP, that is, the set of all optimum solutions, consists of all the feasible solutions in which all variables x_j whose relative cost coefficient $\bar{c}_j$ wrt x_B is strictly >0 have value 0.

As an example, consider the following LP

Original problem

x_1	x_2	x_3	x_4	x_5	x_6	x_7	$-z$	b
0	0	1	1	−1	−5	3	0	7
1	0	0	−1	−1	−3	−8	0	9
0	1	0	−1	−1	0	4	0	1
−1	−1	−1	10	6	8	1	1	0

$x_j \geq 0$ for all j, minimize z

An optimum basic vector for this LP is $x_B = (x_3, x_1, x_2)$. The inverse tableau corresponding to it is given below.

Inverse tableau

Basic var.	Inverse tableau				Basic values
x_3	1	0	0	0	7
x_1	0	1	0	0	9
x_2	0	0	1	0	1
$-z$	1	1	1	1	17

The primal BFS associated with x_B is $\bar{x} = (9, 1, 7, 0, 0, 0)^T$ with an objective value of $\bar{z} = -17$. So, any feasible solution in which $z = -17$ is an optimum solution to this LP. So one representation of the optimum face of this LP is, as the set of feasible solutions of the following system of constraints,

Optimum face representation

x_1	x_2	x_3	x_4	x_5	x_6	x_7	b
0	0	1	1	−1	−5	3	7
1	0	0	−1	−1	−3	−8	9
0	1	0	−1	−1	0	4	1
−1	−1	−1	10	6	8	1	−17

$x_j \geq 0$ for all j.

With respect to the optimum basic vector x_B, the relative cost coefficients of the nonbasic variables are $(\bar{c}_4, \bar{c}_5, \bar{c}_6, \bar{c}_7) = (9, 3, 0, 0)$.

As $\bar{c}_6 = \bar{c}_7 = 0$, any BFS obtained by performing pivot steps with x_6, x_7 as entering variables into x_B is an alternate optimum solution to this LP.

As $\bar{c}_4, \bar{c}_5$ are the only positive relative cost coefficients wrt the present basic vector, we conclude that any feasible solution of the original system in which x_4, x_5 are both 0 is an optimum solution to the problem. So the optimum face of this LP can also be represented as the set of all feasible solutions of

Second representation of the optimum face

x_1	x_2	x_3	x_4	x_5	x_6	x_7	b
0	0	1	1	−1	−5	3	7
1	0	0	−1	−1	−3	−8	9
0	1	0	−1	−1	0	4	1
0	0	0	1	0	0	0	0
0	0	0	0	1	0	0	0

$x_j \geq 0$ for all $j = 1$ to 5.

6.10 The Dual Simplex Algorithm

Consider the LP in standard form

$$\begin{aligned} \text{Minimize} \quad & z = cx \\ \text{subject to} \quad & Ax = b \\ & x \geq 0, \end{aligned} \tag{6.8}$$

where A is a matrix of order $m \times n$ and rank m. Given a basic vector x_B for this problem, let x_D be the corresponding nonbasic vector (i.e., vector of those variables not in x_B). Then the original tableau for this problem can be partitioned into basic, nonbasic parts this way:

Original tableau

x_B	x_D	$-z$	RHS
B	D	0	b
c_B	c_D	1	0

All $x_j \geq 0$, min z

To solve this problem, we have to search for a basic vector x_B satisfying the following two properties:

1. Primal feasibility: $\bar{b} = B^{-1}b \geq 0$,
2. Dual feasibility: $\bar{c}_D = c_D - (c_B B^{-1})D \geq 0$.

The primal simplex algorithm discussed so far begins with a basic vector x_B satisfying primal feasibility, and tries to attain the dual feasibility property while maintaining primal feasibility throughout. It does this by performing a sequence of primal simplex pivot steps, in each of which a column j in which dual feasibility is violated (i.e., $\bar{c}_j < 0$) is selected as the pivot column, and the pivot row is chosen by the primal simplex minimum ratio test to preserve primal feasibility in the next basic vector.

Example 6.7. Consider the following LP for which the canonical tableau wrt the basic vector $x_B = (x_1, x_2, x_3)$ is given below. The basic vector x_B is dual feasible.

Canonical tableau wrt (x_1, x_2, x_3)

BV	x_1	x_2	x_3	x_4	x_5	x_6	$-z$	RHS
x_1	1	0	0	−1	−2	2	0	−6
x_2	0	1	0	1	−1	−2	0	−2
x_3	0	0	1	1	1	−1	0	4
$-z$	0	0	0	3	8	10	1	−100

We do not have a primal feasible basic vector for this LP readily available. The basic vector (x_1, x_2, x_3) available is primal infeasible, but it is dual feasible because the relative cost coefficients of all the nonbasic variables are positive. To solve this LP by the primal simplex method, we have to first carry out a Phase I to find a primal feasible basic vector. During Phase I, the dual feasibility property may be lost, so at the end of Phase I we may have to carry out Phase II to gain dual feasibility again. So this method does not take any advantage of the dual feasible basic vector (x_1, x_2, x_3) that is readily available to us. ■

To solve LPs like this for which a dual feasible basic vector is available, it will be nice to use a method that can be initiated with such

Primal simplex property	Dual simplex property
Initiated with a primal feasible BV	Initiated with a dual feasible BV
Each iteration begins with selecting PC among those violating dual feasibility, i.e., one in which relative cost $\bar{c}_j < 0$	Each iteration begins with selecting PR among those violating primal feasibility, i.e., one in which updated RHS constant $\bar{b}_i < 0$
In each iteration, uses data in PC and updated RHS constants vector, in the primal simplex minimum ratio test, to select PR to keep next BV primal feasible	In each iteration, uses data in PR and relative cost vector, in a dual simplex minimum ratio test, to select PC to keep next BV dual feasible

a basic vector, which maintains dual feasibility throughout, and tries to attain primal feasibility by performing a sequence of pivot steps changing the basic vector by one variable in each step. Such a method was developed by Kantorovich earlier, and again independently by Lemke (1954); it is called the *dual simplex algorithm* (the reason for the name is elaborated on later). Analogous properties of the two algorithms are explained in the table given above (BV = basic vector, PC = pivot column, PR = pivot row) for minimization problems.

We will now describe various steps in a general iteration of the dual simplex algorithm, and how to carry them out using either the canonical tableaus or inverse tableaus (BV = basic vector).

Canonical tableau wrt x_B

BV	x_B	x_D	$-z$	Updated RHS
x_B	I	$\bar{D}$	0	$\bar{b}$
$-z$	0	$\bar{c}_D$	1	$-\bar{z}$

Inverse tableau wrt x_B

BV	Inverse		Updated RHS
x_B	B^{-1}	0	$\bar{b}$
$-z$	$-\bar{\pi}$	1	$-\bar{z}$

Here are the various steps in this iteration.

1. *Check optimality (primal feasibility):* If $\bar{b} \geq 0$, x_B is primal feasible and hence optimal (as it is also dual feasible), and the present BFS is an optimum solution of the original LP, and $\bar{\pi} = c_B B^{-1}$ is an optimum dual solution, terminate. Otherwise continue.
2. *Select pivot row (i.e., dropping basic variable) for a pivot step:* Select one of the rows among $\{i : \bar{b}_i < 0\}$ as the pivot row for the pivot step in this iteration. The current basic variable in this pivot row will be the dropping basic variable in this iteration.

   When there are several i satisfying $\bar{b}_i < 0$, several rules can be used for selecting one of them as the pivot row; for example, the one corresponding to the most negative $\bar{b}_i$, etc.
3. *Check primal infeasibility:* Let the rth row be the pivot row selected. If inverse tableaus are being used, the formula for computing the pivot row is

$$\begin{aligned}
\text{Pivot row} &= \text{updated } r\text{th row} = \bar{A}_{r.}, \\
&= \text{Vector}\times\text{matrix product of } r\text{th row of the inverse tableau} \\
&\quad\ \text{multiplied on the right by the original tableau}, \\
&= (B^{-1})_{r.}A.
\end{aligned}$$

If there is no negative element in the pivot row, then conclude that the primal problem is infeasible, and terminate. The reason for this conclusion is the following: Let $\bar{a}_{rj}$ = entry in pivot row in jth column for $j = 1$ to n. Then the equation corresponding to this pivot row is

$$\sum_{j=1}^{n} \bar{a}_{rj} x_j = \bar{b}_r.$$

From the pivot row selection, we know that $\bar{b}_r < 0$. If $\bar{a}_{rj} \geq 0$ for all $j =$ 1 to n, the above equation cannot have any solution $x \geq 0$, that is, primal is infeasible. As the dual is feasible, by duality theorem the dual objective value is unbounded above in this case.

If this criterion is not satisfied, continue.

4. *Dual simplex minimum ratio test:* The dropping basic variable (the present basic variable in the pivot row) will be a nonbasic variable in the next basic vector, hence it will have a value of 0 in the next primal basic solution, that is, it will satisfy the sign restriction on it.

   The dual simplex algorithm also makes sure that the entering variable (the basic variable in pivot row in next basic vector) will have a strictly positive value in the next basic solution. This will happen iff the pivot element in this pivot step is strictly negative, that is why pivot elements in the dual simplex algorithm are always <0 (compare this to the corresponding property in the primal simplex algorithm in which the pivot element is always a positive element in the pivot column to preserve primal feasibility).

   The pivot row is $\bar{A}_{r.} = (\bar{a}_{r1}, \cdots, \bar{a}_{rn})$, and the updated cost row is $\bar{c} = (\bar{c}_1, \cdots, \bar{c}_n) \geq 0$ by dual feasibility of the present basic vector. Suppose $\bar{a}_{r1} < 0$. If we used $\bar{a}_{r1}$ as the pivot element, after the pivot step, the jth entry in the updated cost row will change to $(\bar{c}_j - (\bar{c}_1\bar{a}_{rj})/(\bar{a}_{r1}))$. So, for the next basic vector to be dual feasible we need therefore

$$(\bar{c}_j - (\bar{c}_1\bar{a}_{rj})/(\bar{a}_{r1})) \geq 0 \quad \text{for all } j,$$
$$\text{i.e.,} \quad \bar{c}_j\bar{a}_{r1} \leq \bar{c}_1\bar{a}_{rj} \quad \text{since } \bar{a}_{r1} < 0.$$

   As $\bar{c} \geq 0$ and $\bar{a}_{r1} < 0$, the above inequality will hold automatically for all j such that $\bar{a}_{rj} \geq 0$. For j such that $\bar{a}_{rj} < 0$, the above inequality will only hold if $\bar{c}_j/(-\bar{a}_{rj}) \geq \bar{c}_1/(-\bar{a}_{r1})$. For this reason, the entering column (i.e., the pivot column) is chosen in this algorithm by the following:

   Dual simplex minimum ratio test: The pivot row is $\bar{A}_{r.} = (\bar{a}_{r1}, \cdots, \bar{a}_{rn})$, and the updated cost row is $\bar{c} = (\bar{c}_1, \cdots, \bar{c}_n) \geq 0$. In all columns j such that $\bar{a}_{rj} < 0$, compute the dual simplex ratio $\bar{c}_j/(-\bar{a}_{rj})$.

   Select the pivot column to be the column that corresponds to the minimum ratio, and the variable associated with that column as the entering variable into the basic vector in this iteration replacing the dropping variable. This selection of the pivot column (entering variable) by the dual simplex minimum ratio test guarantees that the basic vector obtained after the pivot step will be dual feasible.

5. *Pivot step:* Perform the pivot step and go to the next iteration with the new basic vector.

6. *Change in objective value in this iteration:* Let $\bar{z}$ be the objective value wrt x_B and $\bar{b}_r$ the updated RHS constant in the pivot row, at the beginning of this iteration. Suppose the pivot column is the column of the entering variable x_s with relative cost coefficient $\bar{c}_s$. Then the objective value of the new BFS obtained at the end of the pivot step in 5 above is $\bar{z} - \bar{b}_r(\bar{c}_s/(-\bar{a}_{rs})) = \bar{z} - \bar{b}_r\theta$, where θ is the minimum ratio computed in this iteration.

   As $\theta \geq 0$ and $\bar{b}_r < 0$, we see that the objective value either stays the same or increases in this iteration depending on whether $\theta = 0$ or > 0.

Example 6.8. In this example we will solve the problem in standard form for which the canonical tableau wrt the dual feasible basic vector (x_1, x_2, x_3) is given in Example 6.7 (it is the tableau at the top given below) using canonical tableaus. In each step the pivot element selected is boxed, and the PR (pivot row) and PC (pivot column) are those containing this boxed element. The various canonical tableaus obtained are put one below the other. BV indicates "basic variable in row." All x_j are required to be ≥ 0, and z has to be minimized in the problem.

				PC					
BV	x_1	x_2	x_3	x_4	x_5	x_6	$-z$	RHS	
x_1	1	0	0	$\boxed{-1}$	−2	2	0	−6	PR
x_2	0	1	0	1	−1	−2	0	−2	
x_3	0	0	1	1	1	−1	0	4	
$-z$	0	0	0	3	8	10	1	−100	
Ratios				3	4				

					PC				
x_4	−1	0	0	1	2	−2	0	6	
x_2	1	1	0	0	−3	0	0	−8	
x_3	1	0	1	0	$\boxed{-1}$	1	0	−2	PR
$-z$	3	0	0	0	2	16	1	−118	
Ratios					2				

			PC						
x_4	1	0	2	1	0	1	0	2	
x_2	−2	1	$\boxed{-3}$	0	0	−3	0	−2	PR
x_5	−1	0	−1	0	1	−1	0	2	
$-z$	5	0	2	0	0	18	1	−122	
Ratio	5/2		2/3			6			

Optimum canonical tableau

x_4	−1/3	2/3	0	1	0	−1	0	2/3	
x_3	2/3	−1/3	1	0	0	1	0	2/3	
x_5	−1/3	−1/3	0	0	1	0	0	4/3	
$-z$	11/3	2/3	0	0	0	16	1	−370/3	

In the last tableau, all the updated RHS constants have become ≥ 0; hence it is an optimum tableau and the method terminates. The optimum solution of this problem is $\bar{x} = (0, 0, 2/3, 2/3, 4/3, 0)^T$, and the optimum objective value is 370/3. As the relative cost coefficients of the nonbasic variables are all positive in the final tableau, this optimum solution is the unique optimum for this LP. The optimum dual solution is $\pi = (3, 0, 8) \times$ (inverse of basis for the basic vector (x_4, x_3, x_5)), it is the marginal value vector for this problem. ■

Example 6.9. In this example we will solve the problem in standard form for which the canonical tableau wrt the dual feasible basic vector (x_1, x_3, x_4) is given (it is the tableau at the top given below) using canonical tableaus. In each step the pivot element selected is boxed, and PR (pivot row), PC (pivot column), and BV (basic variable in row) are indicated. All the variables are required to be ≥ 0 and z is to be minimized.

					PC					
BV	x_1	x_2	x_3	x_4	x_5	x_6	x_7	$-z$	RHS	
x_1	1	2	0	0	$\boxed{-1}$	−2	−2	0	−6	PR
x_3	0	−1	1	0	1	2	4	0	−14	
x_4	0	0	0	1	1	−1	1	0	4	
$-z$	0	4	0	0	3	8	10	1	−100	
Ratios					3	4	5			
Tableau exhibiting primal infeasibility										
x_5	−1	−1	0	0	1	2	−2	0	6	
x_3	1	1	1	0	0	0	2	0	−8	PR
x_4	1	1	0	1	0	−3	3	0	−2	
$-z$	3	7	0	0	0	2	16	1	−118	

In the final tableau, the second row with updated RHS constant of −8 has been selected as the pivot row.

But there is no negative entry among the updated constraint coefficients in this row, so the primal infeasibility condition is satisfied in this row. Actually the constraint corresponding to this row is $x_1 + x_2 + x_3 + 2x_7 = -8$. As all the variables are required to be ≥ 0 and they have positive coefficients in this equation, it is impossible to make the left-hand-side of this equation negative. So, the system of constraints in this LP has no primal feasible solution; we terminate with this conclusion. ■

Example 6.10. In this example we will solve the following problem in standard form. The vector (x_1, x_2, x_3) is a unit basic vector, and it can be verified to be dual feasible; we will initiate the dual simplex algorithm on this problem with this basic vector. We solve this LP using inverse tableaus. The first inverse tableau wrt (x_1, x_2, x_3) is constructed as discussed earlier. In each step the pivot element selected is boxed, and PR (pivot row, updated row of dropping variable), PC (pivot column, updated column of entering variable), and BV (basic variable in row) are indicated.

Original tableau

x_1	x_2	x_3	x_4	x_5	x_6	$-z$	RHS
1	0	0	4	−5	7	0	8
0	1	0	−2	4	−2	0	−2
0	0	1	1	−3	2	0	2
1	1	1	4	−1	9	1	0

$x_j \geq 0$ for all j, minimize z

Here is the first inverse tableau.

First inverse tableau

BV	Inverse				Updated RHS	PC x_4	
x_1	1	0	0	0	8	4	
x_2	0	1	0	0	−2	−2	PR
x_3	0	0	1	0	2	1	
$-z$	−1	−1	−1	1	−8	1	

The relative cost coefficients $\bar{c}_j$ of the nonbasic variables $x_4,\ x_5,\ x_6$ are given in a table below, and they are all positive. Row 2 with negative updated RHS has been selected as the PR, and the present basic variable in it, x_2, is the dropping basic variable in this step. The following table gives the entries in the PR = updated second row and the $\bar{c}_j$ in nonbasic columns. As the minimum ratio θ occurs in the column of x_4, the entering variable in this step is x_4. Its updated column, the PC in this step, is already entered on the inverse tableau. After the pivot step, we get the second inverse tableau given below.

	x_4	x_5	x_6	
PR	−2	4	−2	
$\bar{c}_j$	1	3	2	
Ratios	1/2		1	$\theta = 1/2$

Second inverse tableau

BV	Inverse				Updated RHS
x_1	1	2	0	0	4
x_4	0	−1/2	0	0	1
x_3	0	1/2	1	0	1
$-z$	−1	−1/2	−1	1	−9

The relative coefficients of the nonbasic variables wrt this basic vector are $(\bar{c}_2,\ \bar{c}_5,\ \bar{c}_6) = (1/2,\ 5,\ 1)$. As the updated RHS constants are all ≥ 0, this basic vector $(x_1,\ x_4,\ x_3)$ is optimal to the problem.

The optimal BFS is $\bar{x} = (4, 0, 1, 1, 0, 0)^T$, and the optimum objective value in the problem is $\bar{z} = 9$. The optimum dual solution is $\pi = (1, 1/2, 1)$. As the optimal primal BFS is nondegenerate, the optimum dual solution is the marginal value vector for this problem. ■

Example 6.11. In this example we will solve the following problem in standard form. Clearly, (x_4, x_5, x_6) is a unit dual feasible basic vector, and we will initiate the dual simplex algorithm on this problem with this basic vector.

We solve this LP using inverse tableaus. In each step the pivot element selected is boxed, and PR (pivot row, updated row of dropping variable), PC (pivot column, updated column of entering variable), and BV (basic variable in row) are indicated. The various inverse tableaus obtained are given below.

Original tableau

x_1	x_2	x_3	x_4	x_5	x_6	$-z$	RHS
1	1	1	1	0	0	0	1
−1	−2	−1	0	1	0	0	−8
1	3	2	0	0	1	0	5
8	24	15	0	0	0	1	0

$x_j \geq 0$ for all j, minimize z

First inverse tableau

BV	Inverse				Updated RHS	PC x_1	
x_4	1	0	0	0	1	1	
x_5	0	1	0	0	−8	[−1]	PR
x_6	0	0	1	0	5	1	
$-z$	0	0	0	1	0	8	

The second row is selected as the PR, and so x_5 is the dropping variable. Updated PR, $\bar{c}_j$ entries in nonbasic columns are given below. x_1 is selected as the entering variable, and the PC is already entered on the inverse tableau. After the pivot step we get the second inverse tableau.

	x_1	x_2	x_3	
PR	−1	−2	−1	
$\bar{c}_j$	8	24	15	
Ratios	8	12	15	$\theta = 8$

Second inverse tableau

BV	Inverse				Updated RHS	PC x_2	
x_4	1	1	0	0	−7	[−1]	PR
x_1	0	−1	0	0	8	2	
x_6	0	1	1	0	−3	1	
$-z$	0	8	0	1	−64	8	

The first row is selected as the PR, so x_4 is the dropping variable. Updated PR, $\bar{c}_j$ entries in nonbasic columns are given below. The variable x_2 is selected as the entering variable, and the PC is already entered on the inverse tableau. After the pivot step we get the third inverse tableau.

	x_2	x_3	x_5	
PR	−1	0	1	
$\bar{c}_j$	8	7	8	
Ratios	8			$\theta = 8$

Third Inverse Tableau

BV	Inverse				Updated RHS
x_2	−1	−1	0	0	7
x_1	2	1	0	0	−6
x_6	1	2	1	0	−10
$-z$	8	16	0	1	−100

Now x_6, the basic variable in the third row, is selected as the dropping variable in this step. The PR $= (B^{-1})_{3.}A = (0, 0, 1, 1, 2, 1)$; it has no negative entry. So, we terminate with the conclusion that the primal is infeasible. ■

6.10.1 Properties of the Dual Simplex Algorithm

Consider the LP in standard form for which the canonical tableau wrt a dual feasible but primal infeasible basic vector x_B is given below. For notational convenience we assume $x_B = (x_1, \cdots, x_m)$. "BV" = "basic variable in row."

Canonical tableau wrt x_B

BV	x_1	…	x_m	x_{m+1}	…	x_j	…	x_n	$-z$	RHS
x_1	1	…	0	$\bar{a}_{1,m+1}$	…	$\bar{a}_{1j}$	…	$\bar{a}_{1n}$	0	$\bar{b}_1$
⋮	⋮	⋱	⋮	⋮		⋮		⋮	⋮	⋮
x_m	0	…	1	$\bar{a}_{m,m+1}$	…	$\bar{a}_{mj}$	…	$\bar{a}_{mn}$	0	$\bar{b}_m$
$-z$	0	…	0	$\bar{c}_{m+1}$	…	$\bar{c}_j$	…	$\bar{c}_n$	0	$\bar{z}$

$x_j \geq 0$ for all j, minimize z.

1. *Current basic solution is the optimum solution for a relaxed problem:* The present primal basic solution is $\bar{x} = (\bar{b}_1, \ldots, \bar{b}_m, 0, \ \ldots, 0)^T$ with objective value $\bar{z}$. However, as x_B is primal infeasible, $\bar{b}_i < 0$ for some i. Let

$$S = \{i : \bar{b}_i < 0\} \text{ and } \bar{S} = \{1, \ldots, m\} \backslash P.$$

Then $\bar{x}$ violates the primal constraints $x_i \geq 0$ for $i \in S$, but satisfies all other primal constraints. Now consider the *relaxed problem* obtained by relaxing the bound constraints "$x_i \geq 0$ for $i \in S$" from the original LP, and let K_S denote its set of feasible solutions. So, $K_S \supset K$, and so the minimum value of z in the relaxed problem is $\leq$ the minimum value of z in the original problem.

$$\begin{aligned} \text{Minimize} \quad & z = cx \\ \text{subject to} \quad & Ax = b \\ & x_j \geq 0, \text{ for } j \in \bar{S} \\ & x_j \quad \text{unrestricted for all } j \in S. \end{aligned} \tag{6.9}$$

From the definition of S, all the variables x_i for $i \in S$ are basic variables in the present basic vector, and so $\{m+1, \cdots, n\} \subset \bar{S}$. The equation corresponding to the objective row in the canonical tableau is

$$z = \bar{z} + \sum_{j=m+1}^{n} \bar{c}_j x_j. \tag{6.10}$$

We know that the present primal basic solution $\bar{x}$ is feasible to the relaxed problem (6.9). Also, as x_B is dual feasible, $\bar{c}_j \geq 0$ for all j, and as $\{m+1, \ldots, n\} \subset \bar{S}$, it is clear that the term $\sum_{j=m+1}^{n} \bar{c}_j x_j$ is ≥ 0 for all feasible solutions of (6.9). So, from (6.10) we see that $\bar{z}$, the objective value at $\bar{x}$, is the minimum objective value in (6.9); that is, the present basic solution $\bar{x}$ is an optimum solution for the relaxed problem (6.9).

As a numerical example consider the LP discussed in Example 6.7 for which the canonical tableau wrt the dual feasible basic vector $x_B = (x_1, x_2, x_3)$ is given. The primal basic solution wrt x_B is $\bar{x} = (-6, -2, 4, 0, 0, 0)^T$ with objective value $\bar{z} = 100$.

The relaxed problem corresponding to x_B is obtained by relaxing the constraints $x_1, x_2 \geq 0$ from the original LP.

The objective equation from the last row of the canonical tableau is $z = 100 + 3x_4 + 8x_5 + 10x_6$. From this equation, as x_4, x_5, x_6 are required to be ≥ 0 in the relaxed problem, we see that the minimum objective value in it is ≥ 100. Therefore, $\bar{x}$ which is feasible to the relaxed problem with objective value 100 is an optimum solution for it.

From this we see that the objective value in any step of the dual simplex algorithm is always a lower bound for the minimum objective value in the original problem. These things show that the dual simplex algorithm approaches the optimum for an LP from outside its feasible region, with the current primal solution always remaining the best solution under the present relaxation, and improving the objective value in every step.

2. *Objective value in the algorithm is a monotonically increasing lower bound converging to the minimum in the original problem from outside:* As in the primal simplex algorithm, we define a pivot step in the dual simplex algorithm to be

- **Degenerate:** if the dual simplex minimum ratio, θ, in it is 0;
- **Nondegenerate:** if the dual simplex minimum ratio, θ, in it is >0.

From the formula discussed earlier for the change in objective value in a dual simplex pivot step, we see that it increases strictly in every nondegenerate step, and stays the same in every degenerate pivot step. Therefore, the objective value in the dual simplex algorithm is a monotonically increasing lower bound converging to the minimum in the original problem from outside.

So, if we plot the objective value of the current basic solution obtained in the various steps of the primal and dual simplex algorithms on the same LP, their paths will be as shown in Fig. 6.2, in which both the paths are shown. Of course, the number of steps that the two algorithms are not guaranteed to be the same. The path at the top is for the primal simplex algorithm that begins with a BFS and walks along BFSs only, with the objective value decreasing monotonically along the path. The path at the bottom corresponds to the dual simplex algorithm, all primal basic solutions in it are primal infeasible excepting the final one which is primal feasible and optimal, and the objective value is monotonically increasing along this path.

3. *Comparison with the primal simplex:* In an LP in standard form with a $m \times n$ coefficient matrix, typically n is much larger than m. On this problem, the dual

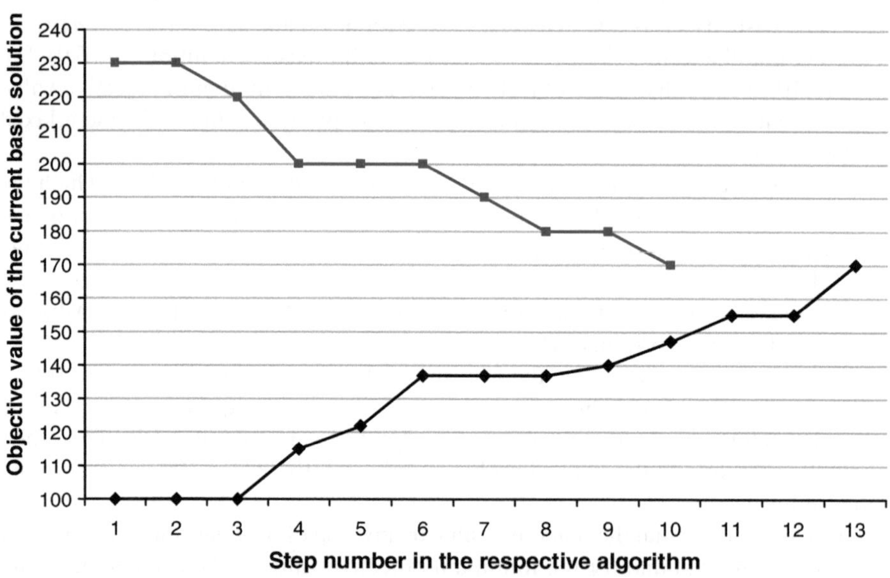

Fig. 6.2 The top line tracks the objective value in various steps of the primal simplex algorithm, and the path at the bottom does the same for the dual simplex. Primal simplex took nine steps (five nondegenerate with strict decrease in objective value, four degenerate with no change) and dual simplex took twelve steps (seven nondegenerate, five degenerate) to solve same LP with optimum objective value of 170. All primal solutions in primal simplex are primal feasible (marked with a little square), in dual simplex all but final primal solution are primal infeasible (these are marked with a little diamond)

simplex minimum ratio test may need up to $O(n)$ comparisons, which is much greater than what the primal simplex minimum ratio step may need, which can be at most $O(m)$ comparisons.

That is why when an LP has both a primal feasible basic vector and a dual feasible basic vector readily available, practitioners show a preference for solving it by the primal simplex algorithm starting with the known primal feasible basic vector. If the LP has neither a primal feasible basic vector, nor a dual feasible basic vector readily available, they prefer solving it by the primal simplex method using Phase I. If it has no primal feasible basic vector readily available, but has a readily available dual feasible basic vector, then they solve it using the dual simplex algorithm starting with that available dual feasible basic vector.

6.11 Importance of the Dual Simplex Algorithm, How to Get New Optimum Efficiently When RHS Changes or New Constraints Are Added to the Model

The dual simplex algorithm has many important applications. It is used very often for the following:

1. To get the new optimum solution when the RHS constants vector changes in an existing LP model with a known optimum basic vector
2. When new inequality constraints have to be added to a model like the above
3. As a tool in a practical method to resolve degeneracy in solving an LP by the primal simplex algorithm

We will now discuss these applications of the dual simplex algorithm. Let the original LP model be

$$\begin{aligned} \text{Minimize} \quad & z = cx \\ \text{subject to} \quad & Ax = b \\ & x \geq 0, \end{aligned} \tag{6.11}$$

where A is a matrix of order $m \times n$ and rank m. Let x_B be the known optimum basic vector for this problem with basis B and basic cost vector c_B. Let x_D denote the vector of nonbasic variables.

1. *Change in the RHS constants vector:* Suppose the RHS constants vector in the model (6.11) has changed from b to b' and there is no other change in the model.

If $B^{-1}b' \geq 0$, then x_B continues to be an optimum basic vector for the modified model associated with the new optimum BFS $(x_B, x_D) = (B^{-1}b', 0)$, optimum objective value $z' = c_B B^{-1}b'$, and the same optimum dual solution $\bar{\pi} = c_B B^{-1}$ as before the change of the RHS vector.

If $B^{-1}b' \ngeq 0$, then x_B is a dual feasible but primal infeasible basic vector for the modified model. The inverse tableau of modified model wrt x_B is obtained by changing the updated RHS vector in the current inverse tableau wrt x_B to $(x_B, -z) = (B^{-1}b', -c_B B^{-1}b')$. Starting with this, solve the modified model by the dual simplex algorithm.

Example 6.12. Consider the LP solved in Example 6.10 in Sect. 6.10. The optimum basic vector for this problem was found to be $x_B = (x_1, x_4, x_3)$, with the second inverse tableau in that example as the inverse tableau corresponding to it. The original RHS vector in this problem is $b = (8, -2, 2)^T$.

Suppose b has to be changed to $b' = (8, -4, 7)^T$, but there is no other change in the data. Then $B^{-1}b' = (0, 2, 5)^T \geq 0$. So, x_B remains primal feasible after the modification of b to b'. Hence x_B continues to be the optimum basic vector for the modified problem too. So, the optimum dual solution to the modified problem remains unchanged as $\pi = (1, 1/2, 1)$ and the optimum primal solution changes to $(x_B, -z) = \text{(inverse tableau)} \times (b', 0) = (0, 2, 5, 13)^T$.

If, on the other hand, the original RHS Vector b changes to $b'' = (6, -4, 1)^T$ instead, we have $B^{-1}b'' = (-2, 2, -1)^T \not\geq 0$. So, with this change the basic vector x_B is not primal feasible, but remains dual feasible. The objective value corresponding to the basic vector x_B is $\pi b'' = (1, 1/2, 1)(6, -4, 1)^T = 5$. So the inverse tableau wrt x_B for the modified model in this case is

Inverse tableau wrt x_B

BV	Inverse				Updated RHS
x_1	1	2	0	0	−2
x_4	0	−1/2	0	0	2
x_3	0	1/2	1	0	−1
$-z$	−1	−1/2	−1	1	−5

Starting with this inverse tableau, the modified model can be solved using the dual simplex algorithm, as before. ■

2. *Adding new constraints to the model:* In practical applications, while reviewing the optimum solution of a model for the problem, one often finds that some important constraints in the problem are missing in the model. These additional constraints have to be included and the modified model solved again.

Solving integer programming models (these are LP models with additional requirements that some variables can only have integer values in the solution) using the *branch and bound* (*B&B*) or *branch and cut* (*B &C*) algorithms is based on this strategy. In these algorithms, an *LP relaxation* (i.e., the original problem without the integer requirements on the variables) is solved first.

If the optimum solution obtained for the LP relaxation violates some integer requirements, then additional linear inequality constraints called *cuts* are generated to cut off some of the nonintegral portions of the feasible set of the LP relaxation, including its present nonintegral optimum solution, and these are included in the model and the augmented model solved. This process may have to be repeated several times before a solution for the original integer program is obtained. So solving augmented models with additional linear inequality constraints is a very important process to solve integer programs. They have applications in many other areas too.

It is possible that the new constraints to be added to the model may be equations or inequalities, but in most applications they turn out to be inequalities. That is why we discuss only that case here. The same technique can easily be adopted for adding new equality constraints to the model and get the new optimum solution.

Suppose x_B is the optimum basic vector for the original LP model in standard form. Let the basic, nonbasic partition of the original tableau wrt x_B for that problem be

Original tableau

x_B	x_D	$-z$	RHS
B	D	0	b
c_B	c_D	1	0

All $x_j \geq 0$, min z

where the original I/O coefficient matrix $(B \vdots D)$ is of order $m \times n$.

After the optimum basic vector x_B has been obtained, suppose we discover that the model has to be augmented with r additional inequality constraints: $Fx_B + Gx_D \leq g$, where $(F \vdots G)$ is of order $r \times n$ (we have expressed all the additional inequality constraints in the "$\leq$" form). Introducing slack variables $x_E = (x_{n+1}, \cdots x_{n+r})^T$, the original tableau for the modified problem is

Original tableau of augmented problem

x_B	x_D	x_E	$-z$	RHS
B	D	0	0	b
F	G	I	0	g
c_B	c_D	0	1	0

All $x_j \geq 0$, min z

Here I is the unit matrix of order r. As x_B is an optimum basic vector for the original model, (x_B, x_E) can be verified to be a dual feasible basic vector for the augmented model. It is associated with the basis whose inverse is

$$\begin{pmatrix} B & 0 \\ F & I \end{pmatrix}^{-1} = \begin{pmatrix} B^{-1} & 0 \\ -FB^{-1} & I \end{pmatrix}. \tag{6.12}$$

So, if

$$\begin{pmatrix} \bar{b} \\ \bar{g} \end{pmatrix} = \begin{pmatrix} B^{-1} & 0 \\ -FB^{-1} & I \end{pmatrix} \begin{pmatrix} b \\ g \end{pmatrix} \geq 0,$$

then (x_B, x_E) is also primal feasible to the augmented problem, and hence an optimum basic vector for the augmented problem. In this case, $(x_B, x_D, x_E) = (\bar{b}, 0, \bar{g})$ is an optimum BFS for the augmented model we are solving, so, terminate.

However, if $\bar{g} \not\geq 0$, then (x_B, x_E) is a dual feasible but primal infeasible basic vector for the augmented problem. The basis inverse corresponding to it can be obtained from (6.12) using the known basis inverse B^{-1} corresponding to x_B for the original model. Then the inverse tableau corresponding to this basic vector can be formed and the augmented problem solved, starting with it, using the dual simplex algorithm.

Example 6.13. Suppose the original LP model is the LP solved in Example 6.10 in Sect. 6.10. The optimum basic vector for this problem was found to be $x_B = (x_1, x_4, x_3)$, with the second inverse tableau in that example as the inverse tableau corresponding to it. The optimum BFS corresponding to this basic vector is $\bar{x} = (4, 0, 1, 1, 0, 0)^T$.

Suppose we need to add the new constraint $2x_1 - 2x_3 - 2x_4 + 2x_5 \leq 4$. The present optimum solution $\bar{x}$ satisfies this new constraint. Hence $\bar{x}$ remains optimal to the augmented problem.

If, on the other hand, the new constraint to be added to the model is $2x_1 - 2x_3 - 2x_4 + 2x_5 \leq 2$. Introducing the nonnegative slack variable x_7, this constraint becomes the equation

$$2x_1 - 2x_3 - 2x_4 + 2x_5 + x_7 = 2,$$

and when we plug in $\bar{x}$ into this equation, we find that the associated value of x_7 becomes $\bar{x}_7 = -2$. We will add this new constraint as the last (i.e., fourth) constraint in the model.

Negative value for the nonnegative slack variable x_7 implies that the current optimum solution $\bar{x}$ violates the new constraint to be added. From the inverse of the basis corresponding to the basic vector (x_1, x_4, x_3) for the original problem from the second inverse tableau in Example 6.10 and using the formula in (6.12), we compute the basis inverse corresponding to the basic vector (x_1, x_4, x_3, x_7) for the augmented model to be the following:

Inverse tableau

BV	Inverse					Updated RHS
x_1	1	2	0	0	0	4
x_4	0	$-1/2$	0	0	0	1
x_3	0	1/2	1	0	0	1
x_7	-2	-4	2	1	0	-2
$-z$	-1	$-1/2$	-1	0	1	-9

The vector (x_1, x_4, x_3, x_7) is a primal infeasible, but dual feasible basic vector for the augmented problem. So, starting with this inverse tableau, the augmented model can be solved using the dual simplex algorithm, as before. ■

3. *A practical method to resolve cycling under degeneracy in the simplex method:* As discussed earlier, when an LP has degenerate BFSs, the simplex method could

in theory cycle while solving it. Cycling is the name given to the phenomenon in which the simplex method moves forever among a set of degenerate basic vectors, making only degenerate pivot steps without ever changing the objective value, or the solution, or satisfying a termination condition. The theoretical possibility of cycling has been shown in specially constructed numerical examples.

Several theoretical techniques have been developed in the literature, with proofs that they will resolve the problem of cycling under degeneracy and guarantee finite convergence of the simplex method if they are used. Some of these require the basis inverse in every step, we discussed earlier that this is difficult to maintain in practice because of the problems of fill-in, roundoff errors, and excessive computer time for each step. Other techniques that do not need the basis inverse for resolving cycling have been found to take far too many steps before satisfying the termination criteria. For this reason, none of the theoretical techniques for resolving cycling are in practical use.

In numerical computation with finite precision on digital computers, when the minimum ratio in a pivot step turns out to be 0, it is hard to tell whether it is really 0, or appears to be so due to accumulated round-off errors. But one often notices that the simplex method goes through several steps in which there is negligible or no change in the objective value. This kind of occurrence is considered as a manifestation of cycling by programmers. To speed up the code when this occurs, many of the software vendors whom I talked to told me that they normally use the following strategy.

Suppose the LP is being solved by the primal simplex algorithm and a degenerate BFS is encountered for the first time in this work. At that time they perturb the updated RHS constants by small amounts so that the values of all the basic variables in the BFS become strictly positive. Continuing with the modified RHS, the next pivot step will be nondegenerate, resulting in a strict decrease in objective value. If the same phenomenon occurs again, repeat a similar process with additional perturbations of the RHS constants. Eventually when termination occurs, check the current primal solution for primal feasibility to the original model. If it satisfies it to within reasonable tolerance, it is accepted as the final solution of the problem. Otherwise, at that stage the current RHS vector is replaced by the original, and the program reverts to the dual simplex algorithm to continue solving the original model starting with the present dual feasible, but primal infeasible basic vector. The dual simplex algorithm is applied with the same kind of perturbation (now of the cost vector) to resolve cycling in it. Again at the end of the dual simplex, one may have to revert to primal simplex and continue the same way. Software vendors told me that after at most two repetitions of this, they converge to an optimum solution within reasonable tolerance and that each of these repetitions usually takes only very few steps.

4. *The primal simplex and the dual simplex as a pair of tools for decision making using LP models:* The primal simplex and the dual simplex as a pair of tools offer a complete tool kit to efficiently handle many types of changes that occur in LP models in applications. Some of the other sensitivity features are discussed in a

later section in this chapter. The development of the dual simplex algorithm and its pairing with the primal simplex algorithm has helped a lot to make LP a powerful and valuable tool to model problems in applications.

6.11.1 The Dual Simplex Method

If an initial primal feasible basic vector for an LP in standard form is not available, to solve it the primal simplex method carries out a Phase I using additional artificial variables, to generate a primal feasible basic vector to initiate the primal simplex algorithm on it in Phase II.

In the same way to solve LP by the dual simplex approach, if an initial dual feasible basic vector is not available, then a dual simplex Phase I to generate a dual feasible basic vector for the original problem has been developed. While the primal simplex approach introduces artificial variables into the model to set up a Phase I with its primal feasible basic vector readily available, the dual simplex approach introduces an artificial constraint to set up a dual simplex Phase I with its dual feasible basic vector readily available.

Once a dual feasible basic vector for the original LP is found, starting with that, the dual simplex Phase II takes over to solve the original LP using the dual simplex algorithm. The combined dual simplex Phase I and II approach is called the *Dual simplex method*; it can solve any LP whether a dual feasible basic vector is readily available or not.

However, to solve a general LP model without known special structure, when neither a primal feasible nor a dual feasible basic vector is readily available, practitioners always seem to prefer solving it by the primal simplex method. For this reason, the dual simplex Phase I is never used in practical applications. For this reason, it seems to be only of theoretical interest, and hence we will not discuss it in this book. The reader is referred to references Dantzig (1963); Dantzig and Thappa (1997) of Chap. 1; Murty (1983) of Chap. 2) for details of the dual simplex method.

6.12 Marginal Analysis

We have seen earlier that each constraint in an LP model for the operations of a system can be interpreted as the material balance equation or inequality associated with an item. From Sect. 5.8, we know that in general, there may be two quantities associated with an item; one called the *positive marginal value* (rate of change in the optimum objective value in the model, per unit *increase* in the RHS constant in the constraint corresponding to the item from its present level), and a *negative marginal value* (same rate as above, but per unit *decrease*). These values are called the marginal values of the item, or of the constraint corresponding to it, or of the

RHS constant in that constraint. Marginal analysis is economic analysis of the various options available to the system based on these marginal values.

When the optimum dual solution for the LP is unique, for each item both the positive and negative marginal values are equal to the value of the dual variable associated with that item in the optimum dual solution; this common value is defined to be the *marginal value of that item, or of the constraint corresponding to it, or of the associated RHS constant in the model.* In this case we say that marginal value is well defined, and for each item its marginal value is the rate of change in the optimum objective value in the model per unit change (positive or negative) in the RHS constant in the constraint corresponding to the item, from its present level. We know that this case occurs if the LP model has a nondegenerate optimal primal BFS. In this case, marginal analysis is simple and nice, and is based on the unique optimum dual solution as the marginal value vector. In Sect. 3.13 in Murty (2005b) of Chap. 1) an example (the fertilizer manufacturer's problem) of this case has been discussed with several planning applications of marginal analysis.

If the LP model does not have a nondegenerate optimal primal BFS, then it may have alternate optimum dual solutions, and in this case the positive and negative marginal values for an item may be different (see Sect. 1.7.1, and Exercise 6.73). In this case, to be correct, marginal analysis has to be based on these positive and negative marginal values. Efficient methods for computing both the positive and negative marginal values do exist and are discussed in Murty (1983) of Chap. 2; they require extra computation on the dual problem. However, these methods are not known widely. None of the commercially available LP software packages have subroutines for them, and practitioners have never demanded that positive and negative marginal values be provided to them in this case.

In this book we consider only marginal analysis in the simpler case when the LP model has a nondegenerate optimal primal BFS, using the unique optimum dual solution as the vector of marginal values.

Example 6.14. Marginal analysis at a company producing three products using four processes. As an example, consider a company that needs three products for its internal use. There are four different processes that the company can use to make these products. When a process is run, it may produce one or more of these products as indicated in the following table. Here b_i = minimum daily requirement for product P_i in units, for $i = 1$ to 3; and c_j = cost (in \$) of running process/unit time.

Product	Output /unit time of				
	Process 1	2	3	4	b_i
P_1	1	2	0	1	17
P_2	2	5	1	2	36
P_3	1	1	0	3	8
c_j	28	67	12	35	

For $j = 1$ to 4, let x_j denote the units of time that process j is run daily. Let x_5, x_6, x_7 denote the slack variables corresponding to P_1, P_2, P_3 (these are the amounts of the product produced in excess of the minimum daily requirement).

Then the model for meeting the requirements of the products at minimum cost is the following LP in standard form.

Tableau: Original tableau

Item	x_1	x_2	x_3	x_4	x_5	x_6	x_7	$-z$	b
P_1	1	2	0	1	−1	0	0	0	17
P_2	2	5	1	2	0	−1	0	0	36
P_3	1	1	0	3	0	0	−1	0	8
	28	67	12	35	0	0	0	1	0

all $x_j \geq 0$; x_5, x_6, x_7 are P_1, P_2, P_3 slacks; minimize z

This problem has been solved by the revised simplex method, yielding the following optimum inverse tableau.

Tableau: Inverse tableau

Basic var.	Inverse tableau				Basic values
x_1	5	−2	0	0	13
x_2	−2	1	0	0	2
x_7	3	−1	−1	0	7
$-z$	−6	−11	0	1	−498

x_7 is P_3 slack

So, the optimum solution is to run processes 1, 2 for 13, 2 units of time daily. This solution attains the minimum cost of \$498, and produces 17, 36, and 15 units of P_1, P_2, and P_3, respectively, meeting the minimum daily requirements of P_1, P_2, exactly, but leaving an excess of seven units of P_3 after meeting its requirement. As the optimal primal BFS is nondegenerate, the vector of marginal values of P_1, P_2, and P_3 is the optimum dual solution $= (6, 11, 0)$.

So, the marginal value of P_3 is 0. This means that small changes in its daily requirement in the neighborhood of its present value of eight units, does not change the cost to the company of meeting its requirements for these three products. At the moment the requirement of P_3 is automatically covered while meeting the requirements of P_1 and P_2, and this actually produces an excess of seven units of P_3 beyond its requirement.

P_2 has the highest marginal value of \$11 among the three products. This means that small changes in its requirement from its present level of 36 units result in a change in the optimum cost at the rate of \$11/unit. And if an outside supplier were to offer to supply P_2, it is worth considering if the rate is $\leq$\$11/unit. As it has the highest marginal value, P_2 is a critical input for the company.

A similar interpretation can be made for P_1 and its marginal value of \$6/unit.

Suppose the company's research lab has come up with a new process, process 5, that produces P_1 and P_2, at the rate of 4 and 9 units per unit time it is run, and does not produce any P_3. Let \$$c_8$ be the cost of running process 5 per unit time. For what values of c_8 is it desirable to run process 5? To answer this question, we evaluate the monetary benefit, in terms of the marginal values, of the output by running this process per unit time. As it is 4 and 9 units of P_1 and P_2, respectively, and the marginal values of P_1 and P_2 are 6 and 11, this monetary benefit is $4\times 6+9\times 11 =$ \$123/unit time. Comparing this with the cost c_8 of running this process we conclude that process 5 is not worth running if $c_8 > 123$; it breaks even with the present optimum solution if $c_8 = 123$; and can save cost if $c_8 < 123$. ∎

Example 6.15. Evaluating the profitability of new products. One major use of marginal values is in evaluating the profitability of new products. It helps to determine whether they are worth manufacturing, and if so at what level they should be priced so that they are profitable in comparison with existing product lines.

We will illustrate this again using the company discussed in the previous example (Example 6.14) that produces products $P_1,\ P_2,$ and P_3 for its internal use with processes 1, 2, 3, and 4.

Suppose the company's research lab has come up with a novel product, code named "NP," that is expected to have good market potential. Its manufacture requires as inputs 1, 2, 1 units of $P_1,\ P_2,\ P_3$, respectively/unit and incurs other costs (in labor, other inputs, etc.) amounting to \$50/unit.

What is the break-even price/unit for selling this NP in the market?

Using the marginal values of $P_1,\ P_2,\ P_3$, derived in Example 6.14, we find that a packet consisting of 1, 2, 1 units of $P_1,\ P_2,\ P_3$ respectively, is equivalent to $1\times 6+2\times 11+1\times 0 =$ \$28 in costs for running the processes to produce these inputs.

So the break-even cost for NP is $(28+50)=78$/unit. Hence the break-even market price for NP is \$78/unit. If the company's expectation is to have a profit margin of $\geq 20\%$, NP is worth manufacturing if it can be sold at a price $\geq \$(1.2)(78) =$ 93.6/unit.

The company can conduct a market survey and determine whether the market will accept NP at this price level. Once this is known, the decision whether to bring NP into production becomes clear. ∎

Marginal analysis is this kind of cost-benefit analysis using the marginal values. It provides very valuable planning information.

It has become a very common practice with practitioners to use this kind of analysis using an optimum dual solution provided by the simplex method, even when the optimal primal solution is degenerate. As pointed out at the beginning of this section, this may lead to wrong conclusions in this case, so one should watch out.

6.12.1 How to Compute the Marginal Values in a General LP Model

The model for a real-world problem may be a general LP model that may not be in standard form originally. Corresponding to each constraint in this model, there will be a dual variable in its dual, which can be interpreted as the marginal value of the RHS constant in that constraint; it is the rate of change in the optimum objective value in the model per unit change in the RHS constant in that constraint.

To solve this model, we first put it in standard form, which is in the form of a minimization problem involving only equality constraints in nonnegative variables. In this process constraints are modified, some may even be eliminated if the original model contains some unrestricted variables.

To distinguish one from the other, let (P) refer to the original model with the objective function and all the constraints as they were stated originally. Let (P') refer to the equivalent LP in standard form *in minimization form* obtained by transforming (P). Suppose (P') has a nondegenerate primal BFS. Then all the marginal values for it are well defined. Knowing the marginal values for (P'), how can we compute the marginal values for the original model (P)? We provide an answer to this question here. Let x_B denote an optimum basic vector for (P'). For (P') we assume that the relative cost coefficients wrt x_B of all nonbasic variables are computed. By definition, the relative cost coefficient of a basic variable is 0.

First suppose there are no unrestricted variables in the original model (P). Inequality constraints in (P) were transformed into equations by introducing appropriate slack variables in the process of getting (P').

The marginal value associated with any "$\geq$" ["$\leq$"] inequality constraint in (P) is the relative cost coefficient of the corresponding slack variable in (P') if the objective function in (P) is in minimization [maximization] form, or its negative if the objective function in (P) is in maximization [minimization] form.

The marginal value associated with any equality constraint in (P) is the same as the marginal value associated with this constraint in (P') if the objective function in (P) is in minimization form, or its negative if the objective function in (P) is in maximization form.

It is possible that there are still some equality constraints in (P) that were used to eliminate the unrestricted variables, marginal values associated with which are not determined by the above rules. If so, go back to the process that transformed the original problem (P) into (P'). In this process, first all the inequality constraints in (P) other than the nonnegativity restrictions on individual variables have been transformed into equations by introducing appropriate slack variables. This is just before eliminating unrestricted variables in the model by performing pivot steps in their columns. When you augment the optimal basic vector x_B for (P'), with all the unrestricted variables in the model at this stage, you get a vector x_F say, which will be a basic vector for the system of equality constraints at this stage. Compute the dual basic solution associated with x_F for this system. The values of the dual variables associated with the eliminated constraints used to eliminate the unrestricted variables to get (P'), in this dual solution, are the marginal values corresponding to these constraints in (P).

6.13 Sensitivity Analysis

In applications of linear programming, the data such as the input–output coefficients, cost coefficients, and RHS constants are normally estimated from practical considerations and may have unspecified errors in them. Given an optimum basic vector for the LP model, the *optimality range* of a data element is the interval within which that element can vary, *when all the other data remain fixed at their current values*, while keeping the present solution or basic vector optimal. In sensitivity analysis there is a technique called *ranging* that determines the optimality range of some of the data very efficiently. The width of the optimality range and the position of the present value of the data element in this range can be used to check the robustness of the present optimum solution or optimum basic vector to possible errors in that data element.

Sensitivity analysis is the collection of simple applications of the optimality conditions for LP to find the optimality ranges for various coefficients in the LP original tableau, and also efficient techniques for finding a new optimum solution beginning with the current one, if a change occurs in the value of one coefficient in the model outside its optimality range. Ranging and these other techniques in sensitivity analysis are all based on the optimality criteria. Here we discuss some sensitivity analysis techniques that proved to be useful in practice.

6.13.1 Introducing a New Nonnegative Variable

Consider solving the original LP

$$\begin{aligned} \text{Minimize} \quad z(x) &= cx \\ \text{subject to} \quad Ax &= b \\ x &\geq 0, \end{aligned} \tag{6.13}$$

where without any loss of generality we assume that A is a matrix of order $m \times n$ and rank m. Here is the original tableau for this problem

Original tableau

x	$-z$	
A	0	b
c	1	$-\alpha$

Suppose we found an optimum basic vector x_B associated with the optimum primal BFS $\bar{x} = (\bar{x}_1, \cdots, \bar{x}_n)^T$ and optimum dual solution $\bar{\pi} = (\bar{\pi}_1, \cdots, \bar{\pi}_m)$ for this original problem.

After solving the original model (6.13), suppose a new activity has become available. Let x_{n+1} denote the decision variable corresponding to it; it is the level at which this new activity is to be carried out. Suppose the original column vector of x_{n+1} is $A_{.n+1}$ and its original cost coefficient is c_{n+1}.

The following questions need to be answered at this stage. Is it worth performing the new activity at a positive level? If so, how can we find the new optimum solution beginning with the inverse tableau of x_B for the original (6.13)? If it is not worth performing the new activity at a positive level, how much should its cost coefficient change before it becomes worthwhile?

Enter the new variable x_{n+1}, with its column vector into the original tableau given above. We will refer to the modified tableau as the *augmented tableau.* From the optimality criterion in the simplex algorithm, and the fact that the basic vector x_B satisfied it before introducing x_{n+1} into the problem, we conclude that x_B remains optimal to the augmented problem if the relative cost coefficient of x_{n+1}, namely $\bar{c}_{n+1} = c_{n+1} - \bar{\pi} A_{.n+1}$, is ≥ 0. Thus if $c_{n+1} \geq \bar{\pi} A_{.n+1}$, it is not worth performing the new activity. In this case $(\bar{x}_1, \cdots, \bar{x}_n, \bar{x}_{n+1} = 0)^T$ is an optimum solution of the augmented problem.

For the new activity to be worth performing, $\bar{c}_{n+1}$ has to be <0, that is, c_{n+1} has to decrease below $\bar{\pi} A_{.n+1}$.

If $\bar{c}_{n+1} < 0$, then x_B does not satisfy the optimality criterion for the augmented problem, and x_{n+1} is the variable eligible to enter it. To get an optimum solution of the augmented problem in this case, select x_{n+1} as the entering variable into x_B and continue the application of the simplex algorithm until it terminates again.

As an example, consider the LP model in Sect. 6.12 of the company trying to produce the required quantities of P_1, P_2, P_3 using four available processes at minimum cost. The optimum basic vector for this problem is $x_B = (x_1, x_2, x_7)$ and its inverse tableau is given in Sect. 6.12.

As discussed in Sect. 6.12, the new process 5 developed by the company's research lab has become available. Let x_8 denote the level at which this process is operated. In Sect. 6.12, the column of x_8 in the model is given to be $(4, 9, 0, c_8)^T$, where c_8 is the cost coefficient. The relative cost coefficient of x_8 wrt x_B is

$$\bar{c}_8 = (-6, -11, 0, 1)(4, 9, 0, c_8)^T = c_8 - 123.$$

If $\bar{c}_8 \geq 0$, that is, $c_8 \geq 123$, it is not worth operating this process, and the present BFS wrt the basic vector (x_1, x_2, x_7) given in Sect. 6.12 remains optimum with $x_8 = 0$.

If $\bar{c}_8 < 0$, that is, $c_8 < 123$, then x_B does not satisfy the optimality criterion for the augmented problem. Suppose $c_8 = 122$. With this value for c_8, here is the original tableau for the augmented problem.

Original tableau

x_1	x_2	x_3	x_4	x_5	x_6	x_7	x_8	$-z$	b
1	2	0	1	−1	0	0	4	0	17
2	5	1	2	0	−1	0	9	0	36
1	1	0	3	0	0	−1	0	0	8
28	67	12	35	0	0	0	122	1	0

all $x_j \geq 0$; x_5, x_6, x_7 are P_1, P_2, P_3 slacks; minimize z

Here is the inverse tableau of $x_B = (x_1, x_2, x_7)$ with the updated column of x_8 selected as the pivot column.

Basic var.	Inverse tableau				Basic values	PC x_8	Ratios
x_1	5	−2	0	0	13	2	13/2
x_2	−2	1	0	0	2	**1** (boxed)	2/1
x_7	3	−1	−1	0	7	3	7/3
$-z$	−6	−11	0	1	−498	−1	Min. $= \theta = 2$
x_1	9	−4	0	0	9		
x_8	−2	11	0	0	2		
x_7	9	−4	−1	0	1		
$-z$	−8	−10	0	1	−496		

PC = pivot column

In the new basic vector, the relative cost coefficients of the nonbasic variables are $(\bar{c}_2, \bar{c}_3, \bar{c}_4, \bar{c}_5, \bar{c}_6) = (1, 2, 7, 8, 10) > 0$. So, this basic vector is optimal to the augmented problem. The optimum BFS is $\hat{x} = (9, 0, 0, 0, 0, 0, 1, 2)^T$. It operates process 1 for nine units of time, and the new process 5 (corresponding to x_8) for two units of time daily, and has a cost of \$496.

6.13.2 *Ranging the Cost Coefficient or an I/O Coefficient in a Nonbasic Column Vector*

Let (6.13) be the original model for which we have an optimum basic vector x_B associated with the optimum primal BFS $\bar{x} = (\bar{x}_1, \cdots, \bar{x}_n)^T$ and optimum dual solution $\bar{\pi} = (\bar{\pi}_1, \cdots, \bar{\pi}_m)$.

Suppose x_j is a nonbasic variable now whose cost coefficient c_j is likely to change, while all the other data remain fixed at present levels. For what range of values of c_j does $\bar{x}$ remain an optimum solution to the problem?

To answer this question, notice that a change in c_j does not affect the primal or dual basic solutions associated with x_B nor does it affect the primal feasibility of $\bar{x}$. However, for $\bar{\pi}$ to remain dual feasible, we need $\bar{c}_j = c_j - \bar{\pi} A_{.j} \geq 0$, that is, $c_j \geq \bar{\pi} A_{.j}$. So, $\bar{x}$ remains an optimum solution to the problem as long as $c_j \geq \bar{\pi} A_{.j}$, this is the optimality range for c_j.

If the new value of c_j is $< \bar{\pi} A_{.j}$, then $\bar{c}_j < 0$, and the basic vector x_B is no longer dual feasible. In this case, x_j is eligible to enter x_B. To get the new optimum solution, correct the value of c_j in the original tableau, bring x_j into the basic vector x_B, and continue the application of the primal simplex algorithm until it terminates again.

As an example, consider the LP model of the company trying to produce the required quantities of P_1, P_2, P_3 using four available processes at minimum cost, discussed in Sect. 6.12. The optimum basic vector for this problem is $x_B = (x_1, x_2, x_7)$. Its inverse tableau is given in Sect. 6.12. Suppose the cost coefficient of x_4, the cost of running process 4 per unit time, is likely to change from its present value of \$35, while all the other data remains fixed. Denote the new value of this cost coefficient by c_4. For what range of values of c_4 does the primal BFS wrt the present basic vector (x_1, x_2, x_7) remain optimal to the problem?

The answer: As long as the relative cost coefficient of x_4, $\bar{c}_4 = (-6, -11, 0, 1)$ $(1, 2, 3, c_4)^T = c_4 - 28$ is ≥ 0, that is, as long as $c_4 \geq 28$. This is the optimality range for c_4.

If the new value of c_4 is <28, say $c_4 = 27$, the basic vector (x_1, x_2, x_7) is no longer dual feasible. x_4 is eligible to enter this basic vector. To get the new optimum solution, correct the original cost coefficient of x_4 to its new value of 27, bring x_4 into the basic vector (x_1, x_2, x_7), and continue the application of the primal simplex algorithm until it terminates again.

Now consider changes in an input–output coefficient in a nonbasic column. These coefficients may change due to changes in the corresponding process or changes in technology. Again consider the LP (6.13) and an optimum basic vector x_B associated with the primal and dual optimum solutions, $\bar{x}$, $\bar{\pi} = (\bar{\pi}_1, \dots, \bar{\pi}_m)$ for it. Let x_j be a nonbasic variable. Its column in the original tableau is $A_{.j} = (a_{1j}, \dots, a_{mj})^T$. Suppose the coefficient a_{rj} in this column is likely to change while all the other data remains fixed at present levels.

Here also, the optimality range for a_{rj} is the set of all values of this coefficient that keeps the relative cost coefficient of x_j, $\bar{c}_j = c_j - \bar{\pi} A_{.j} = c_j - \left(\sum_{\substack{i=1 \\ i \neq r}}^{m} \bar{\pi} a_{ij}\right) - \bar{\pi}_r a_{rj}$ nonnegative.

In the expression for $\bar{c}_j$ here, the only thing that varies is a_{rj}, and all other entries are at their present values. So

If $\bar{\pi}_r = 0$, $\bar{c}_j$ remains at its present nonnegative value whatever a_{rj} is, so the optimality range for a_{rj} is the entire real line. If $\bar{\pi}_r > 0$, the optimality range for a_{rj} is

$$a_{rj} \leq \left(c_j - \left(\sum_{\substack{i=1 \\ i \neq r}}^{m} \bar{\pi} a_{ij}\right)\right) / \bar{\pi}_r,$$

and if $\bar{\pi}_r < 0$, this optimality range is

$$a_{rj} \geq -(c_j - (\sum_{\substack{i=1 \\ i \neq r}}^{m} \bar{\pi} a_{ij}))/\bar{\pi}_r.$$

If the new value for a_{rj} is outside its optimality range, x_j becomes eligible to enter x_B. To get the new optimum solution, bring it in and continue the application of the simplex algorithm until it terminates again.

As an example, again consider the LP model discussed in Sect. 6.12. Suppose by making changes in process 4, we can improve the output of P_3 from this process from its present value of 3 units/unit time. Will this change the optimum solution for the problem? The answer is no, as $\bar{\pi}_3 = 0$ in the dual optimum solution. The explanation for this is not hard to see. The present primal optimum solution that consists of running processes 1, 2 (corresponding to primal variables x_1, x_2) at 13, 2 units of time daily already produces an excess of seven units of P_3 over its requirement, and process 4 will not become economical even if the output of P_3 from it changes.

In the same example, suppose by making changes in process 4 we can improve the output of P_1 from it. Does this change the optimum solution for this problem, and if so how high should this output be before it does? Let a_{14} denote the new value of the output of P_1/unit time of process 4. With this value, the relative cost coefficient of x_4 is $\bar{c}_4 = (-6, -11, 0, 1)(a_{14}, 2, 3, 35)^T = 13 - 6a_{14}$; $\bar{c}_4 \geq 0$ if $a_{14} \leq 13/6$; this is the optimality range for a_{14}. If $a_{14} > 13/6$, the basic vector (x_1, x_2, x_7) is no longer dual feasible to the problem, and x_4 becomes eligible to enter it. To get the new optimum solution, compute the updated column of x_4 with the new value of a_{14} and bring it in, and continue the application of the primal simplex algorithm until it terminates again.

Note on real-world applications of LP: In practical applications, the coefficients in the LP model are estimated from data, and the values of these coefficients like cost coefficients are changing constantly over time. So, when a model is constructed based on the value of this coefficient at some point of time, practitioners realize that in future the value of this coefficient may differ from the value for it used in the model.

So, one worry is: how robust is the solution obtained from the model to these changes taking place continuously over time? Estimating the optimality range for a coefficient for optimality of the current solution is of great help in assessing this robustness. If the interval is wide, and the current value of the coefficient is far away from the bounds of this interval like, for example, if the interval is 1–10 and the current value of the coefficient is 5, then one feels confident that the current optimum solution is likely to remain optimum to the problem even with slight variations in the coefficient from present value. On the other hand if current value of coefficient is close to one of the bounds on this interval, then the value of this coefficient has to be monitored carefully, as small changes in it may lead to a new optimum solution.

6.13.3 Ranging a Basic Cost Coefficient

Let (6.13) be the original model for which we have an optimum basic vector x_B associated with the optimum primal BFS $\bar{x} = (\bar{x}_1, \dots, \bar{x}_n)^T$ and optimum dual solution $\bar{\pi} = (\bar{\pi}_1, \dots, \bar{\pi}_m)$.

Suppose the cost coefficient of a basic variable in x_B, x_1 say, is likely to change, while all the other data remains fixed at present levels. Denote the cost coefficient of x_1 by c_1. For what range of values of c_1 does $\bar{x}$ remain optimal to the problem?

The simple technique used in Sect. 6.13.2 for ranging a nonbasic cost coefficient cannot be used here, because c_1 is now a basic cost coefficient, and any change in it results in a change of the dual basic solution associated with x_B.

Let c'_B denote the row vector of cost coefficients of basic variables in x_B, with that of x_1 recorded as c_1 and that of all others equal to their present values. Then the dual basic solution associated with x_B, as a function of c_1, is $\pi(c_1) = c'_B B^{-1}$, where B^{-1} is the inverse of the basis corresponding to x_B. B^{-1} can be read from the inverse tableau for x_B.

For nonbasic variable x_j, its relative cost coefficient as a function of c_1 is $\bar{c}_j(c_1) = c_j - \pi(c_1)A_{.j}$. Express the condition that each of them should be ≥ 0. This identifies an interval for c_1, which is its optimality range.

Suppose the new value of c_1 is outside its optimality range. Correct the value of c_1 in the original tableau and the dual solution in the last row of the inverse tableau of x_B. Now some nonbasic variables must be eligible to enter x_B. Select one of them to enter x_B and continue applying the primal simplex algorithm until it terminates again.

As an example, consider the optimum inverse tableau associated with the basic vector (x_1, x_2, x_7) for the LP discussed in Sect. 6.12. Suppose the cost coefficient of the basic variable x_1 is likely to change from its present value of 28. Denote its new value by c_1. The dual basic solution of the basic vector $x_B = (x_1, x_2, x_7)$ as a function of c_1 is

$$\pi(c_1) = (c_1, 67, 0) \begin{pmatrix} 5 & -2 & 0 \\ -2 & 1 & 0 \\ 3 & -1 & -1 \end{pmatrix} = (5c_1 - 134, 67 - 2c_1, 0).$$

So, the relative cost coefficient of a nonbasic variable x_j is the dot product of $(-\pi(c_1), 1)$ with the column of x_j in the original tableau. These are $(\bar{c}_3(c_1),$ $\bar{c}_4(c_1), \bar{c}_5(c_1), \bar{c}_6(c_1)) = (2c_1 - 55, 35 - c_1, 5c_1 - 134, 67 - 2c_1)$. To keep all these relative cost coefficients ≥ 0, c_1 must satisfy $55/2 \leq c_1 \leq 67/2$, this is the optimality range for c_1.

If the new value of c_1 falls outside its optimality range, x_B is no

Original tableau

x_1	x_2	x_3	x_4	x_5	x_6	x_7	$-z$	b
1	2	0	1	−1	0	0	0	17
2	5	1	2	0	−1	0	0	36
1	1	0	3	0	0	−1	0	8
27	67	12	35	0	0	0	1	0

$x_j \geq 0$ for all j; minimize z

Inverse tableau

Basic var.	Inverse tableau				Basic values
x_1	5	−2	0	0	13
x_2	−2	1	0	0	2
x_7	3	−1	−1	0	7
$-z$	−1	−13	0	1	−485

longer dual feasible. Suppose this new value is 27. So, the new dual solution corresponding to x_B is $\pi(27) = (1, 13, 0)$. The original tableau for the modified problem and the corrected inverse tableau wrt x_B are given above.

The relative cost coefficient of x_3 now $= \bar{c}_3(27) = -1$, so x_3 is eligible to enter this basic vector. Select x_3 as the entering variable, and continue applying the primal simplex algorithm until it terminates again.

6.13.4 Ranging the RHS Constants

Let (6.13) be the original model for which we have an optimum basic vector x_B associated with the optimum primal BFS $\bar{x}$ and optimum dual solution $\bar{\pi}$.

Suppose one of the RHS constants, b_1, is changing, while all the other data remains fixed at present levels. When b_1 changes, the primal basic solution corresponding to x_B will change with it. So, in this case, the optimality range for b_1 is defined to be the set of values of b_1 for which the basic vector x_B remains optimal to the problem, even though the actual optimal primal BFS changes with b_1 in this interval.

A change in b_1 does not affect the dual feasibility of x_B, but it does affect its primal feasibility. So, x_B remains optimal as long as it is primal feasible. Let b' denote the new RHS constants vector with b_1 left as a parameter, but all the other entries being equal to their current values. Then, the basic vector x_B remains primal feasible if $B^{-1}b' \geq 0$, where B^{-1} is the basis inverse corresponding to x_B read from the inverse tableau. Each entry in $B^{-1}b'$ is an affine function of b_1, express the condition that they must all be ≥ 0. This determines an interval for b_1, which is its optimality range. As long as b_1 is in this range, the optimum solutions are

$$\text{Primal optimum} \begin{cases} \text{all nonbasic variables} = 0 \\ \text{basic vector } x_B = B^{-1}b', \end{cases}$$

$$\text{Optimum dual solution} = \bar{\pi},$$

$$\text{Optimum objective value} = \bar{\pi}b'.$$

Thus in this range, the optimum dual solution remains unchanged, but the basic variables in the primal optimum BFS, and the optimum objective value are linear functions of b_1.

As an example consider the optimum inverse tableau associated with the basic vector (x_1, x_2, x_7) for the LP model discussed in Sect. 6.12. Suppose the first RHS constant in this model, the minimum daily requirement for P_1, is likely to change from its current value of 17. Denote the new value by b_1. As a function of b_1, the primal basic solution wrt the basic vector (x_1, x_2, x_7) is

$$\text{Nonbasic variables } x_3, x_4, x_5, x_6 = 0,$$

$$\text{Basic vector} \begin{pmatrix} x_1 \\ x_2 \\ x_7 \end{pmatrix} = B^{-1}b' = \begin{pmatrix} 5 & -2 & 0 \\ -2 & 1 & 0 \\ 3 & -1 & -1 \end{pmatrix} \begin{pmatrix} b_1 \\ 36 \\ 8 \end{pmatrix}$$

$$= \begin{pmatrix} 5b_1 & -72 \\ -2b_1 & +36 \\ 3b_1 & -44 \end{pmatrix}.$$

For this basic solution to be primal feasible, we need $5b_1 - 72 \geq 0$, which implies $b_1 \geq 72/5$; $36 - 2b_1 \geq 0$, which implies $b_1 \leq 18$; and $3b_1 - 44 \geq 0$, which implies $b_1 \geq 44/3$. The set of values of b_1 satisfying all these conditions is the interval $\max\{72/5, 44/3\} = 44/3 \leq b_1 \leq 18$. This is the optimality range for b_1; for b_1 in this range, the optimum solution of the problem is the one given above, the optimum dual solution is $\bar{\pi} = (6, 11, 0)$ from the inverse tableau, and the optimum objective value is $\bar{\pi} b' = 6b_1 + 396$.

If the new value for b_1 is outside its optimality range, then x_B is a dual feasible but primal infeasible basic vector for the modified model. The inverse tableau of modified model wrt x_B is obtained by changing the updated RHS vector in the current inverse tableau wrt x_B to the correct one corresponding to the new value of b_1. Starting with this, solve the modified model by the dual simplex algorithm as discussed in Sect. 6.10.

6.13.5 Features of Sensitivity Analysis Available in Commercial LP Software

Suppose an LP model is solved using a software package. When the model has an optimum solution, almost all commercially available LP software packages based on the simplex method will output: a list of the basic variables in the final optimum basic vector, together with their values in the primal optimum BFS; the corresponding dual optimum solution; the relative cost coefficients of nonbasic variables; and usually the optimality ranges of all the cost coefficients and RHS constants.

6.13.6 Other Types of Sensitivity Analyses

Many other types of sensitivity analysis techniques are discussed in other LP textbooks, but those discussed in this section and in Sects. 6.11 and 6.12 are the only ones popular with practitioners using LP models for decision making. So, we will not discuss the others.

All sensitivity analysis techniques are based on the optimality conditions for linear programs, so if the reader encounters the need for a new type of sensitivity analysis, she/he can solve their problem using the optimality conditions.

6.14 Revised Primal Simplex Method for Solving Bounded Variable LP Models

So far we have been discussing the primal simplex method for solving LP models in standard form in which the only bound restrictions on the variables are the nonnegativity conditions, which are lower bound conditions. But in many real-world applications of LP, the variables are subject to both lower and upper bound restrictions. The way we have been handling these twin bound restrictions on individual variables, is to transform the lower bound condition into the nonnegativity restriction, and include the upper bound restriction as a linear equality constraint in the model by introducing the appropriate slack variable, a clumsy procedure that increases the order of the basis we have to deal with.

In this section, we will discuss the *bounded variable primal simplex method*, which handles the bound restrictions on individual variables separately from other constraints, each of which involve two or more variables. In this version, all constraints involving two or more variables are converted into equations by introducing the appropriate slack variables, and pivot operations are performed only on this system of equations.

Let the bound restrictions on the decision variable x_j be $\ell_j \leq x_j \leq k_j$. If $\ell_j = -\infty$, $k_j = +\infty$, then x_j is an unrestricted variable in the model. When there are such unrestricted variables in the model, it is possible for the set of feasible solutions to be nonempty, but have no extreme points. As an example for this, let K be the set of feasible solutions of the system in two variables:

- $x_1 + x_2 \leq 2,\ x_1 + x_2 \geq 1$
- In standard form this corresponds to the system: $x_1+x_2+x_3 = 2,\ x_1+x_2-x_4 = 1,\ x_3,\ x_4 \geq 0, x_1,\ x_2$ unrestricted.

The set K is the nonempty unbounded region in R^2 between the parallel lines defined by the two equations $x_1 + x_2 = 1$ and $x_1 + x_2 = 2$. Clearly, K has no extreme points.

However, in applications we encounter only models in which the lower bounds ℓ_j are all finite, so we will consider only this case. In this case, it can be shown easily that if there is a feasible solution, then there is an extreme point solution, and if the LP has an optimum solution, there must be an extreme point optimum. Also, by a simple transformation, we can transform all the lower bounds on individual variables to 0. So, we take the standard form for the bounded variable LP to be

$$\begin{aligned} \text{Minimize} \quad & z(x) = cx \\ \text{subject to} \quad & Ax = b \\ & 0 \leq x \leq u, \end{aligned} \tag{6.14}$$

where A is an $m \times n$ matrix, whose rank we will assume $= m$ without loss of generality, and $u > 0$. Some of the entries in $u = (u_i)$ may be $= \infty$.

Consider the case where u is finite. In this case, by introducing slack variables, we can express the upper bound constraints $x \leq u$ as $x + s = u,\ s \geq 0$, and convert this into the standard form of LP discussed till now in variables $x,\ s$ as

$$\text{Minimize} \quad z(x) = cx$$

$$\text{subject to} \quad \begin{pmatrix} A & \vdots & 0 \\ \cdots & & \cdots \\ I & \vdots & I \end{pmatrix} \begin{pmatrix} x \\ \cdots \\ s \end{pmatrix} = \begin{pmatrix} b \\ \cdots \\ u \end{pmatrix}$$

$$x,\ s \geq 0.$$

The column vectors of the variables s give the coefficient matrix of this problem a special structure. The version of the primal simplex method discussed earlier for a general LP in standard form does not take advantage of this structure, and the special properties of the inverse of bases for it that come from this special structure. Taking advantage of this structure, the method can be simplified and operated using only the portions of the basis inverse corresponding to the rows of the A matrix. We will discuss this simplified version of the primal simplex method for bounded variable LPs in this section.

Given a feasible solution $\bar{x} = (\bar{x}_j)$ of (6.14), variables can be partitioned into three sets wrt it:

x_j satisfying $\bar{x}_j = 0$,

x_j satisfying its upper bound u_j is finite, and $\bar{x}_j = u_j$,

x_j satisfying $0 < \bar{x}_j < u_j$.

From Chap. 4 we know that a feasible solution $\bar{x}$ for (6.14) is a BFS iff $\{A_{.j} : j$ such that $0 < \bar{x}_j < u_j\}$ is linearly independent. Using this, we can define a BFS for (6.14) in terms of a *B-N partitions* (basic, nonbasic partition) of variables in x into three sets denoted by $(x_B,\ x_L,\ x_U)$, where

- x_B = vector of basic variables consisting of m variables associated with a basis B for the coefficient matrix A
- x_L = vector of nonbasic variables whose values in the basic solution are fixed at their lower bound 0

- x_U = vector of nonbasic variables whose values in the basic solution are fixed at their upper bound

B, the submatrix of A whose columns correspond to the basic variables in x_B, must be a square nonsingular submatrix, that is, a basis. Given such a B-N partition (x_B, x_L, x_U), the *basic solution* of (6.14) corresponding to it is defined to be

$$(\bar{x}_B, \bar{x}_L, \bar{x}_U), \text{ where } \bar{x}_L = 0, \ \bar{x}_U = u_U, \ \bar{x}_B = B^{-1}(b - A_{.U}u_U).$$

The B-N partition $(\bar{x}_B, \bar{x}_L, \bar{x}_U)$ is said to be a *(primal) feasible partition* if it is primal feasible, that is, if $0 \leq \bar{x}_B \leq u_B$, (then $\bar{x}$ is said to be a *basic feasible solution* or a *BFS*); otherwise it is a *(primal) infeasible partition.* If $\bar{x}$ is a BFS, it is a

- *Nondegenerate BFS* if $0 < \bar{x}_B < u_B$, that is, every basic variable in the partition has its value in $\bar{x}$ strictly between its bounds
- *Degenerate BFS* otherwise, that is, at least one basic variable in the partition has its value in $\bar{x}$ equal to its lower bound 0, or its upper bound

Some of the entries in the vector u of upper bounds may be $+\infty$, then the corresponding variables do not have an upper bound. Let $J = \{j : u_j < \infty\}$, $\bar{J} = \{j : u_j = +\infty\}$. Associating the dual variables π_i to the ith constraint in $Ax = b$ and the dual variable μ_j for the constraint $x_j \leq u_j$ for $j \in J$, the dual of (6.14) is

$$\begin{aligned} \text{Maximize } \ v = \pi b + \sum_{j \in J} \mu_j u_j & \\ \text{subject to } \ \pi A_{.j} - \mu_j \leq c_j, & \quad \text{for } j \in J \\ \pi A_{.j} \leq c_j & \quad \text{for } j \in \bar{J} \\ \pi \text{ unrestricted}, \ \ \mu_j \geq 0, & \quad \text{for } j \in J, \end{aligned}$$

where $\pi = (\pi_i)$. The CS (complementary slackness) optimality conditions for a primal, dual pair of feasible solutions to (6.14) and its dual above are

$$(u_j - x_j)\mu_j = 0 \text{ for all } j \in J; \text{ and } x_j(c_j - \pi A_{.j} + \mu_j) = 0 \text{ for all } j.$$

Using the special structure of dual constraints, dual feasibility and optimality conditions can be stated in terms of the dual variables $\pi = (\pi_i)$ only without mentioning the dual variables μ_j explicitly. For this reason, usually $\pi = (\pi_i)$ is itself defined as the vector of dual variables for this problem.

The dual basic solution associated with the B-N partition (x_B, x_L, x_U) is defined so as to satisfy the CS conditions with $\bar{x}$, that is, it satisfies as equations the dual constraints corresponding to primal basic variables in x_B:

$$\begin{aligned} \pi A_{.j} &= c_j \text{ for } j \in \bar{J} \text{ s. th. } x_j \in x_B, \\ \pi A_{.j} - \mu_j &= c_j \text{ for } j \in J \text{ s. th. } x_j \in x_B, \\ \mu_j &= 0 \text{ for } j \in J \text{ s. th. } x_j \in x_B. \end{aligned}$$

In other words, the dual basic solution associated with the partition (x_B, x_L, x_U) is defined by $\pi A_{.j} = c_j$ for all j such that x_j is a basic variable, which in matrix notation is $\pi B = c_B$, where B is the corresponding basis and c_B the basic cost vector. Hence $\bar{\pi} = c_B B^{-1}$ is the dual basic solution associated with the partition (x_B, x_L, x_U); the same formula as for the LP in standard form in Chap. 5.

Also, define $\bar{c}_j = c_j - \pi A_{.j}$ as the *relative cost coefficient* of the variable x_j. Hence the definition of the dual basic solution $\bar{\pi}$ implies that the relative cost coefficient $\bar{c}_j$ is 0 for all basic variables x_j in x_B. In fact from the dual constraints, the dual variables μ_j corresponding to $\bar{\pi}$ can actually be obtained from

$$\bar{\mu}_j = \text{minimum}\{\bar{c}_j, 0\} \quad \text{for all } j \in J.$$

This is the reason why the dual solution is stated in terms of $\bar{\pi}$ only.

It can be verified that the dual basic solution $\bar{\pi}$ associated with the B-N partition (x_B, x_L, x_U) is dual feasible iff

$$\bar{c}_j \begin{cases} \geq 0 & \text{for all } j \text{ s. th. } x_j \in x_L, \\ \leq 0 & \text{for all } j \text{ s. th. } x_j \in x_U. \end{cases} \tag{6.15}$$

These are the optimality conditions for a primal feasible B-N partition (x_B, x_L, x_U). A B-N partition (x_B, x_L, x_U) is optimal to (6.14) if it is both primal and dual feasible, in this case the primal, dual basic solutions associated with it are optimum solutions of the respective problems.

6.14.1 The Bounded Variable Primal Simplex Algorithm

To solve (6.14), this algorithm needs an initial B-N primal feasible partition. It goes through a sequence of primal feasible partitions, each obtained by exchanging one variable among a pair of sets in the previous partition. The objective value is monotonic decreasing along the sequence. The method terminates either with an optimum partition or one in which an unboundedness criterion (equivalently, a dual infeasibility criterion) is satisfied. If all the upper bounds in the vector $u = (u_i)$ are finite, then of course the unboundedness criterion will never be satisfied.

To initiate the primal simplex algorithm on an LP like (6.14), we make $b \geq 0$ and look for a unit basis among the columns of A. If a full unit basic vector x_B is found, let x_D be the associated vector of nonbasic variables.

$\bar{x} = (\bar{x}_B, \bar{x}_D) = (b, 0)$ satisfies the equality constraints $Ax = b$ in (6.14) (because the basis corresponding to x_B is the unit matrix). If it also satisfies the bounds on x_B, that is, if $b \leq u_B$, then $\bar{x}$ is feasible to (6.14) and $(x_B, x_L = x_D, x_U = \emptyset)$ is a primal feasible partition, and it can be used to initiate this algorithm.

If $\bar{x}$ is infeasible (i.e., $b \not\leq u_B$), or if a full unit basic vector is not found, then we move to Phase I of the bounded variable primal simplex method whose aim is to find a primal feasible B-N partition first.

Now we will describe a general iteration in the bounded variable primal simplex algorithm.

6.14.2 General Iteration in the Bounded Variable Primal Simplex Algorithm

Let the current primal feasible B-N partition be (x_B, x_L, x_U) associated with the basis B and BFS $\bar{x}$. Let the corresponding inverse tableau be

Inverse tableau

BV	Inverse		Basic values
x_B	B^{-1}	0	$\bar{g}$
$-z$	$-\bar{\pi}$	1	$-\bar{z}$

Partition (x_B, x_L, x_U)

Remember that $\bar{x}_B = B^{-1}(b - A_{.U}u_U) = B^{-1}b - B^{-1}A_{.U}u_U = \bar{g}$. This is what is recorded in the RHS column ("basic values" column) in the inverse tableau.

Here are the various steps in this iteration.

1. *Check optimality:* For all nonbasic variables x_j compute $\bar{c}_j = c_j - \bar{\pi}A_{.j}$. If the optimality criterion (6.15) is satisfied, then (x_B, x_L, x_U) the present partition is an optimum partition, and $\bar{x}$, $\bar{\pi}$ are primal and dual optimum solutions. Terminate.

   Otherwise, let $E = \{j$: either $x_j \in x_L$ and $\bar{c}_j < 0$; or $x_j \in x_U$ and $\bar{c}_j > 0\}$, set of indices of nonbasic variables eligible to enter the basic vector x_B. Go to Step 2.
2. *Select entering variable:* Select a nonbasic variable x_j for some $j \in E$ as the entering variable, say x_s. Go to Step 3.
3. *Minimum ratio test:* Let the updated column of the entering variable x_s be $\bar{A}_{.s} = B^{-1}A_{.s} = (\bar{a}_{is} : i = 1 \text{ to } m)$.

Case 1. $x_s \in x_L$: In this case we have $\bar{c}_s < 0$, so the objective value decreases if the value of x_s is increased from its present value of 0. Let λ represent its new value. All nonbasic variables other than x_s will have their value unchanged in this iteration. So, the new solution $x(\lambda)$ as a function of λ is

$$\begin{array}{ccl} \text{Nonbasics other than } x_s & = & \text{Present value (lower/upper bound)} \\ x_s & = & \lambda, \\ x_B & = & \bar{g} - \lambda \bar{A}_{.s}, \\ z & = & \bar{z} + \lambda \bar{c}_s. \end{array}$$

We need to make sure that x_s and all the variables in x_B remain within their lower and upper bounds in the new solution. The maximum value that λ can have subject to this condition is the *minimum ratio* θ in this step.

$$\theta = \min\left[u_s, \left\{\frac{u_i - \bar{g}_i}{-\bar{a}_{is}} : i \text{ s. th. } \bar{a}_{is} < 0\right\}, \left\{\frac{\bar{g}_i}{\bar{a}_{is}} : i \text{ s. th. } \bar{a}_{is} > 0\right\}\right],$$

where $\bar{g}_i$, u_i represent the present value and upper bound for the ith basic variable, $i = 1$ to m.

- If $\theta = \infty$, z is unbounded below, terminate.
- If $\theta = u_s < \infty$, update the basic values vector by putting $\lambda = \theta$. Let $L' = L\backslash\{x_s\}, U' = U \cup \{x_s\}$. With (B, L', U') as the new partition and new BFS go back to Step 1.
- If $\theta < u_s$ (including $\theta = 0$), make changes in the basic vector and carry out the pivot step to update the inverse tableau, as given below in Step 4.

Examples:

Basic var.	value	Upper bound	Updated entering col. $\bar{A}_{.s}$			
x_i	g_i	u_i				
x_1	0	40	−1	−1	−1	−1
x_2	10	40	−1	−1	−1	−1
x_3	20	40	0	0	0	1
x_4	30	80	1	1	0	0
x_5	0	80	1	−1	−1	0
x_6	15	80	1	−1	−1	0
UB of entering var.			20	10	50	50

Here we give four different examples to illustrate the computation of the minimum ratio and determining the dropping basic variables. All variables in the problem have 0 as their lower bound.

For all these examples, the present basic vector is $x_B = (x_1 \text{ to } x_6)$ and $\bar{g} = \bar{x}_B =$ column vector of values of the basic variables in the present BFS $= (0, 10, 20, 30, 0, 15)^T$ (this is the second column in the above tableau), and $u_B =$ vector of upper bounds for basic variables $= (40, 40, 40, 80, 80, 80)^T$ (this is the third column in the above tableau).

In all the four examples, the entering variable x_s is the nonbasic variable x_7 whose current value in the present BFS is $\bar{x}_7 = 0$. We denote the new value of this variable by the parameter λ and the next B-N partition by $(x_{B'}, x_{L'}, x_{U'})$.

In the first example, $\bar{A}_{.s} = A_{.7} = (-1, -1, 0, 1, 1, 1)^T =$ updated column of the entering variable x_s and $u_7 =$ upper bound on $x_7 = 20$ (these are the entries given in the fourth column in the tableau). So, we need $0 \leq \lambda \leq 20$ to keep x_7 within its bounds in the next solution. Also, the values of the basic variables in the next solution as a function of λ are given by $x_B(\lambda) = (0+\lambda,\ 10+\lambda,\ 20,\ 30-\lambda,\ 0-\lambda,\ 15-\lambda)^T = g + \lambda(-\bar{A}_{.7})$.

To make sure that x_7 and x_B stay within their bounds in the new solution, the maximum value that λ can have is minimum$\{20, 40, (40-10), 30, 0, 15\} = 0$. As the minimum ratio is 0, this is a degenerate pivot step. The entering variable x_7 moves from x_L to $x_{B'}$ replacing the basic variable x_5 from it. x_5 moves to $x_{L'}$.

In the second example, $\bar{A}_{.7},\ u_7$ are given in column 5 of the above tableau. $x_B(\lambda) = (0+\lambda,\ 10+\lambda,\ 20,\ 30-\lambda,\ 0+\lambda,\ 15+\lambda)^T = g + \lambda(-\bar{A}_{.s})$. So, to keep $x_7,\ x_B$ within bounds, the minimum ratio is minimum$\{10 =$ UB of entering variable x_7, 40, 30, 30, 80, (80 -15)$\} = 10$. The entering variable $x_s = x_7$ moves from x_L to x'_U. There is no change in the basic vector. In this example there will be no pivot step to perform, the only change in the inverse tableau is to change the basic values vector to $x_B(10)$, and the value of the nonbasic entering variable x_7 from 0 to 10.

In the third example, $\bar{A}_{.7},\ u_7$ are given in column 6 of the above tableau. So $x_B(\lambda) = (0+\lambda,\ 10+\lambda,\ 20,\ 30,\ 0+\lambda,\ 15+\lambda)^T$. So, to keep $x_7,\ x_B$ within bounds, the minimum ratio is minimum$\{50, 40, 30, 80,\ (80-15)\} = 30$. The entering variable x_7 moves from x_L to $x_{B'}$, replacing x_2 as a basic variable. x_2 moves to $x_{U'}$. In this example, a pivot step has to be performed, and the basic values vector changes to $(30, 40, 20, 30, 30, 45)^T$, and the value of x_2 changes to 40, its upper bound.

In the fourth example, $\bar{A}_{.7},\ u_7$ are given in column 7 (last column) of the above tableau. So $x_B(\lambda) = (0+\lambda,\ 10+\lambda,\ 20-\lambda,\ 30,\ 0,\ 15)^T$. So, to keep $x_7,\ x_B$ within bounds, the minimum ratio is minimum$\{50, 40, 30, 20\} = 20$. The entering variable x_7 moves from x_L to $x_{B'}$, replacing x_3 as a basic variable. x_3 moves to x_L. In this example a pivot step has to be performed, and the basic values vector changes to $(20, 30, 20, 30, 0, 15)^T$, and the value of x_3 changes to 0, its lower bound. ■

Case 2: $x_s \in x_U$: In this case we will have $\bar{c}_s > 0$, so the objective value decreases if the value of x_s is decreased from its present value of u_s. So, we make the new value of $x_s = u_s - \lambda$, leaving all other nonbasic variables at their present values. So, the new solution $x(\lambda)$ is

$$\begin{aligned}\text{Nonbasics other than } x_s &= \text{Present value (lower/upper bound)},\\ x_s &= u_s - \lambda,\\ x_B &= \bar{g} + \lambda \bar{A}_{.s},\\ z &= \bar{z} - \lambda \bar{c}_s.\end{aligned}$$

Based on same arguments as before, in this case the minimum ratio θ in this step is

$$\theta = \min\left[u_s, \left\{\frac{u_i - \bar{g}_i}{\bar{a}_{is}} : i \text{ s. th. } \bar{a}_{is} > 0\right\}, \left\{\frac{\bar{g}_i}{-\bar{a}_{is}} : i \text{ s. th. } \bar{a}_{is} < 0\right\}\right],$$

where $\bar{g}_i$, u_i represent the present value and upper bound for the ith basic variable, $i = 1$ to m.

All remaining work in this case is similar to that in Case 1.

4. *Pivot step:* Select as the dropping basic variable, one whose ratio ties for the minimum in the definition of θ.

   (a) Replace the dropping rth basic variable in the basic vector by x_s. Delete x_s from x_L or x_U where it is currently. Include the dropping variable in x_L or x_U depending on whether it is at its lower bound 0, or upper bound, in $x(\theta)$. This updates all sets in the B-N partition. Let (x'_B, x'_L, x'_U) be the new partition after these changes.
   (b) $x(\theta)$ is the new BFS. Replace the basic values column in the inverse tableau by the values of the new basic vector in $x(\theta)$.
   (c) Update the inverse matrix in the inverse tableau by carrying out the pivot step with the updated column of x_s as the pivot column and row r as the pivot row.

   Begin the next iteration with the new partition and inverse tableau for it.

6.14.3 The Bounded Variable Primal Simplex Method

Consider the bounded variable LP

$$\begin{aligned} \text{Minimize } z(x) &= cx \\ \text{subject to } Ax &= b \\ 0 \leq x &\leq u, \end{aligned} \tag{6.16}$$

where $b \geq 0$ and A is a matrix of order $m \times n$ and rank m. If a primal feasible B-N partition is not available to initiate the bounded variable primal simplex algorithm on this problem, this method goes to a Phase I whose aim is to generate a feasible B-N partition first.

Let I denote the unit matrix of order m. To set up the Phase I problem, for each $i = 1$ to m, check if there is a column $A_{.j}$ of A, which is $= I_{.i}$ and satisfies $u_j \geq b_i$. In that case, make x_j as the basic variable in row i.

If this process generates a full unit basic vector x_B, let x_L be the vector of all nonbasic variables, then the B-L partition $(x_B, x_L, x_U = \emptyset)$ is a primal feasible partition, which can be used to solve (6.16) with the bounded variable primal simplex algorithm, which is called Phase II in this method.

Otherwise, introduce artificial variables associated with the missing unit vectors to make a full unit basic vector x_B. Let x_L be the vector of all the other nonbasic original problem variables. Each artificial variable has a lower bound of 0 and upper bound of ∞. Phase I problem is the problem of minimizing the Phase I objective function $w =$ (sum of all the artificial variables introduced), subject to these constraints; $(x_B, x_L, x_U = \emptyset)$ is a primal feasible B-N partition for it. Solve this Phase I problem by the bounded variable primal simplex algorithm beginning with this feasible partition; this is Phase I of this method. If the minimum value of w in this method is > 0, the original model (6.16) is infeasible, terminate.

If the minimum value of w is 0, let $\bar{x} = (\bar{x}_j)$ denote the BFS at the end of Phase I. Let $\bar{d}_j$ denote the Phase I relative cost coefficient of the original problem variable x_j at this stage for $j = 1$ to n. If any $\bar{d}_j > 0$ [$\bar{d}_j < 0$], then by Phase I termination $\bar{x}_j$ must be 0 (i.e., $x_j \in x_L$) [must be $= u_j$, i.e., $x_j \in x_U$], and $\bar{d}_j > 0$ [$\bar{d}_j < 0$] indicates that $x_j = 0$ [$x_j = u_j$] at every feasible solution of the original problem. So, fix such variables x_j at 0 [at u_j] and delete them from further consideration in the problem [and revise the RHS constants vector at this stage accordingly].

Now you can move to Phase II with the remaining B-N partition. Any artificial variables in the basic vector will have their basic values always = 0 in all future iterations and can be deleted when they leave the basic vector, and they along with their 0 value can be ignored otherwise.

6.15 Exercises

6.1. Solve the following LPs using the revised primal simplex algorithm (both these problems have optimum solutions).

(a) Minimize $z = 3x_1 + 5x_2 + 6x_4 + 3x_5 - 4x_6 - 4x_7$

$$\begin{aligned} \text{subject to} \quad x_1 + x_4 - x_6 - x_7 &= 2 \\ x_2 - x_4 + x_5 - x_7 &= 3 \\ x_3 - x_5 + x_6 + x_7 &= 1 \\ \text{and all } x_j &\geq 0. \end{aligned}$$

(b) Minimize $z = 6x_1 + 4x_3 + 5x_4 + 3x_5 + x_6 + 10x_7$

$$\begin{aligned} \text{subject to} \quad x_1 + x_4 + x_5 + x_6 + x_7 &= 14 \\ x_2 - x_5 - x_7 &= 5 \\ x_3 + x_4 - x_6 + x_7 &= 6 \\ \text{and all } x_j &\geq 0. \end{aligned}$$

6.2. Solve the following LPs using the revised primal simplex algorithm (in both these problems the primal objective function is unbounded below on the set of feasible solutions). Show evidence of infeasibility of the dual problems.

(a) Minimize $z = 7x_1 + 9x_2 + 3x_4 - x_5 + x_6 - 2x_7$

$$\begin{aligned} \text{subject to} \quad x_1 - x_4 - x_5 + x_6 - x_7 &= 1 \\ x_2 + x_5 + x_6 - 3x_7 &= 5 \\ x_3 + x_4 - x_6 + x_7 &= 2 \\ \text{and all } x_j &\geq 0. \end{aligned}$$

(b) Minimize $z = 3x_1 + 9x_3 + 3x_4 - 3x_5 + 3x_6 + 7x_7 + x_8$

$$\begin{aligned} \text{subject to} \quad x_1 + x_5 - x_7 - x_8 &= 4 \\ x_2 - x_6 + x_7 - x_8 &= 3 \\ x_3 + x_5 + x_6 + x_7 - x_8 &= 10 \\ x_4 - x_5 + x_6 + x_8 &= 6 \\ \text{and all } x_j &\geq 0. \end{aligned}$$

6.3. Solve by the revised primal simplex method (all these problems have optimum solutions).

(a) Minimize $z = -4x_1 + 6x_2 - 9x_3 + 2x_4 + 12x_5 + 11x_6 + 12x_7$

$$\begin{aligned} \text{subject to} \quad x_1 + x_3 - x_4 + x_5 + 2x_6 &= -5 \\ 2x_2 - x_3 + x_4 + x_5 - x_6 + x_7 &= 12 \\ x_1 - x_2 + x_3 - x_7 &= -3 \\ \text{and all } x_j &\geq 0. \end{aligned}$$

(b) Minimize $z = 7x_1 - 12x_3 + 14x_4 + 9x_5 - 12x_6 + 11x_7$

$$\begin{aligned} \text{subject to} \quad -x_1 + x_2 + x_3 - x_4 + x_6 + x_7 &= 15 \\ x_2 - x_3 + x_4 + x_5 - x_6 + 2x_7 &= -5 \\ -x_1 - x_4 + x_5 - x_6 - x_7 &= -5 \\ -2x_1 + 2x_2 - x_4 + 2x_5 - x_6 + 2x_7 &= 5 \\ \text{and all } x_j &\geq 0. \end{aligned}$$

(c) Minimize $z = 6x_1 + 4x_2 + 5x_3 + 4x_4 + 4x_6 + 6x_7$

$$\begin{aligned} \text{subject to} \quad x_1 - x_3 + x_4 + x_5 + x_6 + x_7 &= 3 \\ x_1 + x_2 + x_3 - x_5 + x_6 + x_7 &= 6 \\ x_1 + x_2 + x_3 + x_4 - x_7 &= 3 \\ \text{and all } x_j &\geq 0. \end{aligned}$$

6.4. Solve the following problems (both have optimum solutions) by the primal revised simplex method, by setting up Phase I using the smallest number of artificial variables. During the pivot steps of Phase I, whenever there is a tie in the min ratio test for the dropping variable, with an artificial variable and an original problem variable among those tied, always select an original problem variable among those tied as the dropping variable.

(a) Minimize $z = x_1 + x_2 + x_3 + x_4 + x_5$

$$\begin{aligned} \text{subject to} \quad x_1 - x_2 - x_5 &\le 2 \\ -x_1 + x_3 - x_4 + 2x_5 &\le -4 \\ x_1 - 2x_2 + x_3 - x_4 &\ge 0 \\ \text{and all } x_j &\ge 0. \end{aligned}$$

(b) Minimize $z = 3x_1 + 4x_2 + 5x_3 + 2x_4 + x_5 + 3x_6 + 2x_7$

$$\begin{aligned} \text{subject to} \quad x_1 + x_3 + x_5 - x_7 &\le -5 \\ x_2 + x_4 - x_5 + x_6 &\le -1 \\ x_1 - x_6 + x_7 &\le 2 \\ 3x_1 + x_2 + x_3 + x_4 - x_6 + x_7 &\ge 8 \\ \text{and all } x_j &\ge 0. \end{aligned}$$

6.5. Solve the following problems by the dual simplex algorithm (both are primal feasible also).

(a) Minimize $z = -2x_1 - 2x_2 + 10x_3 + 2x_4 - 3x_5$
$-12x_6 + 12x_7 + 14x_8 + 11x_9$

$$\begin{aligned} \text{subject to} \quad x_3 - x_5 - x_6 - x_7 + x_8 + x_9 &= -5 \\ x_1 + x_7 - x_8 - x_9 &= -5 \\ x_4 + x_5 - x_6 - x_9 &= 0 \\ x_2 - x_5 + x_6 - x_8 &= -10 \\ \text{and all } x_j &\ge 0. \end{aligned}$$

(b) Minimize $z = -6x_1 + 2x_3 + 9x_4 - 3x_5 - 4x_6 + 6x_7$

$$\begin{aligned} \text{subject to} \quad x_2 + x_4 - x_6 - x_7 &= 13 \\ x_3 - x_5 + x_7 &= -10 \\ x_1 - x_4 + x_5 + x_6 &= -3 \\ \text{and all } x_j &\ge 0. \end{aligned}$$

6.6. Consider the LP: minimize $z(x) = cx$ subject to $Ax \ge 0$, $x \ge 0$. Show that either 0 is an optimum solution of this problem or that $z(x)$ is unbounded below on its set of feasible solutions.

6.7. Solve the following problems by the dual simplex algorithm (both are primal infeasible).

(a) Minimize $z = -3x_1 - 10x_2 - 15x_3 + 2x_4 - 6x_5 + 14x_6 - 10x_7 + 20x_8 + x_9$

$$\begin{aligned} \text{subject to} \quad x_1 + x_5 - x_7 - x_8 &= 4 \\ x_3 - x_6 + x_7 - x_8 + x_9 &= -2 \\ x_4 - x_5 + x_6 + x_7 + x_8 &= -1 \\ x_2 + 2x_3 + x_5 - x_6 + 2x_7 - x_8 &= -5 \\ \text{all } x_j \geq 0 \text{ for all } j \end{aligned}$$

(b) Minimize $z = -4x_1 + 2x_2 - x_3 - 3x_5 + 5x_6 + 2x_7$

$$\begin{aligned} \text{subject to} \quad x_1 + x_4 + x_5 &= 1 \\ x_3 + x_4 - x_6 - x_7 &= -2 \\ -x_1 + x_2 - x_4 + x_6 &= -2 \\ x_j \geq 0 \text{ for all } j \end{aligned}$$

6.8. Solve the following LP by the Phase I, II primal simplex method. From the final tableau at the end of Phase I, can you conclude whether there are any redundant equality constraints in the original model?

$$\begin{aligned} \text{Maximize } z &= x_1 \\ \text{subject to} \quad x_1 + x_2 + 4x_3 + x_4 &= 20 \\ 2x_1 + x_3 &= 5 \\ -7x_1 + x_2 + x_4 &= 0 \\ x_1 \text{ to } x_5 &\geq 0 \end{aligned}$$

6.9. Solve the following LPs by the bounded variable primal simplex method {(a) is from Dantzig and Thappa (1997), vol. 1, cited in Chap. 1}. (All have optimum solutions.)

(a) Minimize $z = 12x_1 + 2x_2 + 9x_3$

$$\begin{aligned} \text{Subject to} \quad & x_1 + x_2 + x_3 = 5 \text{ and } 2x_1 - x_3 = 2 \\ & 0 \leq x_1 \leq 1, \quad x_2, x_3 \geq 0 \end{aligned}$$

(b) Minimize $z = -2x_1 + 4x_2 + 7x_3 + 6x_4 - 3x_5 + 3x_6 + 6x_7 + 4x_8$

$$\begin{aligned} \text{Subject to} \quad x_2 + x_3 + 2x_5 &= 25 \\ x_2 - x_3 + x_4 - x_7 &= 4 \\ x_1 - x_2 - x_5 + x_6 + x_7 - x_8 &= -2 \\ x_1 \text{ to } x_8 &\geq 0, \text{ and } \leq 10 \end{aligned}$$

(c) Minimize $z = 9x_1 + 11x_2 + 11x_3 + 2x_4 + 11x_5 + 4x_6 + 5x_7 + 2x_8$

$$\begin{aligned} \text{Subject to} \quad -x_1 + x_3 - x_4 + x_5 - x_7 + x_8 &= 15 \\ x_2 - x_3 + x_6 + x_7 - x_8 &= -5 \\ x_3 - 2x_4 - x_6 + 2x_8 &= 5 \end{aligned}$$

$$0 \le x_j \le 15 \text{ for } j \in \{1,3,5,6,8\}; \ 0 \le x_j \le 5 \quad \text{for } j \in \{2,4,7\}.$$

6.10. For the following LPs, find a feasible solution by the bounded variable primal simplex method (both are primal infeasible).

(a)
$$\begin{aligned} & x_2 + x_3 + x_4 - x_5 + 2x_7 = 20 \\ & x_1 + x_3 + x_4 - x_5 + x_7 = 15 \\ & -x_3 + x_5 + x_6 + x_7 = 5 \\ & x_1 \ge 21, \ 0 \le x_j \le 15 \text{ for } j \subset \{2,\dots,7\}. \end{aligned}$$

(b)
$$\begin{aligned} & x_3 + x_4 + x_6 = 5 \\ & x_1 + x_2 + x_4 - x_5 + x_6 = 5 \\ & 2x_1 + x_2 + x_4 - 2x_5 + 2x_6 = 15 \\ & 0 \le x_j \le 10 \text{ for all } j. \end{aligned}$$

6.11. Solve this problem by the primal simplex method. The primal is infeasible; show that the dual objective function is unbounded above in the dual problem.

Minimize $z = -2x_1 - 3x_2 + 10x_3 + 4x_4$

$$\begin{aligned} \text{subject to} \quad x_1 + x_2 - x_3 + x_4 &\le 4 \\ x_2 - x_3 &\ge 8 \\ \text{for all } j; \quad x_j &\ge 0. \end{aligned}$$

Do the same for the following LP: minimize $-x_1 - x_2$

$$\begin{aligned} \text{subject to} \quad x_1 - x_2 &\ge 4 \\ x_1 \le 2 \quad x_1, x_2 &\ge 0 \end{aligned}$$

6.12. Solve the following LP by the revised primal simplex method. If there is a tie for the minimum ratio in any pivot step, always select the bottommost among the rows tied as the pivot row.

$$\begin{aligned} \text{Minimize } z = -2x_3 - 10x_4 \\ \text{s. to } x_1 + x_3 + 2x_4 \geq 2 \\ x_2 + x_3 + x_4 \geq 4 \\ x_1 + x_2 + 2x_3 + 3x_4 \leq 6 \\ 2x_1 + x_2 + 3x_3 + 6x_4 \leq 8 \\ \text{and for all } j, \ \ x_j \geq 0. \end{aligned}$$

6.13. Solve the following LP by the primal simplex method, and show that the objective function is unbounded above on its set of feasible solutions. Show that the dual is infeasible by showing an evidence of its infeasibility.

Maximize $z = x_2$, subject to $x_1 + 2x_2 \geq 2$, $2x_1 + x_2 \geq 2$, $-x_1 + x_2 \leq 2$, and $x_1 - 2x_2 \leq 2$, and $x_1,\ x_2 \geq 0$.

6.14. Consider the LP: minimize $z = cx$ subject to $a \leq x \leq b$, where $a,\ b$ are given column vectors in R^n satisfying $a \leq b$, and x is the vector of variables in the problem. Derive the optimum solution of this problem through direct arguments. Write the dual of this problem and derive its optimum solution using the optimality conditions.

6.15. Consider the LP: minimize $z = cx$ subject to $Ax = b, \quad x \geq 0$. $\bar{x}$ is a nondegenerate optimum BFS for this LP, and the associated dual basic solution is dual degenerate. Then show that the original LP must have alternate optimum solutions.

6.16. Solve the following LP by the revised primal simplex method using Dantzig's entering variable choice rule in every pivot step.

$$\begin{array}{lrrrrcl} \min z = & -x_1 & -10x_2 & +x_3 & & & \\ \text{s. to} & -x_1 & +x_2 & & +x_4 & = & 2 \\ & & -2x_2 & +x_3 & +x_4 & = & 3 \\ & x_1 & +x_2 & -x_3 & & \leq & 2 \\ & & -x_2 & +x_3 & +x_4 & \leq & 3 \\ & -x_1 & +2x_2 & -x_3 & & = & 1 \\ & & \multicolumn{3}{l}{x_j \geq 0 \quad \text{for all } j.} & & \end{array}$$

If the problem is infeasible, show the modification of the original RHS constants vector b to b' indicated by the terminal Phase I inverse tableau to make this model feasible. Also, write the final Phase I inverse tableau for the modified problem, indicating the feasible solution to it contained there.

Explain clearly how to go to Phase II from the final Phase I tableau, in order to get an optimum solution for the modified problem. Carry out Phase II, showing all your work clearly, and obtain an optimum solution for the modified problem.

6.17. For the LP given below, select by inspection, an initial basic vector that is primal feasible but dual infeasible. Solve the LP using the revised primal simplex algorithm, beginning with this basic vector.

Then select by inspection, an initial basic vector that is dual feasible but primal infeasible. Solve the LP using the revised dual simplex algorithm, beginning with this basic vector.

Show that both approaches yield the same solution.

$$\begin{aligned} \text{Minimize} \quad z = -3x_1 - 4x_2 - 4x_3 + 2x_4 + 6x_5 - 6x_6 + 8x_7 - 2x_8 \\ \text{s. to} \quad x_1 + x_4 - x_6 + x_7 + x_8 = 3 \\ x_2 + x_4 + x_6 - x_7 + 2x_8 = 4 \\ -x_3 + x_4 + x_5 - x_6 + x_7 + x_8 = -6 \\ \text{all } x_j \geq 0. \end{aligned}$$

6.18. A is a skew symmetric matrix (i.e., it is square and satisfies $A^T = -A$). Prove that if the LP, minimize yc, subject to $yA \leq c^T$, and $y \geq 0$, has an optimum solution, then the optimum objective value in it must be 0. Here y is the row vector of variables and c is a given column vector.

6.19. While solving the LP minimize $z = cx$ subject to $Ax = b, \;\; x \geq 0$ by the revised primal simplex algorithm, a basic vector x_B was obtained in which x_s was selected as the entering variable. The entering column satisfied the unboundedness criterion. Show that the information in that column can be used to construct a vector y satisfying $Ay = 0, \; y \geq 0, \; cy < 0$.

6.20. Red Brand Canners: Red Brand Canners used to be a medium-size company canning and distributing a variety of fruit and vegetable products in the western part of USA in the 1960s. In one year, they signed an agreement at planting time to purchase the entire tomato crop in a large field at an average delivered price of 6 cents/lb (these are 1960s prices).

At harvest time, Produce Inspection estimates that the total crop will be 3×10^6 lb, of which 20% is expected to be Grade A and the remaining portion expected to be Grade B.

The company makes three different tomato products, and they set the selling prices of these products in light of the long-term marketing strategy of the company. Forecasted demand for these products at these prices are given below.

Product Unit = Case	SP	TU	VC	DF
Whole tomatoes	4	18	2.52	Unlimited
Tomato juice	4.5	20	3.18	50,000
Tomato paste	3.8	25	1.95	80,000

SP selling price ($/case), *TU* tomatoes (lb) used/case,
VC variable costs(excluding tomato costs)$/case,
DF demand forcast in cases at these prices

The company uses a numerical scale to record the quality of both the raw produce and prepared products. The scale runs from 0 to 10 points, higher numbers representing better quality. On this scale, Grade A tomatoes averaged 9 points/lb and Grade B tomatoes 5 points/lb. Minimum average input quality for canned whole tomatoes is 8 for juice 6 points/lb, and paste can be made entirely from Grade B tomatoes.

Remember that tomatoes have already been purchased, so the tomato cost is a fixed cost that has already been incurred by contract. The problem confronting the company now is to decide how to allocate the purchased tomatoes among its three tomato products.

(1) Using as decision variables the number of pounds of each grade of available tomatoes allocated to each of the three products formulate the problem of maximizing the profit contribution from tomato product sales, given the posted selling prices and fixed costs.

Solve the resulting LP model (use any available LP software to answer this question) and find an optimum product mix and the optimum dual solution. What is the average quality/lb of the whole tomatoes and tomato juice if the optimum solution is implemented?

(2) The company has received an offer from another farm to sell additional quantities of Grade A tomatoes at a price of 8.5 cents/lb. Should the company accept this offer – and if so what is the maximum quantity of these tomatoes the company can buy at this price (assuming ample production capacity at the plant)?

Find the new optimum product mix if they buy 80,000 lb of these tomatoes from this farm.

(3) Consider the original problem in (1) again. Suppose the marketing department feels that it can increase demand for tomato juice by 25,000 cases by starting an advertising campaign. How much can the company pay for such a campaign?

(4) How will the optimum product mix change if selling price of tomato juice is increased by 30 cents/case (assume no change in demand)?

(5) If another farm offers to sell 50,000 lbs of Grade B tomatoes, how much can the company pay for this lot?

(6) The estimate of 80,000 cases of demand for tomato paste is not very firm, and it is believed that actually this demand is likely to be between 70,000 and 90,000 cases. Suppose the company decides to apply a holding cost of \$1.8/case of tomato paste remaining unsold into next season and a penalty of \$25/case for each case of tomato paste demand that cannot be met.

Reformulate the LP model taking this uncertainty in tomato paste demand into account, and incorporating these holding and shortage penalty costs into account, to find a new optimum production plan.

(Source: "Red Brand Canners" Case of the Graduate School of Business at Stanford University.).

6.21. Consider the following LP: minimize $z(x) = 3x_1 + 2x_2 + 5x_3$, subject to $x_1 - x_3 = 3$, $3x_2 + 2x_3 = 12$, and $x_j \geq 0$ for all j.

Show that $\bar{x} = (3, 4, 0)^T$ is an optimum solution for this LP.

(1) Suppose the vector of original cost coefficients (3, 2, 5) is going to change, but all the other data in this LP remains fixed at present values. Denote the new cost vector by $c = (c_1, c_2, c_3)$. Describe the set of all c for which $\bar{x}$ remains optimal to the modified LP.

(2) Suppose the RHS constants vector $(3, 12)^T$ in this LP is going to change, but all the other data in the original LP remains unchanged. Denote the new RHS constants vector by $b = (b_1, b_2)^T$. Describe the set of all b for which the basic vector (x_1, x_2) remains optimal to the modified LP.

6.22. Consider the following LP in standard form:

x_1	x_2	x_3	x_4	x_5	x_6	$-z$	RHS
1	2	−1	1	2	2	0	3
0	1	1	2	1	2	0	6
−1	0	2	2	−2	1	0	5
−4	0	9	12	−3	7	1	0

$x_j \geq 0$ for all j, min. z.

Also, you are given that

$$\begin{pmatrix} 1 & 2 & -1 \\ 0 & 1 & 1 \\ -1 & 0 & 2 \end{pmatrix}^{-1} = \begin{pmatrix} -2 & 4 & -3 \\ 1 & -1 & 1 \\ -1 & 2 & -1 \end{pmatrix}.$$

(1) Using the information given, construct the inverse tableau for this problem with respect to the basic vector $(x_1, x_2, x_3, -z)$, and verify that it is an optimum basic vector; showing all your work.

Using this inverse tableau, answer the following questions. **Each of these questions is independent of the others, and each refers to the original LP given above.**

(2) Find the range of values of b_1 (whose present value is 3) for which the basic vector (x_1, x_2, x_3) remains optimal to the problem. Let the lower and upper bounds of this optimality interval be denoted by ℓ_1, u_1, respectively.

(2a) In this range, $\ell_1 \leq b_1 \leq u_1$, give an optimum solution of the LP, optimum objective value, and the optimum dual solution, as functions of b_1.

Does the problem have a marginal value vector when $\ell_1 < b_1 < u_1$ and all the other data in the problem remains unchanged? Why? If so give that marginal value vector.

(2b) Assuming that the new value of $b_1 = (1/2) + u_1$, find the new optimum solution of the problem using sensitivity analysis techniques. Show all the work clearly.

(2c) Instead, if the new value of $b_1 = (-1) + \ell_1$, find the new optimum solution of the problem using sensitivity analysis techniques. Show all the work clearly.

(3) Let $b_1 = 3$, its original value. Find the range of values of c_2 (whose present value is 0) for which the solution from the original inverse tableau obtained above remains optimal to the problem.

Assuming that the new value of c_2 is 1 + (the upper bound of the optimality range computed above), find the new optimum solution using sensitivity analysis techniques.

6.23. A private investor has \$100,000 to invest in various avenues as shown in the table below.

Investment avenue	Code name	Expected annual return	Jones's rating	Liquidity	Risk factor
Savings account	SA	4%	A	Immediate	0
Cert. of deposit	CD	5.2%	A	5-year	0
Atlantic lighting	AL	7.1%	B^+	Immediate	25
Arkansas REIT	AR	10%	B	Immediate	30
Bedrock insurance	BI	8.2%	A	1-year	20
Nocal mining	NM	6.5%	B^+	1-year	15
Mincomp systems	MS	20%	A	Immediate	65
Antony hotel	AH	12.5%	C	Unknown	40

The "liquidity" gives a measure of the waiting period to convert the asset into cash. The "risk factor" is a measure of risk in the investment developed by an experienced agency; the higher the numerical value of the factor, the more risky is the investment.

The investor wants the expected annual return on his total investment to be $\geq$7.5%. He wants at least 50% of his total investment to be in "A" rated investments. To handle emergencies that may show up, he wants $\geq$40% of his total investments to have "immediate" liquidity rating. He wants to limit his investment in SA to be $\leq$\$30,000.

Formulate the problem of finding a portfolio of investments that minimizes the weighted risk factor (sum over avenues of risk factor multiplied by amount invested) subject to the above constraints, as an LP. Write its dual. Solve the model using any available software program.

Is the expected annual return constraint active at the optimum solution? What happens to the optimum solution, and the optimum value of the weighted risk, if the minimum expected annual return is increased from the present 7.5%?

Also, answer the following questions, each part is separate, and they all refer to the original problem given above.

(1) Does this LP have marginal values? Why? Give the practical interpretation of the optimum dual solution obtained.

(2) Suppose the investor wants to include an additional constraint in the model in the form of an upper bound on the amount invested in one stock. Assuming that stock is either SA, or AL, or MS, mention separately for each the range of values of the upper bound in that additional constraint for which the present optimum portfolio will continue to remain optimal.

For each of these stocks, mention what would happen to the optimum weighted risk measure, and annual return, if we decide to invest \$10,000 more in that stock than what is invested in that stock by the present optimum solution? **(3)** Find the optimality range for the risk factors of CD, AR, and AH to maintain the optimum solution unchanged.

Mention how the optimum solution changes if the risk factor for AH changes to $-1+$ (the lower limit of its optimality range computed above).
(Problem composed by Arvind R. Sharma.)

6.24. Consider the system of linear constraints: $Ax = b,\ x \geq 0$, where $A = (a_{ij})$ is a matrix of order $m \times n$ and $b = (b_i)$ is the present RHS constants vector. This system has been found to be infeasible.

We want to make the system feasible by changing the RHS constants vector in the model from the present b to a modified $b' = (b'_i)$ at minimum cost.

For each $i = 1$ to m, if $b'_i > b_i$ $[b'_i < b_i]$, a cost of $c_i^1(b'_i - b_i)$ $[c_i^2(b_i - b'_i)]$ is incurred for making the change in the ith RHS constant, where all the c_i^1, c_i^2 are given positive constants. Formulate the problem of finding the b' to replace b, which will make the original system feasible at minimum cost.

6.25. The vector x_B is a feasible basic vector obtained in the process of solving an LP: minimize cx subject to $Ax = b,\ x \geq 0$ by the primal simplex algorithm. A nonbasic variable x_s has been selected as the entering variable into the basic vector x_B, and there is a tie for the minimum ratio in this step. Then show that the basic vector obtained after this pivot step will be primal degenerate.

6.26. We are given the inverse tableau wrt a basic vector x_B for the LP: minimize $z = cx$, subject to $Ax = b, x \geq 0$. Clearly specify the conditions under which the following conclusions can be drawn: **(1)** x_B is a primal and/or dual feasible basic vector for this problem, **(2)** x_B is the unique optimum basic vector for this LP, **(3)** we can conclude that z is unbounded below on the set of feasible solutions of this LP, **(4)** x_B is a primal, dual, degenerate basic vector for this LP, **(5)** x_B is a nondegenerate primal feasible basic vector for this LP, x_s is eligible to enter the basic vector x_B in the primal simplex algorithm, and if it is chosen as the entering variable, the next basic vector obtained will be primal degenerate, **(6)** x_B is a nondegenerate dual feasible basic vector for this LP, x_s is eligible to be the entering variable into the basic vector x_B in the dual simplex algorithm, and the next basic vector obtained will be dual degenerate, **(7)** x_B is an optimum basic vector for this problem, and we can conclude that this LP has alternate optimum solutions, **(8)** without doing any further pivoting, we can conclude that the optimum objective value in this problem is $\leq \alpha$ (here α is a given number).

6.27. When the primal simplex algorithm is applied to solve the following LPs beginning with the vector of slack variables as the initial primal feasible basic vector, the "most negative $\bar{c}_j$" entering variable choice rule, and the topmost eligible pivot row choice rule in each pivot step, show that the algorithm cycles (H. Kuhn (top problem), M. Beale (second problem)).

x_1	x_2	x_3	x_4	RHS
−2	−9	1	9	≤ 0
1/3	1	−1/3	−2	≤ 0
−2	−3	1	12	$= z$ minimize

$x_j \geq 0$ for all j

x_1	x_2	x_3	x_4	RHS
1/4	−60	−1/25	9	≤ 0
1/2	−90	−1/50	3	≤ 0
0	0	1	0	≤ 1
−3/4	150	−1/50	6	$= z$ minimize

$x_j \geq 0$ for all j

6.28. When Phase I of the simplex method is used to solve the LP: minimize $z = cx$, subject to $Ax = b, x \geq 0$, where A is an $m \times n$ matrix; at the end of Phase I we may find that the minimum value of $w =$ Phase I objective function is 0, but the Phase I final basic vector obtained contains one or more artificial variables as basic variables.

Show that one of the reasons for this to occur is when the statement of the LP given has redundant equality constraints in it. Discuss how we can verify whether there are redundant equality constraints in the statement of the LP given from the final Phase I canonical tableau.

What are some other reasons for this to occur?

6.29. When a general system of linear constraints $Dx = d,\ Ex \geq f$ has at least one feasible solution, an inequality constraint in the system is said to be a *binding inequality constraint* if it holds as an equation at every feasible solution of the system.

When Phase I of the primal simplex method is applied to find a feasible solution of the system $Ax = b,\ x \geq 0$, discuss which conditions at the termination of Phase I of the primal simplex method will lead to identifying binding inequalities in this system. Also, when there are binding inequalities in the system, will all the binding inequality constraints in the system be identified this way by Phase I of the primal simplex method?

6.30. Consider the LP: minimize $z = cx$, subject to $Ax = b,\ x \geq 0$.

We know that multiplying both sides of an equality constraint in the problem by a real number $\alpha \neq 0$ or adding a nonzero multiple of one equality constraint to another does not alter the optimum face of the LP in any way. Discuss what effects these changes to this primal problem will have on the optimum face of the dual problem.

6.31. Consider the LP: minimize $z = cx$, subject to $Ax = b,\ x \geq 0$.

In this system, an inequality constraint $x_1 \geq 0$ is said to be a *redundant inequality constraint* if deleting it from the system of constraints does not change the set of feasible solutions of the system. **(1)** Show that if $x_1 \geq 0$ is a redundant inequality constraint in the system, then there must exist constants $d_j \geq 0$ for all $j = 0, 2, \cdots, n$ such that $x_1 = d_0 + \sum_{j=2}^{n} d_j x_j$ on the set of feasible solutions of this problem. **(2)** Suppose $x_1 \geq 0$ is a redundant inequality constraint in the system of constraints of this LP. Suppose for some i, $a_{i1} \neq 0$. Then from the ith equality constraint in the system, we know that $x_1 = (b_i - \sum_{j=2}^{n} a_{ij} x_j)/a_{i1}$ on the set of feasible solutions of this system. Eliminate the ith equality constraint from the system, and the variable x_1 from the remaining constraints and the objective function using the expression for it in terms of the other variables. If $(x_2^*, \cdots, x_n^*)^T$ is an optimum solution of the remaining problem, show that $x^* = (x_1^*, x_2^*, \cdots, x_n^*)^T$ is an optimum solution of the original problem with $x_1^* = (b_i - \sum_{j=2}^{n} a_{ij} x_j^*)/a_{i1}$. **(3)** Let $((x_2^*, \cdots, x_n^*)^T$ be an optimum solution of the remaining problem discussed in (2) obtained by eliminating the ith equality constraint and the variable x_1 from the original LP. Even if we do not know whether $x_1 \geq 0$ is a redundant inequality constraint in the original LP or not, as long as $x_1 > 0$ on the optimum face of the original LP, show that x^* defined in (2) is an optimum solution of the original LP.

In some applications, we can conclude from practical considerations that some nonnegative variables in the mathematical model for the problem will be positive at every feasible solution or at every optimum solution. We can use the above results to reduce the size of the problem in such instances.

6.32. Consider the LP: minimize $z = cx$, subject to $Ax = b$, $Dx \geq d$, where A, D are $m \times n$, $p \times n$ matrices.

Suppose x^* is an optimum solution of the problem: minimize $z = cx$, subject to $Ax = b$ and $D_{i.}x \geq d_i$, for $i = 2$ to p; obtained from the original LP by ignoring the first inequality constraint in it. Also, suppose $D_{1.}x^* < d_1$. Then prove that if the original LP has an optimum solution, it must have an optimum solution satisfying $D_{1.}x \geq d_1$ as an equation.

6.33. Suppose we know that the LP minimize $z = cx$, subject to $Ax = b$, $x \geq 0$ has a finite optimum objective value. Now if changes are made in the model leaving A, c unchanged, but only changing b to a new $\bar{b}$, prove that the modified LP must be either infeasible or have a finite optimum objective value also (you have to show that under the hypothesis, z cannot be unbounded below on the set of feasible solutions of the modified problem).

6.34. Consider the LP: minimize $z = cx$, subject to $Ax = b$, $x \geq 0$, where A is an $m \times n$ matrix. Suppose we have a primal nondegenerate optimum basic vector x_B for this LP.

A new variable x_{n+1} associated with the column vector $(A_{.n+1}, c_{n+1})$ is introduced into the model. When the augmented model is solved by the primal simplex algorithm beginning with x_B as the initial primal feasible basic vector, show that once x_{n+1} enters the basic vector, it never leaves it. What conclusions can you draw if the optimum basic vector x_B for the original LP is degenerate?

6.35. The LP: minimize $z = cx$, subject to $Ax = b$, $x \geq 0$, where A is an $m \times n$ matrix, is being solved by the primal simplex method, and at the end of Phase I, the conclusion "this LP is infeasible" has been reached.

There are two possibilities in this case: One is that the system of linear equations $Ax = b$ (i.e., with the nonnegativity restrictions on the variables ignored) may itself have no solution. The second is that the system of equations $Ax = b$ may have a solution, but no nonnegative solution. Discuss how the final output from Phase I of the primal simplex method can be used to determine which of these two possibilities occurs for our problem.

Also, under the second possibility, as $Ax = b$ has a solution, but no nonnegative solution, how can we use the output from Phase I to determine a small subset of variables x_j in the model such that when the nonnegativity requirement on them is relaxed, the remaining system becomes feasible?

6.36. Consider the following LP

x_1	x_2	x_3	x_4	x_5	x_6	$-z$	b
0	1	2	−3	−3	0	0	8
2	−1	−1	−1	−2	3	0	12
1	−1	−2	2	1	1	0	1
0	−1	0	5	1	10	1	0

$x_j \geq 0$ for all j, min z

(1) Given the following, find the primal and dual basic solutions of this problem wrt the basic vector (x_1, x_2, x_3). Check whether this basic vector is primal and/or dual feasible. Also, construct the inverse tableau corresponding to this basic vector, using this information.

$$\begin{pmatrix} 0 & 1 & 2 \\ 2 & -1 & -1 \\ 1 & -1 & -2 \end{pmatrix}^{-1} = \begin{pmatrix} 1 & 0 & 1 \\ 3 & -2 & 4 \\ -1 & 1 & -2 \end{pmatrix}.$$

(2) Does this problem have alternate optimum solutions? Why? If so, find the set of all optimum solutions for this problem. Is the set of all optimum solutions bounded or unbounded?

(3) Suppose the cost coefficient of x_5, whose present value is 1, is going to change. Denote its new value by c_5. All the other data in the problem remains unchanged. Find the range of values of c_5 for which the present optimum BFS of this problem remains unchanged. What happens to the problem if c_5 is strictly outside this range, and why?

(4) Suppose c_5 remains equal to 1, but the cost coefficient of x_3 is likely to change from its present value of 0 to c_3. Find the range of values of c_3 for which the present optimum BFS of this problem remains optimal.

Beginning with the basic vector (x_1, x_2, x_3), find an optimum solution of the problem if c_3 changes from 0 to 9.

(5) Suppose $c_5 = 1$ and $c_3 = 0$, but the second RHS constant b_2 is likely to change from its present value of 12. Find the range of values of b_2 for which the basic vector (x_1, x_2, x_3) remains optimal to this problem. In this range, express the optimum primal and dual solutions as functions of the parameter b_2. What is the slope of the optimum objective value in this range?

Find the optimum solution of the problem if b_2 becomes equal to 9.

6.37. A company manufactures products A to G using two types of machines P_1, P_2 and three raw materials R_1, R_2, R_3. Relevant data is given below.

Item	Item input (in units) to make one unit of							Max. available per day
	A	B	C	D	E	F	G	
R_1 (in units)	0.1	0.3	0.2	0.1	0.2	0.1	0.2	500
R_2	0.2	0.1	0.4	0.2	0.2	0.3	0.4	750
R_3	0.2	0.1	0.1	0.2	0.1	0.2	0.3	350
P_1 time (mc. hrs.)	0.02	0.03	0.01	0.04	0.01	0.02	0.04	60
P_2 time	0.04		0.02	0.02	0.06	0.03	0.05	80
Constraint on daily output	≥ 200	≤ 800			≤ 400			
Profit ($/unit)	10	12	8	15	18	10	19	

(a) Let x_1 to x_7 denote the units of products A to G in that order produced/day. Let x_8 to x_{12} denote the slack variables associated with the supply constraint on R_1, R_2, R_3, P_1-time, P_2-time, respectively. Let x_{13}, x_{14} be the slack variables associated with the upper bound constraints on the production of B, E, respectively. Using these decision variables, formulate the product mix problem to maximize total daily profit as an LP, and solve it using one of the available software packages to obtain primal and dual optimum solutions.

(b) Verify that the optimum basic vector for this problem is $(x_2, x_3, x_4, x_5, x_9, x_{10}, x_{13})$ with the values of the basic variables in the optimum BFS as (437.5, 1037.5, 812.5, 400.0, 48.75, 19.0, 362.5).

Let π_1 to π_7 denote the dual variables associated with the supply constraints on R_1, R_2, R_3, P_1-time, P_2-time, and upper bound constraint on the daily production of B, E, respectively, in that order. Using these dual variables, write the dual problem.

Verify that the optimum dual basic solution associated with the primal basic vector $(x_2, x_3, x_4, x_5, x_9, x_{10}, x_{13})$ is $\pi = (12.5, 0, 41.25, 137.5, 0, 0, 10)$. Also, answer each of the following questions about the original problem.

(1) Are the marginal values of the various items well defined in this problem? If so, what are they?

(2) Is it worth increasing the supply of R_1 beyond the present 500 units/day? The current supplier for R_1 is unable to supply any more than the current amount. The procurement manager has identified a new supplier for R_1, but that supplier's price is \$15/unit higher than the current suppliers'. Should additional supplies of R_1 be ordered from this new supplier?

(3) The production manager has identified an arrangement by which 20 h/day of either P_1- or P_2-time can be made available at a cost of \$150/machine hour. Is it worth accepting this arrangement, and if so for which of these machines?

(4) The sales manager would like to know the relative contributions of the various products in the company's total profit. What are they?

(5) The sales manager believes that product C is priced too low for a good image. This manager claims that if the selling price of C were increased by \$2/unit, the demand for it would be 600 units/day. What is the effect of this change?

(6) The production manager claims that the manufacturing process for G can be changed so that its need for P_1-time goes down by 50% without affecting quality, demand, or selling price. What will be the effect of this change on the optimum product mix and total profit?

(7) The production manager believes that by changing specifications, it should be possible to make product B with 33.3% less of R_1, and this would have no effect on the saleability of this product. What will be the effect of this change on the optimum product mix and total profit?

(8) The company's research division has formulated a new product, H, which they believe can yield a profit of \$8–10/unit made. The input requirements to make one unit of this product will be

Item	R_1	R_2	R_3	P_1-time	P_2-time
Input	0.1	0.2	0.1	0.02	0.02

Is this product worth further consideration?

(9) The sales manager feels that the selling price/unit of product F can be increased by \$2 without affecting the demand for it. Would this lead to any changes in the optimum production plan? What is the effect of this change on the total profit?

(Shearn (1984)

6.38. A fertilizer maker can produce three types of fertilizers, call them A, B, C, using three raw materials RM-1, RM-2, RM-3. Present suppliers can supply up to a maximum of 1,700, 500, 1,100 tons of RM-1, 2, 3/day, respectively, at prices of \$50, 70, 60/ton delivered.

Other than the raw materials cost, total all other costs for the company to make any of the three fertilizers is \$20/ton. At present selling prices for the three fertilizers, company gets a net profit of \$50, \$50, \$30/ton of fertilizers A, B, C, manufactured.

The company wants to maximize their total net profit. The LP model for this involving three equality constraints is the following, where x_1, x_2, x_3 represent the tons of A, B, C made daily, and s_1, s_2, s_3 represent the slacks associated with RM-1, 2, 3 supplies.

x_1	x_2	x_3	s_1	s_2	s_3	b
3	2	1	1	0	0	1700
2	1	0	0	1	0	500
1	1	1	0	0	1	1100
50	50	30	0	0	0	$= z$, maximize

All variables ≥ 0.

The basic vector $(x_2,\ x_3,\ s_1)$ is an optimum basic vector for this problem (remember this is a maximization problem). The basis inverse corresponding to this basic vector is

$$\begin{pmatrix} 2 & 1 & 1 \\ 1 & 0 & 0 \\ 1 & 1 & 0 \end{pmatrix}^{-1} = \begin{pmatrix} 0 & 1 & 0 \\ 0 & -1 & 1 \\ 1 & -1 & -1 \end{pmatrix}.$$

(1) Compute the primal and dual basic solutions for this problem associated with the basic vector $(x_2,\ x_3,\ s_1)$, explaining very clearly how you obtained them.

Explain whether the primal and dual basic solutions computed are the unique optimum solutions for the problem, and if so why?

Do marginal values for RM-1, 2, 3 exist in this problem? Why? What are they?

(2) Is it worth it for the company to acquire additional supplies of RM-3? What about RM-1? Why?

What is the maximum price in \$/ton they can pay for additional supplies of RM-3 and still break even? Explain clearly.

(3) A new type of fertilizer, D, has been formulated by the companies chemist. To make one ton of this new fertilizer needs 1, 3, 2 tons of RM-1, 2, 3, respectively, and the other costs will be \$30/ton of D made. Explain at what selling price/ton of D it becomes profitable to make it, explaining clearly how you obtained your answer.

6.39. The vector x_B is an optimum basic vector for an LP in standard form. Show that if x_B is primal (dual) nondegenerate, then the optimum solution of the dual (primal) is unique. Is the converse true? Why?

6.40. The vector $x_B = (x_1, \cdots, x_m)^T$ is an optimum basic vector for the LP: minimize $z = cx$, subject to $Ax = b$, $x \geq 0$, where A is a matrix of order $m \times n$.

Suppose that all the data in the problem remains unchanged except c_1, the cost coefficient of the variable x_1, which is going to decrease from its present value. What will happen to the set of feasible solutions of the primal and dual problems, and the optimum objective value in the dual problem as c_1 decreases? Explain why.

Show that as c_1 decreases from its present value, that it could reach a value $\bar{c}_1$, such that if c_1 decreases any more from $\bar{c}_1$, the objective function z is unbounded below in the primal problem. Model the problem of finding this $\bar{c}_1$ as another linear programming problem, using the dual.

As the value of c_1 varies between its present value and $\bar{c}_1$, can we guarantee that x_B continues to remain as an optimum basic vector for the primal? Why?

Show that the primal has an optimum basic vector containing x_1 as a basic variable for all values of c_1 between its present value and $\bar{c}_1$.

6.41. (i) K is a convex polyhedron specified in its constraint representation as $K = \{x : Ax \geq b\}$, where A, b are given matrices of orders $m \times n,\ m \times 1$ respectively. $\bar{x} \in K$ is given. Discuss how we can find the smallest dimensional face of K containing $\bar{x}$ and determine its dimension.

(ii) Now suppose $\hat{x} \in P$, where P is specified as the convex hull of its set of extreme points $\{x^1, ..., x^r\}$. Discuss whether the smallest dimension face of P containing $\hat{x}$ and its dimension can be determined efficiently in this case.

6.42. Consider the following LP in standard form. The inverse of the basis corresponding to the optimum basic vector $(x_1,\ x_2,\ x_3)$ for it is B^{-1}, where $B^{1}_{1.} = (1, -1, 0)$; $B^{-1}_{2.} = (0, 1, -1)$; $B^{-1}_{3.} = (0, 0, 1)$.

x_1	x_2	x_3	x_4	x_5	x_6	$-z$	Original RHS
1	1	1	1	2	−1	0	17
0	1	1	2	1	1	0	14
0	0	1	1	1	2	0	10
−2	−5	1	3	6	14	1	0

$x_j \geq 0$ for all j, minimize z

Write the optimum solution and the optimum objective value of the primal and the dual.

Is the optimum dual solution unique? Why?

Does this LP have a marginal value vector? If so, what is it?

(1) Suppose the RHS vector changes from the present $(17, 14, 10)^T$ to $(17 + \delta_1,\ 14 + \delta_2,\ 10 + \delta_3)^T$. Find the joint range of variation of $\delta_1,\ \delta_2,\ \delta_3$ that keeps the basic vector $(x_1,\ x_2,\ x_3)$ primal feasible.

(2) The cost coefficient of x_4 whose present value is 3 is changing. Representing its new value by c_4, find its optimality range.

If the new value of c_4 is -3, find the new optimum solution from the present one showing all your work carefully.

(3) Consider the original LP again with $c_4 = 3$ as originally given. Now suppose the RHS constant in row 2, call it b_2 (present value is 14), is changing. Find the optimality range of b_2 showing your work carefully.

In this range express the optimum solution and objective value of the problem as a function of b_2.

Suppose the new value of b_2 is 18. Find the new optimum solution starting from the present one, showing all your work carefully.

6.43. Consider the LP: minimize $z = cx$, subject to $Ax = b,\ x \geq 0$, where A is a matrix of order $m \times n$, which is known to be nondegenerate, and has a bounded feasible solution set. Also, there are feasible solutions in which $x_1 = 0$, and those in which $x_1 > 0$.

All the data in this LP will remain fixed, except c_1, the cost coefficient of x_1, which can vary.

Show that there exists a value $\hat{c}_1$ for c_1 such that for all $c_1 > \hat{c}_1$, $x_1 = 0$ (and so is not contained as a basic variable in any optimum basic vector for this LP), and for all $c_1 < \hat{c}_1$ the variable $x_1 > 0$ in every optimum solution (and hence x_1 is a basic variable in every optimum basic vector for this LP). Formulate the problem of finding this $\hat{c}_1$ as an LP. (From Dantzig and Thappa, vol. 1, 1997, of Chap. 1).

6.44. Construct a numerical example of an LP in which an optimum basic vector x_B for the LP, minimize $z = cx$, subject to $Ax = b$, $x \geq 0$, where A is a matrix of order $m \times n$, is no longer an optimum basic vector when some of the cost coefficients in the basic cost vector c_B are decreased from their values in the original c.

6.45. Consider the following LP in standard form:

x_1	x_2	x_3	x_4	x_5	x_6	$-z$	b
1	1	0	0	-3	-1	0	9
0	-1	2	-5	3	0	0	2
0	1	-1	3	-2	-1	0	1
1	-1	3	-4	8	2	1	0

$x_j \geq 0$ for all j, minimize z

An optimum inverse tableau for this LP is given below.

An optimum inverse tableau

BV	Inverse				$\bar{b}$
x_1	1	-1	-2	0	5
x_2	0	1	2	0	4
x_3	0	1	1	0	3
$-z$	-1	-1	1	1	-10

(1) Write an optimum dual solution. Is it the unique optimum solution of the dual problem? Why?

(2) This LP is the model for minimizing the cost at a company. The company has found out that they can make changes in their operations that will result in a change in the value of b_3 in the model from 1 to 2, leaving everything else unchanged. Is this change worth carrying out? Explain clearly. If this change comes at a price what is the maximum they can pay for it and still break even?

(3) Consider the original LP again. Find the optimality range of the RHS constant b_3 for the present basic vector (x_1, x_2, x_3).

Write the optimum primal and dual solutions and the optimum objective value of this LP as a function of b_3 in its optimality range.

Suppose b_3 changes from 1 to 4. Find the optimum solution of the modified problem starting with the present basic vector (x_1, x_2, x_3) using the revised format (i.e., use inverse tableaus, not canonical tableaus) showing all your work.

(4) Consider the original LP again. The company has found a new activity. This new activity will result in a new nonnegative variable x_7 in the model associated with the column vector $A_{.7} = (-2, -3, 1, c_7)^T$. Find the optimality range for c_7 in which the basic vector (x_1, x_2, x_3) remains optimal to the new model.

Solve the new model assuming that $c_7 = -7$, again using the revised format starting with the present basic vector (x_1, x_2, x_3). Either find the new optimum solution or an extreme half-line along which $z \to -\infty$ if it is unbounded.

6.46. Consider the bounded variable LP:

$$\begin{aligned} \text{minimize } z &= cx \\ \text{subject to } Ax &= b \\ \ell \leq x &\leq u. \end{aligned}$$

Here A is an $m \times n$ matrix, $\ell < u$, and both ℓ and u are finite vectors. Write the dual of this problem.

Looking at the dual very carefully, show that it has a feasible solution.

In fact, assume that $n = 5$, $c = (-2, 3, 0, -8, 10)$. Given only this data about the original problem, write down a feasible solution of the dual problem.

6.47. Let A be an $m \times n$ matrix and $A_{1.}, \cdots, A_{m.}$ its row vectors. Consider the following LP:

$$\begin{aligned} \text{minimize } z &= cx \\ \text{subject to } \ell \leq Ax &\leq u. \end{aligned}$$

Here $\ell = (\ell_1, \cdots, \ell_m) < u = (u_1, \cdots, u_m)$. There are $2m$ constraints in the above LP, they are $\ell_i \leq A_{i.}x \leq u_i$ for $i = 1$ to m. Write the dual of this LP.

(1) Looking at the constraints in the dual problem carefully, determine conditions on A, c to guarantee that the dual is feasible.

(2) We are given a feasible solution $\bar{x}$ to this LP. Prove that there exists a half-line through $\bar{x}$ in the feasible region of this LP, along which $z \to -\infty$ iff the rank of the augmented matrix $\begin{pmatrix} A \\ \cdots \\ c \end{pmatrix} > \text{rank}(A)$.

6.48. Consider the following LP, which we denote by (P).

x_1	x_2	x_3	x_4	x_5	x_6	$-z$	RHS
1	0	0	−1	−1	2	0	2
0	1	0	1	−2	−1	0	1
0	0	1	−2	2	−1	0	−4
−2	−1	2	1	16	1	1	0

$x_j \geq 0$ for all j, min z

Let (RP) denote the problem obtained by relaxing just the one constraint "$x_3 \geq 0$" from (P), but leaving everything else unchanged. Prove that the basic solution of (P) corresponding to the basic vector (x_1, x_2, x_3) is an optimum solution of (RP), even though it is not an optimum solution of (P). Using this, show that the optimum objective value in (P) is ≥ -13.

Solve the original problem (P) starting with the basic vector (x_1, x_2, x_3), using the revised simplex format, showing all your work very clearly.

6.49. We are given the LP:

x_1	x_2	x_3	x_4	x_5	
2	−1	1	1	0	$= 3$
−3	2	−1	1	0	≥ -10
4	1	2	1	−1	≤ 3
4	−1	−1	1	2	≤ 15
−18	2	−1	−4	−4	$= z$ minimize

$x_1, x_2, x_3, x_4, x_5 \geq 0$, x_4 unrestricted

Check whether $\bar{x} = (2, 0, 0, -1, 4)^T$ is an optimum solution to this LP.

6.50. Consider the following LP:

x_1	x_2	x_3	x_4	x_5	x_6		
1	0	−1	2	3	2	$\geq$	6
−2	1	3	−1	2	1	$\geq$	−4
0	2	1	3	1	4	$\geq$	12
3	2	2	3	−1	−2	$\geq$	12
−2	11	22	4	9	6	$= z(x)$	

$x_j \geq 0$ for all j, minimize $z(x)$

It is claimed that $\bar{x} = (0, 0, 0, 4, 0, 0)^T$ is an optimum solution of this LP.

Without actually solving this LP using an algorithm, describe how you can verify whether the claim is true or not using the optimality conditions. Apply this procedure.

Did the procedure lead to a definitive conclusion whether the claim is true or not? Why?

In case the procedure did not lead to a definitive conclusion, write down a system of linear constraints with the property that the claim is true iff that system has a feasible solution.

6.51. Show that the following LP is infeasible.

$$
\begin{aligned}
\text{max.} \quad & 3x_1 - 2x_2 + 10x_3 \\
\text{s. to} \quad & 2x_1 + 5x_2 + 8x_3 \leq 3 \\
& -3x_1 + 8x_2 - 13x_3 \leq -5 \\
& x_1, x_2, x_3 \geq 0.
\end{aligned}
$$

6.52. The following is the LP model for maximizing the *net profit (gross sales − production costs)* at a company. z is the *net profit function* to be maximized.

x_1	$\dots$	x_j	$\dots$	x_n	
a_{11}	$\dots$	a_{1j}	$\dots$	a_{1n}	$\leq b_1$
$\vdots$		$\vdots$		$\vdots$	$\vdots$
a_{m1}	$\dots$	a_{mj}	$\dots$	a_{mn}	$\leq b_m$
p_1	$\dots$	p_j	$\dots$	p_n	$= z =$ net profit, max.

$$x_j \geq 0 \text{ for all } j.$$

For $i = 1$ to m, b_i is the maximum quantity of raw material R_i that the present supplier of R_i can supply daily to the company.

Write the precise definition of the marginal value of R_i in this problem and its practical interpretation.

Given the following information, answer the questions following it:

$b_1 = 100$ tons. Marginal value of R_1 is = \$10/ton. Purchase price of R_1 from present supplier = \$57/ton.

How much of the available 100 tons of R_1 is being used up in the optimum solution of the present model? Explain why clearly.

If a new supplier offers to provide additional quantities of R_1, what is the maximum the company can afford to pay him per ton and still break even? Explain clearly.

6.53. In applications, what is the most commonly used technique for converting an infeasible system of linear constraints into a feasible one? Why?

Consider the following LP model for a problem:

$$\begin{array}{llllllll}
\min z = & -x_1 & -x_2 & +2x_3 & +x_4 & -3x_5 & +4x_6 & \\
\text{subject to} & x_1 & & +x_3 & -x_4 & & -x_6 & \leq 5 \\
 & & x_2 & -x_3 & +x_4 & +2x_5 & -x_6 & \leq 2 \\
 & x_1 & +2x_2 & -x_3 & +x_4 & +3x_5 & -3x_6 & \geq 11
\end{array}$$

$$x_j \geq 0 \quad \text{for all } j.$$

Solve this LP, selecting entering variables that would lead to pivot elements of "1" as far as possible to keep the computation simple. If this model is infeasible, show how the modification discussed above applies to make the system feasible.

Continue the application of the algorithm to obtain an optimum solution of the modified problem.

6.54. Consider the following LP in standard form in which $\alpha,\ \beta,\ \gamma,\ \delta,\ \Delta$ are real-valued parameters.

x_1	x_2	x_3	x_4	x_5	x_6	x_7	$-z$	
1	0	0	-1	1	1	-1	0	4
0	1	0	0	γ	1	-1	0	10
0	0	1	α	δ	-1	-1	0	0
0	0	-1	β	6	Δ	3	1	-10

$x_j \geq 0$ for all j, minimize z

(1) If $\alpha + \beta \geq 0$, $\delta + 6 \geq 0$, $\Delta - 1 \geq 0$, outline a complete proof to show that the minimum value of z in this problem is 10. In this case write an optimum feasible solution.

If $\alpha \leq 0$ and $\alpha + \beta < 0$, show that the minimum value of z in this problem is $-\infty$ by actually producing a set of feasible solutions along which it diverges to $-\infty$.

(2) What conditions should the parameters γ, δ satisfy for x_5 to be eligible to enter the basic vector (x_1, x_2, x_3) when this problem is solved by the primal simplex algorithm and for this operation to lead to x_1 as the unique dropping basic variable? Explain very clearly.

(3) Write the parametric representation of the edge generated when x_6 is selected as the entering variable into the basic vector (x_1, x_2, x_3). As you move away from the BFS corresponding to the basic vector (x_1, x_2, x_3) along this edge, at what rate does the value of z change? Explain very clearly.

6.55. Consider the following LP

x_1	x_2	x_3	x_4	x_5	x_6	x_7	$-z$	b
1	0	0	0	-3	2	-2	0	3
0	1	0	-10	ξ	-1	-3	0	4
0	0	1	β	2	3	0	0	α
0	0	0	γ	δ	Δ	7	1	-10

$x_j \geq 0$ for all j. z to be minimized.

The entries $\alpha, \beta, \gamma, \delta, \Delta, \xi$ in the tableau are real-valued parameters. Let B_1 be the basis for this problem corresponding to the basic vector $x_{B_1} = (x_1, x_2, x_3)$. Write down the range of values of each of the parameters $\alpha, \beta, \gamma, \delta, \Delta, \xi$, which will make the conclusions in the following statements true (each of the following bits is independent of the others).

(1) B_1 is not a primal feasible basis for this problem.

(2) From this tableau it is possible to select an initial primal feasible basic vector and initiate Phase II of the primal simplex method to solve this problem.

(3) B_1 is a primal feasible but nonoptimal basis for this problem, but the updated column vector of x_4 indicates that this problem has no finite optimum solution.

(4) B_1 is a primal feasible basis for this problem, x_5 is a candidate to enter the basic vector x_{B_1}, and when x_5 enters, x_2 is the unique dropping basic variable.

(5) B_1 is a primal feasible basis for this problem, x_6 is eligible to be the entering variable into the basic vector x_{B_1}, but when x_6 enters this basic vector, the objective value remains unchanged.

6.56. Consider the following LP

$$\begin{aligned} \min z &= -x_3 + x_4 \\ \text{s. to} \quad x_1 - x_2 + x_3 - x_4 &\geq 6 \\ 2x_1 - x_2 &\leq 14 \\ x_1 - x_3 + x_4 &\geq 8 \\ x_j &\geq 0 \quad \text{for all } j. \end{aligned}$$

Solve this problem by the revised primal simplex method using the following entering and dropping variable selection rules in each pivot step: the entering variable is the eligible x_j with the smallest subscript j; the dropping variable is chosen so that the pivot row is always the topmost among rows eligible to be pivot rows in this step.

Clearly explain the final conclusion from the algorithm. If the objective function is unbounded below in the problem, write the parametric representation of the extreme half-line along which it is unbounded, write the direction vector of this half-line, and a feasible point on the half-line where the objective function assumes the value −10,000.

6.57. For the following LP construct the inverse tableau corresponding to the basic vector (x_1, x_2, x_3).

x_1	x_2	x_3	x_4	x_5	x_6	$-z$	
1	0	2	3	1	0	0	3
0	1	1	3	0	3	0	3
0	0	1	2	1	1	0	3
1	1	4	18	7	24	1	0

$x_j \geq 0$ for all j, min z

Is the basic vector (x_1, x_2, x_3) primal feasible for this problem? Is it dual feasible?

Solve this LP using (x_1, x_2, x_3) as the initial basic vector.

Suppose the cost coefficient c_2 of x_2 in this problem, whose present value is 1, is likely to change, while all the other data remains fixed.

Find the range of values of c_2 for which the optimum solution obtained above remains optimal to the problem, showing all your work carefully. Also, in this optimality range for c_2, write the optimum dual solution and the optimum objective value as functions of c_2.

6.58. Consider the following two LPs with their optimum solutions given:

x^1 is an optimum solution for: minimize cx, subject to $A^1x = b^1$, $x \geq 0$
x^2 is an optimum solution for: minimize $2cx$, subject to $A^2x = b^2$, $x \geq 0$.

In these problems, $A^1_{m_1 \times n}$, $b^1_{m_1 \times 1}$, $A^2_{m_2 \times n}$, $b^2_{m_2 \times 1}$ are given matrices, and the vector $c_{1 \times n}$ is the same in both the problems. In both these problems, the vector of variables is $x = (x_1, \cdots, x_n)^T$.

Let $x^3 = (x^1 + x^2)/2$, and suppose that x^3 is a feasible solution of the system of constraints $A^1 x = b^1$, $A^2 x = b^2$, $x \geq 0$. Then prove that x^3 is an optimum solution to the problem of minimizing cx subject to these constraints.

6.59. Consider the LP in standard form: minimize $z = cx$, subject to $Ax = b$, $x \geq 0$, where A is a matrix of order $m \times n$ and rank m.

This problem was solved by the primal simplex method. A BFS was obtained at the end of Phase I, and Phase II terminated with the following final canonical tableau

Final canonical tableau

Basic Vars.	x_1	x_2	...	x_m	x_{m+1}	...	x_n	
x_1	1	0	...	0	$\bar{a}_{1,m+1}$	...	$\bar{a}_{1,n}$	$\bar{b}_1$
x_2	0	1	...	0	$\bar{a}_{2,m+1}$	...	$\bar{a}_{2,n}$	$\bar{b}_2$
$\vdots$	$\vdots$	$\vdots$		$\vdots$	$\vdots$		$\vdots$	$\vdots$
x_m	0	0	...	1	$\bar{a}_{m,m+1}$	...	$\bar{a}_{m,n}$	$\bar{b}_m$

in which the unboundedness criterion is satisfied. Write a clear and complete proof to show that the dual problem is infeasible.

6.60. Get the inverse tableau for the following problem wrt the basic vector (x_1, x_2, x_3) and solve the problem using it as the initial feasible basic vector.

x_1	x_2	x_3	x_4	x_5	x_6	$-z$	b
1	1	0	−5	1	1	0	14
1	0	1	−3	0	2	0	15
0	−1	2	2	0	−3	0	6
−6	−3	−2	26	−4	20	1	0

$x_j \geq 0$ for all j, minimize z

6.61. Check whether $\bar{x} = (2, 3, 1, 2, 1, 0, 0)^T$ is on an edge of the set of feasible solutions of the following system. If it is, **(1)** obtain a parametric representation of the edge and all the extreme points on it, mentioning whether each of them is primal degenerate or not; and **(2)** determine whether it is bounded or not.

Also, discuss what role edges play in the simplex algorithm for solving linear programs.

x_1	x_2	x_3	x_4	x_5	x_6	x_7	b
1	2	8	0	1	3	−1	17
0	1	3	1	0	−1	2	8
0	1	3	−1	0	−1	−4	4
−1	2	4	0	1	2	6	9
1	0	2	0	−1	8	9	3

$x_j \geq 0$ for all j

6.62. The symbols $c_{1\times n}$, $A_{m\times n}$, $W_{p\times n}$, $b_{m\times 1}$, $d_{p\times 1}$ denote given matrices. Consider the following linear programs in variables $x = (x_1, \ldots, x_n)^T$.

(1)	(2)	(3)
min cx	min cx	min cx
s. to $Ax = b$	s. to $Wx = d$	s. to $Ax = b$
		$Wx = d$
$x \geq 0$	$x \geq 0$	$x \geq 0$

It is known that $x^* \in R^n$ is an optimum solution to both (1) and (2).

Then, can we conclude that x^* is also optimal to (3)? If so, provide a proof. Otherwise, provide an argument why you think it may be false.

6.63. Let A be a square nonsingular matrix of order n and consider the parallelogram $P = \{x : \ell \leq Ax \leq k\}$ where $\ell < k$. We need to find a maximum [minimum] area rectangle inside P [containing P] whose sides are parallel to the coordinate axes. Give formulations for these problems.

Also, discuss whether these problems can be solved by solving a series of LPs. Discuss a method for solving these problems.

6.64. The vectors c, c^1, c^2 are given row vectors in R^n; A, W are given matrices of order $m \times n$; and b, b^1, b^2 are given column vectors in R^m. Consider the following LPs:

(1)	(2)	(3)
min $c^1 x$	min $c^2 x$	min $(c^1 + c^2)x$
s. to $Ax = b$	s. to $Wx = b$	s. to $(A + W)x = b$
$x \geq 0$	$x \geq 0$	$x \geq 0$

If (1) and (2) are both known to have feasible solutions, can we conclude that (3) must have a feasible solution too? If true, give a rigorous proof; otherwise give a clear explanation why you think it may not be true.

6.65. Consider the bounded variable LP

$$\begin{aligned} \text{Minimize } z &= cx \\ \text{s. to } Ax &= b \\ 0 \leq x &\leq u, \end{aligned}$$

where $u > 0$ is a finite upper bound vector for the variables, and A is a matrix of order $m \times n$ and rank m.

Given a feasible solution $\bar{x}$ to this problem, write the conditions it has to satisfy for being an extreme point of the set of feasible solutions of this problem.

If $\bar{x}$ satisfies the conditions for being an extreme point, state clearly when it is degenerate, nondegenerate.

(2) Consider the following bounded variable LP

x_1	x_2	x_3	x_4	x_5	x_6	x_7	$-z$	b
1	0	0	0	−1	1	−1	0	−1
0	1	0	0	−1	−1	1	0	2
0	0	1	0	−1	0	−1	0	−2
0	0	0	1	0	1	1	0	6
−2	2	1	2	−4	2	6	1	0

$$\min z;\ 0 \leq x_j \leq 10 \text{ for } j = 1 \text{ to } 4;$$
$$0 \leq x_j \leq 5 \text{ for } j = 5, 6, 7$$

Check whether $\bar{x} = (9, 2, 8, 1, 5, 0, 5)^T$ is an extreme point solution to this problem.

Starting with $\bar{x}$, find an optimum solution of this problem using the bounded variable primal simplex algorithm.

For this problem, what are the marginal values of the RHS constants b_1 to b_4, and u_5, u_6, u_7, the upper bounds of x_5, x_6, x_7? Explain carefully.

6.66. If both $\min\{c^1x : Ax = b, x \geq 0\}$ and $\min\{c^2x : Wx = b, x \geq 0\}$ have feasible solutions (here c^1, c^2 may be different vectors, and A, W may be different matrices), can we conclude that $\min\{(c^1 + c^2)x : Ax + Wx = b, x \geq 0\}$ also has a feasible solution? Why?

Further, if x^* is an optimal solution for the problems $\min\{cx : Ax = b^1, x \geq 0\}$ and $\min\{cx : Wx = b^2, x \geq 0\}$, then can we conclude that x^* is also an optimal solution for the problem $\min\{cx : Wx = b^2, Ax = b^1, x \geq 0\}$? Why?

6.67. Consider the following LP involving four variables x_1 to x_4. Show that it is equivalent to a two-variable LP. Using this equivalence, find its optimum solution geometrically.

$$\begin{array}{lrrrrcl}
\text{Maximize } z = & x_1 & & & +x_4 & & \\
\text{s. to} & x_1 & & +2x_3 & -x_4 & = & 1 \\
 & x_1 & +x_2 & & +2x_4 & = & 2 \\
 & & & & & & \\
 & x_1, & x_2, & x_3, & x_4 & \geq & 0.
\end{array}$$

6.68. Verify that (x_1, x_2, x_3) is a unit basic vector, which is infeasible for the following system.

Find a feasible solution to it by Phase I of the primal simplex method using only one artificial variable.

If the system is infeasible, mention a way to modify the RHS data in the original system to make it feasible, and write down a feasible solution of the modified system.

x_1	x_2	x_3	x_4	x_5	x_6	
1	0	0	−1	0	1	−2
0	1	0	0	−1	1	−5
0	0	1	0	1	0	4

$x_j \geq 0$ for all j.

6.69. Consider the following LP in standard form:

x_1	x_2	x_3	x_4	x_5	x_6	x_7	$-z$	b
1	0	0	−1	2	−1	−1	0	4
−1	1	0	2	−3	0	−1	0	2
1	1	1	−1	2	−3	−2	0	16
−1	−1	−1	4	2	13	16	1	0

$x_j \geq 0$ for all j; min z

The optimum inverse tableau for this problem is given below.

Basic var.	Inverse tableau				Basic values
x_1	1	0	0	0	4
x_2	1	1	0	0	6
x_3	−2	−1	1	0	6
$-z$	0	0	1	1	16

(1) Do marginal values exist for this LP? If so, what are they? Explain.

(2) If the company has an opportunity to change only one of the RHS constants, b_is, in the model, which b_i should they change to improve objective function further? Should they increase it or decrease it? Explain.

(3) The first RHS constant, b_1, is changing from its present value of 4. Find the range of values of b_1 for which the marginal value vector found in (1) remains unchanged.

(4) Find an optimum solution for the problem if b_1 changes to 8, beginning with the present basic vector (x_1, x_2, x_3).

6.70. The following LP is being solved by the bounded variable primal simplex algorithm.

x_1	x_2	x_3	x_4	x_5	x_6	x_7	$-z$	b
1	1	1	−3	1	−3	1	0	−4
0	1	1	−2	1	−2	−2	0	0
0	0	1	−1	−1	−2	1	0	−7
1	2	2	−2	0	−12	−3	1	0

$0 \leq x_j \leq 5$ for all j, min z

Here is the present BFS.

Basic var.	Inverse tableau				Basic values
x_1	1	−1	0	0	1
x_2	0	1	−1	0	2
x_3	0	0	1	0	3
$-z$	−1	−1	0	1	−1

Nonbasics $x_4 = x_5 = 5$, nonbasics $x_6 = x_7 = 0$.

Carry out one pivot step with the entering variable as an eligible variable whose present value is at its upper bound.

Now carry out another pivot step with the entering variable as an eligible variable whose present value is at its lower bound.

6.71. Consider the following system of linear constraints.

x_1	x_2	x_3	x_4	x_5		
1	2	1	1	1	=	3
1	1	−1	1	1	=	3
1	3	1	0	2	≤	5
0	0	1	0	1	≤	1

$x_j \geq 0$ for all j

It is required to find a feasible solution to this system. Find it using Phase I of the simplex method. When selecting entering variables, look for an eligible one that will lead to a pivot element of 1 as far as possible to keep arithmetic simple.

Can you conclude that certain variables must be zero in every feasible solution? Explain clearly.

6.72. Consider the following LP

$$
\begin{array}{rrrrrcl}
\min z = & -x_1 & -10x_2 & +x_3 & & & \\
\text{s. to} & -x_1 & +x_2 & & +x_4 & = & 2 \\
 & & -2x_2 & +x_3 & +x_4 & = & 3 \\
 & x_1 & +x_2 & -x_3 & & \leq & 2 \\
 & & -x_2 & +x_3 & +x_4 & \leq & 3 \\
 & -x_1 & +2x_2 & -x_3 & & = & 1
\end{array}
$$

$$x_j \geq 0 \text{ for all } j.$$

Solve this problem by the revised primal simplex method using Dantzig's entering variable choice rule in every pivot step.

If the problem is infeasible, show the modification of the original RHS constants vector b to b' indicated by the terminal Phase I tableau to make this model feasible. Also, write the final Phase I tableau for the modified problem, indicating the feasible solution to it contained there.

Explain clearly how to go to Phase II from the final Phase I tableau in order to get an optimum solution for the modified problem. Carry out Phase II, showing all your work clearly, and obtain an optimum solution for the modified problem.

6.73. Consider the following two-variable LP

$$\begin{aligned} \text{Maximize} \quad & z = x_1 \\ \text{subject to} \quad & x_1 + x_2 \leq b_1 = 4 \\ & 2x_1 - x_2 \leq 2 \\ & x_1 \leq 2 \\ & x_1, \ x_2 \geq 0. \end{aligned}$$

Let $f(b_1)$ denote the optimum objective value in this LP as a function of the RHS constant b_1 whose present value is 4. If you notice any redundant constraints in this LP, do not eliminate them.

Write the dual of this LP. Solve the LP geometrically and find the optimum faces of both the primal and dual. Does either of these problems have a unique optimum solution?

Is there a nondegenerate optimum BFS for the primal?

Compute the ratio $(f(4+\epsilon)-f(4))/\epsilon$, where ϵ is a real number of small absolute value. Does the value of this ratio depend on whether ϵ is ≥ 0 or ≤ 0?

Does this LP have a marginal value wrt b_1? Why? If not, does it have positive and negative marginal values wrt b_1? What are they?

References

Bartels RH (1971) A stabilization of the simplex method. Numerische Mathematik 16:414–434

Bartels RH, Golub GH (1969) The simplex method of linear programming using LU decomposition. Commun Assoc Comput Machinery 12:266–268

Fourer R (2005) Linear programming software survey. ORMS Today 32(3):46–55

Greenberg HJ (1978) Design and implementation of optimization software. Sijthoff and Noordhoff, Alpen aan den Rijn

Harris PMJ (1973) Pivot selection methods of the Devex LP code. Math Program 5:1–28

Lemke CE (1954) The dual method of solving the linear programming problem. Naval Research Logistics Quarterly 1:36–47

Shearn DCS (1984) Postoptimal analysis in linear programming – the right example. IIE Trans 16(1):99–101

Todd MJ (1983) Large scale linear programming: geometry, working bases and factorizations. Math Program 26:1–20

Chapter 7
Interior Point Methods for LP

7.1 Boundary Point and Interior Point Methods

In a linear program, typically there are inequality constraints, and equality constraints, on the variables. In LP literature, a feasible solution is known as a:

- **boundary feasible solution:** if it satisfies at least one inequality constraint in the problem as an equation;
- **interior feasible solution:** if it satisfies all inequality constraints in the problem as strict inequalities.

Methods for solving LPs which move along boundary feasible solutions are called **boundary point methods**; and those that move only among interior feasible solutions are called **interior point methods**.

Variants of the simplex method discussed so far move along edges (1-dimensional boundary faces) of the feasible region K, and hence these are classified as **1-dimensional boundary methods**.

(Murty [1986] and Chang, Murty [1989]) developed newer methods for solving LPs using analogs of the gravitational force. These methods introduce a spherical ball with small radius lying completely in the interior of K, and trace the path of its center as the drop falls under a gravitational force pulling it in the direction of the negative gradient of the objective function to be minimized. The ball cannot cross the boundary of K, so after an initial move it will be blocked by the face of K that it touches; after which it will start rolling down along the faces of K of varying dimensions. Hence the center of the ball will follow a piecewise linear descent path completely contained in the interior of K, but since the drop's radius is small, the center remains very close to the boundary of K. Therefore these methods are essentially boundary methods. However, unlike the simplex method which follows a path strictly along the 1-dimensional boundary of K, these methods are **higher-dimensional boundary methods.**

In this chapter we will discuss interior point methods for LP whose development is primarily fueled by the fundamental work by (Narendra K. Karmarkar [1984] of Chapter 1) which had tremendous impact on algorithms for solving LPs. The boundary methods get entangled with the combinatorial complexity of the boundary of the

K.G. Murty, *Optimization for Decision Making: Linear and Quadratic Models*, International Series in Operations Research & Management Science 137,
DOI 10.1007/978-1-4419-1291-6_7, © Springer Science+Business Media, LLC 2010

feasible region, and the amount of work needed by them to solve an LP typically grows exponentially with the size of the problem in the worst case. Realizing this, Karmarkar developed an interior point method, which he showed to be a polynomial time method and claimed that it will solve large-scale LPs much faster than boundary methods. This claim attracted lot of researchers to the area, and very quickly many other such interior point methods have been developed.

We will use the abbreviation "**IPM**" for "interior point method" in this book.

7.2 Interior Feasible Solutions

In mathematics, if Γ is the set of feasible solutions of a system of linear inequality constraints

$$Dx \geq d \tag{7.1}$$

then a point $\bar{x} \in \Gamma$ is said to be an interior point of Γ iff $D\bar{x} > d$. Some books refer to interior points of Γ as **strict feasible solutions**.

The main characteristic of an interior point $\bar{x} \in \Gamma$ of importance to optimization algorithms is that from $\bar{x}$, one can move in any direction y a positive step length $\alpha > 0$ leading to $\bar{x} + \alpha y$ which is also in Γ. Contrast this to a boundary point $\tilde{x} \in \Gamma$ which satisfies the first constraint in (7.1) as an equation, i.e., $D_{1.}\tilde{x} = d_1$; then in any direction y satisfying $D_{1.}y < 0$, any move of positive length, however small, will take you outside Γ.

If K is the set of feasible solutions of the system

$$\begin{aligned} Ax &= b \\ Dx &\geq d \end{aligned} \tag{7.2}$$

with some equality constraints, then mathematically K has no interior (i.e., no interior points), and every point of K is a boundary point. A point $\bar{x} \in K$ satisfying $D\bar{x} > d$ is called a **relative interior point** of K, its characteristic property is that in any direction y satisfying $Ay = 0$, one can move a positive step length while still remaining inside K.

In linear programming, constraint systems that we have to deal with are typically in the form (7.2), and in LP literature it has become a common practice to adopt the definitions given in Section 7.1 for interior and boundary points, contrary to established mathematical terminology. So, we will use these definitions in the sequel.

7.3 General Introduction to Interior Point Methods

There are many different interior point methods **(IPMs)** now for LP, and the literature on them is very vast. We will discuss some of the common features of all of them.

1. Need an initial interior feasible solution: Most IPMs need an initial interior feasible solution to initiate the method. If an interior feasible solution to the model is not available, the problem can be modified by introducing one artificial variable with a large big-M cost coefficient in the objective function to be minimized, into a Phase I problem for which an initial interior feasible solution is readily available (exactly like the big-M Phase I modification used in the simplex method). We show these modifications first. Suppose the problem to be solved is in the form:

$$\begin{aligned} \text{Minimize} \quad & cx \\ \text{subject to} \quad & Ax \geq b \end{aligned} \tag{7.3}$$

where the coefficient matrix A is of order $m \times n$. For LPs in this form, typically $m \geq n$. Introducing the non-negative artificial variable x_{n+1}, the Phase I modification of the original problem is:

$$\begin{aligned} \text{Minimize} \quad & cx + Mx_{n+1} \\ \text{subject to} \quad & Ax + ex_{n+1} \geq b \\ & x_{n+1} \geq 0 \end{aligned} \tag{7.4}$$

where $e = (1, \dots, 1)^T \in R^m$, and M is a positive number significantly larger than any other number in the problem. Let $x^0_{n+1} > \max \{0, b_1, b_2, \dots, b_m\}$. Then $(0, \dots, 0, x^0_{n+1})^T$ is an interior feasible solution of the Phase I modification which is in the same form as the original problem. If the original problem has an optimum solution, and M is sufficiently large, then the artificial variable x_{n+1} will be 0 at an optimum solution of the Phase I modification.

Now suppose the original problem is in the form:

$$\begin{aligned} \text{Minimize} \quad & cx \\ \text{subject to} \quad & Ax = b \\ & x \geq 0 \end{aligned}$$

where A is a matrix of order $m \times n$. For LPs in this form typically $n > m$, and an interior feasible solution is one which is strictly > 0. Select an arbitrary vector $x^0 \in R^n$, $x^0 > 0$; generally one chooses $x^0 = (1, \dots, 1)^T$, the n-vector of all 1's. If x^0 happens to be feasible to the problem, it is an interior feasible solution, use it as the interior feasible solution to initiate the algorithm. Otherwise, let $A_{.n+1} = b - Ax^0$. The Phase I modification including the non-negative artificial variable x_{n+1} is:

$$\begin{aligned} \text{Minimize} \quad & cx + Mx_{n+1} \\ \text{subject to} \quad & Ax + A_{.n+1}x_{n+1} = b \\ & x,\ x_{n+1} \geq 0. \end{aligned}$$

It is easily confirmed that $(x^0,\ x^0_{n+1})$, where $x^0_{n+1} = 1$ is an interior feasible solution of the Phase I problem which is in the same form as the original problem. Again, if the original problem has an optimum solution and M is sufficiently large, then the artificial variable x_{n+1} will be 0 at an optimum solution of the Phase I modification.

The general LP is in the form

$$\begin{aligned} \text{Minimize} \quad & cx \\ \text{subject to} \quad & Ax = b \\ & Dx \geq d \end{aligned}$$

To construct a Phase I modification of this model with a readily available interior feasible solution, solve the system of linear equations $Ax = b$ and get a solution $\bar{x}$ for it. Then the Phase I modification of this model with an artificial variable x_{n+1} is

$$\begin{aligned} \text{Minimize} \quad & cx \\ \text{subject to} \quad & Ax = b \\ Dx + & ex_{n+1} \geq d \end{aligned}$$

where e is the column vector of all 1s in R^m, assuming D is of order $m \times n$. Let $\beta = 1 + \max\{0,\ d_i - D_{i.}\bar{x}\}$. Then $(\bar{x},\ \bar{x}_{n+1} = \beta)$ is an interior feasible solution for it.

Similar modifications can be made to an LP in any form, to get a Phase I modification in the same form with an interior feasible solution.

2. All interior feasible solutions: All points obtained in the method will be interior feasible solutions. Most of the methods are descent methods in the sense that either the objective function, or some other measure of optimality strictly improves in every iteration. Unlike the simplex method which terminates in finite time, interior point algorithms are continuous optimization algorithms that generate an infinite sequence of solutions converging to an optimum solution.

In practice, when improvements in the objective value per iteration becomes less than a selected tolerance, the method is terminated and the current solution taken as an approximate optimum solution. Arguments can be made that if the approximation is close, then an exact optimum solution can be obtained from it by some procedures which are referred to as **rounding procedures**. The following theorem describes one such rounding procedures.

Theorem 7.1. *Consider the LP in standard form*

$$\begin{aligned} \textit{Minimize} \quad & z = cx \\ \textit{subject to} \quad & Ax = b \\ & x \geq 0 \end{aligned} \tag{7.5}$$

where all the data A, c, b is integer, and A is a matrix of order $m \times n$ and rank m. Let L be the **size** *of problem (7.5), this size is defined to be the total number of binary digits in all the data in (7.5), it is a measure of the storage space needed to store all the data in the problem. Mathematically, size is the measure used to measure how large an LP with integer data is. Suppose $\bar{x}$ is an interior feasible solution at which the objective value is* **sufficiently close** *to the optimum objective value of the problem, meaning that $c\bar{x} \leq 2^{-L}$ + (the optimum objective value in (7.5)).*

Suppose you obtain a BFS of (7.5) by applying the purification routine discussed in Section 4.7 beginning with $\bar{x}$, in each step using the value of the parameter λ in the description of the purification routine in Section 4.7 as the value that leads to a decrease in the objective value. Then the final BFS obtained in this procedure will be an optimum BFS for (7.5).

Proof. Let B denote any square submatrix of

$$\begin{pmatrix} A & b \\ c & 0 \end{pmatrix}. \tag{7.6}$$

Clearly the determinant of B is an integer, and it can be shown that its absolute value is an integer $<2^L/n$ (see Theorem 15.1, Chapter 15 in (Murty [1983] of Chapter 2). Let $(x_{j_1}, \cdots, x_{j_m}, -z)$ be a basic vector for the system of equations

$$\begin{pmatrix} A & 0 \\ c & 1 \end{pmatrix} \begin{pmatrix} x \\ -z \end{pmatrix} = \begin{pmatrix} b \\ 0 \end{pmatrix}.$$

Then by Crammer's rule, the value of any basic variable in the basic solution of this system with respect to the above basic vector is the ratio $\pm(P/Q)$ where P, Q are the values of the determinants of two square submatrices of the matrix in (7.6) and $Q \neq 0$.

These results imply the objective value at any BFS of (7.5) is a ratio of the form $\pm u/v$ where $u \geq 0$, $v>0$ are integers both of which are $<2^L$. This implies that if the objective values at two BFSs of (7.5) are different, then the difference between them is $>2^{-L}$. Since (7.5) has an optimum solution, it must have an optimum BFS.

So any BFS whose objective value is $<2^{-L}+$ (the optimum objective value in (7.5)) must be an optimum BFS for (7.5). So the BFS of (7.5) obtained by applying the purification routine discussed in Section 4.7 beginning with $\bar{x}$, in each step using the value of the parameter λ that leads to a decrease in the objective value, must be an optimum BFS of the original LP. □

In practice 2^{-L} is too small, but it has been noticed that the purification procedure applied on an interior feasible solution $\bar{x}$ produces an optimum BFS even when the difference between $z(\bar{x})$ and the optimum objective value is not this small.

Unfortunately, finding the linear dependence relations required in the purification procedure take computational effort. So programmers have developed even simpler heuristic rounding procedures that are quite successful in practice.

3. General iteration: Each iteration of an IPM begins with the current interior feasible solution (for the first iteration, this is the initial interior feasible solution; for all subsequent iterations it is the final interior feasible solution at the end of the previous iteration), and typically consists of two steps.

Step 1: Determine the search direction: This is the direction to move at the current interior solution.

Step 2: Step length determination: Determine the step length of the move.

If the current solution is x^r, the selected direction is y^r, and step length is λ_r, then the final point obtained in this iteration is $x^{r+1} = x^r + \lambda_r y^r$; it will be the current solution for the next iteration. IPMs can be classified into classes by the strategy they use for selecting the search direction in Step 1. There are two major classes.

Class 1: In these methods the search direction is determined by the solution of a modified approximating problem constructed around the current interior feasible solution. Two of the earliest IPMs developed, the affine scaling method, and Karmarkar's projective scaling method belong in this class. As an example of this class, we will discuss the affine scaling method.

Class 2: These methods work with the optimality conditions for the LP (primal and dual feasibility and the CS conditions), and select the direction of search at the current interior feasible to be the one obtained by applying an appropriate variant of Newton's method for solving systems of nonlinear equations to the system of optimality conditions. Many of the path following IPMs belong in this class. As an example of this class of methods, we will discuss the primal-dual path following IPM.

The various methods differ on whether they work on the primal system only, dual system only, or the system consisting of the primal and dual systems together; on the strategy used to select the search direction d^r; and on the choice of the step length fraction parameter.

To give an idea of the main strategies used by IPMs to select the search directions, we will discuss the two most popular interior point methods.

The first is in fact the first interior point method discussed in the literature, the primal affine scaling method (Dikin [1967]), which predates Karmarkar's work, but did not attract much attention until after Karmarkar made the study of interior point methods popular. This method works on the system of constraints in the original problem (primal) only. To get the search direction at the current interior feasible solution x^r, this method creates an ellipsoid $\bar{E}_r$ centered at x^r inside the feasible region of the original LP. Minimizing the objective function over $\bar{E}_r$ is an easy problem, its optimum solution $\bar{x}^r$ can be computed directly by a formula. The search direction in this method at x^r is then the direction obtained by joining x^r to $\bar{x}^r$.

The second method that we will discuss is a central path-following primal-dual interior point method. It works on the system of constraints of both the primal and dual put together. In this method the search directions used are modified Newton directions for solving the optimality conditions. The class of path-following primal-dual methods evolved out of the work of many authors, see the list of references in (Roos et al. [2006], Saigal [1995], Wright [1997], Ye [1997], Meggido [1989]).

4. Observed number of iterations in practice: Consider LPs in standard form with m constraints in n non-negative variables. The observed average number of iterations in the simplex method to solve these problems keeps growing at a linear or higher rate, as the problem size grows (i.e., as m, n increase). But of course each iteration in the simplex method involves much less work than an iteration of an IPM. In spite of this, the simplex method takes a lot of CPU time to solve large scale LPs, when compared to IPMs.

IPMs have been observed to take much smaller number of iterations to solve LPs, and this number grows very slowly with the size of the LP model being solved; with the result that IPMs have gained the reputation of *taking almost a constant number of iterations* even as the size of the LP grows. Now-a-days, most large scale LPs are typically solved using software based on IPMs.

But the simplex method has some subtle advantages too. It is an ideal method for solving small LPs by hand. The IPMs on the other hand, are not suitable for hand computation even on small LPs. Another is that if the LP being solved has an optimum solution, then at termination the simplex method outputs an optimum BFS even when the LP has alternate optimum solutions. On LPs with alternate optimum solutions, the IPMs will output a relative interior point of the optimum face and not a BFS; to move from that point to an optimum BFS requires purification steps which add to the computational burden. Practitioners typically seem to prefer an optimum BFS over a relative interior point of the optimum face for implementation in real-world applications.

For students of LP, the simplex method and the associated discussion of extreme points and edges of the set of feasible solutions, and optimality conditions, is an ideal way to learn about the geometry of LP, how algorithms solve them, and how to derive useful planning information from the output of algorithms. This is why we have discussed them in detail in this book.

7.4 Center, Analytic Center, Central Path

The **analytic center**, or **center** of a convex polytope; and the **central path** of a linear program, are concepts used to develop interior point methods for solving linear programs and to analyze their convergence properties.

We will give the definitions of these concepts for convex polytopes represented by systems of different types.

1. First consider a convex polytope (i.e., a bounded polyhedron) Γ represented by a system of linear inequalities, let $\Gamma = \{x : v = Ax - b \geq 0\}$, where A is a matrix of order $m \times n$. $\Gamma^0 = \{x : Ax > b\}$ is its interior. In this representation of Γ through a system of linear inequalities, v is the vector of slack variables associated with the inequality constraints. The **analytic center** of Γ in this representation is defined as the point in Γ which maximizes the product of the slack variables associated with the inequality constraints in the representation, equivalently, it is the optimum solution of:

$$\begin{aligned} &\text{Maximize} \sum_{i=1}^{m} \log v_i \\ &\text{subject to} v = Ax - b \geq 0 \end{aligned} \tag{7.7}$$

The analytic center of Γ is well defined if Γ is bounded and $\Gamma^0 \neq \emptyset$, and in this case the optimum solution of (7.7) is unique since Γ is bounded and the objective function to be maximized is a strictly concave function.

If (x^*, v^*) is an optimum solution of (7.7), then by the KKT optimality conditions for it, we know that there exists a row vector y^* which together with x^*, v^* satisfies: $y^* \geq 0, v^* \geq 0$, and

$$\begin{aligned} Ax^* - v^* &= b \\ y^* A &= 0 \\ y^* v^* &= e \end{aligned}$$

The analytic center of Γ depends on the system of inequalities chosen to represent Γ; if redundant constraints are added to or deleted from the system, the analytic center changes.

2. Now consider a convex polytope K represented by a system consisting of linear equations and inequalities. Its analytic center in this representation is defined in a similar way. For example, suppose $K = \{x : Ax = b, x \geq 0\}$, where A is a matrix of order $m \times n$ and rank m. Assume that K is bounded and its set of interior feasible solutions $K^0 = \{x : Ax = b, x > 0\} \neq \emptyset$. Its analytic center in this representation is the unique optimum solution of

$$\begin{aligned} &\text{Maximize} \sum_{j=1}^{n} \log x_j \\ &\text{subject to} \quad Ax = b, \ x \geq 0 \end{aligned} \tag{7.8}$$

3. The central path for an LP: Some interior point methods like the primal-dual path following algorithms discussed later on, are based on finding a sequence of interior feasible solutions along, or in a neighborhood of, a path called the **central path** that converges to the analytic center of the optimum face of the LP. The concept of this central path is a corner stone in the development of interior point algorithms for LP. Consider the LP in standard form:

$$\begin{aligned} &\text{Minimize} \quad z(x) = cx \\ &\text{subject to} \quad Ax = b, \ x \geq 0 \end{aligned} \tag{7.9}$$

where A is a matrix of order $m \times n$ and rank m. Let $t > 0$ be a positive parameter, and consider the problem

$$\begin{aligned} \text{Minimize} \quad & cx - t\left(\sum_{j=1}^{n} \log x_j\right) \\ \text{subject to} \quad & Ax = b, \quad x \geq 0 \end{aligned} \tag{7.10}$$

Problem (7.10) is known as the **logarithmic barrier problem** associated with the original LP. It can be shown that (7.10) has a unique optimum solution for each $t > 0$ iff the LP (7.9) and its dual both have interior feasible solutions. Letting the column vector $y = (y_1, \dots, y_n)^T$ denote the vector of dual variables or Lagrange multipliers associated with the equality constraints in (7.10), the optimality conditions for a feasible solution of (7.10), which we will call $x(t)$, to be optimal to (7.10), is that there exists a $y(t)$ which together with $x(t)$ satisfies the KKT optimality conditions for (7.10), which are: $x > 0,\ s > 0$ and

$$\begin{aligned} Ax &= b \\ -A^T y - s &= -c^T \\ Xs &= te \end{aligned} \tag{7.11}$$

where X denotes diag$(x_1, \dots, x_n)$, i.e., the square matrix of order n with diagonal entries as $x_1, \dots, x_n$, and all off-diagonal entries 0. It can be shown that for all $0 < t < \infty$, the solution $(x(t),\ y(t),\ s(t))$ of (7.11) exists and is unique iff both (7.9) and its dual have interior feasible solutions. The trajectory traced by $(x(t),\ y(t),\ s(t))$ in terms of the parameter $t > 0$ is known as the **primal-dual central path** for the LP (7.9), it is well defined when both this LP and its dual have interior feasible solutions.

Let K denote the set of feasible solutions of (7.9), and assume that both (7.9) and its dual have interior feasible solutions. We also see that if $\mathcal{C}$ denotes the central path for the LP (7.9), and $(x(t),\ y(t),\ s(t))$ is a point on it; then $x(t)$ is the analytic center of $K \cap \{x : cx = cx(t)\}$. Thus every point on the central path is the analytic center of the intersection of K with the objective plane through that point.

7.5 The Affine Scaling Method

This method is due to (Dikin [1967]). We describe the method when the original LP is in the following standard form:

$$\begin{aligned} \text{Minimize} \quad & cx \\ \text{subject to} \quad & Ax = b \\ & x \geq 0 \end{aligned} \tag{7.12}$$

where A is of order $m \times n$ and rank m. Let x^0 be an available interior feasible solution, i.e., $Ax^0 = b$ and $x^0 > 0$, for initiating the method. Starting with x^0, the method generates a sequence of interior feasible solutions $x^0, x^1, \dots$. We will discuss the general step.

Strategy of the General Step

Let $x^r = (x_1^r, \dots, x_n^r)^T$ be the current interior feasible solution. The method creates an ellipsoid with x^r as center, inside the feasible region of the original LP. It does this by replacing:

the non-negativity restrictions "$x \geq 0$" by "$x \in E_r = \{x : \sum_{i=1}^{n}((x_i - x_i^r)/(x_i^r))^2 \leq 1\}$".

E_r is an ellipsoid in R^n with its center at x^r. The ellipsoidal approximating problem is then

$$\begin{aligned} \text{Minimize} \quad & cx \\ \text{subject to} \quad & Ax = b \\ \sum_{i=1}^{n}((x_i - x_i^r)/(x_i^r))^2 & \leq 1 \end{aligned} \tag{7.13}$$

It can be shown that $E_r \subset \{x : x \geq 0\}$. The intersection of E_r with the affine space defined by the system of equality constraints $Ax = b$ is an ellipsoid $\bar{E}_r$ with center x^r inside the feasible region of the original LP. See Fig. 7.1. The ellipsoidal approximating problem given above is the problem of minimizing the objective function cx over this ellipsoid $\bar{E}_r$. Its optimum solution $\bar{x}^r = (\bar{x}_j^r)$ can be computed by a formula. Here we briefly outline the various steps that lead to this formula.

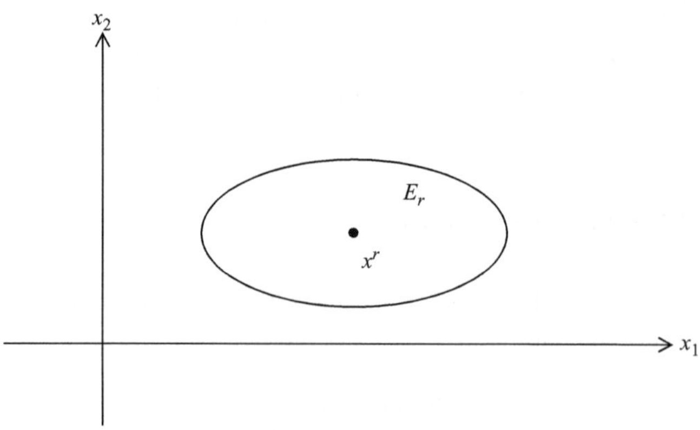

Fig. 7.1 The ellipsoid E^r constructed in affine scaling method when current interior feasible solution is $x^r > 0$. It is inside the non-negative orthant

1. An **ellipsoid** in R^n with center at a point $x^0 \in R^n$ is the set of all points $E = \{x : (x - x^0)^T D(x - x^0) \leq \rho^2\}$; where D is a positive definite matrix, and $\rho > 0$. See Chapter 9 for the definition of positive definiteness of a square matrix, and efficient methods to check whether a given square matrix is positive definite.

The ellipsoid is completely defined by the center x^0, RHS constant ρ^2, and the positive definite matrix D. A **sphere** with center x^0 is a special case of the ellipsoid E with $D = I$, the unit matrix; in this case ρ is the radius of the sphere.

2. We start with the simplest problem of this type, and progressively lead to the ellipsoidal approximating problem above. The sphere is the simplest type of ellipsoid, so we begin with this problem on a sphere first. Consider the problem:

$$\begin{aligned} \text{Minimize} \quad & z = cx \\ \text{s. to} \quad & (x - x^0)^T (x - x^0) \leq \rho^2 \end{aligned}$$

The objective plane through x^0 is $\{x : cx = cx^0\}$. To obtain the optimum solution of the above problem, move this objective plane through x^0, parallel to itself, by decreasing the RHS constant, until it reaches a position where it is touching the sphere at one point and becomes tangent to the sphere at that point. Any further decrease in the RHS constant will make the plane not have any contact with the sphere. So, that touching point, which is $x^0 + \rho(-c^T)/||c||$ is the optimum solution of the above problem. See Fig. 7.2.

3. For any vector $x \in R^n$, let the upper case letter X represent the diagonal matrix in $R^{n \times n}$ whose diagonal entries are the entries in x, i.e., $X = \text{diag}(x)$. This has become standard notation in IPM literature. Correspondingly we will denote $\text{diag}(x^r)$ by X^r for any superscript r.

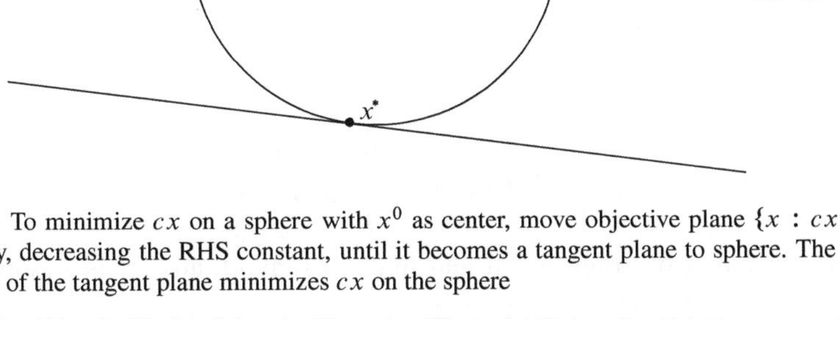

Fig. 7.2 To minimize cx on a sphere with x^0 as center, move objective plane $\{x : cx = cx^0\}$ parallelly, decreasing the RHS constant, until it becomes a tangent plane to sphere. The touching point x^* of the tangent plane minimizes cx on the sphere

Let $x^0 > 0$ be a given point. Consider the problem

$$\begin{aligned} &\text{Minimize} \quad z = cx \\ &\text{s. to} \quad (x - x^0)^T (X^0)^{-2} (x - x^0) \leq \rho^2 \end{aligned} \tag{7.14}$$

This problem can be transformed into the problem discussed under Item 2 by the transformation $y^T = e^T + (x - x^0)^T (X^0)^{-1}$, i.e., $x^T = (y - e)^T X^0 + (x^0)^T = y^T X^0$, where $e \in R^n$ is the column vector in R^n with all entries equal to 1. In terms of the new variables y, the problem (7.14) is

$$\begin{aligned} &\text{Minimize} \quad z' = c'y + \text{constant} \\ &\text{s. to} \quad (y - e)^T (y - e) \leq \rho^2 \end{aligned}$$

where $c' = cX^0$.

From Step 2, we know that the optimum solution of this problem is $\bar{y} = e + \rho(-c'^T / ||c'||)$. So the optimum solution of (7.14) is $\bar{x} = x^0 - \rho (X^0)^2 c^T / (||X^0 c^T||)$.

4. Now consider the problem

$$\begin{aligned} &\text{Minimize} \quad z = cx \\ &\text{s. to} \quad Ax = b \\ &(x - x^0)^T (x - x^0) \leq \rho^2 \end{aligned} \tag{7.15}$$

where A is a matrix of order $m \times n$ and full rank m, and x^0 is a point in the affine space $H = \{x : Ax = b\}$. Denoting the ball $B = \{x : (x - x^0)^T (x - x^0) \leq \rho^2\}$, since the center x^0 of B is in H, $H \cap B$ is another ball which has center x^0 and radius ρ, and is totally contained in H.

Since A is of full row rank, the orthogonal projection of c^T into the subspace $\{x : Ax = 0\}$ is Pc^T where $P = I - A^T (AA^T)^{-1} A$, in which I is the unit matrix of order n. P is the projection matrix corresponding to this subspace.

Verifying that $cPx = cx - cA^T (AA^T)^{-1} b$, we see that solving (7.15) is equivalent to minimizing cPx on $H \cap B$. Using the result in Step 2, we can see that the optimum solution of this problem is $x^0 - \rho(Pc^T)/||Pc^T||$ if $Pc^T \neq 0$. If $Pc^T = 0$, then c is a linear combination of row vectors of A, in this case every feasible solution of (7.15) is optimal to it.

5. Now consider the problem

$$\begin{aligned} &\text{Minimize} \quad z = cx \\ &\text{s. to} \quad Ax = b \\ &(x - x^0)^T (X^0)^{-2} (x - x^0) \leq \rho^2 \end{aligned} \tag{7.16}$$

where A is a matrix of order $m \times n$ and full row rank m, and x^0 is a point in the affine space $H = \{x : Ax = b\}$. See Fig. 7.3.

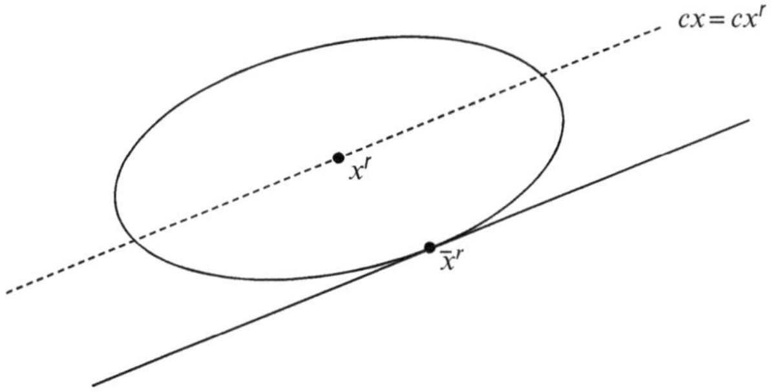

Fig. 7.3 To minimize cx on the ellipsoid $\bar{E}_r$ with center x^r, technique similar to that described in Fig. 7.2 for minimizing cx on a sphere. Formulas given above compute the resulting touching point with $\bar{E}_r$ that minimizes cx on it

As in Step 3, using the transformation of variables $y^T = e^T + (x - x^0)^T (X^0)^{-1}$, i.e., $x^T = (y - e)^T X^0 + (x^0)^T = y^T X^0$, the problem becomes

$$\begin{aligned} \text{Minimize} \quad z = cX^0 y + \text{constant} \\ \text{s. to} \quad AX^0 y = b \\ (y - e)^T (y - e) \leq \rho^2 \end{aligned} \tag{7.17}$$

in terms of the new variables y, using the facts that $x^0 \in H$, and $(X^0)^T = X^0$ since X^0 is diagonal. From Step 4, we know that the optimal y for this problem is $e - \rho(P_0 X^0 c^T)/(||P_0 X^0 c^T||)$ where $P_0 = I - X^0 A^T (A(X^0)^2 A^T)^{-1} A X^0$ is the projection matrix, if $P_0 X^0 c^T \neq 0$. If $P_0 X^0 c^T = 0$, then cX^0 is a linear combination of row vectors in AX^0, i.e., c is a linear combination of rows in A, then every feasible solution is optimal to (7.16). So, an optimal x to this problem is $X^0 \times$ (optimal y to (7.17)) $= X^0(e - \rho(P_0 X^0 c^T)/(||P_0 X^0 c^T||)) = x^0 - \rho(X^0 P_0 X^0 c^T)/(||P_0 X^0 c^T||)$.

Applying the result in Step 5 to the ellipsoidal approximating problem (7.13), we see that its optimum solution is:

$$\bar{x}^r = x^r - [X^r P_r X^r c^T]/(||P_r X^r c^T||) = x^r - [(X^r)^2 s^r]/(||X^r s^r||)$$

where $||.||$ indicates the Euclidean norm, and

$X^r =$ diag$(x_1^r, \dots, x_n^r)$, the diagonal matrix of order n with diagonal entries $x_1^r, \dots, x_n^r$ and off-diagonal entries 0

$I =$ unit matrix of order n

$P_r = (I - X^r A^T (A(X^r)^2 A^T)^{-1} A X^r)$, a projection matrix

$y^r = (A(X^r)^2 A^T)^{-1} A(X^r)^2 c^T$, known as the **tentative dual solution** corresponding to the current interior feasible solution x^r

$s^r = c^T - A^T y^r$, **tentative dual slack vector** corresponding to x^r.

The reason to call s^r the tentative dual slack vector, is that it can be shown to converge to the slack vector corresponding to an optimum dual solution for (7.12) under certain conditions. We have the following theorem:

Theorem 7.2. *Let x^r be the current interior feasible solution of (7.12), and $E_r = E(x^r) = \{x : \sum_{i=1}^{n}((x_i - x_i^r)/(x_i^r))^2 \leq 1\}$. Then $E_r \subset \{x : x \geq 0\}$. So, $\bar{E}_r = \{x : Ax = b,\ x \in E_r\} \subset K$. If $\bar{x}^r = (\bar{x}_j^r)$, an optimum solution for the problem: minimize cx subject to $x \in \{x : Ax = b, x \in E_r\}$ is a boundary feasible solution of K, i.e., $\bar{x}_j^r = 0$ for at least one $j = 1$ to n; then $\bar{x}^r$ is an optimum solution of (7.12), and $y^r = (A(X^r)^2A^T)^{-1}A(X^r)^2c^T$ is an optimum dual solution.*

Proof. Let $x = (x_j) \in E_r$. Suppose there is a j such that $x_j < 0$ in this vector x. As $x^r > 0$, we have $x_j^r > 0$. So, we see that if $x_j < 0$, then $|x_j - x_j^r| > x_j^r$; or $(|x_j - x_j^r|/x_j^r) > 1$; which implies that $x \notin E_r$, a contradiction. So, $E_r \subset \{x : x \geq 0\}$.

From the formulae given above, we verify that $P_rX^rc^T = X^r(I - A^T(A(X^r)^2 A^T)^{-1}A(X^r)^2)c^T = X^r(c^T - A^Ty^r) = X^rs^r$, and $\bar{x}^r = x^r - (X^r)^2s^r/||X^rs^r||$.

Since $x_j^r > 0$, the fact that $\bar{x}_j^r = x_j^r - ((x_j^r)^2s_j^r/||X^rs^r||) = 0$, implies that $x_j^rs_j^r = ||X^rs^r||$. Thus $x_i^rs_i^r = 0$ for all $i \neq j$. Since $x_i^r > 0$, we see that $s_i^r = 0$ for all $i \neq j$. Since $x_j^r > 0$, $x_j^rs_j^r = ||X^rs^r||$ implies that $s_j^r \geq 0$. Therefore from $s^r = c - A^Ty^r$, we see that y^r is dual feasible, and s^r is the corresponding dual slack vector; and that $\bar{x}^r$, y^r together satisfy the CS optimality conditions for (7.12) and its dual respectively. Since they are also primal, dual feasible, respectively, they are optimum solutions to (7.12), its dual, respectively. See Fig. 7.4. □

Theorem 7.3. *Let x^r be the current interior feasible solution of (7.12), and s^r the tentative dual slack vector in this step. If $s^r \leq 0$, then the objective function is unbounded below in the original problem (7.12).*

Proof. Let $x(\lambda) = x^r - \lambda(X^r)^2s^r$. Since $s^r \leq 0$, we see that $x(\lambda) \geq 0$ for all $\lambda \geq 0$; and it can also be verified that $Ax(\lambda) = b$ for all λ. This implies that $x(\lambda)$ is a feasible solution of (7.12) for all $\lambda \geq 0$.

Then $cx(\lambda) = cx^r - \lambda c(X^r)^2s^r = cx^r - \lambda cX^rP_rX^rc^T = cx^r - \lambda||P_rX^rc^T||^2$. So, the coefficient of λ here is < 0, which implies that $cx(\lambda) \to -\infty$ as $\lambda \to \infty$.

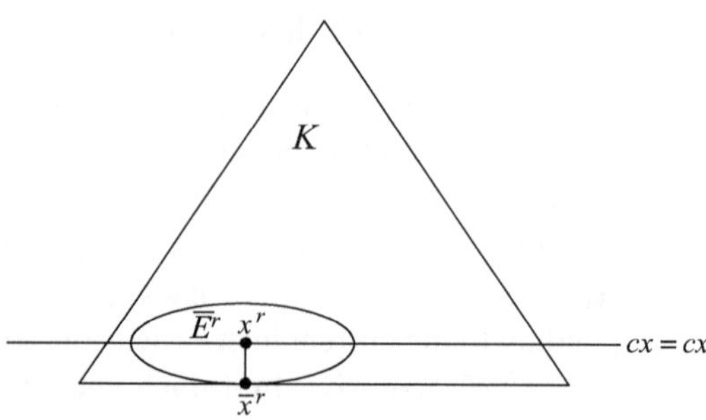

Fig. 7.4 If the point $\bar{x}^r$ minimizing cx on $\bar{E}_r$ is a boundary point of K, then $\bar{x}^r$ is in the optimum face for minimizing cx on K

So $\{x(\lambda) : \lambda \geq 0\}$ is a half-line in K along which $z(x) = cx \to -\infty$, so $z(x)$ is unbounded below on K in this case. □

So, if the conditions in either Theorem 7.2, or Theorem 7.3 are satisfied, the algorithm terminates with the appropriate conclusion. If these termination conditions are not satisfied, then the search direction at x^r is

$$d^r = \bar{x}^r - x^r = -((X^r)^2 s^r)/(||X^r s^r||)$$

is known as the **primal affine scaling direction** at the primal interior feasible solution x^r. Since both x^r, $\bar{x}^r$ are feasible to the original problem, we have $Ax^r = A\bar{x}^r = b$, hence $Ad^r = 0$. So, d^r is a descent feasible direction for the primal along which the primal objective value decreases. The maximum step length θ_r that we can move from x^r in the direction d^r is the maximum value of λ that keeps $x_j^r + \lambda d_j^r \geq 0$ for all j. It can be verified that this is

∞ if $s^r \leq 0$ (this leads to the unboundedness condition stated above); and if $s^r \not\leq 0$ it is equal to
$\theta_r = \min\{(||X^r s^r||)/(x_j^r s_j^r)$: over j such that $s_j^r > 0\}$.

It can be verified that $\theta_r = 1$ if $\bar{x}_j^r = 0$ for some j (in this case $\bar{x}^r$ is an optimum solution of the original LP as discussed above). Otherwise $\theta_r > 1$. In this case the method takes the next iterate to be $x^{r+1} = x^r + \alpha\theta_r d^r$ for some $0 < \alpha < 1$. Typically $\alpha = 0.95$ in implementations of this method. This α is the **step length fraction parameter**. See Fig. 7.5. Then the method moves to the next step with x^{r+1} as the current interior feasible solution. Here is a summary statement of the general step in this method for solving (7.12).

General Step

Substep 1: Let $x^r = (x_1^r, \ldots, x_n^r)^T$ be the current interior feasible solution of the problem. Let $X^r = \text{diag}(x_1^r, \ldots, x_n^r)$.

Substep 2: Compute:

the tentative dual solution $y^r = (A(X^r)^2 A^T)^{-1} A(X^r)^2 c^T$,
the tentative dual slack $s^r = c^t - A^T y^r$,
and the primal affine scaling search direction at x^r which is $d^r = -((X^r)^2 s^r)/(||X^r s^r||)$.

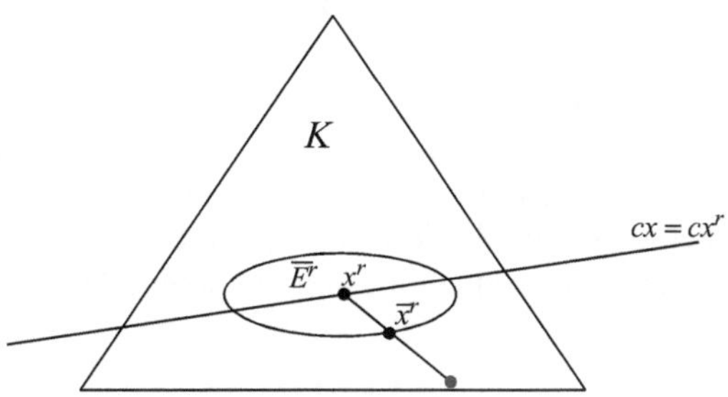

Fig. 7.5 If $\bar{x}^r$ minimizing cx on $\bar{E}_r$ with center x^r, is an interior feasible solution, the affine scaling method uses the direction joining x^r with $\bar{x}^r$ as descent direction at x^r. The bottom point is the output of this descent step, with which the next iteration is initiated

If $s^r \leq 0$, $\{x^r + \lambda d^r : \lambda \geq 0\}$ is a feasible half-line for the original problem along which the objective function $cx \to -\infty$ as $\lambda \to +\infty$, terminate.

Substep 3: If $s^r \nleq 0$, compute the maximum step length that we can move from x^r in the direction d^r, this is the maximum value of λ that keeps $x_j^r + \lambda d_j^r \geq 0$ for all j. It is $\theta_r = \min\{(||X^r s^r||)/(x_j^r s_j^r)$: over j such that $s_j^r > 0\}$. If $\theta_r = 1$, $x^r + d^r$ is an optimum solution of the original LP, terminate.

Otherwise let $x^{r+1} = x^r + \alpha\theta_r d^r$ for some $0 < \alpha < 1$ (typically $\alpha = 0.95$). With x^{r+1} as the current interior feasible solution, go to the next step.

Under some minor conditions it can be proved that if the original problem has an optimum solution, then the sequence of iterates x^r converges to a strictly complementary optimum solution, and that the objective value cx^r converges at a linear or better rate. Also if the step length fraction parameter α is $<2/3$, then the tentative dual sequence y^r converges to the analytic center of the optimum dual solution set. For proofs of these results and a complete discussion of the convergence properties of this method see (Saigal [1995]). So far this method has not been shown to be a polynomial time method.

Versions of this method have been developed for LPs in more general forms like the bounded variable form, and the form in which the LP consists of some unrestricted variables too. When the original LP has unrestricted variables, instead of an ellipsoid, the method creates a hyper-cylinder with an elliptical cross section inside the feasible region centered at the current interior feasible solution. The point minimizing the objective function over this hyper-cylinder can also be computed directly by a formula, and other features of the method remain essentially similar to the above.

A version of this method that works on the constraints in the dual problem only (instead of those of the primal) has also been developed, this version is called the **dual affine scaling method**. There is also a **primal-dual affine scaling method** that works on the system consisting of both the primal and dual constraints together, search directions used in this version are based on Newton directions for the system consisting of the complementary slackness conditions. See (Saigal [1995]) for a complete discussion of all these versions of the affine scaling method with complete proofs of results that can be achieved using them.

7.6 Newton's Method for Solving Systems of Nonlinear Equations

We will now discuss a classical method for solving a square system of nonlinear equations in variables $x = (x_1, \cdots, x_n)^T$. "Square system" means that the number of equations in the system to be solved is the same as the number of variables. The method is called "Newton's method," also as "Newton-Raphson method for solving a system of n equations in n unknowns" in the literature. Consider the system

$$f(x) = (f_1(x), \cdots, f_n(x))^T = 0 \tag{7.18}$$

where each of the functions $f_i(x)$ for $i = 1$ to n is continuously differentiable. The celebrated Newton's method is the most widely used method to solve this system of equations. Let

$$\begin{aligned}\nabla f(x) &= \text{the Jacobian matrix of } f(x) \text{ at } x\\ &= \text{square matrix with } i\text{th row} = \text{row vector of partial}\\ &\quad\ \text{derivatives of } f_i(x)\\ &= \left(\frac{\partial f_i(x)}{\partial x_j} : i = 1, \dots, n; \quad j = 1, \dots, n\right).\end{aligned}$$

Let x^0 be the initial point with which the method is initiated. It generates a sequence of points $x^0, x^1, x^2, \dots$ using the iterative formula

$$x^{r+1} = x^r - (\nabla f(x^r))^{-1} f(x^r)$$

for $r = 0, 1, \dots,$ assuming that $(\nabla f(x^r))^{-1}$ exists.

If for some r, $(\nabla f(x^r))$ is singular, and $||f(x^r)||$ is larger than the specified tolerance for it, the method is unable to proceed further. Under some mild conditions on the functions in $f(x)$, and the initial point x^0, the method can be proved to converge to a solution of the system (7.18). Among several books, (Saigal [1995], Dennis and Schnabel [1983], Fletcher [1981], Murty [1988] of Chapter 2, Ortega and Rheinboldt [1970]) discuss these results in detail.

At the point x^r, if $\nabla f(x^r))$ is nonsingular, then the direction $-(\nabla f(x^r))^{-1} f(x^r)$ is called the **Newton search direction**. Newton's method takes step lengths of 1 in these directions.

Several modifications have been proposed to remedy the situation when the Jacobian matrix turns out to be singular at a point obtained in the sequence; these methods are typically called **modified Newton methods**. We have discussed Newton's method here because the primal-dual path following methods for LP discussed in Section 7.7 use Newton or modified Newton directions.

7.7 Primal-Dual Path Following Methods

The central path-following primal-dual IPMs are some of the most popular methods for LP in software implementations at present. They consider the primal LP:

minimize $c^T x$, subject to $Ax = b$, $x \geq 0$;
and its dual in which the constraints are: $A^T y + s = c$, $s \geq 0$;

where A is a matrix of order $m \times n$ and rank m. The system of primal and dual constraints put together is:

$$\begin{aligned}Ax &= b\\ A^T y + s &= c\\ (x, s) &\geq 0\end{aligned} \tag{7.19}$$

A feasible solution $(x,\ y,\ s)$ to (7.19) is called an **interior feasible solution** if $(x, s) > 0$. Let $\mathcal{F}$ denote the set of all feasible solutions of (7.19), and $\mathcal{F}^0$ the set of all interior feasible solutions. For any $(x,\ y,\ s) \in \mathcal{F}^0$ define $X = \text{diag}(x_1, \dots, x_n)$, the square diagonal matrix of order n with diagonal entries $x_1, \dots, x_n$; and $S = \text{diag}(s_1, \dots, s_n)$.

For each $j = 1$ to n, the pair $(x_j,\ s_j)$ is known as the *jth complementary pair of variables* in these primal, dual pair of LPs. The *complementary slackness conditions* for optimality in this pair of problems are: the product $x_j s_j = 0$ for each $j = 1$ to n; i.e., $XSe = 0$ where e is a vector of all 1s. Since each product is ≥ 0, these conditions are equivalent to $x^T s = 0$.

The Central Path

As defined in Section 7.4, the central path, $\mathcal{C}$ for this family of primal-dual path-following methods is a curve in $\mathcal{F}^0$ parametrized by a positive parameter $\tau > 0$. For each $\tau > 0$, the point $(x^\tau,\ y^\tau,\ s^\tau) \in \mathcal{C}$ satisfies: $(x^\tau, s^\tau) > 0$ and

$$\begin{aligned} A^T y^\tau + s^\tau &= c^T \\ Ax^\tau &= b \\ x_j^\tau s_j^\tau &= \tau, \quad \text{for all } j = 1, \dots, n \end{aligned}$$

If $\tau = 0$, the above equations define the optimality conditions for the LP, shown below in (7.20) (these are obtained by setting $\tau = 0$ in the above system). For each $\tau > 0$, the solution $(x^\tau,\ y^\tau,\ s^\tau)$ is unique, and as τ decreases to 0 the central path converges to the center of the optimum face of the primal, dual pair of LPs.

Optimality Conditions

From optimality conditions, solving the LP is equivalent to finding a solution $(x,\ y,\ s)$ satisfying $(x,\ s) \geq 0$, to the following system of $(2n + m)$ equations in $(2n + m)$ unknowns:

$$F(x, y, s) = \begin{bmatrix} A^T y + s - c \\ Ax - b \\ XSe \end{bmatrix} = 0 \tag{7.20}$$

This is a nonlinear system of equations because of the last equation.

Selecting the Directions to Move

Let the current interior feasible solution be $(\bar{x}, \bar{y}, \bar{s})$. So, $(\bar{x}, \bar{s}) > 0$. Also, the variables in y are unrestricted in sign in the problem.

Primal-dual path-following methods try to follow the central path $\mathcal{C}$ with τ decreasing to 0. For points on $\mathcal{C}$, the value of τ is a measure of closeness to optimality, when it decreases to 0 we are done. Following $\mathcal{C}$ with τ decreasing to 0 keeps all the complementary pair products $x_j s_j$ equal and decreasing to 0 at the same rate.

However, there are two difficulties for following $\mathcal{C}$. One is that it is difficult to get an initial point on $\mathcal{C}$ with all the $x_j s_j$ equal to one another; the second is that $\mathcal{C}$ is a nonlinear curve. At a general solution $(x, y, s) \in \mathcal{F}^0$, the products $x_j s_j$ will not be equal to each other, hence the parameter $\mu = \left(\sum_{j=1}^n x_j s_j\right)/n = x^T s/n$, the *average complementary slackness violation measure*, is used as a measure of optimality for them. Since path-following methods cannot exactly follow $\mathcal{C}$, they stay within a loose but well-defined neighborhood of $\mathcal{C}$ while steadily reducing the optimality measure μ to 0.

Staying explicitly within a neighborhood of $\mathcal{C}$ serves the purpose of excluding points (x, y, s) that are too close to the boundary of $\{(x, y, s) : x \geq 0, s \geq 0\}$, to make sure that the lengths of steps towards optimality remain long.

To define a neighborhood of the central path, we need a measure of deviation from centrality, this is obtained by comparing a measure of deviation of the various $x_j s_j$ from their average μ to μ itself. This leads to the measure

$$(||(x_1 s_1, \dots, x_n s_n)^T - \mu e||)/\mu = (||XSe - \mu e||)/\mu$$

where $||.||$ is some norm. Different methods use neighborhoods defined by different norms.

A parameter θ is used to define a bound for this measure when using the Euclidean norm. A commonly used neighborhood based on the Euclidean norm $||.||_2$, called the 2-norm neighborhood, defined by

$$\mathcal{N}_2(\theta) = \{(x, y, s) \in \mathcal{F}^0 : \ ||XSe - \mu e||_2 \leq \theta\mu\}$$

for some value of the parameter $\theta \in (0, 1)$. Another commonly used neighborhood based on the ∞-norm is the $\mathcal{N}_{-\infty}(\gamma)$ defined by

$$\mathcal{N}_{-\infty}(\gamma) = \{(x, y, s) \in \mathcal{F}^0 : \ x_j s_j \geq \gamma\mu, \quad j = 1, \dots, n\}$$

parametrized by the parameter $\gamma \in (0, 1)$. This is a one-sided neighborhood that restricts each product $x_j s_j$ to be at least some small multiple γ of their average μ. Typical values used for these parameters are $\theta = 0.5$, and $\gamma = 0.001$. By keeping all iterates inside one or the other of these neighborhoods, path-following methods reduce all $x_j s_j$ to 0 at about the same rates.

Since the width of these neighborhoods for a given μ depends on μ, these neighborhoods are conical (like a *horn*), they are wider for larger values of μ, and become narrow as $\mu \to 0$.

Once the direction to move from the current point $(\bar{x}, \bar{y}, \bar{s})$ is computed, we may move from it only a small step length in that direction, and since $(\bar{x}, \bar{s}) > 0$ such a move in any direction will take us to a point that will continue satisfying $(x, s) > 0$. So, in computing the direction to move at the current point, the nonnegativity constraints $(x, s) \geq 0$ can be ignored. The only remaining conditions to be satisfied for attaining optimality are the equality conditions (7.20). So the direction finding routine concentrates only on trying to satisfy (7.20) more closely.

Ignoring the inactive inequality constraints in determining the direction to move at the current point is the main feature of **barrier methods** in nonlinear programming; hence these methods are also known as barrier methods.

Equation (7.20) is a square system of nonlinear equations $2n + m$ equations in $2n + m$ unknowns; it is nonlinear because the third condition in (7.20) is nonlinear. Experience in nonlinear programming indicates that the best directions to move in algorithms for solving nonlinear equations are either the Newton direction, or some modified Newton direction, as discussed in Section 7.6. So, this method uses a modified Newton direction to move. To define that, a centering parameter $\sigma \in [0, 1]$ is used. Then the direction for the move denoted by $(\triangle x, \triangle y, \triangle s)$ is the solution to the following system of linear equations

$$\begin{pmatrix} 0 & A^T & I \\ A & 0 & 0 \\ S & 0 & X \end{pmatrix} \begin{pmatrix} \triangle x \\ \triangle y \\ \triangle s \end{pmatrix} = \begin{pmatrix} 0 \\ 0 \\ -XSe + \sigma\mu e \end{pmatrix} \tag{7.21}$$

where 0 in each place indicates the appropriate matrix or vector of zeros, I the unit matrix of order n, and e indicates the column vector of order n consisting of all 1s.

If $\sigma = 1$, the direction obtained will be a centering direction, which is a Newton direction towards the point $(x^{\mu}, y^{\mu}, s^{\mu})$ on $\mathcal{C}$ at which the products $x_j s_j$ of all complementary pairs in this primal, dual pair of problems are $= \mu$. Moving in the centering direction helps to move the point towards $\mathcal{C}$, but may make little progress in reducing the optimality measure μ. But in the next iteration this may help to take a relatively long step to reduce μ. At the other end the value $\sigma = 0$ gives the standard Newton direction for solving (7.20). Many algorithms choose σ from the open interval $(0, 1)$ to trade off between twin goals of reducing μ and improving centrality.

We will now describe two popular path-following methods.

The Long-Step Path-Following Algorithm (LPF)

LPF generates a sequence of iterates in the neighborhood $\mathcal{N}_{-\infty}(\gamma)$, which for small values of γ (e.g. $\gamma = 0.001$) includes most of the set of interior feasible solutions $\mathcal{F}^0$. The method is initiated with an $(x^0, y^0, s^0) \in \mathcal{F}^0$. In each step the method chooses the centering parameter σ between two selected limits $\sigma_{\min}, \sigma_{\max}$ where $0 < \sigma_{\min} < \sigma_{\max} < 1$. The neighborhood defining parameter γ is selected from $(0, 1)$. Here is the general step in this algorithm.

General Step *k*

Let (x^k, y^k, s^k) be the current interior feasible solution, and $\mu_k = (x^k)^T s^k/n$ the current value of the optimality measure corresponding to it. Choose $\sigma_k \in [\sigma_{\min}, \sigma_{\max}]$. Find the direction $(\Delta x^k, \Delta y^k, \Delta s^k)$ by solving

$$\begin{pmatrix} 0 & A^T & I \\ A & 0 & 0 \\ S^k & 0 & X^k \end{pmatrix} \begin{pmatrix} \Delta x^k \\ \Delta y^k \\ \Delta s^k \end{pmatrix} = \begin{pmatrix} 0 \\ 0 \\ -X^k S^k e + \sigma_k \mu_k e \end{pmatrix} \tag{7.22}$$

Find $\alpha_k =$ the largest value of $\alpha \in [0, 1]$ such that $(x^k, y^k, s^k) + \alpha(\Delta x^k, \Delta y^k, \Delta s^k) \in \mathcal{N}_{-\infty}(\gamma)$.

Setting $(x^{k+1}, y^{k+1}, s^{k+1}) = (x^k, y^k, s^k) + \alpha_k(\Delta x^k, \Delta y^k, \Delta s^k)$ as the new current interior feasible solution; go to the next step.

The Predictor-Corrector Path-Following Method (PC)

Path-following methods have two goals, one to improve centrality (closeness to the central path while keeping optimality measure unchanged), and the other to decrease the optimality measure μ. The PC method takes two different steps alternately to achieve each of these twin goals. The PC uses two $\mathcal{N}_2$ neighborhoods nested one inside the other. They are $\mathcal{N}_2(\theta_1), \mathcal{N}_2(\theta_2)$ for selected $0 < \theta_1 < \theta_2 < 1$. For example $\theta_1 = 0.25, \theta_2 = 0.5$. In some versions of this method values of θ larger than 1 are also used successfully.

Every second step in this method is a "predictor" step, its starting point will be in the inner neighborhood. The direction to move in this step is computed by solving the system (7.22) corresponding to the current solution with the value of $\sigma = 0$. The step length in this step is the largest value of α that keeps the next point within the outer neighborhood. The gap between the inner and outer neighborhoods is wide enough to allow this step to make significant progress in reducing μ.

The step taken after each predictor step is a "corrector" step, its starting point will be in the outer neighborhood. The direction to move in this step is computed by solving the system (7.22) corresponding to the current solution with the value of $\sigma = 1$. The step length in this step is $\alpha = 1$, which takes it back inside the inner neighborhood to prepare for the next predictor step.

It has been shown that the sequence of interior feasible solutions obtained in this method converges to a point in the optimum face. All these path-following methods have been shown to be polynomial time algorithms.

Each step of these interior point methods requires a full matrix inversion, a fairly complex task in solving large-scale problems. This involves much more work than a step of the simplex method. But the number of steps required by these interior point methods is smaller than the number of steps needed by the simplex method.

7.8 Summary of Results on the Primal-Dual IPMs

Here we summarize the results on the solutions obtained in primal-dual IPMs, and related results. We consider the following primal, dual LPs; (7.23) is the primal, and (7.24) is the dual.

$$\begin{aligned} \text{Minimize} \quad z(x) &= cx \\ \text{subject to} \quad Ax &= b \\ x &\geq 0 \end{aligned} \tag{7.23}$$

$$\begin{aligned} \text{Maximize} \quad v(\pi) &= \pi b \\ \text{subject to} \quad \pi A + \bar{c} &= c \\ \bar{c} &\geq 0 \end{aligned} \tag{7.24}$$

where $x = (x_1, \ldots, x_n)^T$, $\pi = (\pi_1, \ldots, \pi_m)$, $\bar{c} = (\bar{c}_1, \ldots, \bar{c}_m)$, A is a matrix of order $m \times n$ and rank m; and $c = (c_1, \ldots, c_n)$, $b = (b_1, \ldots, b_m)^T$. If $x^* = (x_j^*)$, $(\pi^* = (\pi_i^*)$, $\bar{c}^* = (\bar{c}_j^*))$ are primal, dual solution pairs; they are optimal to the respective problems iff $x_j^* \bar{c}_j^* = 0$ for all j. For $j = 1$ to n, $(x_j, \bar{c}_j)$ are the various complementary pairs in this primal, dual pair of LPs.

We assume that this primal, dual pair of LPs both have optimum solutions. For any vector $y = (y_j) \geq 0$, we denote $\sigma(y) = \{j : y_j > 0\}$, the index set of positive components in y.

1. Strictly complementary optimum solution pairs: There exists at least one optimum pair x^*, $(\pi^*, \bar{c}^*)$ that is strictly complementary, i.e., satisfying the property that in every complementary pair $(x_j^*, \bar{c}_j^*)$ exactly one quantity is 0, and the other is strictly positive. In other words, $\sigma(x^*)$, $\sigma(\bar{c}^*)$ satisfy: $\sigma(x^*) \cap \sigma(\bar{c}^*) = \emptyset$, and $\sigma(x^*) \cup \sigma(\bar{c}^*) = \{1, \ldots, n\}$ (see Ye [1992], Mehrotra and Ye [1993] for proofs).

Also, the sets $\sigma(x^*)$, $\sigma(\bar{c}^*)$ for a strictly complementary optimum solution pair are invariant; i.e., if $(x^*, (\pi^*, \bar{c}^*))$, $(\hat{x}, (\hat{\pi}, \hat{\bar{c}}))$ are two strictly complementary optimum solution pairs, then $\sigma(x^*) = \sigma(\hat{x})$, and $\sigma(\bar{c}^*) = \sigma(\hat{\bar{c}})$. So, we will denote these sets for strictly complementary optimum solution pairs by $\sigma^* = \sigma(x^*)$, and $\bar{\sigma}^* = \sigma(\bar{c}^*) = \{1, \ldots, n\} \backslash \sigma(x^*)$, in the sequel.

2. If $\tilde{x}$, $(\tilde{\pi}, \tilde{\bar{c}})$ is any optimum solution pair for this primal, dual pair of LPs, then $\sigma(\tilde{x}) \subset \sigma^*$ and $\sigma(\tilde{\bar{c}}) \subset \bar{\sigma}^*$.

Therefore, given any strictly complementary optimum solution pair, then the optimum faces of the two problems are:

- Ω_P = Optimum face of the primal $= \{x : Ax = b, \ x \geq 0, \ x_j = 0$ for all $j \in \bar{\sigma}^*\}$.
- Ω_D = Optimum face of the dual $= \{\pi : \pi A + \bar{c} = c, \ \bar{c} \geq 0, \ \bar{c}_j = 0$ for all $j \in \sigma^*\}$.

3. When (7.23) is solved by the primal-dual path following algorithms, let x^k, $(\pi^k, \bar{c}^k)$ be the kth solution in the sequence generated. Then $\max\{x_j^k, \bar{c}_j^k\} > \xi$ for all k, where ξ is a fixed positive number independent of k, the iteration count.

Also, let $\sigma^k = \{j : x_j^k \geq \bar{c}_j^k\}$. Then for all k sufficiently large we will have $\sigma^k = \sigma^*$. See Ye [1992], Mehrotra and Ye [1993] for proofs of these results. So, these algorithms converge to a strictly complementary optimum solution pair.

4. As in 3, let x^k, $(\pi^k, \bar{c}^k)$ be the kth solution in the sequence generated, and let $\sigma^k = \{j : x_j^k \geq \bar{c}_j^k\}$.

Let $\mathcal{B}$ be the matrix consisting of columns $A_{.j}$ for $j \in \sigma_k$, and $\mathcal{N}$ be the matrix consisting of the remaining columns of A. Let $x_{\mathcal{B}}$, $x_{\mathcal{N}}$ be the corresponding partition of the variables in x. Let

- $x_{\mathcal{B}}^{k*}$ = orthogonal projection of $x_{\mathcal{B}}^k$ on the affine space $\{x_{\mathcal{B}} : \mathcal{B}x_{\mathcal{B}} = b\}$
- π^{k*} = orthogonal projection of π^k on the affine space $\{\pi : \pi\mathcal{B} = c_{\mathcal{B}}\}$.

If $x_{\mathcal{B}}^{k*} > 0$, and $c_{\mathcal{N}} - \pi^{k*}\mathcal{N} \geq 0$, then $x^{k*} = (x_{\mathcal{B}}^{k*}, 0)$, $(\pi^{k*}, (0, \bar{c}_{\mathcal{N}}^{k*}))$ is an optimum solution pair for this pair of LPs, terminate the algorithm. These conditions will hold for some k finite. Otherwise continue the algorithm.

The orthogonal projection step here provides another way to obtain a true optimum solution for the original LP when it is solved using these IPMs.

7.9 Exercises

7.1. Let $f(x) = 4x_1^2 + 3x_2^2 - 4x_1x_2 + x_1$. Write the necessary optimality conditions for finding the global minimum of $f(x)$ over $x \in R^2$.

At the point $x^0 = (1, 1)^T$ what is the Newton direction for solving these necessary optimality conditions? Find the next point in the sequence generated by Newton's method to solve this system.

7.2. Consider the system of nonlinear equations

$$f_1(x) = x_1^2 - 2x_1x_2 + x_2^4 - x_1 - 2 = 0$$
$$f_2(x) = x_1^4 + 3x_1x_2 - x_2^4 + 3x_2 - 3 = 0.$$

Starting with the point $x^0 = (1, 1)^T$ apply one iteration of Newton's method to solve this system. (R. Saigal)

7.3. Consider the problem: maximize $\sum_{j=1}^{n} log(x_j)$, subject to $Ax = b, \ x \geq 0$, where A is a matrix of full row rank.

Show that if this problem has an optimum solution, then all the variables x_j will be > 0 in it. Using the necessary optimality conditions, and the nature of the objective function being maximized, find the conditions under which this problem has a unique optimum solution.

When the problem has a unique optimum solution, that optimum solution is called the **analytic center** of the polytope $\{x : Ax = b, \ x \geq 0\}$. Find it using the KKT necessary optimality conditions.

References

Chang SY, Murty KG (1989) The steepest descent gravitational method for linear programming. Discrete Applied Mathematics 25:211–239

Dennis JE Jr, Schnabel RB (1983) Numerical Methods for Unconstrained Optimization and Nonlinear Equations. Prentice-Hall, Englewood Cliffs, NJ

Fletcher R (1981) Practical Methods of Optimization. Vols. 1, 2. Wiley-Interscience, New York

Kojima M, Mizuno S, Yoshise A (1989) A primal-dual interior point algorithm for linear programming. Progress in Mathematical Programming: Interior Point and Related Methods. Megiddo N (ed), Springer-Verlag, New York, 2:29–47

Megiddo N (1989) Pathways to the optimal set in linear programming. Progress in Mathematical Programming: Interior Point and Related Methods. Meggiddo N (ed), Springer-Verlag, New York, 8:131–158

Mehrotra S (1992) On the implementation of a primal-dual interior point method. SIAM Journal on Optimization 2:575–601

Mehrotra S, Ye Y (1993) Finding an interior point in the optimal face of linear programs. Mathematical Programming 62:497–515

Mizuno S, Todd M, Ye Y (1993) On adaptive step primal-dual interior point algorithms for linear programming. Mathematics of Operations Research 18:964–981

Monteiro RDC, Adler I (1989) Interior path-following primal-dual algorithms, Part I: Linear programming. Mathematical Programming 44:27–41

Murty KG (1986) The Gravitational method for LP. Opsearch 23(4):206–214

Ortega JM, Rheinboldt WC (1970) Iterative Solutions of Nonlinear Equations in Several Variables. Academic Press, NY

Roos C, Terlaky T, Vial J-Ph (2006) Interior Point Methods for Linear Optimization. Springer, New York

Saigal R (1995) Linear Programming A Modern Integrated Analysis. Kluwer Academic Publishers, Boston, MA

Sonnevend G, Stoer J, Zhao G (1989) On the complexity of following the central path of linear programming by linear extrapolation. Mathematics of Operations Research 62:19–31

Wright SJ (1997) Primal-Dual Interior-Point Methods. SIAM, Philadelphia, PA

Ye Y (1992) On the finite convergence of interior-point algorithms for linear programming. Mathematical Programming 57:325–335

Ye Y (1997) Interior Point Algorithms, Theory and Analysis. Wiley-Interscience, New York

Chapter 8
Sphere Methods for LP

8.1 Introduction

For solving an optimization problem in which an objective function $z(x)$ is to be minimized subject to constraints, starting with an initial feasible solution, a method that works by generating a sequence of feasible solutions along which the objective value $z(x)$ strictly decreases is known as a *descent algorithm*. Strict improvement in the objective value in each step is an appealing property, and hence descent algorithms are the most sought after.

If the original objective function is to be maximized instead, algorithms that maintain feasibility and improve the objective value (i.e., here, increasing its value) in every step are also referred to as descent algorithms.

For LPs we discussed the simplex algorithm and some interior point methods (IPMs), all of which are descent algorithms. There are several other algorithms too for LP in literature which have this desirable property. However, all these algorithms are based on matrix inversion operations. In every step of the simplex algorithm, the inverse of the basis needs to be updated by a single pivot step. Even though IPMs need far fewer steps than the simplex algorithm, in each step they need more complex full matrix inversion to find the search direction in that step.

Also, in every step of these methods, the matrix inversion operations involve every constraint in the problem including any redundant constraints in the model. The process of identifying and eliminating the redundant constraints, if any, from the model using the methods available in theory is prohibitively expensive (in fact the effort needed by these methods is many times more than that needed to solve the original model with all the redundant constraints in it), and hence is not practical. Presolvers based on various heuristics are commonly employed in LP solvers to identify and eliminate redundant constraints in the model; they are reported to be effective, albeit not optimal (see Cartis and Gould 2006). Instead of depending on these presolvers, methods that have the ability to avoid redundant constraints in their computational work by themselves have a definite advantage over existing methods.

K.G. Murty, *Optimization for Decision Making: Linear and Quadratic Models*, International Series in Operations Research & Management Science 137,
DOI 10.1007/978-1-4419-1291-6_8,

Density of an LP Model

The *density* of an LP (or its coefficient matrix) refers to the percentage of nonzero entries in the data. Problems in which this percentage is small (typically $\leq 1\%$) are said to be *sparse*, as this percentage increases, the problem becomes more and more dense. Many LP models in real-world applications tend to be very sparse. In matrix inversion operations of the simplex method and IPMs discussed in Chap. 7, programmers have developed techniques to exploit this sparsity. Using these they produced implementations of these methods with reasonable memory requirements that can solve large-scale models fast. However, the effectiveness of these techniques fades as the density of the coefficient matrix increases, that is why solving of large-scale dense LPs is difficult using existing methods.

Dense LP models do arise in several important application areas. For example, a typical LP model for a data envelopment analysis (DEA) problem has a coefficient matrix that is essentially 100% dense (see Cooper et al. 2006). Also, many LP models arising in location problems, distribution problems, and supply chain problems are typically dense. That is why there is a lot of research being carried out targeting dense LP models. In solving LPs that are not very sparse, it is definitely advantageous to have to deal with smaller number of constraints.

It seems that practitioners are quite content with obtaining solutions not necessarily optimal, but close to being so, but they want a method that can obtain this approximate solution faster than existing methods. In many applications, this requires algorithms that can give good performance on models that may not be very sparse. For this, we need to investigate fast methods that satisfy the following properties:

1. Should be a descent method (i.e., starting with a feasible solution the method should maintain feasibility throughout and the objective value should improve monotonically in every step).
2. Should be implementable with no matrix inversions, or using matrix inversion operations only sparingly.
3. If matrix inversion operations are used, they should never involve any redundant constraints in the model and should only involve a small subset of the other constraints selected intelligently.

The future of algorithmic research in LP Is In this area.

Sphere Methods for LP

The *sphere methods* that we will discuss in this chapter are interior point methods that are descent methods. In contrast to all other IPMs discussed in Chap. 7, these methods can be implemented with no matrix inversions or using them only sparingly (see Murty (2006a, b) of Chap. 1; Murty and Oskoorouchi 2008a, b; Murty 2009a). Another advantage of these methods is that if matrix inversion operations are used in any iteration of the algorithm, only a small subset of constraints (called the *touching set of constraints* in that iteration) will be involved in those matrix inversion operations. Also, amazingly, redundant constraints, if any in the model, never enter into the touching set. So, they offer the prospect of extending the superior performance of existing software systems for LP to models that do not have the property of being very sparse. Hence, they seem to be well suited to the goals mentioned above. The sphere methods belong to the class of predictor–corrector type IPMs discussed in Sect. 7.7. They consider LPs in the form

$$\begin{aligned} \text{minimize} \quad & z(x) = cx \\ \text{subject to} \quad & Ax \geq b \end{aligned} \tag{8.1}$$

where A is an $m \times n$ data matrix with a known initial interior feasible solution x^0 (i.e., satisfying $Ax^0 > b$). The constraints in (8.1) include all bound constraints on individual variables. We assume that the rows of A, denoted by $A_{i.}$ for $i = 1$ to m, have been normalized, so that $||A_{i.}|| = 1$ ($||.||$ denotes the Euclidean norm) for all $i = 1$ to m.

If $c = 0$, every feasible solution is optimal, so we can change the objective function in the problem to that of minimizing $A_{1.}x$ instead. So without any loss of generality we can assume that $c \neq 0$. We will assume that c is also normalized so that $||c|| = 1$.

If the set of feasible solutions does not have an interior or if it does but an interior feasible solution is not known, we modify the problem with the usual big-M augmentation involving one artificial variable as follows:

$$\begin{aligned} \text{minimize} \quad & z(x) = cx + Mx_0 \\ \text{subject to} \quad & Ax + ex_0 \geq b \\ & x_0 \geq 0 \end{aligned}$$

where e is the column vector of all 1s in R^m and M is a large positive penalty parameter. The vector $(x^0 = 0, x_0^0)$, where x_0^0 is a sufficiently large positive number, gives an initial interior feasible solution to this Phase I problem. This Phase I problem is in the same form as (8.1); we solve it instead. For other such Phase I strategies, see (Cartis and Gould 2006).

How to Implement These Algorithms for Solving a General LP?

As mentioned in Chap. 7, at present the most popular IPM for software implementations is the primal–dual interior point method, because: (1) it gives optimum solutions to both the primal and the dual when both have feasible solutions and (2) it provides a lower bound that serves as an indicator to check how far left to go to reach the optimum. Also, in many algorithms, specifying a good termination condition to be used in practice is not easy. The lower bound in the primal–dual format provides an automatic practical termination condition.

We will show how to convert the sphere methods into primal–dual algorithms for LPs in general form. Consider an LP in general form in which there may be equality constraints on the variables, inequality constraints, and bounds on individual variables. By combining the bounds on individual variables with the inequality constraints, the problem is in the form

$$\begin{aligned} \text{Minimize} \quad & f\xi \\ \text{subject to} \quad & F\xi = h \\ & G\xi \geq g \end{aligned} \tag{8.2}$$

where F is a matrix of order $p \times q$, say. Let $\pi,\ \mu$ be dual vectors corresponding to the constraints in the two lines in (8.2). Solving (8.2) and its dual involves finding a feasible solution to the following system

$$\begin{aligned} F\xi &= h \\ \pi F + \mu G &= f \\ (G\xi, \mu, -f\xi + \pi h + \mu g) &\geq (g, 0, 0) \end{aligned} \tag{8.3}$$

.

Solving (8.3) is the same as solving the LP

$$\begin{aligned} \text{Minimize} \quad \sum_{i=1}^{p}(F_{i.}\xi - h_i) + \sum_{j=1}^{q}(\pi F_{.j} + \mu G_{.j} - f_j) \\ \text{subject to} \\ (F\xi, G\xi) \geq (h, g) \\ (\pi F + \mu G, \mu, -f\xi + \pi h + \mu g) \geq (f, 0, 0) \end{aligned} \tag{8.4}$$

The LP (8.4) is same form as (8.1). Also, since we are applying the algorithm using matrix inversion operations sparingly, having all these additional constraints over those in the original LP (8.2) in the model may not make it numerically difficult to handle. The objective value in (8.4) will be 0 at the optimum if both (8.2) and its dual have feasible solutions, so this provides a convenient lower bound to judge how far is left to go. Hence, in the sequel we will discuss the methods for solving the LP in the form (8.1).

Notation

The orthogonal distance from a point $y \in R^n$ to the facetal hyperplane $\{x : A_{i.} x = b_i\}$ of the set of feasible solutions of (8.1) is minimum $\{||x-y|| : A_{i.}x = b_i\} = A_{i.}y - b_i$, because $||A_{i.}|| = 1$.

Fig. 8.1 $x^0 \in K^0$, and the ball shown is $B(x^0, \delta(x^0))$, the largest ball inside K with x^0 as center. Facetal hyperplanes of K corresponding to indices 1, 2 are tangent planes to this ball, so $T(x^0) = \{1, 2\}$

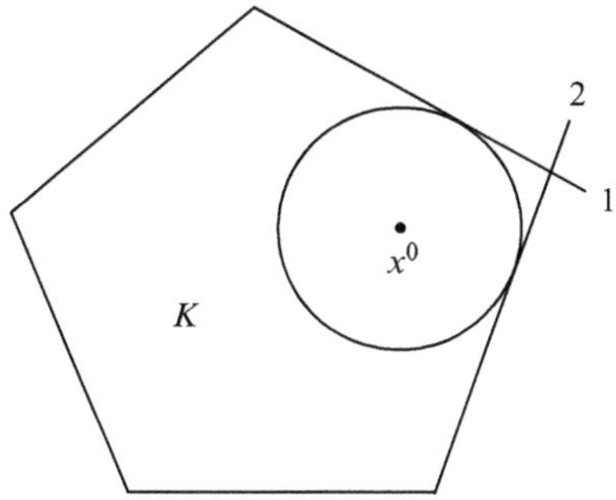

In this chapter, we will use the following notation.

K = Set of feasible solutions of (8.1).

$K^0 = \{x : Ax > b\}$ = interior of K.

$\delta(x) = \text{Min}\{A_{i.}x - b_i : i = 1 \text{ to } m\}$, defined for $x \in K^0$, it is the radius of the largest ball inside K with x as its center, since $||A_{i.}|| = 1$ for all i.

$B(x, \delta(x))$ = Defined for $x \in K^0$, it is the largest ball inside K with x as its center.

$T(x)$ = Defined for $x \in K^0$, it is the set of all indices i satisfying: $A_{i.}x - b_i = \text{Minimum}\{A_{p.}x - b_p : p = 1 \text{ to } m\} = \delta(x)$. The hyperplane $\{x : A_{i.}x = b_i\}$ is a tangent plane to $B(x, \delta(x))$ for each $i \in T(x)$; therefore, $T(x)$ is called the *index set of touching constraints in (8.1)* at $x \in K^0$ (see Fig. 8.1).

$t_{\min}, t_{\max}$ = Minimum, maximum values of $z(x)$ over K, respectively.

$\Gamma_1 = \{A_{i.}^T, -A_{i.}^T : i = 1 \text{ to } m\}$. This is a set of directions normal to facetal hyperplanes of K.

$\Gamma_2 = \{P_{.1}, \cdots, P_{.m}, -P_{.1}, \cdots, -P_{.m}\}$, where $P_{.i} = (I - c^T c)A_{i.}^T$, the orthogonal projection of $A_{i.}$ (the direction normal to the facet of K defined by the ith constraint in (8.1)) on the hyperplane $\{x : cx = 0\}$, for $i = 1$ to m.

$\delta[t]$ = It is the Maximum$\{\delta(x) : x \in \{x : cx = t\}\}$, i.e., the maximum radius of the ball that can be inscribed inside K with its center restricted to $\{x : cx = t\}$. Notice the difference between $\delta(x)$ defined over K^0, and this $\delta[t]$ defined over the interval $[t_{\min}, t_{\max}]$ of the real line.

t^* = Value of $t \in [t_{\min}, t_{\max}]$ that maximizes $\delta[t]$.

$c^i = (I - A_{i.}(A_{i.})^T)c^T$, the orthogonal projection of c^T on $\{x : A_{i.}x = 0\}$ for $i = 1$ to m.

Each iteration of sphere methods consists of only two steps, a centering step and a descent step. The centering step is a corrector step; it tries to move the current interior feasible solution into another one with higher value for $\delta(x)$ without sacrificing objective quality. The descent step is a predictor step that results in a strict decrease in objective value.

In earlier chapters, we discussed all the previous methods for solving LPs in a theoretical style. In this chapter, we will discuss the sphere methods for LP in the "implementation style," i.e., as they are implemented.

8.2 Ball Centers: Geometric Concepts

In this section, we first give the theoretical definition of the ball center of the given polytope K and the ball center of K on a given objective plane intersecting K in its interior; these are important concepts used in the sphere methods. Then, we discuss methods for computing these ball centers approximately in the next section.

Earlier versions of a sphere method were discussed in (Murty (2006a, b) of Chap. 1). In the centering step in those versions, the concept of the "center" of a polytope is defined as a point that is the center of the largest radius ball inscribed inside the polytope; i.e., for K it will be a point x that is optimal to the LP (8.5) discussed later. But the optimum face for (8.5) in the x-space may contain multiple points, in that case the description of the algorithm in these papers did not clearly specify which of those multiple optimum points of (8.5) in the x-space should be selected as the "center" of the polytope K defined by (8.1). Here, we will specify rules that will make that choice "precise" from (Murty and Oskoorouchi 2008a, b). For this we introduce the concepts of *the ball center of a polytope* and *the ball center of a polytope on a given objective plane* $\{x : cx = t\}$.

The definition of the "center" of a polytope used in the sphere method is very different from that used in earlier IPMs discussed in Chap. 7. For clarity between the two, since the "center" that we need in the sphere method is the center of the largest ball inside the polytope, we will use the word *ball center* of the polytope for the center that is used in sphere methods.

Our definitions guarantee that every polytope K (i.e., a bounded convex polyhedron) has a well defined and unique ball center for the polytope as a whole. Also, when H is a hyperplane having a nonempty intersection with the polytope K in its interior, the ball center for K on H is again well defined and unique.

These concepts are well defined only for convex polytopes (i.e., bounded convex polyhedra) and may not really work for unbounded convex polyhedra, we discuss this issue at the end of this section. For this reason, we will first give the definitions under the assumption that the set of feasible solutions of the LP (8.1), K, is a polytope.

What happens to the LP (8.1) if we do not know whether its set of feasible solutions is bounded or not, or if we know that it is unbounded?

In the unbounded case, even though the concepts of ball centers may not be well defined, the implementation of the algorithm based on the approximate computation of ball centers can be carried out as usual. In this case, in one of the descent steps, the step length for the move may turn out to be $+\infty$. That is an indication that the objective value in (8.1) is unbounded below on its set of feasible solutions, and the algorithm terminates if this happens.

The Ball Center of a Polytope

Now we will continue with the definition of the ball center of a polytope. A polytope of dimension 1 is a line segment, its ball center is its unique midpoint (see Fig. 8.2).

Now consider the polytope K of dimension n represented by (8.1). Its ball center x^* is a point in K^0, which is the center of the largest radius ball inscribed inside K. Letting $\delta^* = \delta(x^*)$, (x^*, δ^*) is therefore an optimum solution of the LP

$$\begin{aligned} &\text{Maximize} \quad \delta \\ &\text{subject to} \quad \delta \leq A_{i.}x - b_i, \quad i = 1 \text{ to } m \end{aligned} \tag{8.5}$$

If the optimum solution of this LP (8.5) is unique, it will be $(x^*, \ \delta^*)$, and x^* is *the ball center* of K (see Fig. 8.3).

If the optimum solution of (8.5) is not unique, all of its optimum solutions are of the form $(x, \ \delta^*)$ for $x \in S$, where S is the optimum face of (8.5) in the x-space. In this case, the ball center of K is defined recursively by dimension to be the ball center of the lower-dimensional polytope S. This definition guarantees that every polytope has a unique ball center. See Fig. 8.4 for an illustration.

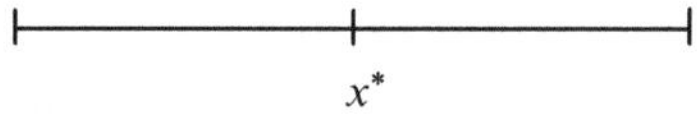

Fig. 8.2 The ball center of a 1-dimensional polytope (a line segment) is its mid-point x^*

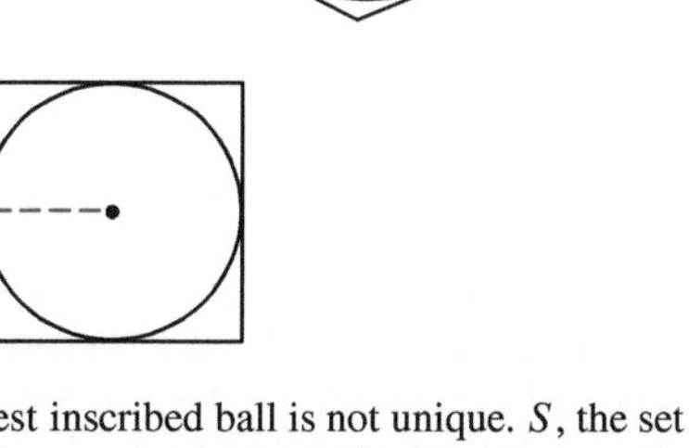

Fig. 8.3 The polytope K and the largest ball inside it are shown in the figure. When the largest inscribed ball in K is unique as here, its center x^* is the ball center of K

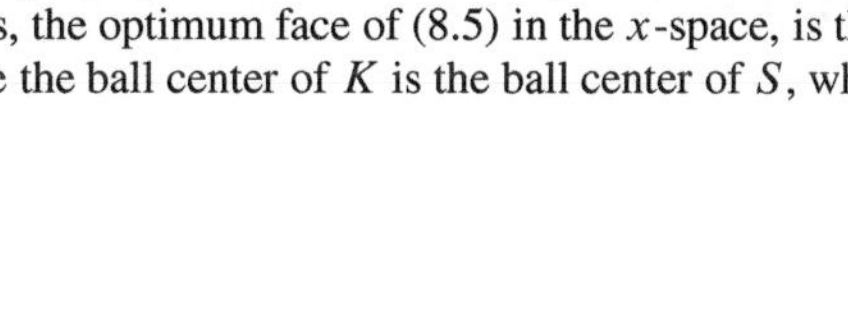

Fig. 8.4 A 2-dimensional polytope K for which the largest inscribed ball is not unique. S, the set of centers of all such balls, the optimum face of (8.5) in the x-space, is the *dashed line* segment in this polytope. Hence, here the ball center of K is the ball center of S, which is its mid-point x^*

The Ball Center for (8.1) on the Objective Plane $\{x : cx = t\}$ for Given t

The ball center of (8.1), or K, on the objective plane $\{x : cx = t\}$ for given t is defined as the center of the largest ball subject to the constraint that the center lie on the objective plane $\{x : cx = t\}$. It is an optimum solution of the LP

$$\begin{aligned} &\text{Maximize } \delta \\ &\text{subject to } \delta - A_{i.}x \leq -b_i, \quad i = 1 \text{ to } m \\ &\qquad\qquad cx = t \end{aligned} \tag{8.6}$$

If the optimum solution x for (8.6) is unique, denote it by $x(t)$ it is called *the ball center for (8.1) (or for its set of feasible solutions K) on the current objective plane* $\{x : cx = t\}$. In this case, the unique optimum solution of (8.6) is $(x(t), \ \delta[t] = \delta(x(t)))$ (see Fig. 8.5).

In general, even though the optimum δ in (8.6) is always unique, there may be multiple points x which are optimal. Hence, let $S(t)$ denote the optimum face of (8.6) in the x-space. In this case, the ball center for (8.1) on the objective plane $\{x : cx = t\}$ is defined to be the ball center of the polytope $S(t)$ as defined earlier.

This definition guarantees that for each $t \in (t_{\min}, \ t_{\max})$, the ball center for (8.1) on the objective plane $\{x : cx = t\}$ is unique.

The Case of Unbounded Convex Polyhedra

Suppose the set of feasible solutions, K of (8.1) is an unbounded polyhedron. Typically, the radius of a maximum inscribed ball in K will be ∞ (i.e., δ is unbounded

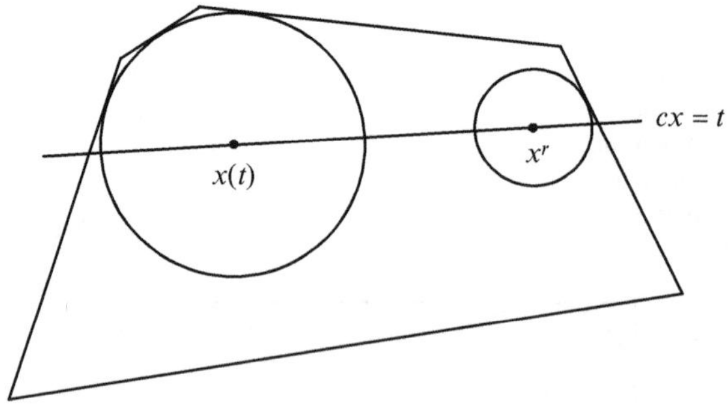

Fig. 8.5 K is the polytope in this figure. When the optimum solution for (8.6) is unique and the largest ball inside K with center on the current objective plane $\{x : cx = t\}$ is unique (like here, it is the large ball in the figure), its center is $x(t)$, the ball center for (8.1) corresponding to the present objective plane $\{x : cx = t\}$

above in the LP (8.5) typically). Even if the radius of the maximum inscribed ball is finite, the optimum face of (8.5) in the x-space may be an unbounded polyhedron itself, so the ball center of K is not defined.

Even when K is an unbounded convex polyhedron, the hyperplane H may be such that $K \cap H$ is bounded, i.e., is a polytope; in this case the ball center of K on H is well defined. If $K \cap H$ is also unbounded, then the ball center of K on H is also not defined.

8.3 Approximate Computation of Ball Centers

In Sect. 8.2, we defined *the ball center of a polytope* K and *the ball center of* K *on a hyperplane* H *intersecting* K *in its interior*, recursively by dimension.

For computation of these ball centers, we will refer to any point that is the center of the largest inscribed ball inside K (i.e., any optimum solution of (8.5)) as *a ball center* of K, and discuss here how to compute one such point approximately. In practice, the sphere methods are based on such a point computed approximately. When the optimum solution of (8.5) is not unique, the schemes discussed here will only approximate such a ball center, and not *the ball center* defined conceptually in Sect. 8.2.

Several techniques have been discussed in (Murty (2006a, b) of Chap. 1; and Murty and Oskoorouchi 2008a, b; Murty 2009a) to compute these ball centers approximately. In the following Sects. 8.3.1 and 8.3.2 we give mathematical descriptions of some techniques used for computing approximately a ball center of the polytope K and a ball center of K on the objective plane $\{x : cx = t\}$ for given t.

8.3.1 Approximate Computation of Ball Centers of Polyhedra

We consider (8.5), the problem of computing a ball center of the whole polytope K approximately beginning with an initial interior feasible solution. We summarize these techniques and discuss ways to implement them.

Our approach involves a series of line search steps. The search direction for each line search step is selected such that it is a *profitable direction*, i.e., the move in that direction at the current point is guaranteed to lead to a point that increases the radius of the inscribed ball with that point as center. Also, we will show that the optimum step length in this line search step can be found by solving a 2-variable LP.

Let $\bar{x}$ be the current interior feasible solution. Hence, a direction $y \neq 0$ is defined to be a *profitable direction* at $\bar{x}$ if $\delta(\bar{x} + \alpha y)$ strictly increases as α increases from 0; *unprofitable direction* to move at $\bar{x}$ if the maximum value of $\delta(\bar{x} + \alpha y)$ over $\alpha \geq 0$ is attained at $\alpha = 0$. We have the following result.

Theorem 8.1. *A given direction y is a profitable direction to move at the current interior feasible solution $\bar{x}$ iff $A_{i.}y > 0$ for all $i \in T(\bar{x})$. Also, $\bar{x}$ is a ball center of K iff the system $A_{i.}y > 0$ for all $i \in T(\bar{x})$ has no solution in y.*

Proof. For any α we have $\delta(\bar{x} + \alpha y) = \text{minimum}\{A_{i.}(\bar{x} + \alpha y) - b_i : i = 1 \text{ to } m\}$, and $\delta(\bar{x} + \alpha y) = \delta(\bar{x})$ when $\alpha = 0$. From the definition of $T(\bar{x})$, we have $\delta(\bar{x}) = A_{i.}x - b_i$ for all $i \in T(\bar{x})$. So, we have the following facts.

Fact 1. *For all $i \notin T(\bar{x})$, we have $A_{i.}\bar{x} - b_i > \delta(\bar{x})$, and hence when α is positive but sufficiently small $A_{i.}(\bar{x} + \alpha y) - b_i > \delta(\bar{x})$ what ever the sign of $A_{i.}y$ may be.*

Fact 2. *If $A_{i.}y > 0$ for all $i \in T(\bar{x})$, then each of these $A_{i.}(\bar{x} + \alpha y)$ is strictly monotonically increasing as α increases from 0.*

Facts 1 and 2 together imply that when α is positive but sufficiently small and y satisfies $A_{i.}y > 0$ for all $i \in T(\bar{x})$; then $\delta(\bar{x} + \alpha y) = \text{minimum}\{A_{i.}(\bar{x} + \alpha y) - b_i : i = 1 \text{ to } m\}$ is $> \delta(\bar{x})$; i.e., $\delta(\bar{x} + \alpha y)$ strictly increases as α increases from 0. This shows that y is a profitable direction to move at $\bar{x}$ under the conditions stated in the theorem.

If there is at least one index, say $i_1 \in T(\bar{x})$ satisfying $A_{i_1.}y \leq 0$, then for positive but sufficiently small values of α, we have $A_{i_1.}(\bar{x} + \alpha y) - b_{i_1} \leq A_{i_1.}\bar{x} - b_{i_1} = \delta(\bar{x})$ since $i_1 \in T(\bar{x})$. From the definition of $\delta(\bar{x} + \alpha y)$ this implies that in this case $\delta(\bar{x} + \alpha y)$ either stays the same or strictly decreases as α increases from 0, implying that y is an unprofitable direction to move at $\bar{x}$.

For $x \in K$, $\delta(x) = \text{minimum}\{A_{i.}x - b_i : i = 1 \text{ to } m\}$, is a piecewise linear function that is the pointwise infimum of a finite set of affine functions, and so by Theorem 2.5 of Chap. 2, we know that $\delta(x)$ is a concave function. If there is no profitable direction at $\bar{x}$, then $\bar{x}$ is a local maximum for the concave function $\delta(x)$ over K, and by Exercise 2.3.9 in Sect. 2.3 we know that it is the global maximum for $\delta(x)$ over K, i.e., $\bar{x}$ is a ball center of K. □

Since $\delta(\bar{x} + \alpha y) = \text{minimum}\{A_{i.}(\bar{x} + \alpha y) - b_i : i = 1 \text{ to } m\}$, we have $\delta(\bar{x} + \alpha y) \leq A_{i.}\bar{x} + \alpha A_{i.}y - b_i$ for all $i = 1$ to m.

So, for a given direction y, suppose $\bar{\delta}$ = maximum value of $\delta(\bar{x} + \alpha y)$ over $\alpha \geq 0$ and $\bar{\alpha} \geq 0$ is the value of α where this maximum is attained in $\delta(\bar{x} + \alpha y)$. Then, $(\bar{\delta}, \bar{\alpha})$ is the optimum solution of the following 2-variable LP in which the variables are δ and α.

$$\begin{aligned} &\text{Maximize } \delta \\ &\text{subject to } \delta - \alpha A_{i.}y \leq A_{i.}\bar{x} - b_i, \quad i = 1 \text{ to } m \\ &\delta, \alpha \geq 0. \end{aligned} \tag{8.7}$$

Therefore, the optimum step length for the line search problem of maximizing $\delta(\bar{x} + \alpha y)$ over $\alpha \geq 0$ in the given direction y can be found by solving this 2-variable LP (8.7).

LSFN, Using Line Search Steps in Facetal Normal Directions

One advantage of this technique is that it needs no matrix inversions. Beginning with the initial $x^{r,0} = x^r$, it generates a sequence of points $x^{r,k}$, $k = 1, 2, \cdots$ along which the radius of the inscribed ball δ is strictly increasing.

Whether a given direction y is profitable or not at the current point $x^{r,k}$ can be checked very efficiently using the conditions proved in Theorem 8.1 (these are: $A_{i.}y > 0$ for all $i \in T(x^{r,k})$). Since the goal in this centering step is to increase the minimum distance of x from each facetal hyperplane of K, this procedure LSFN uses the directions normal to the facetal hyperplanes of K for the line searches, i.e., directions in $\Gamma_1 = \{A_{i.}^T, -A_{i.}^T : i = 1 \text{ to } m\}$. This is the set of directions normal to facetal hyperplanes of K. This procedure continues as long as profitable directions for line search are found in Γ_1 and terminates with the final point as an approximate center of K, which is denoted by $\bar{x}^r$ (see Fig. 8.6).

2-Variable LP to Compute the Step Length in Profitable Direction in the Centering Step: Once a profitable direction y at the current point $x^{r,k}$, has been found, the optimum step length α in this direction that maximizes $\delta(x^{r,k} + \alpha y)$ over $\alpha \geq 0$ is $\bar{\alpha}$, where $(\bar{\delta}, \bar{\alpha})$ is the optimum solution of the 2-variable LP

$$\begin{aligned} &\text{Maximize} \quad \delta \\ &\text{subject to} \quad \delta - \alpha A_{i.}y \leq A_{i.}x^{r,k} - b_i \qquad i = 1, \ldots, m \\ &\qquad\qquad\quad \delta,\ \alpha \geq 0 \end{aligned} \tag{8.8}$$

and $\bar{\delta}$ is the optimum objective value $\delta(x^{r,k} + \bar{\alpha} y)$. Hence, the line search for the maximum value of δ in the direction y involves solving this 2-variable LP, which can be carried out efficiently (e.g., by the simplex algorithm) as discussed above.

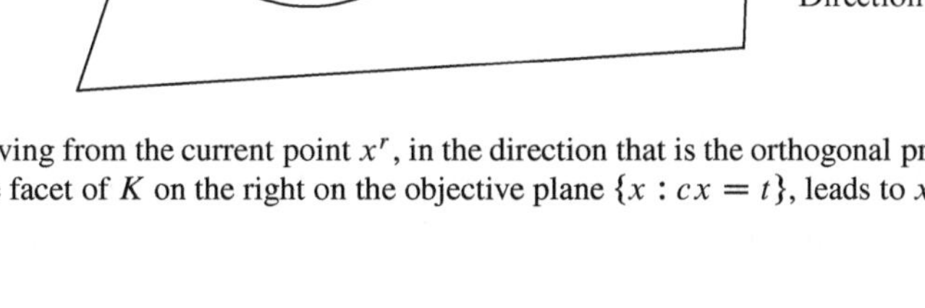

Fig. 8.6 Moving from the current point x^r, in the direction that is the orthogonal projection of the normal to the facet of K on the right on the objective plane $\{x : cx = t\}$, leads to $x(t)$

LSCPD, Sequence of Line Search Steps Using Computed Profitable Directions

This technique, discussed in (Murty 2006b of Chap. 1), consists of a sequence of at most n line search steps in profitable directions computed by solving a system of linear equations. After each step in the sequence, the index set of touching constraints at the current solution x keeps growing by at least one more constraint, and we will stop the sequence when the set of coefficient vectors of the touching set of constraints becomes linearly dependent. That's why the number of steps in the sequence is at most n. The entire sequence needs a single matrix inversion, carried out in stages adding one row and column to the matrix at a time; hence, it uses matrix inversion operations sparingly.

The sequence begins with the initial point in the sequence $x^{r,0} = x^r$. When $x^{r,k}$ is the current solution, from Theorem 8.1, we know that any solution y of the system

$$A_{i.}y = d_i \quad \text{for all } i \in T(x^{r,k}) \tag{8.9}$$

where $d = (d_i : \ i \in T(x^{r,k})) > 0$ is any positive vector, is a profitable direction to move at $x^{r,k}$.

We will use (8.9) with $d = e$, the column vector of all 1s of appropriate dimension, to generate a profitable direction which is a basic solution of (8.9) to move. Once a profitable direction y is determined, the step length to move in this direction is determined by solving the 2-variable LP (8.8) as discussed under LSFN.

Hence, this sequence begins with the initial point $x^{r,k}$ for $k = 0$, solves (8.9) with $d = e$ for a basic solution. If this system has no solution, then this technique is terminated with the initial point as the approximation of a ball center of K obtained in it.

If it does yield a solution y^k, suppose it is the basic solution with respect to a basic vector y_{B^k} and basis B^k for the system. If α^k is the optimum step length maximizing $\delta(x^{r,k} + \alpha y^k)$, notice that as α increases from 0, $A_{i.}(x^{r,k} + \alpha y^k) = A_{i.}x^{r,k} + \alpha$ increases at the same rate as α for all $i \in T(x^{r,k})$. This implies that at the solution $x^{r,k+1} = x^{r,k} + \alpha^k y^k$ obtained after this line search step, we will have $T(x^{r,k+1}) \supset T(x^{r,k})$.

The sequence will now continue the same way with $x^{r,k+1}$ as the initial solution for the next step. Suppose $T(x^{r,k}) = \{1, \cdots, s\}$ and $T(x^{r,k+1}) = \{1, \cdots, s, s+1\}$. Then the profitable direction y^{k+1} to be used at $x^{r,k+1}$ is computed by solving the system

$$A_{i.}y = 1 \quad \text{for all } i \in \{1, \cdots, s+1\} \tag{8.10}$$

Let $\mathcal{A}^k = (B^k \vdots D^k)$ denote the coefficient matrix of (8.9) with rows $\{A_{i.} : i \in T(x^{r,k})\}$, with columns partitioned into the basic, nonbasic parts with respect to the basic vector y_{B^k} for it, and $(A^{B^k}_{s+1.}, A^{D^k}_{s+1.})$ the corresponding partition of $A_{s+1.}$.

The set of touching constraint coefficient vectors $\{A_{i.} : i \in T(x^{r,k+1})\}$ is linearly independent iff $A^{D^k}_{s+1.} - A^{B^k}_{s+1.}(B^k)^{-1} D^k \neq 0$.

Therefore, if $A^{D^k}_{s+1.} - A^{B^k}_{s+1.}(B^k)^{-1} D^k = 0$, terminate the sequence with $x^{r,k+1}$ as the final approximation of a ball center of K obtained in it. On the other hand, if this vector is $\neq 0$, select a nonzero entry in it, suppose it is in the column of the variable y_j, then let $\mathcal{A}^k_{.j}$ be the column corresponding to y_j in $\mathcal{A}^k$, the coefficient matrix of (8.9). Then $y_{B^{k+1}} = (y_{B^k}, y_j)$ is a basic vector for (8.10). The corresponding basis for (8.10) is

$$B^{k+1} = \begin{pmatrix} B^k & \vdots & \mathcal{A}^k_{.j} \\ \dots & & \dots \\ A^{B^k}_{s+1.} & \vdots & a_{s+1,j} \end{pmatrix}.$$

where $a_{s+1,j}$ is the coefficient corresponding to y_j in $A_{s+1.}$. Hence

$$\left(B^{k+1}\right)^{-1} = \begin{pmatrix} P & \vdots & Q \\ \dots & & \dots \\ R & \vdots & S \end{pmatrix}$$

where

$$S = 1/\left(a_{s+1,j} - A^{B^k}_{s+1.}(B^k)^{-1}\mathcal{A}^k_{.j}\right)$$

$$R = \left(-A^{B^k}_{s+1.}(B^k)^{-1}\right) / \left(a_{s+1,j} - A^{B^k}_{s+1.}\left(B^k\right)^{-1}\mathcal{A}^k_{.j}\right)$$

$$Q = -\left(B^k\right)^{-1}\mathcal{A}^k_{.j} S$$

$$P = \left(B^k\right)^{-1} + QR/S$$

Thus, $\left(B^{k+1}\right)^{-1}$ can be obtained by updating $\left(B^k\right)^{-1}$ using the above formulas.

The sequence repeats the same way with $x^{r,k+1}$ until it terminates at some stage. Thus, in this sequence, whenever system (8.10) is augmented by a new constraint in a step, the basic vector and basis inverse in this step can be updated quite efficiently for the next step as discussed above.

To compute a ball center of K approximately starting with the initial interior feasible solution we first apply the LSFN sequence of steps as described above selecting profitable directions from the set Γ_1 until the improvement in the radius of the ball δ per step decreases below some selected tolerance. At that time we switch and initiate an LSCPD sequence starting with the final point obtained at the end of the LSFN sequence of steps. When the LSCPD sequence terminates, the final point obtained in that sequence is an approximate ball center for K.

8.3.2 Computing An Approximate Ball Center of K on the Current Objective Plane

Consider the current objective value t and the current objective plane $\{x : cx = t\}$. The model for finding the ball center on this objective plane is (8.6); the only difference between this and (8.5) is the additional constraint $cx = t$ in (8.6).

Hence, an approximate solution to (8.6) can be obtained directly by adopting the line search techniques discussed in earlier sections to this situation.

The only change required to adopt the line search techniques LSFN to this problem is to look for profitable directions from the set $\Gamma_2 = \{P_{.1}, \cdots, P_{.m}, -P_{.1}, \cdots, -P_{.m}\}$, where $P_{.i} = (I - c^T c)(A_{i.})^T$, the orthogonal projection of $A_{i.}$ (the direction normal to the facet of K defined by the ith constraint in (8.1)) on the hyperplane $\{x : cx = 0\}$, for $i = 1$ to m, instead of the set Γ_1. Since Γ_2 is the set of directions that are orthogonal projections of the directions in Γ_1 on the plane $\{x : cx = 0\}$, any step length from a point in the current objective plane, in a direction from Γ_2, will keep the point on the current objective plane (see Fig. 8.6).

To adopt the line search sequence LSCPD to this problem, right from the initial step in this sequence, we include the additional constraint $cy = 0$ in all systems of the form (8.10) in the sequence, and continue as discussed under LSCPD.

8.3.3 Ball Centers of Some Simple Special Polytopes

An n-dimensional simplex in R^n is a convex polytope S which has exactly $(n + 1)$ extreme points $\{x^1, \cdots, x^{n+1}\}$ satisfying the property that for each i, the set $\{x^j - x^i : j = 1 \text{ to } n + 1, j \neq i\}$ is a basis for R^n. It can be represented as the set of feasible solutions of a system of linear inequalities

$$Dx \geq d$$

where D is an $(n+1)\times n$ matrix satisfying the properties that every square submatrix of it of order $n \times n$ is nonsingular, and for each i, $D_{i.}$ is a linear combination of the other rows of D with strictly negative coefficients.

As examples, a triangle is a 2-dimensional simplex, and a tetrahedron (e.g., the one with the set of vertices $\{(0, 0, 0), (1, 0, 0), (0, 1, 0), (0, 0, 1)\}$) is a 3-dimensional simplex.

Assume that in the above system, all rows of D have been normalized so that $||D_{i.}|| = 1$ for all i. In Murty (2009a, b) it has been shown that this simplex S has a unique ball center, which is the unique solution of

$$Dx - \delta e = d$$

where e is a column vector of all 1s, and (x, δ) in the solution are the ball center and the radius of the largest ball inside S. This can be proved easily using the conditions derived in Theorem 8.1. See Fig. 8.7 illustrating the ball center of a 2-dimensional simplex.

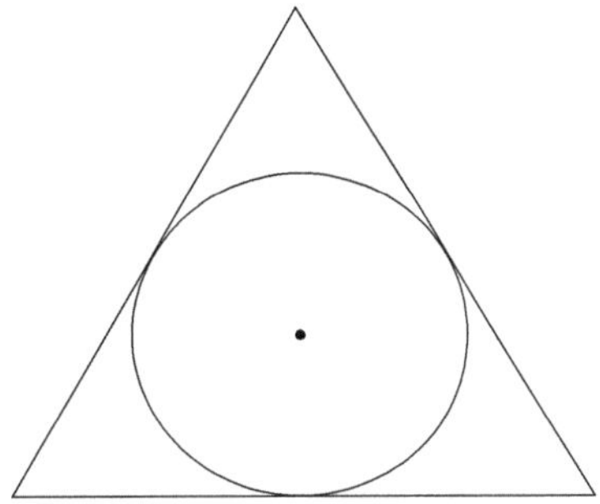

Fig. 8.7 A 2-dimensional simplex (a *triangle* in R^2) shown with the largest radius ball inside it; see that it touches all the facets of the simplex. The center of this ball marked in the figure is the ball center of this simplex

In the same way they show that the ball center of S can be computed directly even when it is represented as the convex hull of its set of extreme points $\{x^1, \cdots, x^{n+1}\}$. They also show that the ball center of the convex hull of a linearly independent set of vectors $\{x^1, \cdots, x^r\}$ can be computed directly in the same way.

Murty (2009a, b) also discuss the application of these results to improve the performance of sphere methods.

The Sphere Methods

There are two versions of the sphere method which we will call sphere methods 1 and 2. We will discuss them in the following sections. To solve (8.1) both methods need an initial interior feasible solution. In both methods, each iteration consists of two steps: a centering step and a descent step.

In concept, the centering step is a corrector step; it tries to move the current interior feasible solution into another with a higher value for $\delta(x)$ while keeping the objective value either constant or decreasing. Of the two steps in each iteration, the centering step is computationally the most expensive. Once the center is computed in an iteration, the effort needed for a descent step in a given descent direction is just one minimum ratio computation, which is cheap in comparison. That is why, in contrast to earlier methods, we carry out descent steps in six different descent directions (all obtained directly without additional computation) in each iteration and take the best result obtained from all of them as the output of this iteration with which the method goes to the next iteration.

8.4 Sphere Method 1

This is the version discussed in (Murty and Oskoorouchi 2008a). The general iteration $r + 1$, begins with the current point x^r, the interior feasible solution obtained at the end of the previous iteration (or with the available interior feasible solution if $r = 0$).

Centering step: Starting with the current point x^r, in concept the aim of this step is to find a ball center for (8.1) on the current objective plane $\{x : cx = t\}$ approximately, where $t = cx^r$, using line search steps discussed in Sects. 8.3.1 and 8.3.2. The reason for getting a ball with the maximum possible radius with center on the current objective plane is this: from the center of a ball with radius δ, we can move a step length of at least δ in any direction. Maximizing δ helps to make longer steps towards optimality in the descent steps in each iteration.

But in practice, if the same step that increases $\delta(x)$ can also decrease the value of cx instead of keeping it constant, that is even better for solving the LP (8.1). For this reason, in this step we select line search directions which are both profitable and are also descent directions for cx if possible.

First, apply LSFN as discussed in Sect. 8.3.1, selecting profitable directions for the line search from the set Γ_2 (instead of Γ_1, to keep the objective value unchanged) until the improvement in the radius of the ball δ per step decreases below some selected tolerance. At that time suppose the point in the sequence being generated is $x^{r,k}$.

With $x^{r,k}$, switch and initiate an LSCPD sequence with $x^{r,k}$ as the initial point of the sequence. We will now follow a different procedure than what is described in Sect. 8.3.2. Let $x^{r,g}$ denote a general point in the sequence being generated in this sequence.

When $x^{r,g}$ is the current point, we look for a profitable direction to move at $x^{r,g}$ which is a basic solution of the system

$$A_{i.}y = 1 \quad \text{for all } i \in T(x^{r,g}). \tag{8.11}$$

Notice that the additional constraint "$cy = 0$" as discussed in Sect. 8.3.2 is not included in (8.11). Suppose we obtain a basic solution y^0 for this system, with respect to the basic vector y_{B^0} and basis B^0. There are two cases to consider.

Case 8.1. If $cy^0 \leq 0$, then y^0 is not only profitable at $x^{r,g}$ (i.e., $\delta(x^{r,g} + \alpha y^0)$ increases as α increases from 0), but it is also a descent direction for cx. So, we carry out a line search step at $x^{r,g}$ in the direction y^0 exactly as described in Sect. 8.3.1 and continue. This move increases $\delta(x)$ and may also decrease cx.

Case 8.2. If $cy^0 > 0$, there are two subcases to consider here. Let $(B^0 \vdots D^0)$ be the partition of the coefficient matrix of (8.11) into basic and nonbasic parts with respect to the basic vector y_{B^0} for it. Let $(c_{B^0} \vdots c_{D^0})$ be the corresponding partition of the row vector c.

Subcase 2.1: $c_{D^0} - c_{B^0}(B^0)^{-1}D^0 = 0$. In this subcase, cy = a constant = $cy^0 > 0$ in every solution of (8.11). So, in this subcase using any solution of (8.11) as the direction for the move helps increase $\delta(x)$, but also increases cx. So we terminate the sequence at this stage and take the current point $x^{r,g}$ as the approximation to the ball center on the objective plane through $x^{r,g}$ and go to the descent step discussed next.

Subcase 2.2: There is a nonzero entry in $c_{D^0} - c_{B^0}(B^0)^{-1}D^0$, suppose it occurs in the column of the nonbasic variable y_j and that nonzero entry is denoted by $\bar{c}_j$. Let the column vector of this nonbasic variable y_j in D^0 be denoted by $(D^0)_{.j}$. Then the solution y^1 of (8.11) given by: all the variables except y_j in the nonbasic vector y_{D^0}, are $= 0$, and

$$y_{B^0} = y^0_{B^0} - \theta(B^0)^{-1}(D^0)_{.j}$$
$$\text{Nonbasic } y_j = \theta.$$

where $\theta = (-1 - cy^0)/\bar{c}_j$, satisfies $cy^1 = -1$. Now carry out a line search step at the current point $x^{r,g}$ in the direction y^1 exactly as in Sect. 8.3.1 and continue. As under Case 1, this move not only increases $\delta(x)$ but also decreases cx.

This seems to output a good approximation to a ball center with objective value $\leq$ the objective value at the starting point x^r. With that go to the Descent steps discussed next.

Descent Steps: Let $\bar{x}^r$ denote the approximate ball center obtained in the centering step.

Minimum Ratio Test to Compute the Step Length in Descent Steps: Each descent step carried out in this iteration requires one minimum ratio computation. For example, consider a descent step from the current center $\bar{x}^r$ in the descent direction y (i.e., satisfying $cy < 0$). If the step length is λ, the move leads to the point $\bar{x}^r + \lambda y$. Select a small positive number ϵ_1 as the tolerance for minimum $\{A_{i.}x - b_i : i = 1 \text{ to } m\}$ for the point x to be in the interior of K. Then we will take the step length from $\bar{x}^r$ in the direction y to be: $(-\epsilon_1) +$ (the maximum step length possible while remaining inside K), which is

$$\gamma = \text{minimum}\left\{\frac{-A_{i.}\bar{x}^r + b_i + \epsilon_1}{A_{i.}y} : i \text{ such that } A_{i.}y < 0\right\}$$

and then the point obtained at the end of this descent step will be $\bar{x}^r + \gamma y$ if γ is finite.

If $\gamma = \infty$, the objective function $z(x)$ is unbounded below in (8.1) and $\{\bar{x}^r + \lambda y : \lambda \geq 0\}$ is a feasible half-line along which $z(x)$ diverges to $-\infty$ on K. Terminate the method if this occurs.

We now list the various descent steps carried out in this iteration. After each descent step, include the point obtained at the end of it, along with its objective value, in a *List*.

D1, Descent Step 1: From the ball center $\bar{x}^r$ take a descent step in the direction $d^1 = -c^T$.

D2, Descent Step 2: From the ball center $\bar{x}^r$ take a descent step in the direction $d^2 = \bar{x}^r - \bar{x}^{r-1}$, where $\bar{x}^{r-1}$ denotes the ball center computed in the previous iteration r. Hence, this direction is the direction of the path of ball centers generated in the algorithm.

D3, Descent Steps 3: The directions $-c^i$ for $i \in T(\bar{x}^r)$ are called *GPTC* (gradient projection on touching constraint) directions at this stage. Carry out descent steps from the ball center $\bar{x}^r$ in each of these GPTC directions.

D4, Descent Step 4: From the ball center $\bar{x}^r$ take a descent step in the direction $d^4 = (\sum(-c^i : \text{for } i \in T(\bar{x}^r))/|T(\bar{x}^r)|$, the average direction of all the GPTC directions.

D5.1, Descent Steps 5.1: For $i \in T(\bar{x}^r)$, let x^{ir} denote the orthogonal projection of the center $\bar{x}^r$ on the touching facetal hyperplane $\{x : A_{i.}x = b_i\}$; it is the point where this facetal hyperplane touches the ball $B(\bar{x}_r, \delta(\bar{x}^r))$. The points x^{ir} for $i \in T(\bar{x}^r)$ are called the *touching points (TPs)* of the ball $B(\bar{x}_r, \delta(\bar{x}^r))$ with its touching facetal hyperplanes of K.

Let $0 < \epsilon < 1$ be a small positive tolerance ($\epsilon = 0.1$ works well). Then for $i \in T(\bar{x}^r)$, the point on the line segment joining $\bar{x}^r$ and x^{ir} close to the TP x^{ir}, $\hat{x}^{ir} = \epsilon\bar{x}^r + (1-\epsilon)x^{ir}$ is called the *near touching point (NTP)* corresponding to the tangent plane $\{x : A_{i.}x = b_i\}$ of the ball $B(\bar{x}_r, \delta(\bar{x}^r))$.

Carry out a descent step in the GPTC $-c^i$ from the NTP $\hat{x}^{ir}$, for each $i \in T(\bar{x}^r)$ (see Fig. 8.8).

At the end of all these steps, let $\tilde{x}^{r1}$ denote the best point (i.e., the one corresponding to the least objective value) among the $|T(\bar{x}^r)|$ points obtained in these descent steps.

D5.2, Descent Step 5.2: This descent step is from Murty and Kabadi (2008).

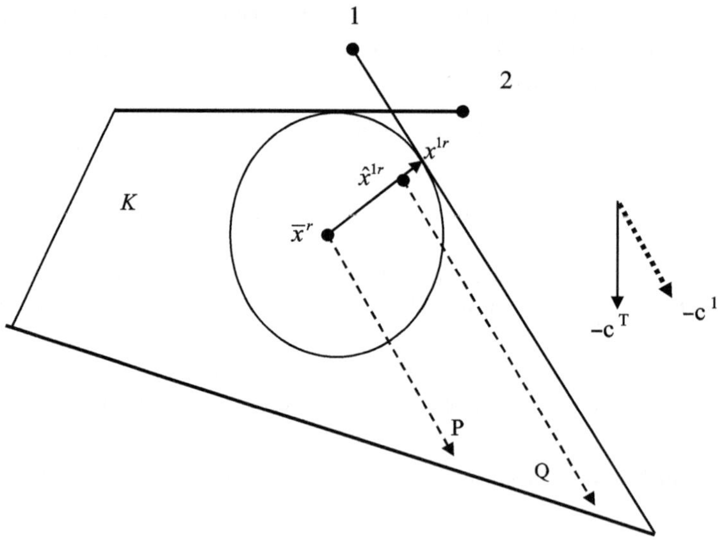

Fig. 8.8 Descent steps in a GPTC direction. Here, $\bar{x}^r$ is the current center, $T(\bar{x}^r) = \{1, 2\}$. Directions $-c^T$ pointing down south, $-c^1$ = orthogonal projection of $-c^T$ on facetal hyperplane of constraint 1, are shown. x^{1r} = TP of constraint 1, $\hat{x}^{1r}$ = NTP corresponding to constraint 1. Descent step from $\bar{x}^r$ [$\hat{x}^{1r}$] in direction $-c^1$ are shown, leading to points P [Q], respectively. Here Q is a much better point than P

By the way the descent steps are carried out, it is clear that $\tilde{x}^{r1}$, the best point obtained in Descent Step 5.1, is close to the boundary of K and $\delta(\tilde{x}^{r1}) \leq \epsilon$. Find the touching set $T(\tilde{x}^{r1})$ = set of all i that tie for the minimum in $\{A_{i.}\tilde{x}^{r1} - b_i : i = 1 \text{ to } m\}$.

For each $i \in T(\tilde{x}^{r1})$, from $\tilde{x}^{r1}$ take a descent step in the GPTC direction $-c^i$ and include the resulting point along with its objective value in a new List 5.2.

At the end, let $\tilde{x}^{r2}$ denote the best point by objective value, in List 5.2. If $c\tilde{x}^{r1} - c\tilde{x}^{r2}$ is:

$\leq$ some selected tolerance for objective value reduction, take $\tilde{x}^{r2}$ as the output of this Descent Step 5.2, put $\tilde{x}^{r2}$ along with its objective value in the List, and go to the next iteration.

$>$ the selected tolerance for objective value reduction, with $\tilde{x}^{r2}$ as the initial interior feasible solution repeat this Descent Step 5.2, and continue the same way.

When all these descent steps are carried out, the best point x^{r+1}, among the output points of all the descent steps, is the output of this iteration, with that point the method goes to the next iteration. Just as with other IPMs, this method also terminates when the change in the final points obtained in successive iterations is smaller than some tolerance (i.e., it terminates at the end of iteration $r+1$ if $||x^{r+1} - x^r||/||x^r|| < \epsilon$, concluding that x^{r+1} is an optimum solution of (8.1)).

8.4.1 Summary of Computational Results on Sphere Method 1

In this section, we present some computational results of implementing sphere method 1, using MATLAB 7.0 to implement the algorithm and test it on some randomly generated problems from (Murty and Oskoorouchi 2008a). For details on how the test problems are generated and computational issues (like various tolerances used in the code, etc.), see this paper. Also, in these tests, Descent Step 5.2 is not included, but all the others are.

Like other IPMs discussed in Chap. 7 we noticed in these experiments that the number of iterations needed to solve an LP in n variables with m constraints, is very small, and grows very slowly as n, m increase. In effect, like those other IPMs, sphere method 1 also seems to need almost a constant number of iterations to solve an LP as n, m increase.

The following table summarizes the results from solving an LP in $n = 50$ variables, with 150 constraints and bounds on individual variables, from (Murty and Oskoorouchi 2008a), to minimize an objective function $z(x) = cx$, starting with an initial interior feasible solution, at which the objective value is 0. Sphere method 1 needed 10 iterations to solve this problem. In the following table we give: Itr = iteration number r, δ_r = radius of the ball obtained at the end of the centering step in iteration r, T_r = number of touching constraints for this ball, z_r = objective value of the output point obtained at the end of iteration r, and D_r = the descent step that yielded the output point in iteration r.

Itr r	δ_r	T_r	z_r	D_r
1	2.052	22	−3.31	D5.1
2	2.053	30	−14.80	D5.1
3	1.40	50	−22.88	D5.1
4	0.60	50	−28.04	D5.1
5	0.27	49	−30.12	D5.1
6	0.11	47	−31.06	D5.1
7	0.04	46	−31.43	D2
8	0.003	49	−31.62	D5.1
9	Small	48	−31.64	D5.1
10	Small	48	−31.65	D5.1

When x^*, x^r, x^{r+1} are an optimum solution and the current feasible solutions at the beginning and end of iteration r, the quantity $\frac{100(cx^r - cx^{r+1})}{(cx^r - cx^*)}$ is known as the percent moves towards optimality, in this iteration r in the algorithm. On problems with $n = 100$, $m = 300$, and densities ranging from 10 to 100%, this percent move towards optimality per iteration in sphere method 1 varied from 10 to 20 and from 0.22 to 0.50 in the simplex method.

In terms of CPU time, even this preliminary code based on MATLAB 7.0 routines for each individual step separately in sphere method 1 outperforms production versions of the Simplex method on these problems.

8.5 Sphere Method 2

Consider the LP in the form (8.1), which we reproduce below for convenience.

$$\begin{aligned} \text{Minimize} \quad & z = cx \\ \text{subject to} \quad & Ax \geq b \end{aligned} \tag{8.12}$$

where A is a matrix of order $m \times n$ with $||c|| = ||A_{i.}|| = 1$ for all $i = 1$ to m; its set of feasible solutions K and an initial interior feasible solution x^0.

In sphere method 1 the set of feasible solutions considered remains unchanged (i.e., remains the original K) throughout the algorithm, but the current objective plane $\{x : cx = t\}$ keeps on sliding in a parallel fashion towards decreasing values of t from one iteration to the next. The centering step in this method in each iteration has the aim of finding a ball center on the current objective plane, at least in principle. Even though line search directions y used in LSCPD in the centering step in sphere method 1 may satisfy $cy < 0$; all the search directions used in LSFN satisfy $cy = 0$, and hence leave the objective value unchanged.

In sphere method 2, in contrast, the set of feasible solutions K considered is updated by the current objective value after each iteration, and hence gets smaller. Thus, to distinguish, we will denote by K^r, the set of feasible solutions considered in Step r and we will have $K^r \subset K^{r-1} \subset K$ for all r. And in the centering step of sphere method 2, all line search directions used (in both the LSFN and LSCPD sequences) will both be profitable and descent directions for the original objective function $z = cx$.

The first iteration of sphere method 2 begins with the initial interior feasible solution x^0. We will now describe the general iteration, iteration $r+1$, in this method.

General Iteration $r + 1$

The initial point for this iteration is x^r, the interior feasible solution obtained at the end of the previous iteration. Define the set of feasible solutions to be considered for this iteration to be K^{r+1}, where

$$K^{r+1} = \{x : Ax \geq b, \text{ and } \; cx \leq cx^r + \epsilon\}$$

where ϵ is a small positive tolerance parameter. Go to the centering step in this iteration.

Centering Step: The aim of this step is to find a ball center of K^{r+1} approximately as described in Sect. 8.3.1 (see Fig. 8.9).

LSFN: The set of facetal normal directions of K^{r+1} is $\Gamma_1^{r+1} = \{\pm c^T, \pm A_{i.} : i = 1 \text{ to } m\}$. Apply LSFN to find a ball center for K^{r+1} as in Sect. 8.3.1 using profitable directions from Γ_1^{r+1}.

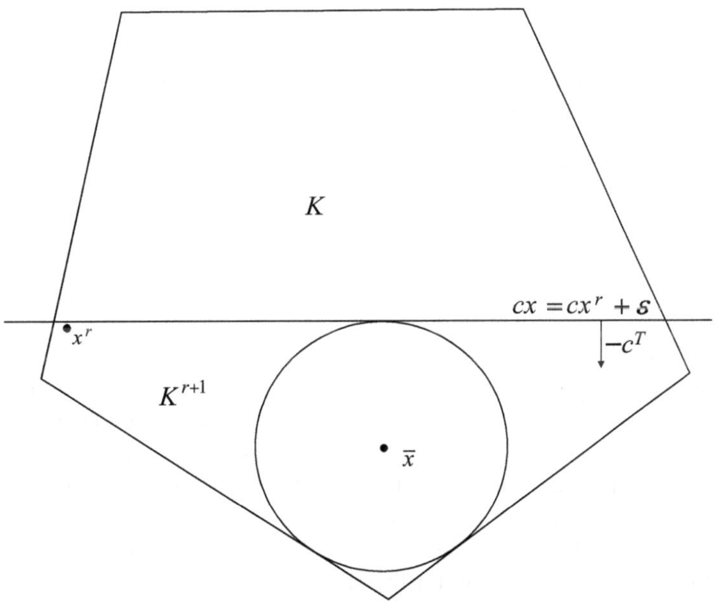

Fig. 8.9 K is the original set of feasible solutions of the LP being solved. The current set of feasible solutions in an iteration when x^r is the initial interior feasible solution, is K^{r+1}. The ball shown is the largest ball inside K^{r+1} and its center $\bar{x}$ is a ball center obtained in the centering step in this iteration

LSCPD: This sequence begins with the interior feasible solution obtained at the end of LSFN.

Let $\hat{x}$ denote the interior feasible solution in a step of this sequence. The touching constraint set at $\hat{x}$ for K^{r+1} will typically include the objective constraint in the definition of K^{r+1}. If it does not, then apply this sequence as discussed in Sect. 8.3.1.

On the other hand, if the touching constraint set includes the objective constraint, let $T^{r+1}(\hat{x})$ denote the touching constraint index set for K^{r+1}. Solve the system

$$\begin{aligned} A_{i.}y &= 1 \quad \text{for all} \quad i \in T^{r+1}(\hat{x}) \\ -cy &= \beta \end{aligned} \tag{8.13}$$

where β is a non-negative parameter. In Sect. 8.3.1, we used only $\beta = 1$. But here we will leave it as a parameter that is restricted to take positive values only and obtain a solution of (8.13) as a function of this parameter β. Let this solution be denoted by $p + \beta q$.

As in Sect. 8.3.1, if B is a basis associated with the basic vector y_B obtained for (8.13) let y_D denote the vector of remaining nonbasic variables in (8.13) associated with the basic vector y_B. Let $p = (p_B, p_D)$, $q = (q_B, q_D)$ be the partition of the vectors p, q corresponding to the partition of y into basic and nonbasic parts (y_B, y_D). Then $q_D = p_D = 0$, and q_B is the last column of B^{-1} and p_B is the sum of the remaining columns of B^{-1}.

3-Variable LP to Find Step Length in This Step: Hence, for all $\beta > 0$, $p + \beta q$ is a profitable direction at $\hat{x}$ for K^{r+1}. With $p + \beta q$ as line search direction, the optimum step length α (maximizing $\delta(\hat{x} + \alpha(p + \beta q))$, the radius of the maximum radius ball inscribed in K^{r+1} with $\hat{x} + \alpha(p + \beta q)$ as center) is determined by solving the 3-variable LP in variables δ, α, γ.

$$\begin{aligned} &\text{Maximize} \quad \delta \qquad \text{subject to} \\ \delta - \alpha A_{i.}p - \gamma A_{i.}q &\leq A_{i.}\hat{x} - b_i, \quad i = 1, \cdots, m \\ \delta - \alpha(-c)p - \gamma(-c)q &\leq (-c)\hat{x} - ((-c)\hat{x} - \epsilon) \\ \delta, \alpha, \gamma &\geq 0. \end{aligned}$$

Here, α, γ will both be > 0 at optimum. Actually, this γ is $(\alpha)(\beta)$.

If $(\bar{\delta}, \bar{\alpha}, \bar{\gamma})$ is an optimum solution of this 3-variable LP, then the point obtained at the end of this step is $\hat{x} + \bar{\alpha}p + \bar{\gamma}q$. With that the next LSCPD step is applied again as here, and so on until the LSCPD sequence is completed,

Let $\bar{x}$ denote the point obtained at the end of LSCPD, it is the approximate ball center of K^{r+1} obtained in this iteration (see Fig. 8.9).

With the point $\bar{x}$ obtained at the end of the centering step, the iteration moves to the descent steps in this iteration for the current set of feasible solutions K^{r+1}, to apply descent steps D1 to D5.2 as described in Sect. 8.4.

Instead of giving β the specific value 1 as in earlier methods, leaving it as a positive parameter in (8.13) improves the performance of the centering step.

With the point obtained at the end of the iteration, the algorithm moves to the next iteration. Termination criteria are the same as in sphere method 1.

Computational Performance of Sphere Method 2: Results from computational experiments reported in Murty and Oskoorouchi (2008b) comparing both sphere methods 1 and 2, and comparing these with the simplex method indicate that sphere method 2 (including all six descent steps D1–D4, D5.1, D5.2) is about 40% faster than sphere method 1; particularly on large scale LPs. These computational tests also confirmed that implementing sphere method 2 by a low-level programming language would make it competitive with other IPMs discussed in Chap. 7.

8.6 Improving the Performance of Sphere Methods Further

The most computationally expensive step in the sphere methods is the centering step in each iteration, so reducing the number of times this step has to be carried out will improve the performance of the methods.

The Descent Step D5.1 in sphere methods presents an opportunity to reduce the number of times the centering step has to be used in them.

Let $\{\hat{x}^1, \cdots, \hat{x}^s\}$ denote the set of all points obtained at the end of the various descent steps in D5.1 in an iteration. We will have $s \leq m$, and typically $s \leq n+1$, so s is not a large number in comparison to m, n.

Let K_2 denote the convex hull of $\{\hat{x}^1, \cdots, \hat{x}^s\}$. Typically, $\hat{x}^1, \cdots, \hat{x}^s$ are spread out in different directions all around K, each one in the interior of K but close to the boundary of K. So, intuitively, it seems that a ball center for K_2 may be close to a ball center for K on the objective plane through it.

We checked whether the average of $\hat{x}^1, \cdots, \hat{x}^s$ is a reasonable approximation to a ball center of K_2, but typically it is not.

When $s = n+1$, we find that typically K_2 is a simplex. In this case, the ball center of K_2 can be computed directly using the method discussed briefly in Sect. 8.3.3, and in greater detail in Murty (2009a, b). When $s \leq n$, we found that typically the set $\{\hat{x}^1, \cdots, \hat{x}^s\}$ is linearly independent. Again in this case, the ball center of K_2 is unique which can be computed directly using again the method mentioned above. Using these results, a new descent step labeled D5.3 to be carried out after D5.1 in sphere method 1 has been developed that promises to significantly reduce the number of times the centering step has to be used (see Murty 2009b). At the time this book is going to press, this new step has not been tested computationally; but it will be tested soon and the results reported in Murty and Oskoorouchi (2008b).

Conclusions

We presented some preliminary computational results on implementing sphere methods 1 and 2 by solving each step in these methods using MATLAB 7.0 routines separately and compared this performance with that of MATLABs finished

LP code "linprog" based on the simplex method. The results show that even this implementation of the sphere methods performs much better than "linprog."

To compare the sphere methods with existing IPMs will require developing a low-level programming language code for them using advanced techniques of numerical linear algebra and updating the basis inverse in LSCPD steps as the matrix grows by a row and column as described above, which we have not done in these preliminary experiments. But these preliminary results and the fact that the work in each iteration of the sphere methods is much simpler than an iteration of other IPMs, indicates that sphere methods will have advantage over them for solving large-scale models, in particular when the models may have redundant constraints, or a coefficient matrix that is not very sparse.

8.7 Some Open Theoretical Research Problems

Let $K = \{x : Ax \geq b\}$, where A is a matrix of order $m \times n$, and K^0 its interior. Assume that K is a polytope, i.e., it is bounded.

Let $z = cx$ be an objective function for which the minimum and maximum values in K are $t_{\min}$ and $t_{\max}$. Also, let the symbols $\delta(x)$, t^*, $B(x, \delta(x))$, $T(x)$ have the same meanings as defined in the notation at the beginning of this chapter.

For $t_{\min} \leq t \leq t_{\max}$ let $x(t)$ denote the ball center of K on the objective plane $\{x : cx = t\}$, as defined conceptually in Sect. 8.2. Then for each $t_{\min} \leq t \leq t_{\max}$, $B(x(t), \delta(x(t)))$ is a maximum radius inscribed ball in K with its center on the objective plane $\{x : cx = t\}$.

For simplicity we denote $\delta(x(t))$ by $\delta[t]$, it is the radius of the maximum radius ball that can be inscribed inside K with its center restricted to the objective plane $H = \{x : cx = t\}$. Let t^* be the value of t in the interval $t_{\min} \leq t \leq t_{\max}$ where $\delta[t]$ attains its maximum value.

In Murty (2006a) of Chap. 1, it has been shown that $\delta[t]$ is a piecewise linear concave function defined on the interval $t_{\min} \leq t \leq t_{\max}$ and that $\delta[t]$ is monotonic increasing in the interval $t_{\min} \leq t \leq t^*$ and monotonic decreasing in the interval $t^* \leq t \leq t_{\max}$.

1. In Murty (1980), it has been shown that in general the number of slope changes in the optimum objective value function of a parametric RHS LP can grow exponentially with the size (i.e., the number of constraints m or variables n in the LP) in the worst case.

   We know that the piecewise linear concave function $\delta[t]$, is the optimum objective value in (8.6), which is a parametric right-hand-side (RHS) LP problem in which t is the parameter. The LP (8.6) is of course a special parametric RHS LP with special structure, determine whether the number of slope changes in $\delta[t]$ can grow also exponentially in m, n like the optimum objective value function of a general parametric RHS LP, or whether it can be proved to be bounded above by a polynomial in m, n. In particular, is the number of slope changes in $\delta[t]$ bounded above by a linear function of m, n?

For the following problem 2, make the following

Assumption 1: for each $t_{\min} \le t \le t_{\max}$, the maximum radius inscribed ball in K with its center on the objective plane $\{x : cx = t\}$, is unique; i.e., that (8.6) has a unique optimum solution for all t.

Under this assumption it has been shown in (Murty (2006a) of Chap. 1) that $x(t)$ is a piecewise linear function in t. Also that the interval $t_{\min} \le t \le t_{\max}$ can be partitioned into a finite number of subintervals, such that in each subinterval both $x(t)$ and $\delta(x(t))$ are linear functions, and $T(x(t))$ remains unchanged. As t moves from one of these subintervals to an adjacent one, $T(x(t))$ changes by one or more entries.

2. Suppose a specific constraint, say constraint 1, drops out of the set $T(x(t))$ at $t = t_1$ as the parameter t is increasing from a value slightly less than t_1 to t_1.

   As t continues to keep on increasing, is it possible for constraint 1 to come back into the set $T(x(t))$ again? If so, what is the maximum number of times that constraint 1 can go out and come back into the set $T(x(t))$ as t increases from $t_{\min}$ to $t_{\max}$?

   What is the maximum number of changes that the set $T(x(t))$ can undergo as t increases from $t_{\min}$ to $t_{\max}$?

   Note: In Murty (2006a) of Chap. 1, I included a proof that the total number of changes that the set $T(x(t))$ can undergo as t increases from $t_{\min}$ to $t_{\max}$ is at most $2m$, but Andy Mirzaian (2007) has shown me that this proof has an error in it. So, right now whether the total number of changes that the set $T(x(t))$ can undergo as t increases from $t_{\min}$ to $t_{\max}$ grows linearly with m is an open question.
3. Now, suppose that assumption 1 does not hold. In this case we defined $x(t)$ as the ball center of K on the objective plane $\{x : cx = t\}$, which is uniquely defined as in Sect. 8.2. In this general case also, is $x(t)$ a piecewise linear function of t? Discuss what happens to the questions posed in 2 in this general case.
4. For each $t_{\min} \le t \le t_{\max}$ assume that the largest radius ball inscribed in the set $K(t)$ of feasible solutions of

$$\begin{aligned} Ax &\ge b \\ A_{m+1.} &\ge b_{m+1} \end{aligned} \tag{8.14}$$

   where $A_{m+1.} = -c$ and $b_{m+1} = -t$ is unique, and let $x\{t\}$ denote its center and $\delta\{t\}$ its radius. Let $T_t(x\{t\})$ denote the index set of touching constraints in (8.14) at $x\{t\}$ for $K(t)$. For all $t^* + \delta[t^*] \le t \le t_{\max}$, clearly $\delta\{t\} = \delta[t^*]$ and $x\{t\} = x(t^*)$.

   (1) What is the set of all t satisfying $m + 1 \in T_t(x\{t\})$? (2) As t is decreasing from $t_{\max}$ to $t_{\min}$ when an $i \in \{1, \cdots, m + 1\}$ leaves $T_t(x\{t\})$, can it come back into this set later on? How many times can an index $i \in \{1, \cdots, m+1\}$ leave and enter $T_t(x\{t\})$ as t decreases from $t_{\max}$ to $t_{\min}$?

5. For the original polytope K represented by the system of linear constraints in (8.1) and (8.5), and given an interior point x^0 of K, we can determine the radius of the largest sphere with center x^0 that can be inscribed in K very efficiently as described by the formula for $\delta(x^0)$ given in Sect. 8.1. Also, the results discussed in Sect. 8.3 offer a good characterization for a ball center for K. Using this characterization, the problem of obtaining a ball center (center for a largest radius ball inside K) requires solving the LP (8.5), and in Sect. 8.3, we even developed a procedure for doing this approximately.

   Suppose K_1 is a polytope represented as the convex hull of its extreme points, say K_1 = convex hull of $\{x^1, \cdots, x^N\}$, where the set of extreme points of K_1 is given and $K_1 \subset R^n$ is of full dimension. Given an interior point x^0 of K_1, can we determine efficiently the radius of the largest sphere with center x^0 that can be inscribed in K_1? Is there a simple characterization for a ball center of K_1 similar to that of K? Can an approximation to a ball center of K_1 be computed by a fast algorithm? At the moment these are all open research questions.

8.8 Future Research Directions

The sphere methods are new and are still under development. We are also investigating additional directions for line search to include in the procedures used for computing approximate ball centers to accelerate their convergence rate and to improve the quality of the approximation to an optimum ball center, and additional descent steps to use in each iteration. Also, we have so far been able to test the numerical performance of some alternatives in the algorithm using a preliminary MATLAB code. More extensive testing is needed to determine the best alternatives and then prepare a good code with these alternatives for tests using available large scale test problems.

The sphere methods also open up important new research topics in computational linear algebra. To get the best results from an implementation of the simplex method, researchers developed efficient factorization techniques to update the inverse of a matrix as one of its column vectors changes from one step to the next. An important component of the sphere method is the LSCPD sequence of steps in which a matrix grows by a row and a column from one step to the next. To get the best results from the sphere method, we need to develop appropriate techniques to update the inverse as the matrix grows in this way.

8.9 Exercises

8.1. Consider the polytope, K, defined by the following system of linear constraints in two variables x, y: $x - 2 \leq y \leq 2x + 1$, $y \leq x + 2$, $0 \leq x$, $y \leq 4$.

Draw K geometrically, and by trial and error determine the largest inscribed ball B inside K. Is B unique? Why? Formulate an LP model for determining B. Solve this LP and check whether it tallies with the B that you generated by trial and error.

Also, let $H = \{(x,\ y) : y = 2\}$. Determine by trial and error the largest ball B_H, with its center on H that is inscribed inside K. Formulate an LP model for determining B_H. Solve this LP and check whether it tallies with the B_H that you generated by trial and error.

8.2. Give a simple numerical example of a system of linear inequalities in two variables for which the set of feasible solutions K is unbounded, but it has a ball center and a maximum radius inscribed ball with finite radius.

8.3. Give a simple numerical example of a system of linear inequalities in two variables for which the set of feasible solutions K is unbounded and does not have a ball center because the maximum inscribed ball in K is unbounded and a hyperplane H intersecting K in its interior, such that the ball center of K on H is well defined.

8.4. Let $K = \{x : Ax \geq b\}$, where A is a matrix of order $m \times n$ and K^0 its interior.

Let $z = cx$ be an objective function for which the minimum and maximum values are $t_{\min}$ and $t_{\max}$. Also, let the symbols $\delta(x),\ \delta[t], t^*, B(x, \delta(x)), T(x)$ have the same meanings as defined in Sect. 8.2 or in the notation at the beginning of this chapter.

Let $x(t)$ denote the ball center of K on the objective plane $\{x : cx = t\}$, as defined conceptually in Sect. 8.2. Then for each $t_{\min} \leq t \leq t_{\max}$, $B(x(t),\ \delta(x(t)))$ is a maximum radius inscribed ball in K with its center on the objective plane $\{x : cx = t\}$.

(1). Prove that $\delta[t]$ is a piecewise linear concave function. Also show that $\delta[t]$ is monotonic increasing in the interval $t_{\min} \leq t \leq t^*$ and monotonic decreasing in the interval $t^* \leq t \leq t_{\max}$.

(2). Under assumption 1 of Sect. 8.7, show that the interval $t_{\min} \leq t \leq t_{\max}$ can be partitioned into a finite number of subintervals, such that in each subinterval both $x(t), \delta(x(t))$ are linear functions, and $T(x(t))$ remains unchanged. As t moves from one of these subintervals to an adjacent one, $T(x(t))$ changes by one or more entries.

(3). Prove that (1) and (2) hold even when Assumption 1 does not hold.

8.5. Let $K = \{x : Ax \geq b\}$, where $||A_{i.}|| = 1$ for all i. Suppose $\bar{x} \in \{x : Ax > b\}$ and $T(\bar{x})$ = set of all indices i that tie for the minimum in: minimum$\{A_{i.}\bar{x} - b_i : i = 1 \text{ to } m\}$. Show that $\bar{x}$ is the center of a largest sphere inscribed inside K iff $y = 0$ is the unique feasible solution for the system: $A_{i.}y \geq 0$ for all $i \in T(\bar{x})$.

8.6. Let $K = \{x : Ax \geq b\}$, where A is a given matrix of order $m \times n$; and $S \subset K$, an inscribed sphere. Both K *and* S are full dimensional. The point $\bar{x} \in S$ minimizes the linear objective function cx on S. If $\bar{x}$ is also a boundary point of K prove that $\bar{x}$ minimizes cx over K.

References

Cartis C, Gould NIM (2006) Finding a point in the relative interior of a polyhedron, RAL Technical Report 2006-016, Available from Optimization Online. www.optimization-online.org/
Cooper WW, Seiford LM, Tone K (2006) Data envelopment analysis: a comprehensive text with models, applications, references and DEA-Solver software, 2nd edn. Springer, NY
Mirzaian A (2007) Private communication
Murty KG (1980) Computational complexity of parametric linear programming. Math Program 19:213–219
Murty KG (2009a) Ball centers of polytopes. Department of IOE, University of Michigan, Ann Arbor
Murty KG (2009b) New sphere methods for LP. Tutorials in OR, INFORMS
Murty KG, Kabadi SN (2008) Additional descent steps in the sphere method. Department of IOE, University of Michigan, Ann Arbor
Murty KG, Oskoorouchi MR (2008a) Note on implementing the new sphere method for LP using matrix inversions sparingly. Optim Lett 3(1):137–160
Murty KG, Oskoorouchi MR (2008b) Sphere methods for LP. Department of IOE, University of Michigan, Ann Arbor

Chapter 9
Quadratic Programming Models

9.1 Introduction

Quadratic programming (QP) deals with a special class of mathematical programs in which a quadratic function of the decision variables is required to be optimized (i.e., either minimized or maximized) subject to linear equality and/or inequality constraints.

Let $x = (x_1, \dots, x_n)^T$ denote the column vector of decision variables. In mathematical programming, it is standard practice to handle a problem requiring the maximization of a function $f(x)$ subject to some constraints by minimizing $-f(x)$ subject to the same constraints. Both problems have the same set of optimum solutions. Because of this, we restrict our discussion to minimization problems.

A quadratic function of decision variables x is a function of the form

$$Q(x) = \sum_{i=1}^{n} \sum_{j=i}^{n} q_{ij} x_i x_j \ + \sum_{j=1}^{n} c_j x_j \ + c_0.$$

Define $c = (c_1, \dots, c_n)$ and a square symmetric matrix $D = (d_{ij})$ of order n, where

$$\begin{aligned} d_{ii} &= 2q_{ii} \quad \text{for all } i = 1 \text{ to } n \\ d_{ij} = d_{ji} &= q_{ij} \quad \text{for } j > i \end{aligned}$$

Then in matrix notation, $Q(x) = \frac{1}{2}x^T Dx + cx + c_0$. Here, D is the Hessian matrix (i.e., the matrix of second order partial derivatives) of $Q(x)$.

As an example, consider $n = 3$, $x = (x_1, x_2, x_3)^T$, and $h(x) = 81x_1^2 - 7x_2^2 + 5x_1x_2 - 6x_1x_3 + 18x_2x_3$. This quadratic function $h(x) = \frac{1}{2}x^T Dx$ where

$$D = \begin{pmatrix} 162 & 5 & -6 \\ 5 & -14 & 18 \\ -6 & 18 & 0 \end{pmatrix}$$

K.G. Murty, *Optimization for Decision Making: Linear and Quadratic Models*, International Series in Operations Research & Management Science 137,
DOI 10.1007/978-1-4419-1291-6_9, © Springer Science+Business Media, LLC 2010

A quadratic function is the simplest nonlinear function, and hence they have always served as model functions for approximating general nonlinear functions by local models (through Taylor series and other such approximations). Hence, quadratic programming models serve as a bridge between linear programming and nonlinear programming models. Also, some algorithms for nonlinear programming are based on solving quadratic approximations of them using quadratic programming techniques.

A square matrix D of order n is said to be

Positive semidefinite (PSD): if $x^T Dx \geq 0$ for all x
Positive definite (PD): if $x^T Dx > 0$ for all $x \neq 0$

Negative semidefinite (NSD): if $x^T Dx \leq 0$ for all x
Negative definite (ND): if $x^T Dx < 0$ for all $x \neq 0$

Indefinite: if it is neither PSD nor NSD.

Hence, the square matrix D is NSD, ND iff $-D$ is PSD, PD, respectively. Also, remember that all these concepts are only defined for square matrices. These matrix theoretic concepts are important in the study of QP because the quadratic function $Q(x) = \frac{1}{2}x^T Dx + cx + c_0$ is a convex function over R^n iff the matrix D is PSD.

Result 9.1. *Let* $Q(x) = \frac{1}{2}x^T Dx + cx + c_0$*, where* D *is a symmetric matrix of order* n*. Then the Hessian matrix (the* $n \times n$ *matrix of second partial derivatives) of* $Q(x)$ *at* x *is* $H(Q(x)) = \left(\frac{\partial^2 Q(x)}{\partial x_i \partial x_j}\right) = D$*.* $Q(x)$ *is a convex [concave] function defined over* R^n*, iff the matrix* D *is PSD [NSD].*

Result follows directly from Theorem 2.4 of Sect. 2.2.

9.2 Superdiagonalization Algorithm for Checking PD and PSD

Let M be a square matrix of order n which may or may not be symmetric. By definition, M is PD or PSD iff $D = M + M^T$ is PD or PSD, respectively. To check whether the symmetric matrix $D = (d_{ij})$ is PD or PSD, we can use the algorithms given below. They are based on the following results.

Result 9.2. *If* D *is PD, all its diagonal entries* d_{ii}*,* $i = 1$ *to* n *must be* >0*. If* D *is not PD, but PSD, all its diagonal entries* d_{ii}*,* $i = 1$ *to* n*, must be* ≥ 0*.*

Result 9.3. *If the symmetric matrix* $D = (d_{ij})$ *is PSD and a diagonal entry in it* $d_{ii} = 0$*, then all the entries in its row and column, i.e., row* i *and column* i*, must be zero.*

Theorem 9.1. *Let $D = (d_{ij})$ be a symmetric matrix with its first diagonal entry $d_{11} \neq 0$. Subtract suitable multiples of row 1 from each of the other rows to convert all entries in column 1 in rows 2–n to 0, i.e., transform*

$$D = \begin{pmatrix} d_{11} & \dots & d_{1n} \\ d_{21} & \dots & d_{2n} \\ \vdots & & \vdots \\ d_{n1} & \dots & d_{nn} \end{pmatrix} \quad \text{into} \quad D_1 = \begin{pmatrix} d_{11} & d_{12} & \dots & d_{1n} \\ 0 & \tilde{d}_{22} & \dots & \tilde{d}_{2n} \\ \vdots & \vdots & & \vdots \\ 0 & \tilde{d}_{n2} & \dots & \tilde{d}_{nn} \end{pmatrix}$$

let E_1 be the matrix of order $(n-1) \times (n-1)$ obtained by deleting column 1 and row 1 from D_1. Then E_1 is also symmetric; and D is PD (PSD) iff $d_{11} > 0$ and E_1 is PD (PSD).

For proofs of Results 9.2 and 9.3, and Theorem 9.1, see (Murty 1988 of Chap. 2), for example. The following algorithms for checking PD and PSD are based on repeated use of these results.

Algorithm 1: Superdiagonalization Algorithm for Checking Whether M is PD
Let $M = (m_{ij})$ be the matrix of order $n \times n$ being tested for positive definiteness. Let $D = M$ if M is symmetric, otherwise $D = M + M^T$.

Step 1: If any of the principal diagonal elements in D are ≤ 0, D and hence M is not PD, then terminate. Otherwise go to Step 2 with D as the current matrix.

Step 2: Subtract suitable multiples of row 1 of D from all the other rows, so that all the entries in column 1 and rows 2–n of D are made into 0; i.e., transform D into D_1 as in Theorem 9.2.2. If any diagonal element of D_1 is ≤ 0, D and hence M is not PD, then terminate. Otherwise go to Step 3 with the matrix D_1 as the current matrix.

General Step $r+2$: At this stage the current matrix will be D_r of the following form.

$$D_r = \begin{pmatrix} d_{11} & d_{12} & & & & \dots & d_{1n} \\ 0 & \tilde{d}_{22} & & & & \dots & \tilde{d}_{2n} \\ 0 & 0 & \ddots & & & \dots & \vdots \\ & & & \bar{d}_{rr} & & \dots & \bar{d}_{rn} \\ & & & 0 & \hat{d}_{r+1,r+1} & \dots & \hat{d}_{r+1,n} \\ \vdots & \vdots & & \vdots & \vdots & & \vdots \\ 0 & 0 & & 0 & \hat{d}_{n,r+1} & \dots & \hat{d}_{nn} \end{pmatrix}$$

Subtract suitable multiples of row $r+1$ in D_r from rows i for $i > r+1$, so that all the entries in column $r+1$ and rows $i > r+1$ are transformed into 0. This transforms D_r into D_{r+1}. If any element in the principal diagonal of D_{r+1} is ≤ 0, D and hence M is not PD, terminate. Otherwise, if $r+2 < n$ go to the next step with D_{r+1} as the current matrix. If $r+2 = n$, the current matrix is D_{n-1} and it is of the form

$$D_{n-1} = \begin{pmatrix} d_{11} & d_{12} & \dots & d_{1n} \\ 0 & \tilde{d}_{22} & \dots & \tilde{d}_{2n} \\ 0 & 0 & & \\ \vdots & \vdots & & \vdots \\ 0 & 0 & \dots & \bar{d}_{nn} \end{pmatrix}$$

If no termination has occurred earlier and all the diagonal entries in D_{n-1} are positive, D and hence M is PD, terminate.

If the method goes through Step $n-1$, the final matrix D_{n-1} is upper triangular. All the work in this method has transformed the original matrix D into the upper triangular matrix D_{n-1}. That is why this method is called the *superdiagonalization algorithm.*

Example 9.1. Test whether the following matrix M is PD.

$$M = \begin{pmatrix} 3 & 1 & 2 & 2 \\ -1 & 2 & 0 & 2 \\ 0 & 4 & 4 & 5/3 \\ 0 & -2 & -13/3 & 6 \end{pmatrix}$$

$$D = M + M^T = \begin{pmatrix} 6 & 0 & 2 & 2 \\ 0 & 4 & 4 & 0 \\ 2 & 4 & 8 & -8/3 \\ 2 & 0 & -8/3 & 12 \end{pmatrix}$$

All the entries in the principal diagonal of D are > 0. So, apply Step 1 in superdiagonalization getting D_1. Since all elements in the principal diagonal of D_1 are > 0, continue. The matrices obtained in the order are

$$D_1 = \begin{pmatrix} 6 & 0 & 2 & 2 \\ 0 & 4 & 4 & 0 \\ 0 & 4 & 22/3 & -10/3 \\ 0 & 0 & -10/3 & 34/3 \end{pmatrix}, \quad D_2 = \begin{pmatrix} 6 & 0 & 2 & 2 \\ 0 & 4 & 4 & 0 \\ 0 & 0 & 10/3 & -10/3 \\ 0 & 0 & -10/3 & 34/3 \end{pmatrix},$$

$$D_3 = \begin{pmatrix} 6 & 0 & 2 & 2 \\ 0 & 4 & 4 & 0 \\ 0 & 0 & 10/3 & -10/3 \\ 0 & 0 & 0 & 8 \end{pmatrix}.$$

The algorithm terminates now. Since all the diagonal entries in D_3 are >0, D and hence M is PD.

Example 9.2. Check whether

$$\overline{M} = \begin{pmatrix} 1 & 0 & 2 & 0 \\ 0 & 2 & 4 & 0 \\ 2 & 4 & 4 & 5 \\ 0 & 0 & 5 & 3 \end{pmatrix}$$

is PD. $\overline{M}$ is already symmetric and its diagonal entries are > 0. Carrying out Step 1 on $D = \overline{M}$ leads to

$$D_1 = \begin{pmatrix} 1 & 0 & 2 & 0 \\ 0 & 2 & 4 & 0 \\ 0 & 4 & 0 & 5 \\ 0 & 0 & 5 & 3 \end{pmatrix}$$

Since the third diagonal entry in D_1 is 0, the matrix $\overline{M}$ here is not PD.

Algorithm 2: Algorithm for Checking Whether M Is PSD
Let $M = (m_{ij})$ be the matrix of order $n \times n$ being tested for positive semidefiniteness. Let $D = M$ if M is symmetric, otherwise $D = M + M^T$.

Step 1: If any of the principal diagonal elements in D are < 0, D and hence M is not PSD, then terminate. Otherwise, continue.

If any diagonal entries in D are 0, all the entries in the row and column of each 0 diagonal entry must be 0. Otherwise, D and hence M is not PSD, so terminate. If termination has not occurred, reduce the matrix D by striking off the 0-rows and columns of 0 diagonal entries. We will call the remaining matrix by the same name D. Go to Step 2 with D as the current matrix.

Step 2: Start off by performing row operations as in Step 1 of Algorithm 1 discussed above, i.e., transform D into D_1. If any diagonal entry in D_1 is < 0, D and hence M is not PSD, terminate. Otherwise, let E_1 be the submatrix of D_1 without its row 1 and column 1. If a diagonal entry in E_1 is 0, all entries in its row and column in E_1 must be 0 too; otherwise, D and hence M is not PSD, terminate. Continue if termination did not occur. With D_1 as the current matrix, go to Step 3.

General Step $r + 2$: At this stage the current matrix will be D_r of the same form as in the above algorithm. Let E_r be the submatrix of D_r with its rows 1 to r and columns 1 to r struck off. If any diagonal entry in E_r is < 0, D and hence M is not PSD, terminate. If any diagonal element of E_r is 0, all the entries in its row and column in E_r must be 0 too; otherwise, D and hence M is not PSD, terminate. If termination did not occur, continue.

Let D_{ss} be the first nonzero (and hence positive) diagonal element in E_r. Subtract suitable multiples of row s in D_r from row i for $i > s$, so that all the entries in column s and rows $i > s$ in D_r are transformed into 0. This transforms D_r into D_s, which is the new current matrix. Go to Step $s + 2$ with D_s as the current matrix.

If termination does not occur until D_{n-1} is obtained, and if all the diagonal entries in D_{n-1} are ≥ 0, D and hence M is PSD, terminate.

In the process of obtaining D_{n-1}, if all the diagonal elements in D and in all the matrices D_r obtained during the algorithm are > 0, D and hence M is not only PSD but actually PD.

Example 9.3. Check whether the following matrix M is PSD.

$$M = \begin{pmatrix} 0 & -2 & -3 & -4 & 5 \\ 2 & 3 & 3 & 0 & 0 \\ 3 & 3 & 3 & 0 & 0 \\ 4 & 0 & 0 & 8 & 4 \\ -5 & 0 & 0 & 4 & 2 \end{pmatrix}, \quad D = M + M^T = \begin{pmatrix} 0 & 0 & 0 & 0 & 0 \\ 0 & 6 & 6 & 0 & 0 \\ 0 & 6 & 6 & 0 & 0 \\ 0 & 0 & 0 & 16 & 8 \\ 0 & 0 & 0 & 8 & 4 \end{pmatrix}$$

$D_{.1}$ and $D_{1.}$ are both zero vectors. So we eliminate them, but will call the remaining matrix by the same name. All diagonal entries in D are ≥ 0. So we apply Step 1 of superdiagonalization. This leads to

$$D_1 = \begin{pmatrix} 6 & 6 & 0 & 0 \\ 0 & 0 & 0 & 0 \\ 0 & 0 & 16 & 8 \\ 0 & 0 & 8 & 4 \end{pmatrix}, \quad E_1 = \begin{pmatrix} 0 & 0 & 0 \\ 0 & 16 & 8 \\ 0 & 8 & 4 \end{pmatrix}$$

The first diagonal entry in E_1 is 0, and the column and row of this entry in E_1 are both zero vectors. And all the diagonal entries in D_1 are ≥ 0. So continue with superdiagonalization. Since the second diagonal element in D_1 is 0, move to the third diagonal element of D_1. This step leads to

$$D_3 = \begin{pmatrix} 6 & 6 & 0 & 0 \\ 0 & 0 & 0 & 0 \\ 0 & 0 & 16 & 8 \\ 0 & 0 & 0 & 0 \end{pmatrix}$$

All diagonal entries in D_3 are ≥ 0, and it is upper triangular. So, D and hence M are PSD. M is not PD because of the 0-diagonal entries encountered in the algorithm.

Example 9.4. Is the matrix $\bar{M}$ in Example 9.2 PSD? We have seen there that it is not PD. Referring to Example 9.2, after Step 1 in superdiagonalization we have

$$E_1 = \begin{pmatrix} 2 & 4 & 0 \\ 4 & 0 & 5 \\ 0 & 5 & 3 \end{pmatrix}$$

The second diagonal entry in E_1 is 0, but its row and column in E_1 are not zero vectors. So, $\overline{M}$ is not PSD.

These algorithms for checking whether a given square matrix is PD or PSD have many uses in NLP. For example, given a quadratic program (QP) in which the objective function $f(x) = x^T Dx + cx$ is to be minimized subject to linear constraints, to check whether this problem is a convex QP, we need to check whether the matrix D defining the quadratic form in $f(x)$, is PSD. Also, given a nonlinear twice continuously differentiable function $g(x)$, a sufficient condition for it to be locally convex at a point $\bar{x}$, is that its Hessian matrix at $\bar{x}$ is PD. Also, these algorithms are needed to check whether a given point satisfies the optimality conditions for being a solution to an NLP (nonlinear program), as well.

9.3 Classification of Quadratic Programs

Considering the discussion in Sect. 5.9, QPs can be classified into the following types.

Unconstrained quadratic minimization problem is one that requires the minimization of a quadratic function $Q(x)$ over the whole space R^n with no constraints.

Equality constrained quadratic minimization problem is one that requires the minimization of a quadratic function $Q(x)$ subject to linear equality constraints on the variables, $Ax = b$. These equations can be used to eliminate some variables by expressing them in terms of the others, and thereby transform the problem into an unconstrained one in the remaining variables. Thus, these problems are mathematically equivalent to (and can be solved by techniques similar to those of) unconstrained quadratic minimization problems.

Inequality constrained quadratic minimization problem is one that requires the minimization of a quadratic function $Q(x)$ subject to linear inequality constraints $Bx \geq d$, and possibly bounds on individual variables $\ell \leq x \leq u$, and may be some equality constraints $Ax = b$.

Bound constrained quadratic minimization problem is one that requires the minimization of a quadratic function subject only to bounds (lower and/or upper) on the variables.

Convex quadratic program (CQP) any of the above problems in which the objective function to be minimized, $Q(x)$, is convex.

Nonconvex quadratic program any of the above problems in which the objective function to be minimized, $Q(x)$, is nonconvex.

Linear complementarity problem (LCP) a special problem dealing with a system of equations in non-negative variables in which the variables are formed into various pairs called *complementary pairs*. A feasible solution in which at least one variable in each pair is zero is desired. There is no objective function to be minimized in this problem. The first-order necessary optimality conditions for a QP are in the form of an LCP. And in turn every LCP can be posed as a QP.

Among optimization problems, those in which a convex function is to be minimized over a suitably defined convex set are called *convex programming problems*. For convex programming problems, optimum solution(s) are well characterized, and mathematical theory has derived necessary and sufficient optimality conditions characterizing the optimum solution set, and efficient algorithms have been developed for solving these problems based on those optimality conditions.

Since linear functions are both convex and concave, all linear programs are convex programming problems, and we discussed optimality conditions, and efficient algorithms based on them for LPs in earlier portions of the book.

However, for nonconvex programming problems the situation is very different. Talking in terms of minimization problems, in a nonconvex programming problem there may be several types of minimum solutions. Those that attain the minimum possible value for the objective function are called *global minima*. But there may also be *local minima* that are not global minima; these local minima are minimum solutions in a small neighborhood about them. For all convex programming problems, every local minimum is a global minimum; that is why when discussing LPs, we only refer to "optimum solutions" and never worry about them being only local minima.

Also, for nonconvex programming problems in general, so far there are no useful necessary and sufficient optimality conditions for a feasible solution to be a "global minimum" or even a "local minimum." All we have available at this time are some necessary conditions (called *Karush–Kuhn–Tucker (KKT) optimality conditions*, or *first-order necessary optimality conditions*) for being a local minimum discussed in Sect. 5.9. Unfortunately, these conditions may not be sufficient, so a solution satisfying them cannot be guaranteed to be even a local minimum. That is why a feasible solution satisfying the KKT optimality conditions is called a *KKT point*, or *KKT solution*, or *stationary point*. Algorithms developed for nonconvex programming problems can at best be proved to converge to a KKT point. That is why nonconvex programming problems are considered hard problems and practitioners are usually happy if they can find a descent algorithm for them which converges to a KKT point.

Among QPs we have two classes: convex QPs and nonconvex QPs. There are efficient algorithms to find global optima for convex QPs. To solve nonconvex QPs, we apply the same algorithms on them, but we cannot guarantee that the solutions obtained by them are even local minima.

9.4 Types of Solutions and Optimality Conditions

In linear programming (LP), we talk about optimum solutions but not about different types of optima such as local and global optima. That is because every local optimum and every point satisfying the first-order necessary optimality conditions for an LP, are also a global optimum. Unfortunately, this is not the case in general QPs, as discussed in Sect. 5.9.

For a QP, or any mathematical program in which an objective function $\theta(x)$ is required to be minimized, a

Local minimum is a feasible solution $\bar{x}$ for which there exists an $\epsilon > 0$ such that $\theta(x) \geq \theta(\bar{x})$ for all feasible solutions within a Euclidean distance of ϵ from $\bar{x}$.
Global minimum is a feasible solution $\hat{x}$ satisfying $\theta(x) \geq \theta(\hat{x})$ for all feasible solutions x.

Stationary point or KKT point is a feasible solution satisfying the first-order necessary optimality conditions (also called the KKT optimality conditions) for the problem. We will state these conditions now.

Consider the general QP with equality and inequality constraints:

$$\begin{aligned} \text{Minimize } Q(x) &= \frac{1}{2}x^T Dx + cx \\ \text{subject to } Ax &= b \\ Ex &\geq f \\ x &\geq 0 \end{aligned} \tag{9.1}$$

where A, E are matrices of orders $m \times n$, $p \times n$, respectively. We assume that D is a symmetric matrix; if not, just replace it by $(D + D^T)/2$ and this leaves $Q(x)$ unchanged. To state the first-order necessary optimality conditions for this QP, we introduce the Lagrange multiplier vectors associated with the equality constraints, $\pi = (\pi_1, \cdots, \pi_m)$ and inequality constraints, $\mu = (\mu_1, \cdots, \mu_p)$ and $\delta = (\delta_1, \cdots, \delta_n)$ associated with the nonnegativity restrictions (these π, μ, δ are all row vectors) and write the Lagrangian for the problem which is $L(x, \pi, \mu, \delta) = Q(x) - \pi(Ax - b) - \mu(Ex - f) - \delta x$. Then the first-order necessary optimality conditions for (9.1) are the following:

$$\begin{aligned} \frac{\partial L}{\partial x} &= Dx + c^T - A^T\pi^T - E^T\mu^T - \delta = 0 \\ Ax = b,\ Ex &\geq f,\ x \geq 0,\ \mu,\ \delta \geq 0 \\ \text{For all } i \in \{1, \cdots, p\},\ \mu_i(E_{i.}x - f_i) &= 0 \\ \text{For all } j \in \{1, \cdots, n\},\ \delta_j x_j &= 0 \end{aligned} \tag{9.2}$$

From the first and fourth lines in this system, we see that the Lagrange multipliers δ_j for $j = 1$ to n can easily be eliminated from this system and it can be expressed as:

$$\begin{aligned} \frac{\partial L}{\partial x} &= Dx + c^T - A^T\pi^T - E^T\mu^T \geq 0 \\ Ax = b,\ Ex &\geq f,\ x \geq 0,\ \mu \geq 0 \\ \text{For } i \in \{1, ..., p\},\ \mu_i(E_{i.}x - f_i) &= 0 \\ \text{For } j \in \{1, ..., n\},\ [(D_{j.}x + c_j - \pi A_{.j} - \mu E_{.j})]x_j &= 0 \end{aligned} \tag{9.3}$$

These conditions are often given in the form (9.3) in some textbooks. A feasible solution $\bar{x}$ to (9.1) is called a KKT point or stationary point for this problem, iff there exist Lagrange multiplier vectors $\bar{\mu}, \bar{\pi}$, which satisfy the above KKT conditions (9.3) together with $\bar{x}$.

For the unconstrained minimization problem: minimize $Q(x) = \frac{1}{2}x^T Dx + cx$ over $x \in R^n$, the first-order necessary optimality conditions are:

$$\frac{\partial L}{\partial x} = Dx + c^T = 0.$$

Here again, $\bar{x}$ is called a KKT point or stationary point for this problem if it satisfies this condition.

For the equality constrained minimization problem: minimize $Q(x) = \frac{1}{2}x^T Dx + cx$, subject to equality constraints $Ax = b$ only, where A is a matrix of order $m \times n$ we write the Lagrangian $L(x, \pi) = Q(x) - \pi(Ax - b)$, where $\pi = (\pi_1, \cdots, \pi_m)$ is the row vector of Lagrange multipliers associated with the equality constraints in the problem.

Then the first-order necessary optimality conditions for this problem are

$$\frac{\partial L}{\partial x} = Dx + c^T - A^T \pi^T = 0$$
$$Ax = b \tag{9.4}$$

As before, a feasible solution $\bar{x}$ to this problem is called a KKT point or stationary point for this problem, iff there exist Lagrange multiplier vectors $\bar{\pi}$ that satisfy the above KKT conditions together with $\bar{x}$.

In a convex QP, every stationary point (KKT point), or a local minimum, is a global minimum; hence, all these concepts converge in a convex QP. The same may not be true in nonconvex QPs, i.e., there may be local minima that are not global minima, and stationary points which are neither global nor local minima. Also, the problem may have some local minima even when the objective function is unbounded below on the set of feasible solutions.

We will refer to the first-order (KKT) necessary optimality conditions for a QP as its *KKT system.*

Many textbooks have proofs of these optimality conditions; see, for example, (Murty 1988 of Chap. 2; Bazaraa et al. 2006 of Chap. 5).

9.5 What Types of Solutions Can Be Computed Efficiently by Existing Algorithms?

Like LPs, QPs have the property that when the set of feasible solutions is nonempty either a global minimum exists or the objective function is unbounded below on the set of feasible solutions. And for both convex and nonconvex QPs, there exist finite algorithms for checking whether the objective function is unbounded below on the set of feasible solutions, and for computing a global optimum solution when one exists, but these finite algorithms may not be practical.

For convex QPs there are very efficient algorithms for computing a global minimum when it exists, and very high quality software systems implementing these algorithms are available commercially.

For nonconvex QPs, even though finite algorithms for computing a global minimum are available, they are impractical, because the computational effort needed by them grows exponentially with the size of the problem being solved. Nonconvex QP is NP-hard, and so far there is no algorithm known that is guaranteed to find a global minimum for it within a reasonable time.

Can we at least compute a local minimum for a nonconvex QP efficiently? Unfortunately, even the problem of checking whether a given feasible solution is a local minimum for a nonconvex QP may be a hard problem. In (Murty and Kabadi 1987), it has been shown that the problem of checking whether 0 is a local minimum in the following simple QP:

$$\begin{array}{ll} \text{Minimize} & x^T D x \\ \text{subject to} & x \geq 0 \end{array}$$

is a co-NP-complete problem when D is not PSD. In this paper, it has been explained that when dealing with a nonconvex QP, a reasonable goal is to look for an algorithm that produces a descent sequence (i.e., a sequence of feasible points along which the objective value strictly decreases) converging to a KKT point. Practitioners seem to accept this goal at the moment. Some of the algorithms discussed below achieve this goal.

9.6 Some Important Applications of QP

Here we discuss some important practical applications of QP models in different areas.

Finance: Analysis using QP models is an established part of selecting optimum investment strategies. Perhaps (Markowitz 1959) is the first published book in this area. With x as the vector of stock investments, the Markovitz model employs the variation in return as measured by the quadratic function $x^T Dx$, where D is the variance/covariance matrix of returns for measuring the risk. This risk is the objective function to be minimized. Constraints in the model guarantee conservation on the flow of funds and a lower bound on the expected returns from the portfolio. There may also be bounds placed on the investments in particular sectors of the economy (such as pharmaceuticals, utilities, etc.) to make sure that the model does not put too many eggs in any one basket, thus achieving diversification. Many other practical aspects of investing can easily be included by either adding appropriate constraints or modifying the objective function by including quadratic penalty terms.

For selecting the best investment strategy, several publications measure risk by different objective functions (see Murty 2008a, b).

Many authors (e.g., Crum and Nye 1981; Mulvey 1987) have designed similar multiperiod quadratic generalized network flow models in which interest, dividends, and loans are modeled by means of arc multipliers.

Taxation: QP models play a very important role these days in the analysis of tax policies. Political leaders at the national and state levels are relying more and more on such analyses to forecast growth rates in tax revenues and to set various taxes at levels that are likely to ensure growth at desired rates. White (1983) gives a detailed description of such an analysis carried out for the state of Georgia.

National and state government taxes, such as sales tax, motor fuels tax, alcoholic beverages tax, personal income tax, etc., are all set at levels to ensure a healthy economic growth. Government finance is based on the assumption of predictable and steady growth of each tax over time.

If s is the tax rate for a particular tax and S_t the expected tax revenue for this tax in year t, then a typical regression equation used to predict S_t as a function of s and t is $\log_e S_t = a + bt + cs$ where a, b, c are parameters to be estimated from past data to give the closest fit by the least squares method, a QP technique. The annual growth rate in this tax revenue is then the regression coefficient b multiplied by 100 to convert it to percent.

The decision variables in the model are s_j = the tax rate for tax j in the base year (0th year) as a fraction. From the known tax base for tax j in the 0th year, the revenues from tax j in this year can be obtained as s_j (tax base for tax j) $= x_j$. The instability or variability in this revenue is measured by the quadratic function $Q(x) = x^T Vx$, where V is the variance/covariance matrix estimated from past data. $Q(x)$ is to be minimized. The constraints in the model consist of bounds on the x_j and a condition that $\sum x_j = T$, the total expected tax revenue in the 0th year. And there is an equation that the overall growth rate which can be measured by the weighted average of the growth rates of the various taxes j, $\sum(x_j b_j)/T$ should be equal to the desired growth rate λ. Any other linear constraints that the decision variables are required to satisfy can also be included. In fact, λ can be treated as a parameter and the whole model solved as a parametric QP model. Exploring the optimum solution for different values of λ in the reasonable range yields information for the political decision makers to determine good values for the various tax rates that are consistent with expected growth in tax revenues.

For an example, see Exercise 9.13.

Equilibrium Models: Economists use equilibrium models to analyze expected changes in economic conditions, predict prices, inflation rates, etc. These models often involve QPs. As an example, in (Glassey 1978), a simple equilibrium model of interregional trade in a single commodity is described. He considers N regions and the following data elements and variables.

Data: $a_i > 0$ the equilibrium price in the ith region in the absence of imports and exports.

$b_i > 0$ the elasticity of supply and demand in the ith region.

c_{ij} the cost/unit to ship from i to j.

Variables: p_i equilibrium price in the ith region.
y_i net imports into the ith region (may be > 0, or 0, or < 0)
x_{ij} actual exports from region i to region j.

If $p_i > a_i$, supply locally exceeds demand in the ith region, the difference being available for export. From this we have $p_i = a_i - b_i y_i$. Also, the y_i and x_{ij} are linked through flow conservation equations. The interregional trade equilibrium conditions are

$$p_i + c_{ij} \geq p_j \quad \text{for all } i,\ j$$
$$(p_i + c_{ij} - p_j)x_{ij} = 0 \quad \text{for all } i,\ j$$

If the first condition above does not hold, exports from i to j will increase until the elasticity effects in markets i and j rise, and prices will adjust so that additional profit from export no longer exists. Also, if $x_{ij} > 0$, we must have $p_i + c_{ij} - p_j = 0$.

It can be verified that these conditions are the first-order necessary optimality conditions for a quadratic network flow problem in which the quadratic objective function can be interpreted as a net social payoff function. Using this observation (Glassey 1978) describes a procedure for computing the equilibrium prices and flows based on solving the QP.

In the same way traffic engineers use traffic equilibrium models solved by quadratic network flow algorithms for road and communication network planning. These traffic equilibrium models typically have hundreds of thousands of variables and constraints and are probably the largest QP models solved on a regular basis.

Electrical Networks: Even during the physicist J.C. Maxwell's time in the second half of the 19th century, it has been well recognized that the equilibrium conditions of an electrical or a hydraulic network are attained at the point where the total energy loss is minimized. Dennis (1959) has formally shown that the sum of the energy losses in the resistors and at the voltage sources in an electrical network, is a quadratic function of the branch currents, if all devices in the network are of a linear (i.e., ohmic) nature. Using this he formulated the problem of determining the branch currents at equilibrium in an electrical network connecting various devices, voltage sources, diodes, and resistors, as a QP. He then showed that the optimality conditions for this QP are precisely the Kirchoff laws governing the equilibrium conditions of the network, with the Lagrange multipliers representing node potentials. In the distribution of electrical power, this QP model is used to solve the load flow problem concerned with the flow of power through the transmission network to meet a given demand.

Power System Scheduling Problem: The economic dispatch problem in an electrical power system operation deals with the problem of allocating the demand for power – or system load – among the generating units in operation at any point of time. The optimal allocation of load among the units to achieve a least cost allocation depends on the relative efficiencies of the units and can be modeled as a QP (see Wood 1984). In power system operation, this model is usually solved many times during the day with appropriate load adjustments.

Application in Solving General Nonlinear Programs: At the moment, one of the most popular algorithms for solving general nonlinear programming problems is the SQP (sequential or recursive quadratic programming) method. It is an iterative method that in each iteration solves a convex QP to find a search direction and a line search problem (1-dimensional minimization problem for a merit function) in that direction. The original concepts of this method are outlined in (Wilson 1963; Han 1976; Powell 1978), but it has been developed into a successful approach through the work of many researchers (see Eldersveld 1991; Bazaraa et al. 2006 of Chap. 5; Murty 1988 of Chap. 2). The success of these methods has made QP a very important topic in mathematical programming. A nice software package for nonlinear programs based on this approach is FSQP (Zhou and Tits 1992).

9.7 Unconstrained Quadratic Minimization in Classical Mathematics

Historically, quadratic functions became prominent because they provide simple local models for general nonlinear functions. A quadratic function is the simplest nonlinear function and when used as a local approximation for a general nonlinear function it can capture the important curvature information that a linear approximation cannot.

The use of quadratic approximations to handle general nonlinear functions goes back a very long time. We discuss some important instances of this.

1. *Newton's method:* Newton used it when he developed the celebrated *Newton's method* for finding an unconstrained minimum of a twice continuously differentiable function, $f(x)$. This method constructs the local model for $f(x^r + y)$ at the current point x^r to be the quadratic function $Q(y) = f(x^r) + \nabla f(x^r)y + \frac{1}{2}y^T H(f(x^r))y$, where $\nabla f(x^r)$ is the row vector of the first-order partial derivatives of $f(x)$ at x^r and $H(f(x^r))$ is the Hessian matrix of $f(x)$ at x^r. $Q(y)$ is the second-order Taylor series approximation for $f(x)$ at x^r. The method computes y^r, the minimizer of the model function $Q(y)$ (assuming that $H(f(x^r))$ is PD, we have $y^r = -H(f(x^r))^{-1}(\nabla f(x^r))^T$) and takes the next point to be $x^{r+1} = x^r + y^r$.

   Thus, Newton's method solves an unconstrained quadratic minimization problem in each step. Starting from an initial point x^0 it generates the sequence $\{x^r\}$, which under certain conditions can be shown to converge to the minimum of the original function $f(x)$.

   To treat the case where the Hessian $H(f(x^r))$ may not be PD, several modified Newton methods based on quadratic models different from the second order Taylor series approximation at x^r have been developed.

   Also, the mathematically beautiful theory of quasi-Newton methods for unconstrained minimization has also been developed through the study of quadratic models (see Dennis and Schnabel 1983; Fletcher 1987; Bazaraa et al. 2006 of Chap. 5).

2. *Conjugate gradient method:* There is the very efficient Gaussian elimination method for solving a square nonsingular system of linear equations, $Ax = b$ say, of order n. However, when n is very large, this method becomes unwieldy and difficult to implement. The least squares formulation of this system of equations is the unconstrained quadratic minimization problem

$$\text{Minimize} \quad (Ax - b)^T(Ax - b)$$

and (Hestenes and Stiefel 1952) developed the conjugate gradient method for solving this problem. Subsequently, through the study of the quadratic model, several researchers have extended this method directly into a variety of conjugate gradient methods for the unconstrained minimization of general nonlinear functions.
3. *Linear least squares:* Suppose we have a large system of linear equations (typically overdetermined, i.e., where the number of equations exceeds the number of variables), say $Ax = b$, which has no exact solution. A common approach for handling such a system is to look for a least squares solution, i.e., an optimum solution of the unconstrained quadratic minimization problem

$$\text{Minimize} \quad (Ax - b)^T(Ax - b)$$

This problem is known as the *linear least squares problem*. Powerful numerical linear algebra techniques such as singular value decomposition (SVD) have been developed to solve large scale versions of this special class of QPs (see Dennis and Schnabel 1983).

Statisticians have been using the linear least squares model for computing the estimates of the coefficients in a linear regression model for a long time.

9.8 Summary of Some Existing Algorithms for Constrained QPs

1. *Frank–Wolfe method:* One of the first methods for QP developed in recent times is that of (Frank and Wolfe 1956). It is an iterative method which in each iteration solves an LP to find a search direction, and a line search problem in that direction. It produces a descent sequence such that every limit point of this sequence is a KKT point. However, the method has slow convergence, and is not popular except on problems with special structure that makes it possible to solve the LP in each iteration by an extremely fast special method taking advantage of the structure.
2. *Reduced gradient methods:* The simplex method for LP has been extended to solve problems involving the minimization of a quadratic (or in general a smooth nonlinear) function subject to linear constraints. The method is called the *reduced gradient method* and is discussed in (Wolfe 1959). The name reduced gradient method refers to any method that uses the equality constraints to eliminate some

variables (called the dependent or basic variables) from the problem and treats the remaining problem in the space of the independent (or nonbasic variables) only, either explicitly or implicitly. The reduced gradient is the gradient of the objective function in the space of independent variables. The method is quite popular; see, for example (Murty 1988 of Chap. 2; Bazaraa et al. 2006 of Chap. 5). The OSL software package uses this method for solving QPs. The MINOS 5.4 software package uses this method for minimizing a smooth nonlinear function subject to equality constraints.

This method has been generalized directly into the GRG (generalized reduced gradient) method for solving nonlinear programs involving nonlinear constraints (Abadie and Carpentier 1969). The GRG is a popular method on which several successful nonlinear programming software packages are based.

3. *Methods based on the LCP:* In the 1950s and 1960s, several researchers proposed schemes for solving the QP by solving its KKT system. Lemke (1965) formulated the KKT system for a QP as an LCP and developed a beautiful algorithm for it called the *complementary pivot algorithm.* The data for an LCP of order n consists of a square matrix M of order n and a column vector $q \in R^n$, and it is to find a $w = (w_j) \in R^n$ and a $z = (z_j) \in R^n$ satisfying

$$\begin{aligned} w - Mz &= q \\ w, z &\geq 0 \\ w_j z_j &= 0 \quad \text{for all } j \end{aligned}$$

Checking whether the general LCP has a solution is an NP-complete problem and there are no efficient algorithms known for it. But the complementary pivot algorithm is a finite path following method for finding a solution, when one exists, to a class of LCPs which includes the KKT systems corresponding to convex QP.

The development of the complementary pivot method is a nice theoretical breakthrough for which Lemke received the Von Neumann theory award of ORSA/TIMS in 1978.

However, the complementary pivot method and several other methods developed for the LCP are not currently popular for solving even convex QPs, because a QP involving m inequality constraints in n non-negative variables leads to an LCP of order $m + n$, blowing up the size.

For tackling nonconvex QPs, the complementary pivot approach is clearly unsuitable, as it focusses attention purely on the KKT system and never even computes the objective value, and if it leads to a KKT point at termination that point may not even be a local minimum.

But the theoretical contribution of the formulation of the LCP and the complementary pivot method for it is great. The LCP has a fascinating geometrical interpretation. The study of the geometry of LCP is initiated in (Murty 1972) and continues to be a very active area of research. And the mathematical principle behind the complementary pivot method has been used to develop simplicial methods (which are also called complementary pivot methods) to solve systems of nonlinear equations and fixed point problems (see Murty 1988; Cottle et al. 1992).

4. *Active set methods:* A popular method for solving QP is based on a combinatorial approach to iteratively determine the set of active constraints at the optimum. This type of strategy for handling inequality constrained optimization problems is called the *active set strategy*. The method solves a sequence of equality constrained QPs by treating some of the inequality constraints as equations (the active set) and temporarily ignoring the others. Several rules are employed to modify the active set from one iteration to the next, to guarantee finite convergence of the procedure (see Theil and van de Panne 1961). Several researchers have extended this method to minimize a smooth nonlinear function subject to linear equality and inequality constraints.
5. *Interior point methods:* Since the development of a very successful interior point method for LP by Karmarkar in 1984, a variety of interior point methods have been developed for convex QPs and the LCPs associated with them. These methods are polynomially bounded, and some versions of them give excellent computational performance on large sparse problems. The monograph (Kojima et al. 1991) establishes the theoretical foundations for primal–dual interior point methods for LP and LCP. The authors won the 1992 Lanchaster award for this monograph (some other references on these methods are Ye 1991; Fang and Puthenpura 1993).

   We do not discuss details of these methods here, as detailed treatments of them are already available in severable books among the references at the end of this chapter and in (Bazaraa et al. 2006 of Chap. 5).

   In the next section, we discuss the extension to QP of the sphere method discussed in Chap. 8 for LP. This new method still in development shows a lot of promise for large-scale problem solving for both convex and nonconvex QPs.

9.9 The Sphere Method for QP

We consider the quadratic program (QP) in the following form

$$\begin{aligned} \text{Minimize} \quad & Q(x) = cx + (1/2)x^T Dx \\ \text{subject to} \quad & Ax \geq b \end{aligned} \tag{9.5}$$

where the objective coefficient matrix D is a symmetric matrix of order n, the constraint coefficient matrix A is of order $m \times n$, and b, c are column and row vectors of appropriate orders. Let K denote the set of feasible solutions. We also assume that an interior point x^0 of K (i.e., a point satisfying $Ax^0 > b$) is available.

First we will assume that D is positive definite, i.e., that $Q(x)$ is strictly convex. Strategies for relaxing this assumption are discussed briefly later on.

Let $K^0 = \{x : Ax > b\}$, it is the interior of K. We assume that the row vectors of A, denoted by $A_{i.}$ for $i = 1$ to m, are normalized so that their Euclidean norm $||A_{i.}|| = 1$ for all i. For each $x \in K^0$, we define $\delta(x) = \min\{A_{i.}x - b_i : i = 1$ to $m\}$, $\delta(x)$ is the radius of the largest ball that can be inscribed within K with its center at s x.

In this section, we will discuss the sphere method for QP, which is the extension of the sphere method 1 for LP discussed in Chap. 8. It is an interior point method. Just as the sphere methods for LP, each iteration in this method consists of a centering step and then some descent steps. When the current interior feasible solution at the beginning of an iteration is x^r, the centering step in that iteration tries to move that point to a different feasible solution, $\bar{x}$ say, that maximizes $\delta(x)$ subject to the constraint that $Q(x) \leq Q(x^r)$.

The Strategy of the Centering Step

When x^r is the current interior feasible solution for (9.5), the problem of finding the largest inscribed sphere inside K with center at a point where the objective value $Q(x)$ is $\leq Q(x^r)$ is the following constrained max–min problem:

$$\begin{aligned} &\text{Maximize} \quad \delta \\ &\text{subject to} \quad \delta - A_{i.}x \leq -b_i, \quad i = 1, \cdots, m \\ &\qquad\qquad Q(x) \leq Q(x^r) \end{aligned} \tag{9.6}$$

If $(\bar{x},\ \bar{\delta})$ is an optimum solution of this problem, then $\bar{\delta} = \delta(\bar{x})$, and the ball $B(\bar{x},\ \bar{\delta})$ with $\bar{x}$ as center and $\bar{\delta}$ as radius is the largest inscribed sphere required. This problem (9.6) is itself a quadratic program. This type of model may have to be solved several times before we get a solution for our original QP (9.5) and for implementing our algorithm an exact solution of (9.6) is not essential; hence, solving (9.6) exactly will be counterproductive. Using the special max–min structure of (9.6), we now develop an efficient procedure for getting an approximate solution to (9.6), similar to the one developed in Chap. 8 for the corresponding centering problem in sphere method 1 discussed there for LP.

9.9.1 Procedure for Getting an Approximate Solution for (9.6)

Let x^r be the current interior feasible solution. This procedure generates a sequence of interior feasible solutions $x^{r,0} = x^r,\ x^{r,1}, \cdots$ all of them feasible to (9.6), along which $\delta(x)$ is monotonic increasing. When $x^{r,k}$ is the current solution, a direction y is said to be

- **Descent direction for $Q(x)$ at $x^{r,k}$:** if $\nabla Q(x^{r,k})y \leq 0$;
- **Profitable direction to move at $x^{r,k}$:** if $\delta(x^{r,k} + \alpha y)$ increases as α increases from 0.

From the result in Sect. 8.3.1 we know that a direction y is a profitable direction to move at the current interior feasible solution $x^{r,k}$ iff $A_{i.}y > 0$ for all $i \in T(x^{r,k})$ = set of all p that tie for the minimum in $\{A_{p.}x^{r,k} - b_p : p = 1$ to $m\}$.

So, this procedure consists of a series of line search steps. The interior feasible solution obtained at the end of a step is the initial interior feasible solution for the next step. The first step begins with $x^{r,0} = x^r$.

Each step consists of two substeps. Substep 1 selects a profitable direction to move that is also a descent direction for $Q(x)$ at the current interior feasible solution. In the step when $x^{r,k}$ is the initial interior feasible solution and y is the profitable direction to move selected at it in Substep 1, Substep 2 carries out a line search step to find the optimum step length α maximizing $\delta(x^{r,k} + \alpha y)$ over values of α between 0 and α_1 = value of α that minimizes $Q(x^{r,k} + \alpha y)$ over $\alpha \geq 0$. We will first describe this Substep 2.

Substep 2: Line search substep: When the current point is $x^{r,k}$ and the profitable direction selected to move at it is y. Find

- α_1 = the value of α that minimizes $Q(x^{r,k} + \alpha y)$ over $\alpha \geq 0$. Finding α_1, therefore, requires minimizing a quadratic function in the single variable α, which can be solved easily.
- α_2 = the value of α that maximizes $\delta(x^r + \alpha y)$ over $\alpha \geq 0$. In Sect. 8.3.1, it has been shown that this can be found by solving the following 2-variable linear program in which the variables are θ, α.

$$\begin{aligned} &\text{Maximize} \quad \theta \\ &\text{subject to} \quad \theta - \alpha A_{i.}y \leq A_{i.}x^{r,k} - b_i \quad i = 1, \ldots, m \\ &\theta, \alpha \geq 0 \end{aligned}$$

which can be found with at most $O(m)$ effort.

Once α_1, α_2 are determined, let α = minimum$\{\alpha_1, \alpha_2\}$, take the next point in the sequence to be $x^{r,k+1} = x^{r,k} + \alpha y$, and continue the procedure in the same way with $x^{r,k+1}$.

Substep 1: Selecting profitable direction to move: For this we use the two methods LSFN and LSCPD similar to those discussed in Sect. 8.3.1. In steps at the beginning of the procedure, we use LSFN and select the profitable directions to move from the set $\Gamma_1 = \{A_{i.}, -A_{i.} : i = 1 \text{ to } m\}$ which also satisfy the property of being descent directions for $Q(x)$ at the current interior feasible solution. This is continued until at some stage the improvement per step in the value of the radius of the inscribed ball becomes smaller than some selected tolerance.

Then we begin a cycle of LSCPD which selects profitable directions to move as in LSCPD of Sect. 8.3.1. In this cycle, in the step when $x^{r,k}$ is the current point, the profitable direction to move is selected exactly as in LSCPD of Sect. 8.3.1 with c replaced by $\nabla Q(x^{r,k})$.

The current point in the sequence at the end of this cycle, $\bar{x}$ say, is taken as the approximate center and call it the *ball center* selected by this procedure. As can be seen, the procedure used in this approximate centering strategy uses matrix inversion steps only in LSCPD and uses them only sparingly and only solves a series of 2-variable LPs and single variable quadratic function minimization problems, which can be solved very efficiently. Hence, this approximate centering strategy can be expected to be efficient.

What Is the Purpose of Maximizing the Radius of the Inscribed Ball in This Centering Step?

Our goal is to find an optimum solution to the original quadratic program (9.5). Then, why are we focussing on the seemingly unrelated problem of maximizing the radius of the inscribed ball in this centering step? The reason is the following.

Let $B(\bar{x}, \bar{\delta})$, the ball with center $\bar{x}$ and radius $\bar{\delta}$ be the ball constructed in this centering step. Then, in this iteration the algorithm uses the direction $\hat{x} - \bar{x}$ as a descent direction for a line search step to minimize $Q(x)$ over $\{\bar{x}+\lambda(\hat{x}-\bar{x}) : \lambda \geq 0$, and λ such that $\bar{x} + \lambda(\hat{x} - \bar{x}) \in K\}\}$, where $\hat{x}$ is a point that minimizes $Q(x)$ over the ball $B(\bar{x}, \bar{\delta})$.

The problem of minimizing $Q(x)$ over the ball $B(\bar{x}, \bar{\delta})$ is known in the literature as the *trust region problem*. Efficient algorithms for solving it, and software implementations of them are available, whether $Q(x)$ is convex or not (e.g., see the links: http://www2.imm.dtu.dk/~mr/lstrs.html, http://galahad.rl.ac.uk/galahad-www/, for software for this problem, one containing a MATLAB program and the other a Fortran program). But solving this trust region problem is perhaps the most expensive computational operation in this algorithm. Maximizing $\bar{\delta}$, the radius of the ball found in this centering step helps to reduce the number of times this expensive step has to be used in this algorithm.

9.9.2 Descent Steps

Just as in the Sphere method for LP discussed in Chap. 8, we take several descent steps and take the best point among the output points obtained from all of them as the initial interior feasible solution for the next iteration.

DQ1: Descent Step Using a Descent Direction

Let $B(\bar{x}, \bar{\delta}) = \{x : (x - \bar{x})^T(x - \bar{x}) \leq \bar{\delta}^2\}$ be the ball with center $\bar{x}$ and radius $\bar{\delta}$ obtained in the centering step. In this step, we solve the problem

$$\begin{aligned} \text{Minimize } Q(x) &= cx + (1/2)x^T Dx \\ \text{subject to } (x - \bar{x})^T(x - \bar{x}) &\leq \bar{\delta}^2 \end{aligned} \tag{9.7}$$

This is the well-known *trust region problem* discussed in nonlinear programming (see Conn et al. 2000), a problem of minimizing a quadratic function inside a ball, for which efficient polynomial time algorithms exist. Associating the Lagrange multiplier $\lambda \in R^1$ with the constraint, the KKT optimality conditions for this problem are

$$\begin{aligned} c^T + Dx + 2\lambda(x - \bar{x}) &= 0 \\ \lambda \geq 0, \quad \bar{\delta}^2 - (x - \bar{x})^T(x - \bar{x}) &\geq 0 \\ \lambda(\bar{\delta}^2 - (x - \bar{x})^T(x - \bar{x})) &= 0 \end{aligned}$$

Since $\lambda \in R^1$, this problem can be solved efficiently (in polynomial time) using the KKT conditions; see (Conn et al. 2000) for complete details of this algorithm. The algorithm becomes simpler when D is positive definite or semidefinite, but even if D is not positive semidefinite, it can be solved efficiently using the KKT conditions.

Let $\hat{x}$ be the optimum solution computed for (9.7). If $\hat{x}$ is an interior point of $B(\bar{x}, \bar{\delta})$, or if it is a boundary point of both $B(\bar{x}, \bar{\delta})$ and K, or if $\nabla Q(\hat{x}) = 0$, then $\hat{x}$ is an optimum solution of (9.5), then terminate.

Otherwise, using $\hat{x} - \bar{x}$ as the descent direction for $Q(x)$ at $\bar{x}$, do a line search to minimize $Q(x)$ on the line segment $\{\bar{x} + \lambda(\hat{x} - \bar{x}) : \lambda \geq 0$, and λ such that $\bar{x} + \lambda(\hat{x} - \bar{x}) \in K\}$. Let λ_1 be the optimum step length for this line search. If $\bar{x} + \lambda_1(\hat{x} - \bar{x})$ is an interior point of K, then terminate if $\nabla Q(x) = 0$ at this point, otherwise define this point as the output of this step.

If, however, $\bar{x} + \lambda_1(\hat{x} - \bar{x})$ is a boundary point of K, let $I = \{i : i$-th constraint in (9.5) is satisfied as an equation by $\bar{x} + \lambda_1(\hat{x} - \bar{x})\}$. If the following system in Lagrange multipliers $\pi_I = (\pi_i : i \in I)$

$$\begin{aligned} c + (\bar{x} + \lambda_1(\hat{x} - \bar{x}))^T D - \sum_{i \in I} \pi_i A_{i.} &= 0 \\ \pi_i \geq 0 \quad \text{for all} \quad & i \in I \end{aligned} \tag{9.8}$$

has a feasible solution, then $\bar{x} + \lambda_1(\hat{x} - \bar{x})$ is an optimum solution of (9.5), so terminate. However, it may not be productive to check if system (9.8) is feasible every time this step ends up at this stage. If this operation of checking the feasibility of (9.8) is not carried out or if (9.8) turns out to be infeasible, then take the output of this step as $\bar{x} + (\lambda_1 - \epsilon)(\hat{x} - \bar{x})$, where ϵ is some preselected positive tolerance for the current point to be an interior point of K.

DQ2: Descent Step Using the Touching Constraints

We will first provide the motivation for this step. Assume that the centering step is carried out exactly and suppose $B(\bar{x}, \bar{\delta}) = \{x : (x - \bar{x})^T(x - \bar{x}) \leq \bar{\delta}^2\}$ is the ball with center $\bar{x}$ and radius $\bar{\delta}$ obtained in the centering step in this iteration. $T(\bar{x}) = \{i : A_{i.}\bar{x} = b_i + \bar{\delta}\}$ is the index set of *touching constraints* in this iteration this is the index set of facetal hyperplanes of K that are touching the ball $B(\bar{x}, \bar{\delta})$ and hence are tangent hyperplanes for it. Actually, $T(\bar{x})$ is the index set of linear constraints in (9.6) that are active at its optimum solution, all other linear constraints in (9.6) are inactive at its optimum solution and the same thing is also true for the problem obtained by replacing x^0 in (9.6) by $\bar{x}$. So, $(\bar{x}, \bar{\delta})$ is an optimum solution for (9.6) when x^0 there is replaced by $\bar{x}$, i.e., for

$$\begin{aligned} &\text{Maximize } \delta \\ &\text{subject to } \delta - A_{i.}x \leq -b_i, \quad i = 1, \cdots, m \\ &\qquad Q(x) \leq Q(\bar{x}) \end{aligned} \tag{9.9}$$

It often happens the index set of touching constraints for the ball obtained from an optimum solution of (9.9) with $Q(\bar{x})$ replaced by $Q(\bar{x}) - \gamma$ remains the same as $T(\bar{x})$, for a range of values of γ, say $0 \leq \gamma \leq \gamma_1$. In this range $0 \leq \gamma \leq \gamma_1$, let $\delta(\gamma)$ denote the optimum radius of the ball and $x(\gamma)$ the center. Beginning with $\delta(0) = \bar{\delta}$, clearly, $\delta(\gamma)$ decreases as γ increases to γ_1. From these facts we see that in the range $\delta(0) \geq \delta(\gamma) \geq \delta(\gamma_1)$, $x(\gamma)$ is the optimum solution of

$$\begin{aligned} &\text{Minimize } Q(x) \\ &\text{subject to } A_{i.}x = b_i + \delta(\gamma), \quad i \in T(\bar{x}) \end{aligned} \tag{9.10}$$

Replacing the parameter $\delta(\gamma)$ by the symbol s, an optimum solution for (9.10) can be obtained by solving

$$\begin{aligned} c^T + Dx - \sum_{i \in T(\bar{x})} \pi_i A_{i.} &= 0 \\ A_{i.}x = b_i + s, \quad i &\in T(\bar{x}) \end{aligned} \tag{9.11}$$

where $\pi_{T(\bar{x})} = (\pi_i : i \in T(\bar{x}))$ is the vector of Lagrange multipliers for (9.10). If $(x(s), \pi_{T(\bar{x})}(s))$ is a solution of (9.11) as a function of the parameter s, then $x(s)$ defines a straight line in R^n in terms of the parameter s. The above argument shows that by carrying out a line search step on this straight line, we can decrease the value of $Q(x)$ to reach $Q(x(\gamma_1))$; and any further decrease in the value of $Q(x)$ below this will lead to an optimal touching constraint index set for the ball different from $T(\bar{x})$.

Even when (9.6) is solved approximately, we may improve the objective value by carrying out this work with the ball obtained. That is what this step does.

Denoting the ball obtained in the centering step by the same symbol $B(\bar{x}, \bar{\delta}) = \{x : (x - \bar{x})^T(x - \bar{x}) \leq \bar{\delta}^2\}$, denote the touching constraint index set by the same symbol as above $T(\bar{x}) = \{i : A_{i.}\bar{x} = b_i + \bar{\delta}\}$. With this $T(\bar{x})$, get the solution $(x(s), \pi_{T(\bar{x})})$ for system (9.11). Then, do a line search to minimize $Q(x)$ over the line segment $\{x(s) : s$ such that $x(s) \in K\}$. Suppose $s = s_1$ gives the optimum $x(s)$ in this line search step.

If $x(s_1)$ is an interior point of K, then terminate if $\nabla Q(x) = 0$ at this point, otherwise define this point as the output of this step.

If, however, $x(s_1)$ is a boundary point of K, let $I = \{i : i\text{-th constraint in (9.5)}$ is satisfied as an equation by $x(s_1)\}$. If the following system in Lagrange multipliers $\pi_I = (\pi_i : i \in I)$

$$c + x(s_1)^T D - \sum_{i \in I} \pi_i A_{i.} = 0 \tag{9.12}$$

$$\pi_i \geq 0 \quad \text{for all} \quad i \in I$$

has a feasible solution, then $x(s_1)$ is an optimum solution of (9.5), so terminate. However, it may not be productive to check if system (9.12) is feasible every time this step ends up at this stage. If this operation of checking the feasibility of (9.12) is not carried out, or if (9.12) turns out to be infeasible, then take the output of this descent step as a point on the line segment $\{x(s) : s \in R^1\}$ close to $x(s_1)$ but in the interior of K.

Other Descent Steps

For each $i \in T(\bar{x})$, find the NTP $\hat{x}^i$ as defined in Sect. 8.3. Let $-c^i$ denote the orthogonal projection of $-\nabla Q(\hat{x}^i)^T$ on $\{x : A_{i.}x = 0\}$. For the QP (9.5) the descent steps corresponding to D5.1 in Chap. 8 for LP, would be the descent steps for $Q(x)$ from $\hat{x}^i$ in the descent direction $-c^i$ for each $i \in T(\bar{x})$.

And if $\tilde{x}^{r1}$ is the best point obtained in Descent Step D5.1, the descent steps corresponding to D5.2 in Chap. 8 for LP would be the descent steps for $Q(x)$ from $\tilde{x}^{r1}$ in the descent direction $-c^i$ for each $i \in T(\tilde{x}^{r1})$, and repeating this as long as good reductions in objective value are occuring/repetition.

In Chap. 8, we saw that the Descent Steps D5.1 and D5.2 gave excellent results for solving LP. It is not clear that the corresponding steps for the QP would be equally effective. If computational experiments indicate that they are, then these descent steps can also be included in the algorithm.

9.9.3 The Algorithm

The algorithm consists of repetitions of the following iteration beginning with an initial interior point of K. We will now describe the general iteration. In each iteration, Steps 2.1 and 2.2 are parallel steps, both of which begin with the ball obtained in the centering step in the iteration.

A General Iteration

Let x^0 be the current interior feasible solution.

1. *Centering Strategy:* Apply the centering strategy described in Sect. 9.9.1 beginning with the current interior feasible solution. Let $B(\bar{x}, \bar{\delta})$ denote the ball obtained with ball center $\bar{x}$ and radius $\bar{\delta}$. Let $T(\bar{x}) = \{i : A_{i.}\bar{x} = b_i + \bar{\delta}\}$ is the index set of touching constraints for this ball.

2.1. *DQ1: Descent Step Using a Descent Direction:* Apply this strategy described in Sect. 9.9.2 beginning with the ball $B(\bar{x}, \bar{\delta})$. If termination did not occur in this step, let x^1 denote the interior feasible solution of (9.5) which is the output point in this step.

2.2. *DQ 2: Descent Step Using the Touching Constraints:* Apply this strategy described in Sect. 9.9.2 beginning with the ball $B(\bar{x}, \bar{\delta})$. If termination did not occur in this step, let x^2 denote the interior feasible solution of (9.5) which is the output point in this step.

Similarly, if other descent steps discussed in the previous section are used, let x^3 denote the best point obtained at the end of these descent steps.

3. *Move to Next Iteration:* Define the new current interior feasible solution as the point among x^1, x^2, x^3 obtained in Steps 2.1 and 2.2, which gives the smallest value for $Q(x)$. With it, go to the next iteration.

9.9.4 The Case when the Matrix D is not Positive Definite

Relaxing the positive definiteness assumption on the matrix D leads to a vast number of applications for the model (9.5). For example, an important model with many applications is the following 0–1 mixed integer programming (MIP) model:

$$\begin{aligned} \text{Minimize} \quad & cx \\ \text{subject to} \quad & Ax \geq b \\ & x \geq 0 \\ x_j = \quad & 0 \text{ or } 1 \text{ for each } j \in J \end{aligned} \tag{9.13}$$

where J is the subscript set for variables that are required to be binary. Solving this problem is equivalent to finding the global minimum in the quadratic program

$$\begin{aligned} \text{Minimize} \quad & cx + M \sum_{j \in J} x_j(1 - x_j) \\ \text{subject to} \quad & Ax \geq b \\ & x_j \geq 0 \quad \text{for } j \notin J \\ & 0 \leq x_j \leq 1 \quad \text{for each } j \in J \end{aligned} \tag{9.14}$$

where M is a large positive penalty coefficient, which is in the form (9.5) with D negative semidefinite. Unlike the model (9.5) when D is positive definite, (9.14) may have many local minima, and we need to find the global minimum for (9.14).

Some of the steps in this algorithm can still be carried out exactly. The approximate centering procedure can be carried out. Also, Step 2.1 can be carried out exactly. For Step 2.2, the system of (9.11) may typically have a unique solution. Even when (9.11) has many feasible solutions, a solution to (9.12) may not even be a local minimum for (9.11); in fact, it may be a local maximum for (9.11). Hence, the value of including Step 2.2 in the algorithm is not clear in this case.

However, since the ball minimization problems in Step 2.1 can be solved exactly, there is reason to hope that by adjusting the value of the penalty cost coefficient M during the algorithm, the algorithm can be made to lead to a good local minimum, and thereby offer a good heuristic approach. For this general case, these and other issues need to be pursued.

9.10 Commercially Available Software

MINOS 5.4 – available from Stanford Business Software or from The Scientific Press as part of either of the algebraic modeling systems AMPL or GAMS and OSL – available from IBM, or from the Scientific Press as part of AMPL, are two of the commercially available software packages for solving QPs.

AMPL (Fourer et al. 1993) is a modeling language for mathematical programming that provides a natural form of input for linear, integer, and nonlinear mathematical models besides QP models. The book is accompanied by a PC student version of AMPL and representative solvers, enough to easily handle problems of a few hundred variables and constraints. Versions that support much larger problems are available from the publisher. AMPL uses either the MINOS 5.4 solver or the OSL solver for solving QP models.

GAMS (Brooke et al. 1988) is a high-level language that is designed to make the construction and solution of large and complex mathematical programming models straightforward for programmers, and more comprehensible to users of models. It uses the MINOS solver for solving QPs, it also has solvers for linear, integer, and nonlinear programming problems. A student version and a professional version are available.

IBM's OSL is a collection of high-performance mathematical subroutines for solving linear, integer, and quadratic programming models.

MINOS 5.4 (Murtagh and Saunders 1987) is a Fortran-based computer system designed to solve large-scale linear, quadratic, and nonlinear models.

9.11 Exercises

9.1. A is a given matrix of order $m \times n$. Consider the following two problems: (1) Maximize $f(x) =$ Minimum $\{A_{i.}x : i = 1 \text{ to } m\}$, subject to $||x|| = 1$ and (2) Minimize $y^T y$, subject to $A_{i.}y \geq 1, i = 1$ to m.

Show that these two problems are equivalent (Avi-Itzak 1994).

9.2. Let $f(x) = 3x_1 + 4x_2 - x_1x_2 + x_1^2 + 2x_2^2$. Consider the QP: minimize $f(x)$ subject to $x_1 + x_2 = 1$.

Check whether this is a convex QP and write the KKT optimality conditions for it.

Find a KKT point for this problem and check whether it is a global minimum.

Suppose the additional constraints: $x_1, x_2 \geq 0$ are added to the problem. Write the KKT conditions for the augmented problem. Find all the KKT points for this problem and comment on whether they are global minima (R. Saigal).

9.3. Consider the QP: minimize $cx + (1/2)x^T Dx$, subject to $Ax = b$ where D is a PD matrix.

Write the KKT necessary optimality conditions for this problem. Under what conditions is this KKT system guaranteed to have a solution? Under these conditions, is the solution of the system unique? Find a solution of the system.

Now consider the same problem with additional constraints $x \geq 0$. Write the KKT necessary optimality conditions for the augmented problem. Obtain the conditions under which this system will have a solution. Under these conditions, show that the solution to this system must be unique (R. Saigal).

9.4. Let $K \subset R^n$ be a convex set with a nonempty interior, and $Q(x)$ a quadratic function. If $Q(x)$ is a convex function over K, show that it is actually convex over the whole space R^n.

9.5. Consider the QP: minimize $Q(x) = cx + (1/2)x^T Dx$ subject to $Ax \geq b$, $x \geq 0$. If D is not PSD prove that an interior feasible solution $\bar{x}$, i.e., one satisfying $A\bar{x} > b$, $\bar{x} > 0$, cannot be an optimum solution of this QP using the necessary optimality conditions.

9.6. Write the optimality conditions for the following QP. Given that it has an optimum solution in which all of x_1, x_2, x_3 are > 0, find that optimum solution.

$$\begin{aligned} \text{Minimize} \quad & -6x_1 - 4x_2 - 2x_3 + 3x_1^2 + 2x_2^2 + (1/3)x_3^2 \\ \text{subject to} \quad & x_1 + 2x_2 + x_3 \leq 4 \\ & x_1, x_2, x_3 \geq 0 \end{aligned}$$

9.7. Test whether the following matrices are either PD, PSD, or not.

$$\begin{pmatrix} 0 & 1 & -1 \\ 0 & 0 & -2 \\ 1 & 2 & 1 \end{pmatrix}, \begin{pmatrix} 4 & 3 & -7 \\ 0 & 0 & -2 \\ 4 & 0 & 6 \end{pmatrix}, \begin{pmatrix} 4 & 100 & 2 \\ 0 & 2 & 10 \\ 4 & 0 & 4 \end{pmatrix}, \begin{pmatrix} 5 & -2 & -2 \\ -5 & 5 & -2 \\ 5 & 0 & 5 \end{pmatrix}.$$

9.8. $Q(x) = cx + (1/2)x^T Dx$. If D is PD, prove that $Q(x)$ is bounded below.

9.9. Prove that $Q_\lambda(x) = x^T(\lambda I + D)x + cx$ is a convex function when λ is sufficiently large, whatever D may be.

9.10. Given a square matrix D, the superdiagonalization algorithm discussed in Sect. 9.2 can be used to check if it is PSD. If this algorithm terminates with the conclusion that D is not PSD, show how to find a vector $y \in R^n$ satisfying $y^T Dy < 0$.

9.11. Sylvester's Problem. $A_{.1}, A_{.2}, \cdots, A_{.n}$ are given points in R^m. It is required to find the smallest diameter sphere in R^m containing all these points inside it. Formulate this problem as a QP. Show this formulation for the set of points $\{(1,1), (-3,2), (1,-5), (-2,4)\}$ in R^2.

9.12. Consider the QP: minimize $cx + (1/2)x^T Dx$, subject to $Ax = b$, where D is a PD matrix.

Write the KKT necessary optimality conditions for this problem. Under what conditions is this KKT system guaranteed to have a solution? Under these conditions, is the solution of the system unique? Find a solution of the system.

Now consider the same problem with additional constraints $x \geq 0$. Write the KKT necessary optimality conditions for the augmented problem. Obtain the conditions under which this system will have a solution. Under these conditions show that the solution to this system must be unique (R. Saigal).

9.13. Quadratic Programming Model to Determine State Taxes. It is required to determine optimum levels for various state government taxes that minimizes instability while meeting constraints on growth rates over time. Seven different taxes are considered: sales, motor fuel, alcoholic beverages, tobacco, motor vehicle, personal income, and corporate taxes. State government finance is based on the assumption of predictable and steady growth of each tax over time. Instability in tax revenue is measured by the degree to which the actual revenue differs from predicted revenue.

Using past data, a regression equation can be determined to measure the growth in tax revenue over time. Let s be the tax rate for a particular tax and S_t the expected tax revenue from this tax in year t. Then the regression equation used is

$$\log_e S_t = a + bt + cs$$

where a, b, c are parameters to be determined using past data to give the closest fit. Data for the past 10 years from a state is used for this parameter estimation. Clearly, the parameter c can only be estimated, if the tax rate s for that tax has changed during this period, this has happened only for the motor fuel and the tobacco taxes. The best-fit parameter values for the various taxes are given below in Table 9.1 (for all but the motor fuel and tobacco taxes, the tax rate has remained the same over the 10-year period for which the tax data is available, and hence the parameter a given below for these taxes is actually the value of $a + cs$, as it was not possible to estimate a and c individually from the data).

Table 9.1 Regression coefficient values

j	Tax j	a	b	c
1	Sales	12.61	0.108	
2	Motor fuel	10.16	0.020	0.276
3	Alcoholic beverages	10.97	0.044	
4	Tobacco	9.79	0.027	0.102
5	Motor vehicle	10.37	0.036	
6	Personal income	11.89	0.160	
7	Corporate	211.09	0.112	

The annual growth rate is simply the regression coefficient b multiplied by 100 to convert it to percent.

For 1984, the tax revenue from each tax as a function of the tax rate can be determined by estimating the tax base. This data, available with the state, is given below.

j	Tax j	Tax base (in $\$10^6$)
1	Sales	34,329
2	Motor fuel	3,269
3	Alcoholic beverages	811
4	Tobacco	702
5	Motor vehicle	2,935
6	Personal income	30,809
7	Corporate	4,200

If s_j is the tax rate for tax j in 1984 as a fraction, x_j = tax revenue to be collected in 1984 in millions of dollars for the jth tax is expected to be: (tax base for tax j)s_j.

Choosing the decision variables to be x_j for $j = 1$–7, let $x = (x_1, \dots, x_7)^T$. The total tax revenue is $\sum_{j=1}^{7} x_j$. Then the variability or instability in this revenue is measured by the quadratic function $Q(x) = x^T V x$ where V the variance–covariance matrix estimated from past data is

$$\begin{pmatrix} 0.00070 & -0.00007 & 0.00108 & -0.00002 & 0.00050 & 0.00114 & 0.00105 \\ & 0.00115 & 0.00054 & -0.00002 & 0.00058 & -0.00055 & 0.00139 \\ & & 0.00279 & 0.00016 & 0.00142 & 0.00112 & 0.00183 \\ & & & 0.00010 & 0.00009 & -0.00007 & -0.00003 \\ & & & & 0.00156 & 0.00047 & 0.00177 \\ & & & & & 0.00274 & 0.00177 \\ & & & & & & 0.00652 \end{pmatrix}.$$

Since V is symmetric, only the upper half of V is recorded above.

The problem is to determine the vector x that minimizes $Q(x)$, subject to several constraints. One of the constraints is that the total expected tax revenue for 1984 should be $T = 3300$ in millions of dollars. The second constraint is that a specified growth rate of λ in the total tax revenue should be maintained. It can be assumed that this overall growth rate is the function $\sum_{i=1}^{7} \frac{x_j b_j}{T}$ which is a weighted average of the growth rates of the various taxes. We would like to solve the problem treating λ as a non-negative parameter. Of particular interest are values $\lambda = 9\%$ and 13%.

The other constraints are lower and upper bounds on tax revenues x_j; these are of the form $0 \le x_j \le u_j$ for each j, where u_j is twice the 1983 revenue from tax j. The vector $u = (u_j)$ is (2,216, 490, 195, 168, 95, 2,074, 504) in millions of dollars.

Formulate this problem as a QP. Using the tax base information given above determine the optimal tax rates for 1984 for each tax (White 1983).

9.14. To Determine Optimum Mix of Ingredients for Moulding Sand in a Foundry. In a heavy casting steel foundry, moulding sand is prepared by mixing sand, resin (phenol formaldehyde), and catalyst (para toluene sulfonic acid). In the mixture, the resin undergoes a condensation polymerization reaction resulting in a phenol formaldehyde polymer that bonds and gives strength. The bench life of the mixed sand is defined to be the length of the time interval between mixing and the starting point of setting of the sand mix. In order to give the workers adequate time to use the sand and for proper mould strength, the bench life should be at least 10 min. Another important characteristic of the mixed sand is the dry compression strength which should be maximized. An important variable which influences these characteristics is the resin percentage in the mix. Extensive studies have shown that the optimum level for this variable is 2% of the weight of sand in the mix; hence, the company has fixed this variable at this optimal level. The other process variables that influence the output characteristics are:

$$x_1 = \text{Temperature of sand at mixing time}$$
$$x_2 = \text{Percent of catalyst, as a percent of resin added}$$
$$x_3 = \text{Dilution of catalyst added at mixing.}$$

The variable x_3 can be varied by adding water to the catalyst before it is mixed. An experiment conducted yielded the following data.

Dry compression strength

	$x_3 = 0$				$x_3 = 10$			
x_1	$x_2 = 25$	30	35	40	25	30	35	40
20^c	31.4	32.4	33.7	37.3	32.7	33.7	36.3	34.0
30^c	33.4	34.1	34.9	32.6	30.1	31.1	35.0	35.2
40^c	33.8	31.4	38.0	32.4	31.6	32.3	34.7	34.8

Bench life

	$x_3 = 0$				$x_3 = 10$			
x_1	$x_2 = 25$	30	35	40	25	30	35	40
20^c	13.3	11.5	10.8	10.3	15.8	14.0	12.8	11.8
30^c	10.3	9.0	8.0	6.8	12.3	11.0	10.3	9.3
40^c	7.0	6.3	5.0	4.3	11.8	10.5	7.3	5.8

Bench life can be approximated very closely by an affine function in the variables x_1, x_2, x_3; dry compression strength can be approximated by a quadratic function in the same variables. Find the functional forms for these characteristics that provide the best approximation. Using them, formulate the problem of finding the optimal values of the variables in the region $0 \leq x_3 \leq 10$, $25 \leq x_2 \leq 40$, $20 \leq x_1 \leq 40$, so as to maximize the dry compression strength subject to the additional constraint that the bench life should be at least 10, as a QP. Find its optimum solution. (*Hint*: For curve fitting use either the least squares method or the minimum absolute deviation methods based on linear programming discussed in Chap. 2.) (Bharat Heavy Electricals Ltd., Hardwar, India).

9.15. A stock broker has been following the stock price of a company by measuring the weekly average stock price denoted by W. The company is well established, and its stock price distribution has been stable for a long time. W varied between 50 to 80 in the past with following discretized distribution.

Interval of W	Probability
I_1 (= 50 to 53)	p_1
I_2 (= 53 to 56)	p_2
$\vdots$	$\vdots$
I_{10} (= 77 to 80)	p_{10}

Let $p = (p_1, \ldots, p_{10})$ denote the given probability vector. In Fall 2002 the company acquired another company through a merger. This may have changed the probability distribution of W. Since than W varied between 56 to 86. From the observations for one year after the merger, in Fall 2003 we estimated the following discretized distribution.

Interval of W	Probability
I_3	q_3
I_4	q_4
$\vdots$	$\vdots$
I_{10} (= 77 to 80)	q_{10}
I_{11} (= 80 to 83)	q_{11}
I_{12} (= 83 to 86)	q_{12}

Let q denote the given probability vector corresponding to these latest observations.

Let $x = (x_1, ..., x_{12})$ denote the unknown true probability vector corresponding to the intervals I_1 to I_{12} in the current discretized distribution of W. Since q is based on too few observations, it is not a good estimate for x by itself.

Let S_1, S_2 denote the sum of squared deviations of corresponding entries between x and p, x and q respectively.

The standard quadratic model takes the estimate of x to be a probability vector that minimizes the weighted average $\alpha S_1 + (1 - \alpha)S_2$ where $0 < \alpha < 1$ is a numerical parameter whose value is taken to be larger than $(1 - \alpha)$ because q is based on too few observations (for example $\alpha = 0.7$).

Formulate the model for estimating x. Derive an optimum solution of this model showing clearly how you obtained it. Can you conclude that the optimum solution of this model is unique, and why?

References

Abadie J, Carpentier J (1969) Generalization of the wolfe reduced gradient method to the case of nonlinear constraints. In: Fletcher R (ed) Optimization. Academic Press, NY

Avi-Itzak (1994) High-accuracy correlation-based pattern recognition. Ph.D. thesis, EE, Stanford University, Stanford, CA; Dantzig, Thappa (1997) vol 1

Brooke A, Kendrick D, Meeraus A (1988) GAMS: a user's guide. Scientific Press, San Francisco

Conn AR, Gould NIM, Toint PL (2000) Trust-region methods. MPS-SIAM Series on Optimization

Cottle RW, Pang JS, Stone RE (1992) The linear complementarity problem. Academic Press, NY

Crum RL, Nye DL (1981) A network model of insurance company cash flow management. Math Program Stud 15:86–101

Dennis JB (1959) Mathematical programming and electrical networks. Wiley, NY

Dennis JE Jr, Schnabel RB (1983) Numerical methods for unconstrained optimization and nonlinear equations. Prentice Hall, NJ

Eldersveld SK (1991) Large scale sequential quadratic programming, SOL91. Department of OR, Stanford University, CA

Fang SC, Puthenpura S (1993) Linear optimization and extensions: theory and algorithms. Prentice Hall, NJ

Fletcher R (1987) Practical methods of optimization, 2nd edn. Wiley, NY

Fourer R, Gay DM, Kernighan BW (1993) AMPL: a modeling language for mathematical programming. Scientific Press, San Francisco

Frank M, Wolfe P (1956) An algorithm for quadratic programming. Nav Res Logist Q 3:95–110

Glassey CR (1978) A quadratic network optimization model for equilibrium single commodity trade flows. Math Program 14:98–107

Han SP (1976) Superlinearly convergent variable metric algorithms for general nonlinear programming problems. Math Program 11:263–282

Hestenes MR, Stiefel E (1952) Method of conjugate gradients for solving linear systems. J Res Natl Bur Stand 49:409–436

IBM (1990) OSL- Optimization subroutine library guide and reference. IBM Corp, NY

Kojima M, Megiddo N, Noma T, Yoshise A (1991) A unified approach to interior point algorithms for linear complementarity problems, Lecture Notes in Computer Science 538. Springer, NY

Lemke CE (1965) Bimatrix equilibrium points and mathematical programming. Manag Sci 11:681–689

Markovitz HM (1959) Portfolio selection: efficient diversification of investments. Wiley, NY

Mulvey JM (1987) Nonlinear network models in finance. Adv Math Program Finan Plann 1: 253–271

Murtagh BA, Saunders MA (1987) MINOS 5.4 user's guide, SOL 83-20R. Department of OR, Stanford University, CA

Murty KG (1972) On the number of solutions of the complementarity problem and spanning properties of complementary cones. Lin Algebra Appl 5:65–108

Murty KG (2008a) Forecasting for supply chain and portfolio management, Chap 3. In: Neogy SK, Bapat RB, Das AK, Parthasarathy T (eds) Mathematical programming and game theory for decision making, vol 1, pp 231–255. World Scientific, Singapore

Murty KG (2008b) A new practically efficient IPM for convex quadratic programming, Chap 3. In: Neogy SK, Bapat RB, Das AK, Parthasarathy T (eds) Mathematical programming and game theory for decision making, vol 1, pp 21–31. World Scientific, Singapore

Murty KG, Kabadi SN (1987) Some NP-complete problems in quadratic and nonlinear programming. Math Program 39:117–129

Powell MJD (1978) Algorithms for nonlinear constraints that use Lagrangian functions. Math Program 14:224–248

Theil H, van de Panne C (1961) Quadratic programming as an extension of conventional quadratic maximization. Manag Sci 7:1–20

Vavasis SA (1992) Local minima for indefinite quadratic knapsack problems. Math Program 54:127–153

White FC (1983) Trade-off in growth and stability in state taxes. Natl Tax J 36:103–114

Wilson RB (1963) A simplicial algorithm for convex programming, Ph.D. dissertation, School of Business Administration, Harvard

Wolfe P (1959) The simplex method for quadratic programming. Econometrica 27:382–398

Wood AJ (1984) Power generation, operation, and control. Wiley, NY

Ye Y (1991) Interior point algorithms for quadratic programming. In: Kumar S (ed) Recent developments in mathematical programming, pp 237–261. Gordon and Breach, PA

Zhou JL, Tits AL (1992) User's guide to FSQP Version 3.1. SRC TR-92-107r2, Institute for Systems Research, University of Maryland, College Park

Epilogue

As the Greek proverb says

$$\Omega\sigma\pi\varepsilon\rho\ \xi\varepsilon\nu o\iota\ \chi\alpha\iota\rho o\upsilon\sigma\iota\ \pi\alpha\tau\rho\iota\delta\alpha\ \beta\lambda\varepsilon\pi\varepsilon\iota\nu,$$

$$O\upsilon\tau\omega\varsigma\ \kappa\alpha\iota\ \tau o\iota\varsigma\ \kappa\alpha\mu\nu o\upsilon\sigma\iota\ \beta\iota\beta\lambda\iota o\upsilon\ \tau\varepsilon\lambda\emptyset\varsigma.$$

(which could loosely be translated as: "As travelers rejoice to see their destination, so too is the end of a book to those who labor to learn from it"), I hope the readers will find their efforts rewarding.

Katta Gopalakrishna Murty

Index

L

M

N

O

P

Q

R

S

MIX
Papier aus verantwortungsvollen Quellen
Paper from responsible sources
FSC® C105338

If you have any concerns about our products,
you can contact us on
ProductSafety@springernature.com

In case Publisher is established outside the EU,
the EU authorized representative is:
Springer Nature Customer Service Center GmbH
Europaplatz 3, 69115 Heidelberg, Germany

Printed by Libri Plureos GmbH
in Hamburg, Germany